BASIC APPLIED BIOINFORMATICS

Basic Applied Bioinformatics

Chandra Sekhar Mukhopadhyay
Ratan Kumar Choudhary
Mir Asif Iquebal

With contributions from

Ravi GVPPS Kumar, Sarika, Dinesh Kumar, Aditya Prasad Sahoo, Amit Kumar, Saurabh Jain, Surbhi Panwar, Ashwani Kumar, Harpreet Kaur Manku

WILEY Blackwell

This edition first published 2018

Registered Office(s)
John Wiley & Sons, Inc., 111 River Street, Hoboken, NJ 07030, USA

Editorial Office
111 River Street, Hoboken, NJ 07030, USA

For details of our global editorial offices, customer services, and more information about Wiley products visit us at www.wiley.com.

Wiley also publishes its books in a variety of electronic formats and by print-on-demand. Some content that appears in standard print versions of this book may not be available in other formats.

Library of Congress Cataloging-in-Publication Data

Names: Mukhopadhyay, Chandra Sekhar, author. | Choudhary, Ratan Kumar, author. | Iquebal, Mir Asif, author.
Title: Basic applied bioinformatics / by Chandra Sekhar Mukhopadhyay, Ratan Kumar Choudhary, Mir Asif Iquebal.
Description: 1st edition. | Hoboken, NJ : Wiley, [2017] | Includes bibliographical references and index. |
Identifiers: LCCN 2017015387 (print) | LCCN 2017019742 (ebook) | ISBN 9781119244370 (pdf) | ISBN 9781119244417 (epub) | ISBN 9781119244332 (hardback)
Subjects: LCSH: Bioinformatics–Textbooks. | BISAC: MEDICAL / Biostatistics.
Classification: LCC QH324.2 (ebook) | LCC QH324.2 .M85 2017 (print) | DDC 572.80285–dc23
LC record available at https://lccn.loc.gov/2017015387

Cover Design: Wiley
Cover Images: (top) @ mashuk/Gettyimages; (middle) @ alice-photo/Shutterstock; (bottom) © kentoh/Shutterstock

Set in 10.5/13pt Times by SPi Global, Pondicherry, India

Printed in Singapore by C.O.S. Printers Pte Ltd

10 9 8 7 6 5 4 3 2 1

Dedicated to students, researchers and professionals

Contents

Appendices

Preface

Bioinformatics, a discipline that attempts to make predictions about biological functions using data from molecular sequence (nucleotide and protein) analysis and involves application of information science to biology has, over the years, evolved exponentially in the genomics era. Today it has become an indispensable component of biological science, including its application in a number of applied areas.

There has long been a need among students and researchers for a primer book on the application of bioinformatics tools in various spheres of veterinary and agricultural sciences. This being the era of multi-tasking, research workers who do not possess a background in computer or bioinformatics often stumble over *in silico* analysis of molecular data. This book provides practical know-how for graduate students of bioinformatics, biotechnology and other streams of biological science, and also for those who need to learn bio-computational analyses of the large volume of molecular data that is being generated in thousands of laboratories throughout the world.

The topics considered in this book are the basic ones that a student or researcher of the fields above should know. In addition, this book covers the syllabi of the graduate or undergraduate course called "Introduction to Bioinformatics" (or course name similar to that) that is offered in several universities. Some of the chapters, covering areas such as genome annotation in prokaryotes and eukaryotes, an overview of microarray data analysis, use of MISA for microsatellite sequence identification and SNP mining, have also been introduced for out-of-the-box applications.

In general, the book serves as a reference book for those working in biocomputational research and studies. The contents of this book cover wider areas of bioinformatics. Several freely available software tools (online or offline) are available, and students and researchers can use them for *in silico* analysis. However, in some instances, students become stuck while optimizing parameters for data analysis and drawing appropriate inferences. Also, they are often not familiar with several terminologies. This book explains steps for parameter optimization of the tools being used, as well as the basic terminologies. The results obtained have been explained, to demonstrate how inferences are drawn.

This book can also serve as a practical manual for the elucidation of critical steps, with annotation and explanation. It begins with basic aspects of bioinformatics, including frequently used terminology, concept development, handling molecular sequences,

BLAST analyses, primer designing, phylogenetic tree construction, prediction of protein structures and genome annotation. In the last few chapters, some advanced topics of bioinformatics have been covered, such as analysis of transcriptome data, identification of differentially expressed genes and prediction of microRNA targets.

Each chapter demonstrates the steps with an example, which involves stepwise elucidation of the procedures and explanation of the obtained results. The practical methodology is depicted with screenshots of the software being used, along with legends to explain the screenshot view. New terminologies introduced in some chapters have been provided. Additionally, four or five questions are given at the end of each chapter, with any hints which are deemed to be required for some questions.

We believe that there could be some unintentional mistakes remaining in this book. We sincerely request the reader to apprise the editors for typographical or other errors, if found. It is very common that the version of molecular sequences in public repository is updated over time, or sometimes one or more sequence entries are deleted. The readers are requested to update the editors about such changes. Similarly, the uniform resource locators (URLs) of the websites containing bioinformatics tools or databases can change suddenly. We will be careful to update these changes in the next edition of the book. Readers are also requested to assist us in this regard.

It is hoped that this book will be a useful primer for beginners of this fast-expanding field.

Acknowledgements

We thank Ms. Mindy Okura-Marszycki and Mr. Vishnu Narayanan of Wiley for their timely help and encouragements. We extend a special note of thanks to Prof. G.S. Brah, Founder Director, and Prof. Ramneek Verma, Director, of the School of Animal Biotechnology, GADVASU, Ludhiana, for providing the conducive working environment and for inspiring us to contribute to the field of bioinformatics. They evaluated some of the chapters and raised critical questions for improving the quality. The authors thankfully appreciate Miss J.K. Dhanoa and Ms. H.K. Manku for thoroughly checking the syntax of the manuscript, helping in editing and framing the diagrams in the proper format. The chapters were also evaluated for lucidness and ease of understanding by the graduate students of the Iowa State University, Mrs. Supreet Kaur (MSc Biochemistry) and Shravanti Krishna (PhD Biochemistry). Sincere thanks are extended to Dr. Nikhlesh Singh (Assistant Professor, Physiology, the University of Tennessee Health Science Center (UTHSC), Memphis, USA), Dr Monson Melissa (Postdoc Research Associate, Animal Science, Iowa State University) and Dr. Sangita Singh (Post Doc., Department of Food Science and Human Nutrition, Iowa State University) for their constructive criticisms to improve the chapters. Dr. Shivani Sood, Assistant Prof. (Biotechnology, Mukand Lal National College, Yamuna Nagar, Haryana, India), deserves special mention for critically checking the chapters and giving constructive input. All the freely available software and databases covered in this book are duly acknowledged. We are obliged to all those who have directly or indirectly contributed to writing this book.

Our sources of inspiration have been our families, colleagues and students. Nevertheless, we bow before the Almighty and Mother Nature for giving us the potential to accomplish the task.

List of Abbreviations

AFLP	amplified fragment length polymorphism
ASCII	American Standard Code for Information Interchange
BAC	bacterial artificial chromosome
BAM	binary alignment/map
BIC	Bayesian information criterion
BLAST	basic local alignment search tool
BWA	Burrows–Wheeler algorithm
BWT	Burrows–Wheeler transformation
cDNA	complementary DNA
CINEMA	color interactive editor for multiple alignments
cRNA	complementary RNA
dbGaP	database of genotypes and phenotypes
dbVar	database of variation
DDBJ	DNA data bank of Japan
DEG	differentially expressed genes
DNA	deoxyribonucleic acid
DP	dynamic programming
EMBL	European Molecular Biology Laboratory
EST	expressed sequence tag
ExPASy	expert protein analysis system
F81 model	Felsenstein (1981) model
FASTA	fast all
FDQN	fully quantified domain name
FPKM	fragment per kilobase of exon per million mappable reads
GATK	genome analysis toolkit
gi or GI	gene identity
GO	gene ontology
GOR	Garnier, Osguthorpe, and Robson
GSS	genome survey sequence
GTF	gene transfer format
GTR	generalized time-reversible
GUI	graphical user interface

GWAS	genome-wide association studies
HKY85 model	Hasegawa, Kishino and Yano (1985) model
IBL	internal branch length
InDels	insertion and deletions
INSDC	International Nucleotide Sequence Database Collaboration
IUPAC	International Union of Pure and Applied Chemistry
JALVIEW	Java alignment viewer
JC69 model	Jukes and Cantor (1969) model
K80 model	Kimura (1980) model
MACAW	multiple alignment construction and analysis workbench
MAFFT	multiple alignment using fast Fourier transform
ME	minimum evolution
MEGA	molecular evolution and genetic analysis
MISA	microsatellite identification tool
ML	maximum likelihood
MP	maximum parsimony
MSA	multiple sequence alignment
MSRE	methylation sensitive restriction enzymes
MUSCLE	multiple sequence comparison by log-expectation
mYa	million years ago
NBRF	National Biomedical Research Foundation
NCBI	National Center for Biotechnology Information
NGS	next-generation sequencing
NJ	neighbor joining
NWA	Needleman–Wunsch algorithm
ORF	open reading frame
OTU	operational taxonomic unit
PDB	protein data bank
pI/MW	isoelectric point to molecular weight ratio
PIR	protein information resource
PSD	protein sequence database
PWMs	position weight matrices
RCSB	Research Collaboratory for Structural Bioinformatics
RE	restriction enzyme
RF	reading frame
RFLP	restriction fragment length polymorphism
RPKM	read per kilobase of exon per million mappable reads
rRNA	ribosomal RNA
SAM	sequence alignment/map
SCOP	structural classification of protein
SIB	Swiss Institute of Bioinformatics
SNPs	single nucleotide polymorphisms
SPR	subtree pruning regrafting
SSR	simple sequence repeats
STS	sequence-tagged site

SWA	Smith–Waterman algorithm
T92 model	Tamura (1992) model
TBR	tree bisection reconnection
T-Coffee	tree-based consistency objective function for alignment evaluation
TFBS	transcription factor binding sites
TFs	transcription factors
TIS	translation initiation sites
TN93 model	Tamura and Nei (1993) model
T-P	transversion-parsimony
TPA	third-party annotation
TPM	transcripts per million
TRANSFAC	transcription regulatory factors
tRNA	transfer RNA
TTS	triplex-forming oligonucleotide target sequences
uGDT	unnormalized global distance test
UniProt	universal protein resource
UPGMA	unweighted pair group method with arithmetic mean
URL	uniform resource locator
VCF	variant call format
VNTR	variable number tandem repeat
WGS	whole genome shotgun
wwPDB	worldwide protein data bank
YAC	yeast artificial chromosome

SECTION I

Molecular Sequences and Structures

Retrieval of Sequence(s) from the NCBI Nucleotide Database

CHAPTER 1

CS Mukhopadhyay and RK Choudhary
School of Animal Biotechnology, GADVASU, Ludhiana

1.1 INTRODUCTION

The NCBI nucleotide database (http://www.ncbi.nlm.nih.gov/nucleotide/) is an archive of gene, transcript, and fragments of genomic DNA sequences. It combines several online public repositories, including *GenBank* (the genetic sequence database of NIH), *RefSeq* (annotated, non-redundant reference sequence from genomic, transcript and protein), *TPA* (third-party annotated data on nucleotide sequences), and *PDB* (protein databank: a repository of 3D structures of proteins and nucleic acids). The International Nucleotide Sequence Database Collaboration (INSDC) maintains the liaison between the three major molecular data repositories – namely, NCBI, DDBJ, and EMBL – to share the nucleotide data present in any of those databanks.

A brief description of the NCBI databases has been given in Appendix A "NCBI Database: A Brief Account" at the end of this book.

1.2 COMPONENTS OF THE NCBI NUCLEOTIDE DATABASE

- **GenBank**: An annotated collection of all publicly available nucleotide and *in silico* translated protein sequences.
- **EST database**: Maintains expressed sequence tags (ESTs) and short, single-pass reads (the sequence-fragments/reads obtained by loading the reaction in a lane only once and, hence, obtained after analyzing the input sequence by the sequencer only once) from mRNA (cDNA).
- **GSS database**: A database of genome survey sequences (GSS), or short single-pass genomic sequences (TTS, Exon Trapped, BAC/YAC, etc.)

Basic Applied Bioinformatics, First Edition. Chandra Sekhar Mukhopadhyay, Ratan Kumar Choudhary and Mir Asif Iquebal.

1.3 OBJECTIVES

To search and download nucleotide sequences from NCBI Nucleotide database and save as a text file (*.txt). The sequence of interest for downloading could be complete or partial gene/mRNA/coding sequence, non-coding RNA (rRNA, tRNA), non-coding and repeat sequences (VNTR) in the genome, partial genomic DNA sequences, and so on.

1.4 PROCEDURE

1.4.1 Nucleotide sequence search

a. Open the NCBI nucleotide page: http://www.ncbi.nlm.nih.gov/nucleotide/

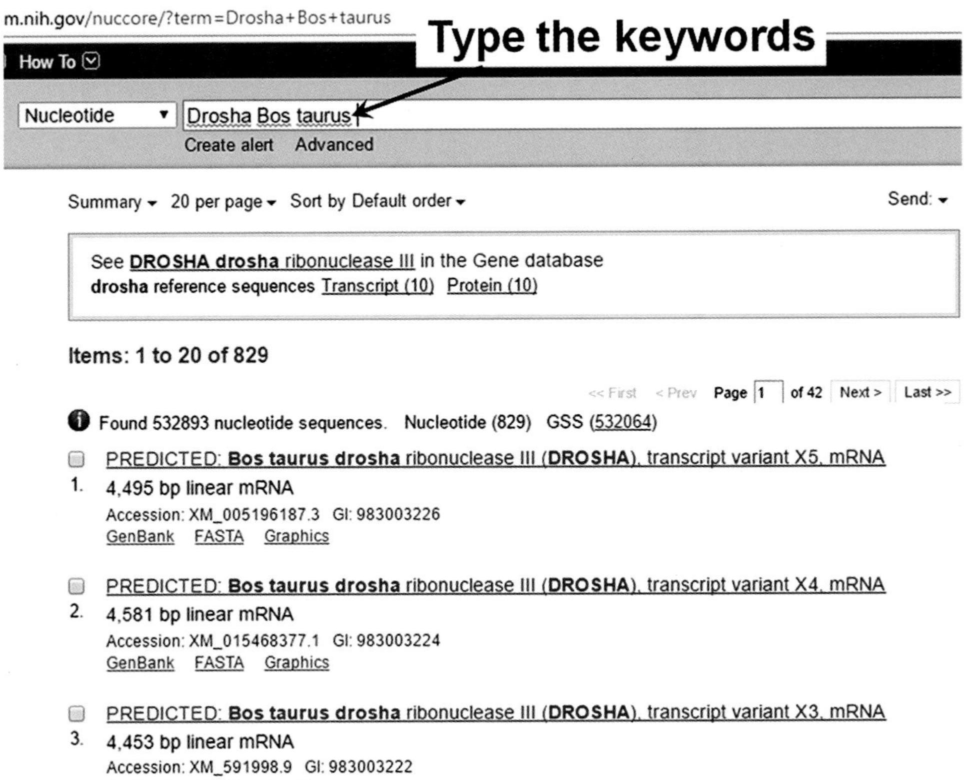

FIGURE 1.1 Main search window of NCBI Nucleotide page and list of hits for nucleotide sequences of taurine *Drosha* (gene/mRNA). (*See insert for colour representation of the figure.*)

b. Search the target sequence by providing the name of the gene and keywords – say, for example, the *Drosha* gene sequence in taurine cattle (*Bos taurus*) (Figure 1.1). Thus, the keywords are: "Drosha Bos taurus" (type your keywords without quotes, or else the quotes will instruct the search engine to find out the exact phrase within quotes, which ultimately limits your search). Then click on the "Search" button.

c. The nucleotide sequence(s) can also be searched by specifying the *accession number*(s), separated by a space (or comma). Please note that from September 2016 onwards, NCBI has phased out the sequence gi numbers. The accession numbers are *unique codes* assigned as an *identifier* to each nucleotide sequence in the database.

1.4.2 Downloading the selected sequences

a. Now, for example, select the first three sequences (depending on your requirement) by checking the small checkboxes on the left-hand side of each of the sequences.
b. Click on the "Send to" button, located at the top-right side of the page (Figure 1.2). Choose the destination of the selected sequences (to a *.txt file or to the clipboard for copying and pasting to a separate file, or collection in your NCBI account). Register yourself to NCBI and get your account-Id and password. Select the sequence format (Summary, GenBank, FASTA, etc.), the items per page and mode of sorting the selected sequences from the drop-down menus before downloading in a text file.
c. Finally, click on the "Create File" button to download the nuccore_result.txt file (default name) (see Figure 1.2, below). Open the file to obtain the sequences in the specified format and order.

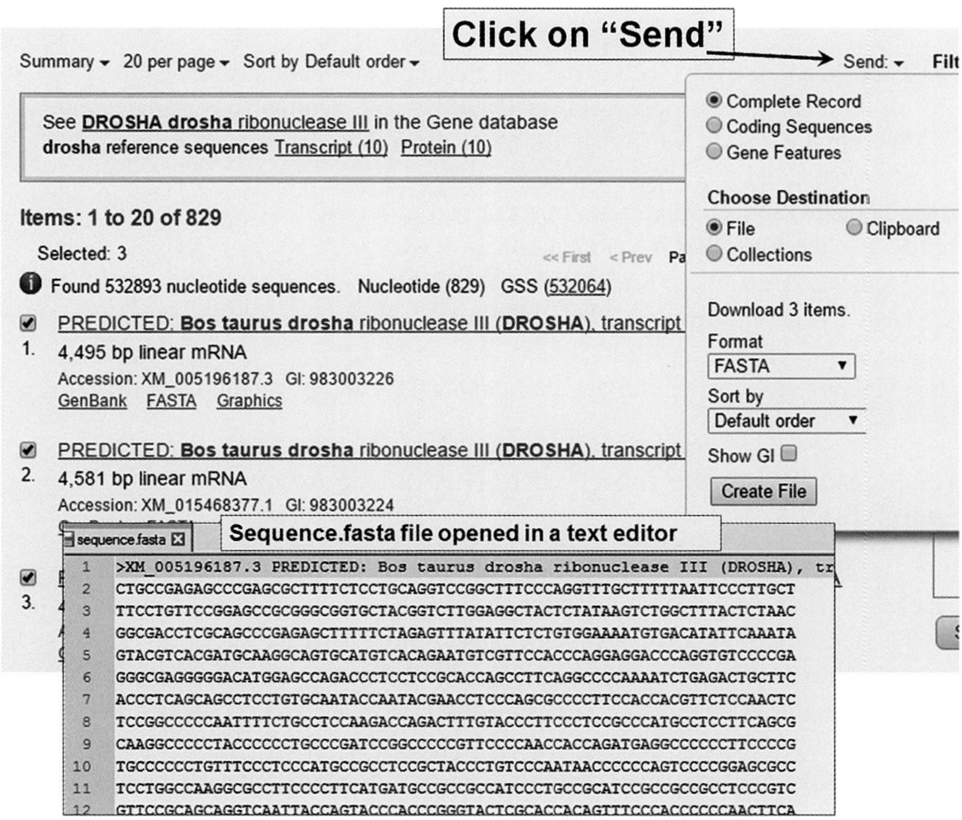

FIGURE 1.2 Click on the "Send to" button to download and save (in a text file) the first three Drosha mRNA sequences in "Summary" format. (*See insert for colour representation of the figure.*)

1.5 SOME USEFUL NUCLEOTIDE SEQUENCE DATABASES OF NCBI

One can search other NCBI databases that archive nucleotide sequences:

a. species-wise or chromosome assembly search (WGS or other assembly of chromosome or genome, likeBos_taurus_UMD_3.1.1);
b. clone (clones associated with genomics, cDNA and cell-based libraries, viz. BTDAEX-80K11, HWYUBAC-1-028-04-H12);
c. dbGaP (interaction of genotype and phenotype, viz. phs000287.v4.p1Cardiovascular Health Study (CHS) Cohort);
d. dbVar (large-scale genomic variation, nsv836042, nsv836041 etc), SNP, etc., among many other databases. The process of downloading the data as a text file is the same.

1.5.1 Modifying the search with the "Limits" option (currently not available)

The user can *narrow down the search* by using the parameters available after clicking on the hyperlink "Limits". However, NCBI has removed this option nowadays. The available options are:

a. Published in the last (specify the available days or mention date range).
b. Modified in the last (specify the available days or mention your own date range).
c. Search *Field Tags* (different fields of GenBank flat file Accession, Author, Bioproject, etc.).
d. Segmented sequences (master of set or part of set).
e. Source database (RefSeq or GenBank or EMBL or DDBJ or PDB).
f. Molecule (Genomic DNA/RNA, mRNA, rRNA or cRNA).
g. Gene Location (Genomic DNA/RNA or Mitochondrion, Chloroplast or "any" of the above types).
h. Exclude (STSs and/or working draft and/or TPA and/or patents).

1.5.2 Modifying the search with "Nucleotide Advanced Search Builder"

Click on the hyperlinked word "Advanced" just below the text box. The new page enables you to build your search settings.

Please note that the *search builder* enables us to specify the keywords according to their type (i.e., accession, assembly, author, journal and so on); in turn, this instructs the search engine to pinpoint the keywords from the database, depending on its feature or type.

Let us take our previous example: "Drosha Bos taurus". In the search builder, click on the drop-down menu (shown as "All Fields") and select "*Gene Name*" and type "Drosha". The role of "Show index list" is discussed in the next paragraph. Next, click on the drop-down list of the second-row field and select "*Organism*" and then type "Bos taurus" (without quotes). If you have more keywords, then add more rows accordingly, and select the specific field before typing the keyword(s).

The "Show index list" button will show the list of indexes from which you can specify your index. To move further along the index, you can use "Previous 200" or "Next 200" options. The "+" and "–" symbols beside each of the text boxes allow you to add (new) or delete the corresponding text boxes.

Please note that we can also do the advanced search without using the "Advanced Search Builder". Type the keywords in the form as given below: "term [field] OPERATOR term [field]". So, in the case of our current example, it will be:

Drosha[Gene name] AND Bos taurus[Organism]

1.6 QUESTIONS

1. Download the following sequences in FASTA format, and save these sequences in a single text file: NCBI Nucleotide Accession numbers are AB909393.1; AB909392.1; AB909391.1; AB906338.1; AB906337.1; KF773864.1; AB898237.1
2. Suppose you need to download the "Drosha" full-length sequence of taurine cattle. How will you proceed with the "Advanced" option of the Nucleotide search?
3. Download the following sequences in NCBI (full length) sequences and save them in a text file: NCBI Nucleotide Accession numbers: KF021228.1; KC831578.1; KC822646.1; KC758965.1; KC758964.1; KC424594.1; KC424593.1
4. How will you search and save the bubaline SRY coding sequence in NCBI, Nucleotide and check only the full-length cds or mRNA, and then save the FASTA sequences in a text file?
5. What are the differences between Genome Survey Sequence (GSS) and Nucleotide sequences in the NCBI database? In your search result, if you get both types of sequences, which one will you download to use as a template for primer designing?

Retrieval of Protein Sequence from UniProtKB

CHAPTER 2

CS Mukhopadhyay and RK Choudhary
School of Animal Biotechnology, GADVASU, Ludhiana

2.1 INTRODUCTION

The Universal Protein Resource (UniProt) is a database of protein sequence and function, created by combining the Protein Information Resource-Protein Sequence Database (PIR-PSD), Swiss-Prot, and TrEMBL databases. UniProt(www.uniprot.org/) has two sections: the Swiss-Prot knowledgebase (it harbors fully annotated records) and the TrEMBL protein database (contains computationally analyzed records on proteins).

2.1.1 Features of UniProtKB/Swiss-Prot

- Non-redundancy of records.
- High level of integration of data deposited in different related databases (NCBI-GenBank, EMBL, DDBJ for translated coding sequences).
- High level of manual curation.
- Contains more than 0.25 million entries.

2.1.2 Features of UniProtKB/TrEMBL

- Translations of nucleotide coding sequence (cds) in EMBL/NCBI-GenBank/DDBJ.
- Automatic annotation.
- Contains more than 3.3 million entries.

2.2 OBJECTIVE

To download the amino acid sequence of protein (say, taurine sex-determining region, Y-encoded (SRY) peptide).

Basic Applied Bioinformatics, First Edition. Chandra Sekhar Mukhopadhyay, Ratan Kumar Choudhary and Mir Asif Iquebal.

2.3 PROCEDURE

a. Open the Expert Protein Analysis System (ExPASy) homepage: http://www.expasy.org/
b. Locate the browser on the drop-down menu "Query all databases" at the upper center portion of the page, and click on "Proteomics" (to obtain information from all relevant databases, such as Prosite, String, ENZYME, UniProtKB etc), or else select UniProtKb (Figure 2.1 below).

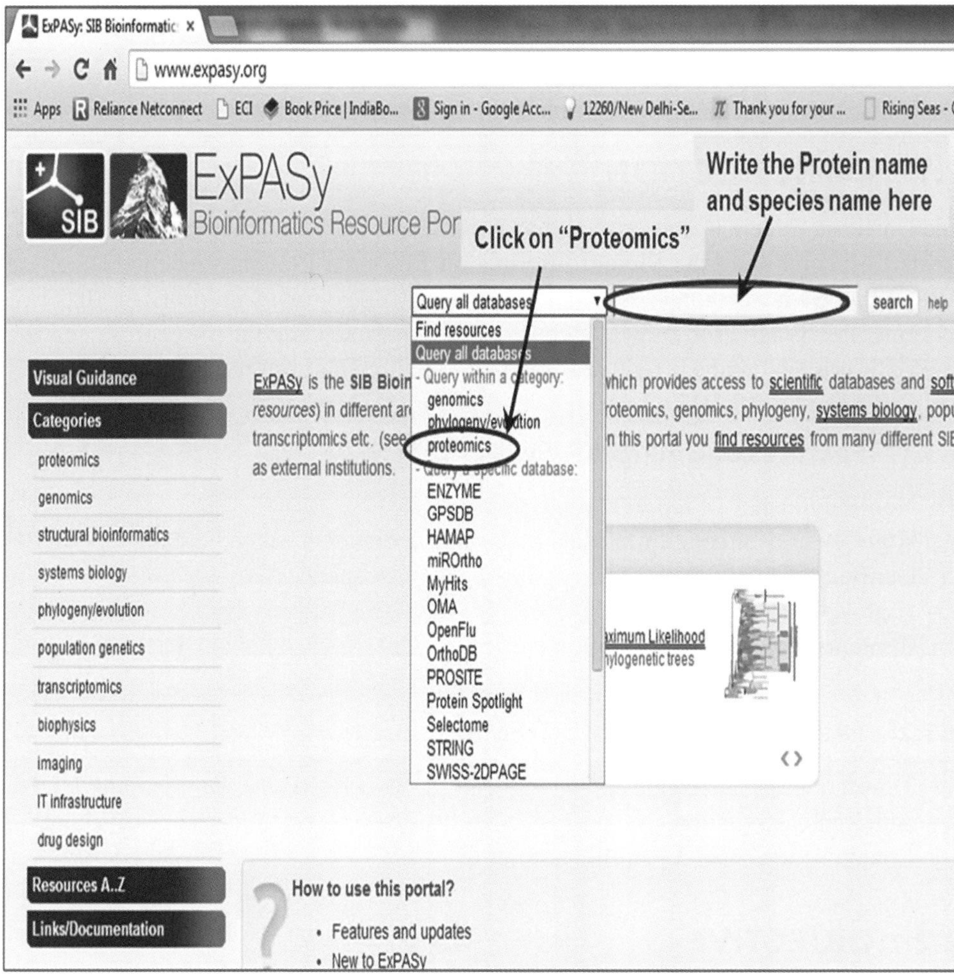

FIGURE 2.1 Homepage of ExPASy server: select the "proteomics" option from the drop-down menu for databases, and enter your protein name along with other keywords to begin search.

c. Write the name of the protein and the species: "SRY Bos taurus" in the blank text box just beside "Find resources".
d. Click on "Search".
e. A list of search results is obtained in a table. Select the specific result: here it is "SRY_Bovin."

f. Click on the Entry (here "Q03255") to get the detail of the sequence (see Figure 2.2).

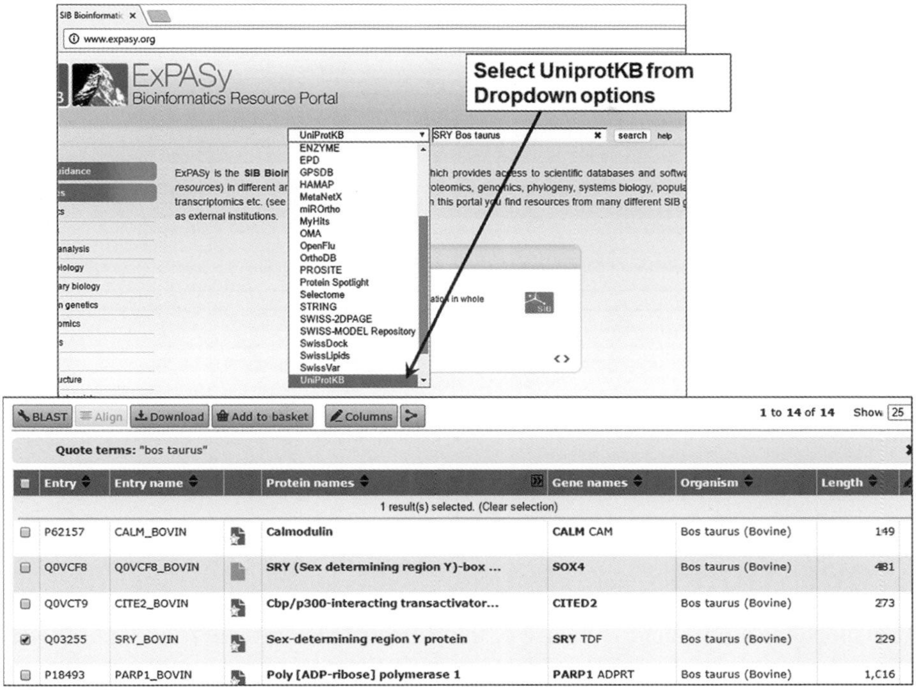

FIGURE 2.2 Click on the specific entry to open it in a separate window. (*See insert for colour representation of the figure.*)

g. The newly opened page shows detailed information on the target protein, including names and origin, protein attributes, general annotation (i.e., comments), ontologies, sequence annotation (features), sequences, references, and so on.
h. Click on FASTA to obtain the sequence in FASTA format.
i. Select the sequence in FASTA format and copy and paste in a text file (see Figure 2.3).

One can also *BLAST* the sequence or do the computation of physical as well as chemical parameters of the protein being studied (by ProtParam, i.e., Protein Parameters), compute pI/MW ratio and peptide mass to explore the molecular features of the protein.

Use the "Align" tab to align the above entry to its isoform (if the isoform is available). Note that the current entry (Q03255 (SRY_BOVIN)) does not have any isoform, so alignment is not possible.

The "Add to Basket Tab" is available to enable the user to select the entry and place it in a separate place (called "Basket") for later use.

The "History" tab is meant for checking the history (dates of initial version, revised version, etc.) of entry.

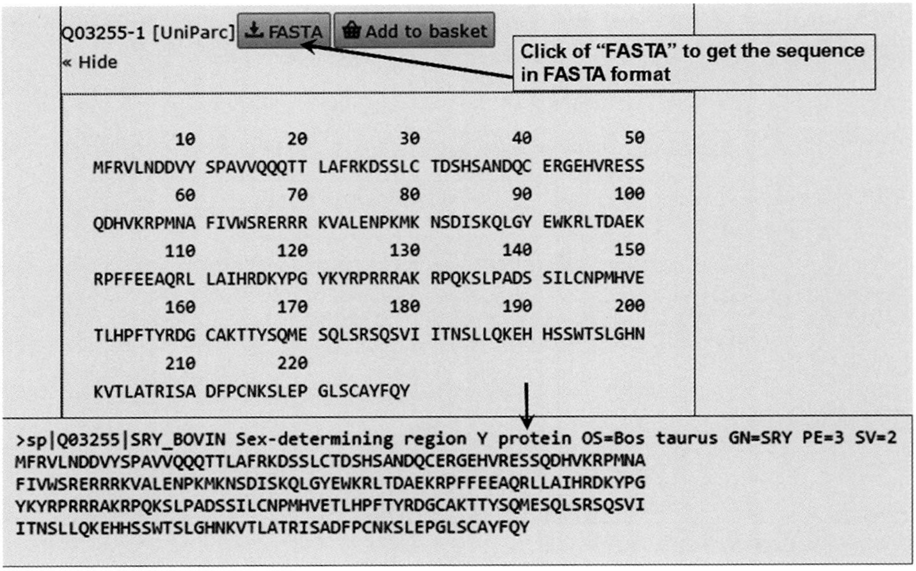

FIGURE 2.3 Peptide sequence of taurine SRY in FASTA format.

2.4 QUESTIONS

1. Download the amino acid sequence of the Human TSPY protein from UniProtKB.
2. Write down the protein feature of the bovine SRY-HMG-box, using the ProtParam tool.
3. Enumerate the uses of UniProtKB/Swiss-Prot *vis-à-vis* UniProtKB/TrEMBL.
4. Use the following sequence to find out the name of the protein (NCBI Protein Accession Number NP_001032554.1):

   MAAADGDDSLYPIAVLIDELRNEDVQLRLNSIKKLSTI
   ALALGVERTRSELLPFLTDTIYDEDEVLLALAEQLGTFT
   TLVGGPEYVHCLLPPLESLATVEETVVRDKAVESLRAIS
   HEHSPSDLEAHFVPLVKRLAGGDWFTSRTSACGLFSVCY
   PRVSSAVKAELRQYFRNLCSDDTPMVRRAAASKLGEFAK
   VLELDNVKSEIIPMFSNLASDEQDSVRLLAVEACVNIAQ
   LLPQEDLEALVMPTLRQAAEDKSWRVRYMVADKFTELHK
   AVGPEITKTDLVPAFQNLMKDCEAEVRAAASHKVKEFCE
   NLSADCRENVIMTQILPCIKELVSDANQHVKSALASVIM
   GLSPILGKDSTIEHLLPLFLAQLKDECPEVRLNIISNLD
   CVNEVIGIRQLSQSLLPAIVELAEDAKWRVRLAIIEYMP
   LLAGQLGVEFFDEKLNSLCMAWLVDHVYAIREAATSNLK
   KLVEKFGKEWAHATIIPKVLAMSGDPNYLHRMTTLFCIN
   VLSEVCGQDITTKHMLPTVLRMAGDPVANVRFNVAKSLQ
   KIGPILDNSTLQSEVKPVLEKLTQDQDVDVKYFAQEALTVLSLA

What is the SwissProt-Id of the above sequence? Write down the common name of the protein, gene encoding the protein, and molecular weight of the protein.

5. Compare the above protein for the available information in NCBI (Protein Accession Number NP_001032554.1) and SwissProt (UniProtKB Id Q03255) databases.

Downloading Protein Structure

CHAPTER 3

CS Mukhopadhyay and RK Choudhary
School of Animal Biotechnology, GADVASU, Ludhiana

3.1 INTRODUCTION

The tertiary structure of a protein can be downloaded from the Research Collaboratory for Structural Bioinformatics Protein Databank (RCSB-PDB). It is a repository for the three-dimensional structural data of large biological molecules, such as proteins and nucleic acids. The PDB is managed by the Worldwide Protein Data Bank (wwPDB).

3.2 OBJECTIVE

To download the structure of taurine beta-lactoglobulin peptide from the protein databank (PDB).

3.3 PROCEDURE

a. Open RCSB-PDB: http://www.rcsb.org/pdb/home/home.do
b. Enter the name (or sequence or ID) of the target protein in the search text box provided at the top of the page. In this example, it is <Beta-lactoglobulin AND “Bos taurus”> (Figure 3.1).
c. One can further filter the search results by clicking on “Refine Search”. This process will narrow down the search, based on the following classifications: “IDs and Keywords,” “Structure Annotations,” “Structure Features,” “Deposition,” “Chemical Components,” and so on.
d. Click on “Unliganded form of bovine beta-lactoglobulin, ambient pressure” to get the structure of the target protein (Figure 3.2).
e. Download the protein structure in PDB file format. This is a file format where the file has a *.pdb extension.
f. The three-dimensional image can be viewed by clicking on the “3D View”. Java™ must be installed in the system to view the 3D image (Figure 3.2).

Basic Applied Bioinformatics, First Edition. Chandra Sekhar Mukhopadhyay, Ratan Kumar Choudhary and Mir Asif Iquebal.

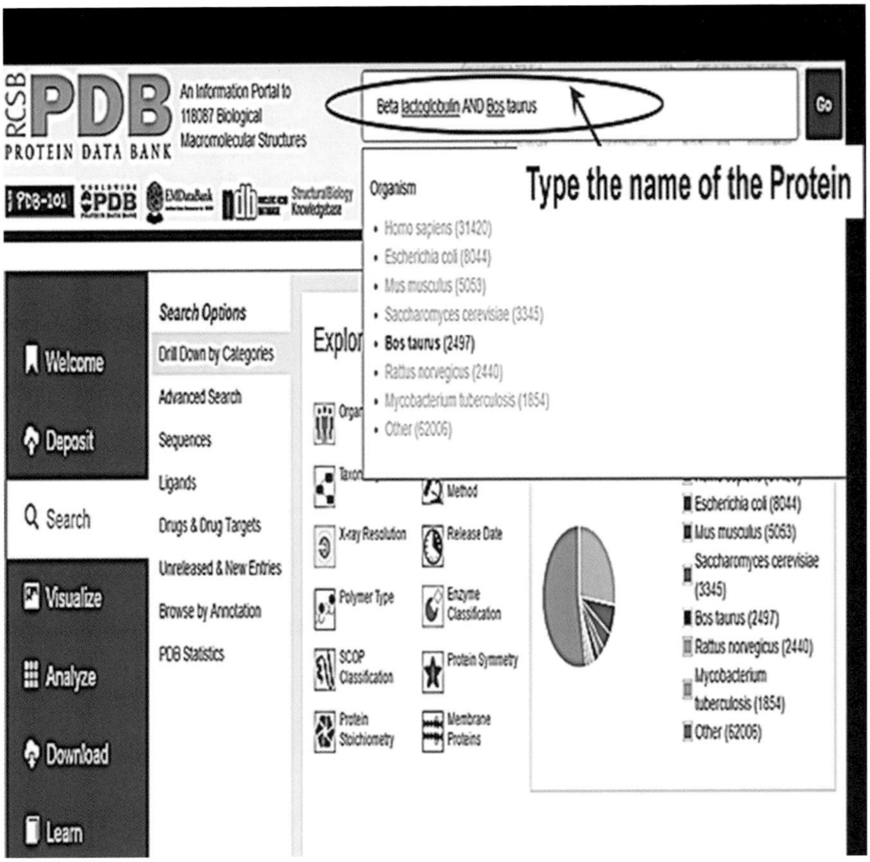

FIGURE 3.1 Homepage of RCSB-PDB. Specify the name of the protein and the species in the given box, and click on the search button (denoted by the symbol of a lens).

- One can narrow down the search (when huge numbers of results are obtained from a search based on "Everything") by clicking on the bases of search, such as "Author", or "Macromolecule" or "Sequence of molecule" or "name of ligand". A help option on search is also available, indicated by a question mark "?" just above the search text box.
- The PDB-ID (a unique identifier comprising four alphanumeric characters), which is assigned randomly to each entry, can be used to obtain the structure of a specific protein.
- Double-quoted keywords (viz. "Ribonuclease III") will enable you to search all entries specific to that protein only (here, Ribonuclease III only, not any other kinds of Ribonucleases).
- Use the search operators "AND", "OR" and "NOT" to enable a more refined search.
- Use of parenthesis and wildcards (i.e. "?" and "*") are also permitted.

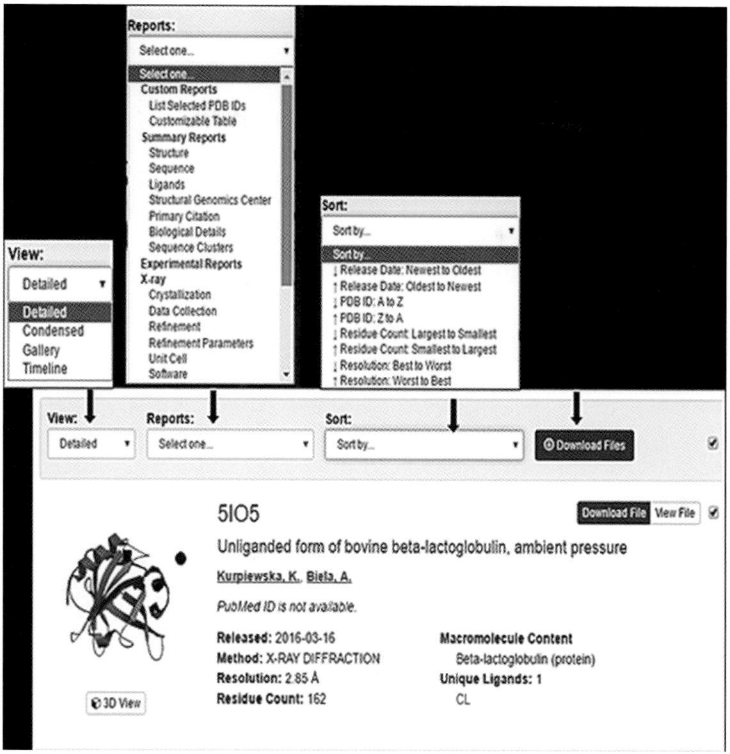

FIGURE 3.2 Visualization of 3D peptide structure obtained following PDB search.

3.4 QUESTIONS

1. Download the structure of bovine insulin and save it in PDB format.
2. Search and download the structure of bovine pancreatic ribonuclease from PDB.
3. What will you do if a queried structure is not available in PDB? Suppose you need the structure for homology modeling (see Chapter 29) and, thereby, to predict the structure of a novel peptide of the same protein family.

> **Hint**: Structure of the protein from phylogenetically close species, or structure of protein from the same family (with identity more than 70%) could be searched.

4. What are the different file formats in which a protein structure can be downloaded? How to select the desired file format?
5. Search and save the structures of human and mice keratin.

Visualizing Protein Structure

CHAPTER 4

CS Mukhopadhyay and RK Choudhary
School of Animal Biotechnology, GADVASU, Ludhiana

4.1 INTRODUCTION

RasMol, developed by Roger Sayle, is freely downloadable interactive software that is used for visualizing molecular graphics. The name "RasMol" is an abbreviation of "*Raster of a Molecule*", which means a computer display (Raster) of the solid surface molecule. RasMol displays the 3D format of the atom coordinates of a molecule, written in PDB file format.

4.2 OBJECTIVE

To visualize the structure of a peptide (we will use "bovine beta-lactoglobulin (isoform A) in complex with dodecyl trimethyl ammonium chloride (DTAC)") using RasMol software.

4.3 PROCEDURE

a. Download Rasmol: http://rasmol.org/ and run the executive file. Please get registered when you use RasMol for the first time.
b. Open RasMol by double clicking the "RasWin" icon on the desktop. This simultaneously opens two windows:
   i. The 3D visualization window(the GUI with black appearance). RasMol can be operated through the menu bar drop-down options.
   ii. The RasMol command line prompt. This window is used to write the commands (instead of operating RasMol through keyboard and GUI) to visualize the structure.
c. Go to "File >Open" and select the particular PDB file. We need to download the file 4IB7.pdb (i.e., Bovine beta-lactoglobulin (isoform A)), as stated above, from PDB or UniProtKB (Figure 4.1). Presently, the following display patterns are

Basic Applied Bioinformatics, First Edition. Chandra Sekhar Mukhopadhyay, Ratan Kumar Choudhary and Mir Asif Iquebal.

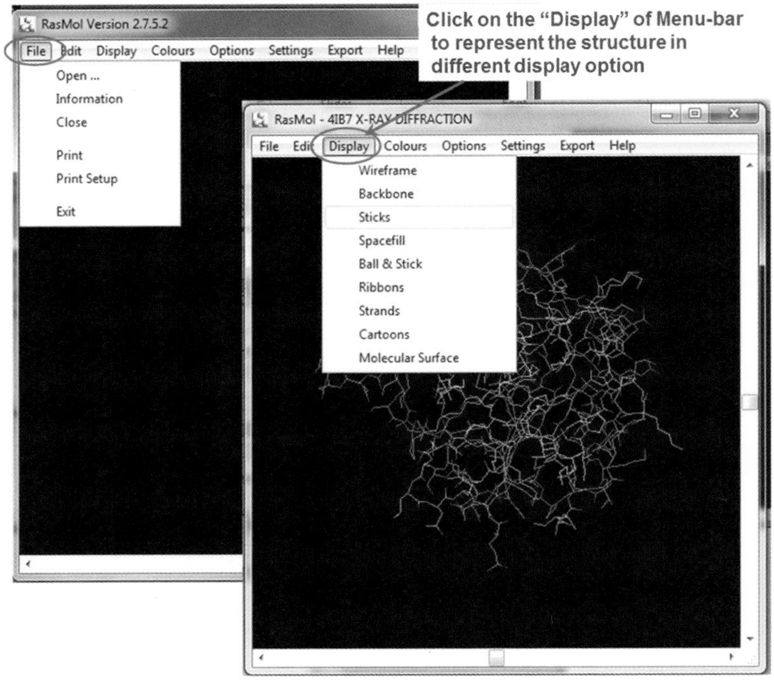

FIGURE 4.1 Graphical user interface (GUI) of RasMol and the drop-down menu to open, modify or alter the display of the peptide. (*See insert for colour representation of the figure.*)

available in the Display menu: "Wireframe" (default), "Backbone", "Sticks", "Spacefill", "Ball and Stick", "Ribbons", "Strands", "Cartoons", and "Molecular Surface" (Figure 4.2).

d. The menu "Colours" provides several alternative color combinations, as well as the monochromatic view.
e. You may also view in slab mode, with (default) or without hydrogen atom, display of hetero atoms (non-DNA, non-Protein atoms), display of labels, etc.
f. Rotating the structure: The displayed image can be rotated or enlarged using keyboard only. Table 4.1 shows the keys used in the *Windows* operating system.

Students are advised to visit the site http://ww2.chemistry.gatech.edu/~williams/bCourse_Information/4581/labs/tbp/rasmol/rasmol_tbp_fset.html for the detailed elaboration of the RasMol Menus and Commands.

The presentation of the molecular structure of protein or protein–DNA complex can be manipulated according to requirements, through the RasMol command line interface. The RasMol commands are easy and user-friendly:

- **Control of the rendering**: This enables the user to restrict the visualization to protein (command: *restrict protein*) or DNA (command: *restrict DNA*), change the color of the molecules (command: *color red*, *color green*, etc.), or select the

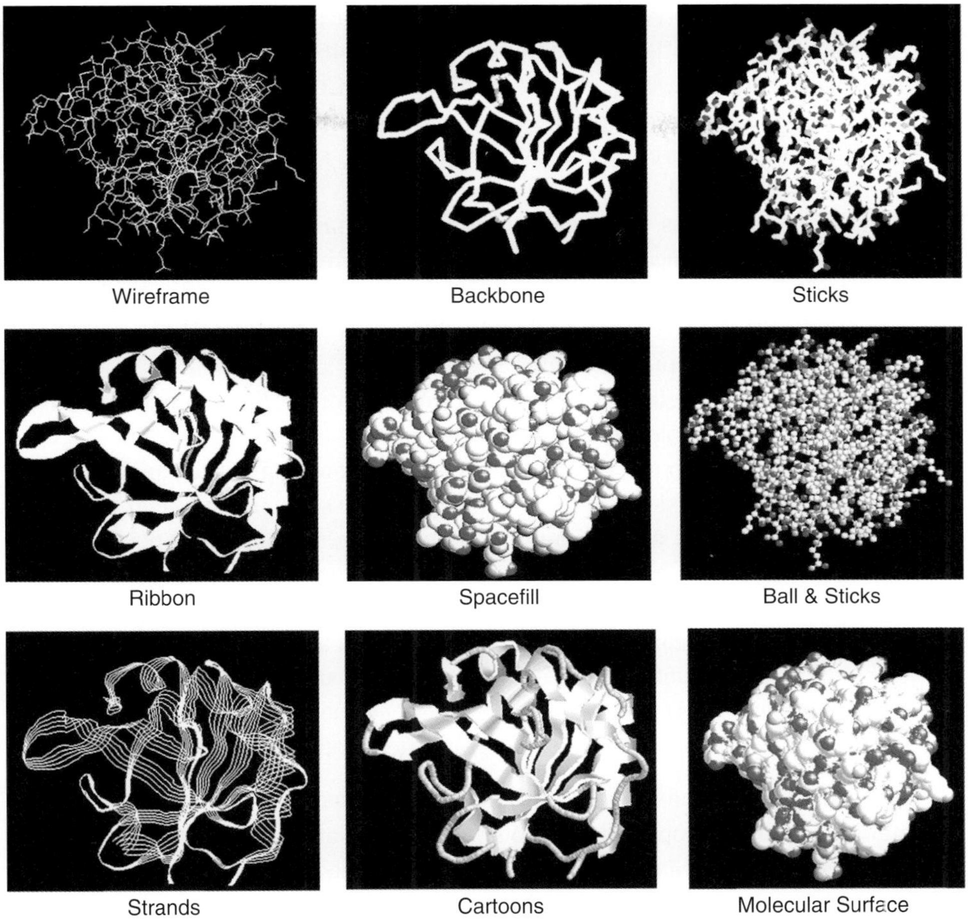

FIGURE 4.2 A single peptide, displayed in 'Wireframe', 'Backbone', 'Sticks', 'Spacefill', 'Ball and Stick', 'Ribbons', 'Strands', 'Cartoons' and 'Molecular surface' patterns. (*See insert for colour representation of the figure.*)

TABLE 4.1 Computer short-cuts to work on the image displayed by RasMol (http://www.openrasmol.org/doc/).

Action on image	Press key
Rotate X,Y	Left
Translate X,Y	Right
Rotate Z	Shift-Right
Zoom	Shift-Left
Slab Plane	Ctrl-Left

specific nucleotides (A/T/G/C/combination of pairing bases) using commands like "*select a or t*". The presentation of the molecule is also required to be modified accordingly, through the "Display" and "Color" menus in the menu bar. This is needed to identify specific types of molecules in different colors, in order to analyze the structure. Please note that the prime symbol (′) is indicated by an asterisk (*) in Rasmol.

- **Measuring distances between atoms**: The command "*set picking distance*" switches on the distance measuring option. Alternatively, this can also be done in the molecules window, by selecting the "Pick Distance" option from the "Setting" menu in the menu bar. Single- or double-click on two atoms to get the distance (in Angstroms) between those two atoms.
- **Measurement of angles**: The method is same for either the command line ("*set picking angle*") or molecules window, and to select the atoms between which the angle is to be determined. The command line window displays the angle.
- **Measurement of torsion angles**: Try this using the same procedure as above.
- **Measurement of Phi and Psi angles**: Visualize the structure as "Stick" and type "select all". Deselect everything except for the target portions (alpha-helix, beta sheet or, say, atoms from 100–110 (command: "*restrict backbone and 100–110*")). If a black screen appears, reselect the display option as "Stick", select the types of atoms and color them differently. Turn on the dihedral button in the Molecules window to measure the Phi angle (clicking on the atoms one after another C' ->Ca ->N ->C') and the Psi angle (clicking on the atoms one after another N ->C' ->Ca ->N).

As well as the aforementioned necessary applications, DNA–protein interaction, hydrophobicity, the hydrophilicity of polar and non-polar residues, amphiphilicity of alpha-helices and so on can also be studied using Rasmol. The techniques are similar to that mentioned above, and have been elaborated step by step in the Rasmol tutorial.

4.4 QUESTIONS

1. Open the following PDB files (you need to download them first from a suitable database such as UniProtKB) and display the peptide structures as ribbon, space-fill, and cartoons:
   i. 4TLJ
   ii. 4MTV
   iii. 2GJM
   iv. 1B1Y
   v. 2Z5Z
2. Display the antimicrobial domain of fowlicidin: 2AMN (PDB file).
3. How do you measure the angle of torsion using RasMol? Consider actin protein (PDB Id: 1ATN) as an example, and determine the Phi-Psi angles for Tyrosine (position 133), Alanine (210th) and Isoleucine (274th) amino acids.
4. Can you determine the distance between the Arginine (95th) and Glutamine (156th)?

Sequence Format Conversion

CHAPTER 5

CS Mukhopadhyay and RK Choudhary
School of Animal Biotechnology, GADVASU, Ludhiana

5.1 INTRODUCTION

A computer file format is a distinct way of encoding data to store in a file. Biological sequence format is an assemblage of distinct file formats, with the aim of rendering the files legible to specific programs.

Note: Biological sequences are generally written in Courier New font. This enables us to arrange the sequences uniformly in each line of the text

Sequence formats are manipulated or inter-converted by the system in the base level through ASCII (American Standard Code for Information Interchange – i.e. binary code) text – that is, A–Z characters are encoded by 65–90; a–z characters by 97–122. Thus, the sequence formats are the required arrangement of characters, symbols, and keywords that specify the sequence, ID name, comments, and so on.

The sequence formats are needed for two purposes:

a. Different programs recognize different types of formats. We need to convert one format to an other to use the sequence for that program.
b. Presentations of the molecular sequence are sometimes required in a particular format.

Commonly used sequence formats.

1. IG/Stanford
2. GenBank/GB
3. NBRF
4. EMBL
5. GCG
6. DNAStrider
7. Fitch
8. Pearson/Fasta
9. Zuker (in-only)
10. Olsen (in-only)
11. Phylip3.2
12. Phylip
13. Plain/Raw
14. PIR/CODATA
15. MSF
16. ASN.1
17. PAUP
18. Pretty (out-only)

Basic Applied Bioinformatics, First Edition. Chandra Sekhar Mukhopadhyay, Ratan Kumar Choudhary and Mir Asif Iquebal.

5.2 OBJECTIVE

To convert the format of a given molecular sequence to other sequence formats like NCBI, EMBL, PIR, etc.

5.3 PROCEDURE

The online program *ReadSeq* (by Don Gilbert) will be used to convert the sequence formats. *ReadSeq* accepts the following formats: FASTA, Abstract Syntax Notation (ASN.1), National Biomedical Research Foundation (NBRF), EMBL, Fitch (phylogenetic analysis), GenBank, GCG, DNA Strider, Intelligenetics, Multiple sequence format, Protein Information Resource (PIR), and eight additional specialised formats.

a. Open the online *ReadSeq* sequence conversion tool using the URL: http://www-bimas.cit.nih.gov/molbio/readseq/
b. A molecular sequence (nucleotide or amino acid sequence) in any format is pasted into the text box. The software can determine the input sequence automatically (Figure 5.1).

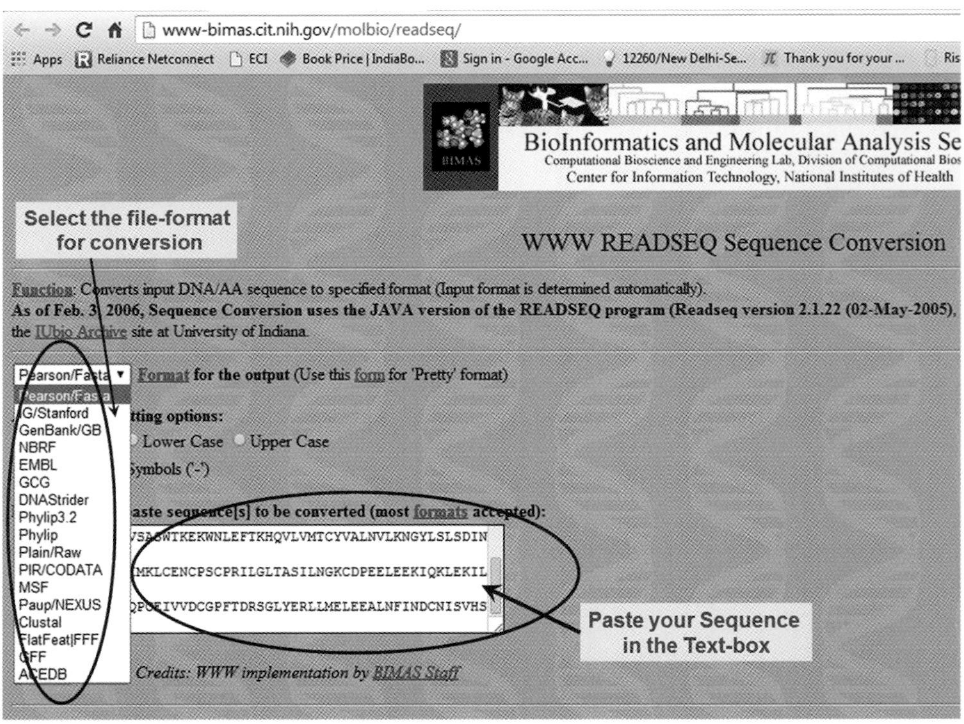

FIGURE 5.1 Homepage of the *ReadSeq* biosequence format conversion tool. (*See insert for colour representation of the figure.*)

c. Click on the drop-down menu, just above the text box (on the left side) and select the desired output format.
d. There are additional formatting options:
   i. *Altering the case of the output sequence*: click on one of the radio buttons "MiXeD case", "UPPER" or "lower" case.

   ii. *Removal of the gaps*: click on the check box to remove existing gaps in the input sequence.

e. Click on the "Submit" button to get the output.
f. The "reset" button is there to erase all the input data and start afresh with default settings.

The International Union of Pure and Applied Chemistry (IUPAC) nucleic acid code has been adopted to specify a single or a group of nucleotide(s) by a single alphabet:

A=adenine	U=uracil	M=A or C (amino)	D=G or A or T
C=cytosine	R=G or A (purine)	S=G or C	H=A or C or T
G=guanine	Y=T or C (pyrimidine)	W=A or T	V=G or C or A
T=thymine	K=G or T (keto)	B=G or T or C	N=A or G or C or T (any)

IUPAC amino acid codes:

A=Alanine	G=Glycine	M=Methionine	S=Serine
C=Cysteine	H=Histidine	N=Asparagine	T=Threonine
D=Aspartic Acid	I=Isoleucine	P=Proline	V=Valine
E=Glutamic Acid	K=Lysine	Q=Glutamine	W=Tryptophan
F=Phenylalanine	L=Leucine	R=Arginine	Y=Tyrosine

5.3.1 Other online sequence conversion tools

a. **FMTSeq** – This is an elaborative version of *ReadSeq*. It is furnished with data manipulation for ClustalW, Zuker, ELEX (I/O files) and so on. URL: http://www.bioinformatics.org/JaMBW/1/2/
b. **Emboss**: This has several features, including cutseq, pasteseq, nthseq, extractseq, and so on. URL: http://emboss.sourceforge.net/docs/themes/SequenceFormats.html
c. **EMBOSS Seqret**: This is another sequence format conversion tool available online, offering several output formats for conversion. The URL is as follows: http://www.ebi.ac.uk/Tools/sfc/emboss_seqret/

5.4 QUESTIONS

1. Identify the sequence format given below:

   A >DL;readseq-43434_tmp_1
   readseq-43434_tmp_1 100 bases
   cagacggaaaagctggagcgcaggcgcaagccccacctggaccgcagagg
   cgccatcatccggggcatccccggcttctgggccaatgccattgcgaacc*

   B LOCUS readseq-13129_tmp_1 100 bp
   ORIGIN
   1 cagacggaaaagctggagcgcaggcgcaagccccacctggaccgcagaggcgccatcatc
   61 cggggcatccccggcttctgggccaatgccattgcgaacc
   //

C >readseq-14738_tmp_1 100 bp
cagacggaaaagctggagcgcaggcgcaagccccacctggaccgcagaggcgccatcatc
cggggcatccccggcttctgggccaatgccattgcgaacc

D ID readseq-10695_tmp_1 standard; DNA; UNC; 100 BP.
SQ Sequence 100 BP;
cagacggaaaagctggagcgcaggcgcaagccccacctggaccgcagaggcgccatcatc 60
cggggcatccccggcttctgggccaatgccattgcgaacc 100

E readseq-946_tmp_1 cagacggaaaagctggagcgcaggcgcaagccccacctggaccgcagagg cgccatcatc
readseq-946_tmp_1 cggggcatccccggcttctgggccaatgccattgcgaacc

F 1 100
readseq-26 cagacggaaaagctggagcgcaggcgcaagccccacctggaccgcagagg
cgccatcatccggggcatccccggcttctgggccaatgccattgcgaacc

2. Download a nucleotide sequence of your interest from NCBI Nucleotide. Then convert it to the following formats:
   a. Clustal **b.** EMBL **c.** Phylip
3. Given below is an amino acid sequence (GenBank: BAA36473.1) in lower case. Convert it to upper case and show in PIR format:
   QTEKLERRRKPHLDRRGAIIRGIPGFWANAIANHPQMSALITDQDE
4. Suppose you have custom sequenced a cloned product. How will you open the sequence file, and to which format will you convert it to do basic biocomputational analysis (i.e., using BLAST, Alignment, *in silico* translation (if applicable), etc.)?
5. What are the uses of sequence format conversion? A DNA sequence has been presented in some of the commonly used formats. Please write the name of the formats.

(A)
>readseq-26104_tmp_1 204 bp
ccatgaacgccttcattgtgtggtctcgtgaacgaagacgaaaggtggctctagagaatc
ccaaaatgaaaaactcagacatcagcaagcagctgggatatgagtggaaaaggcttacag
atgctgaaaagcgcccattctttgaggaggcacagagactactagccatacaccgagaca
aatacccgggctataaatatcgac

(B)
LOCUS readseq-11577_tmp_1 204 bp
ORIGIN
1 ccatgaacgccttcattgtgtggtctcgtgaacgaagacgaaaggtggctctagagaatc
61 ccaaaatgaaaaactcagacatcagcaagcagctgggatatgagtggaaaaggcttacag
121 atgctgaaaagcgcccattctttgaggaggcacagagactactagccatacaccgagaca
181 aatacccgggctataaatatcgac

(C)
ID readseq-2117_tmp_1 standard; DNA; UNC; 204 BP.
SQ Sequence 204 BP;
ccatgaacgccttcattgtgtggtctcgtgaacgaagacgaaaggtggctctagagaatc 60
ccaaaatgaaaaactcagacatcagcaagcagctgggatatgagtggaaaaggcttacag 120

atgctgaaaagcgcccattctttgaggaggcacagagactactagccatacaccgagaca 180
aatacccgggctataaatatcgac 204
//

(D)
\\\
ENTRY readseq-18456_tmp_1
TITLE readseq-18456_tmp_1 204 bases
SEQUENCE
5 10 15 20 25 30
1 c c a t g a a c g c c t t c a t t g t g t g g t c t c g t g
31 a a c g a a g a c g a aa g g t g g c t c t a g a g a a t c
61 c c a aaa t g a aaaa c t c a g a c a t c a g c a a g c
91 a g c t g gg a t a t g a g t g g a aaa g g c t t a c a g
121 a t g c t g a aaa g c g c cc a t t c t tt g a g g a g g
151 c a c a g a g a c t a c t a g c c a t a c a c c g a g a c a
181 a a t a c cc g gg c t a t a aa t a t c g a c
///

5.5 BRIEF DESCRIPTION OF SOME OF THE IMPORTANT MOLECULAR SEQUENCE FORMATS

1. **FASTA/Pearson format**: This is the simplest and the most common form of representing biological sequences. It was developed by Pearson and Lipman (1996).

   Features:
   - It starts with a ">" *sign*, followed by a sequence identifier that designates name, description, identity number of the sequence.
   - One-letter symbols represent the sequence entities.
   - The sequence is written continuously (without gaps or numbering).
   - In the end, the asterisk (i.e., "*") symbol indicates the end of the sequence (this is optional).
2. **PHYLIP format**: This is the format of the Phylip package for phylogenetic analysis.

   Features:
   - The first line of the input file contains the *number of species* and *number of characters* in that sequence (with space, no comma).
   - The information for each sequence starts with a ten-character-long species name (any alphabetic characters, with or without space and dots).
   - This is followed by a string of sequences.
   - The sequences may be *interleaved or sequential.*
3. **CLUSTAL/.ALN format**: This format originated from the *CLUSTAL* program for sequence alignment. The alignment is written in blocks of 60 characters, and the sequence is written in either UPPER CASE or lower case.

Features:

- Every block starts with sequence name (of any length), followed by at least one space.
- "-" denotes gap (for InDel) in multiple or pair-wise sequence alignment.
- *Residue* number is shown at the terminus of each line (optional).
- The last line bearing asterisks (*) at the end of each block indicates the conservation (in sequence alignment) (Figure 5.2).

FASTA or Pearson Format

```
>readseq-26104_tmp_1 204 bp
ccatgaacgccttcattgtgtggtctcgtgaacgaagacgaaaggtggctctagagaatc
ccaaaatgaaaaactcagacatcagcaagcagctgggatatgagtggaaaaggcttacag
atgctgaaaagcgcccattctttgaggaggcacagagactactagccatacaccgagacaa
atacccgggctataaatatcgac*
```

Phylip Format

```
Phylip
1 204
readseq-13    ccatgaacgc cttcattgtg tggtctcgtg aacgaagacg aaaggtggct
              ctagagaatc ccaaaatgaa aaactcagac atcagcaagc agctgggata
              tgagtggaaa aggcttacag atgctgaaaa gcgcccattc tttgaggagg
              cacagagact actagccata caccgagaca aatacccggg ctataaatat
              cgac
```

Clustal or .aln Format

```
FinalBbu MKSPALQPLSMAGLQLMTPASSPMGPFFGLPWQQEAIHDNIYTPRKYQVELLEAALDHNT 60

AY386968 MKSPALQPLSMAGLQLMTPASSPMGPFFGLPWQQEAIHDNIYTPRKYQVELLEAALDHNT 60
         ************************************************************
```

FIGURE 5.2 Three sequence formats – namely, FASTA, Phylip and Clustal.

4. **GCG format**: The programs in the Genetics Computer Group (GCG) suite use the GCG format of molecular sequences.

   Features:

   - The sequence begins with either of the following lined (mandatorily *all uppercase*)
     - <for nucleic acid sequences>: !!NA_MULTIPLE_ALIGNMENT 1.0
     - <for amino acid sequences>: !!AA_MULTIPLE _ALIGNMENT 1.0
   - The next line is a description line that holds sequence information.
   - There is one dividing line that shows the number of molecular elements (residues) in the sequence, date and time of creation of file, a checksum (a number that indicates the total number of correct digits in a digital data, in order to compare data corruption or data loss in a process of storage or data transmission).
   - Two dots (..) act as a divider between the descriptive line and the sequence. These dots are not optional.

5. **GenBank format**: This format is used to display the sequences in the GenBank flat file at NCBI. The format has three parts:
   - **The header**: contains the locus field (locus name, sequence length, molecule type, GenBank division, modification date, definition, accession, etc.), Geneinfo identifier, key words, source, organism, reference, authors, title, and so on.
   - **Features**: information about genes and gene products, regions of biological significance in the sequence, a sequence that code for proteins and RNA molecules.
   - **Sequence**: contains the sequence with row numbering. Each row contains 60 entities/residues sub-divided into six blocks (each of 10 residues).
6. **NBRF format**: The National Biomedical Research Foundation (NBRF) format is read and written by Multalign Viewer. It is also known as *Protein Information Resources* (PIR) format.

   Features:
   - The first line starts with ">", followed by the sequence code.
   - The second line displays sequence information.
   - The third line onwards of the sequence is presented in the form of blocks of 10 entities, with five blocks in each row.
   - When multiple sequences are presented in NBRF format, individual sequences are concatenated together to make them of equal length, using leading or trailing characters, and gap positions.
   - Any non-alphanumeric characters except the asterisk (*) can be used to make the sequence legible to Multalign Viewer.
   - Spaces within the sequence are ignored.
7. **Rich Sequence Format (rsf) format**: These files harbor one or more sequences which could be either related or unrelated to each other. To create a file in "rsf format," GCG's NetFetch program can be used to download the flat file from NCBI and save it in *.rsf format. The rsf files are particularly useful for "*Seqlab*" (the graphical user interface version of GCG).

   Features:
   - The sequence is presented in a manner similar to EMBL format.
   - The annotations of the sequence are rich, and the important points are:
     - Author: list of authors related to sequence.
     - Sequence weight.
     - Date of creation.
     - Description line of the sequence.
     - Number of leading gaps (called Offset) in the sequence corresponding to an alignment or assembly of fragments.
     - Other sequence features.

Nucleotide Sequence Analysis Using Sequence Manipulation Suite (SMS)

CHAPTER 6

CS Mukhopadhyay and RK Choudhary
School of Animal Biotechnology, GADVASU, Ludhiana

6.1 INTRODUCTION

Nucleotide and amino acid sequences are analyzed to understand their hidden features, to discover patterns, and to determine function, structure and their evolution. The frequently used *in silico* analyses using molecular sequences are: sequence alignment; determining conserved regions; identification of low-complexity region of nucleotides; gene prediction; nucleotide sequence assembling; exploring biochemical and immunogenic properties of amino acid sequences; protein structure prediction, and so on. After completing this chapter, you will learn how to use some of the sequence analytical techniques using free online software called *Sequence Manipulation Suite*. The original examples cited in the software suit (as "help" for explaining the programs) have been used here. In some places, the explanations may be verbatim.

6.1.1 Sequence Manipulation Suite (SMS)

This is a collection (which is why it is called a "suite") of software, written in JavaScript1.5, for generating, formatting, and analyzing short DNA and protein sequences. Paul Stothard (of the University of Alberta, Canada) wrote the software suite (Stothard, 2000). The off-line suite can be downloaded from the link http://www.bioinformatics.org/sms2/mirror.html/.

Sequences are submitted to the sequence box of SMS and then analyzed according to the particular query.

6.2 OBJECTIVE

To learn the use of the different programs within SMS for analyzing nucleotide and amino acid sequences.

Basic Applied Bioinformatics, First Edition. Chandra Sekhar Mukhopadhyay, Ratan Kumar Choudhary and Mir Asif Iquebal.

6.3 PROCEDURE

a. Open the home page of Sequence Manipulation Suite using the URL: http://lion.img.cas.cz/sms2/index.html/. Check its compatibility with your web browser.
b. Various analytical tools, categorized into five groups, are available in the panel on the left-hand side of the page:
    i. Format Conversion
    ii. Sequence Analysis
    iii. Sequence Figures
    iv. Random Sequences
    v. Miscellaneous
c. Click on the required tool in the left-hand side panel: for example, if the first tool "Combine FASTA" is clicked, the specific page is opened.
d. Every tool has been explained for "how to use", along with an example, on its original page.
e. The common steps are to paste sequence(s) in the sequence box, and then click "Submit". The result is returned in a separate window. The "Clear" button will erase all data in the sequence box. The last button "Reset" will delete any document pasted in the sequence box and will reset the parameter(s).

6.4 FORMAT CONVERSION

6.4.1 Combine FASTA

This converts multiple FASTA sequences (either nucleotide or amino acid records) into a single sequence (Figure 6.1). The software imposes a restriction of input to 500 000 characters in total (inclusive of description line and input sequences).

6.4.2 EMBL feature extractor

This program extracts the salient features (according to the annotations) of one or more EMBL file(s), and returns the sequences in FASTA format in a new window. The program thus returns the whole nucleotide sequence, the mRNA and the cDNA sequence as separate FASTA format files as output. This is useful if the user wants to extract only the cds (coding sequence) or mRNA sequence out of the whole gene sequence (containing exons and introns, as well).

This program has a limit of 200 000 characters as input. There are two options for the output sequence features:

- "separated" – only the specific portions, viz. mRNA or cDNA parts out of the whole sequence, will be shown in lower case; or

- "UPPER-case" – the specified stretches of the mRNA or cDNA will be high-lighted in upper case while the rest of the source sequence will be in un-high-lighted upper case.

6.4.3 EMBL Trans Extractor

This program accepts one or more EMBL files and extracts the translated amino acid sequence in the result window (Figure 6.2). This program has a limit of 200 000 characters of input.

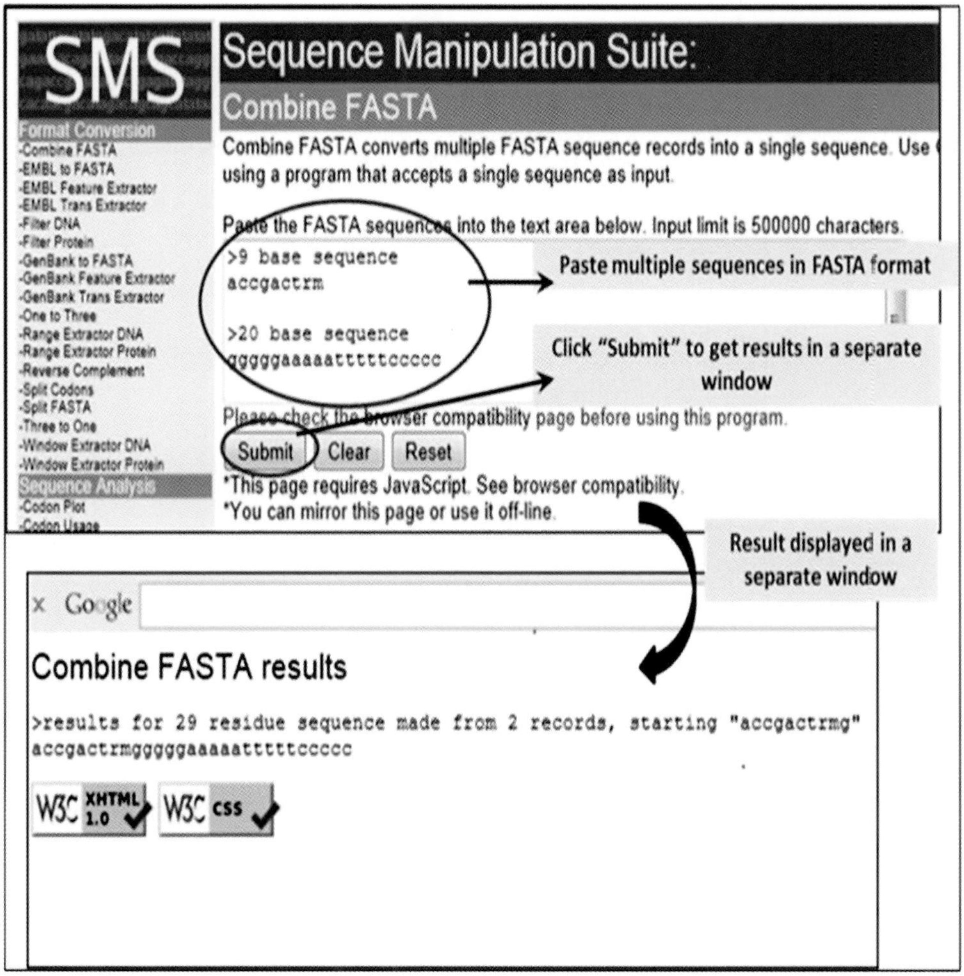

FIGURE 6.1 "Combine FASTA" input page to provide input data, and the corresponding output page with the result.

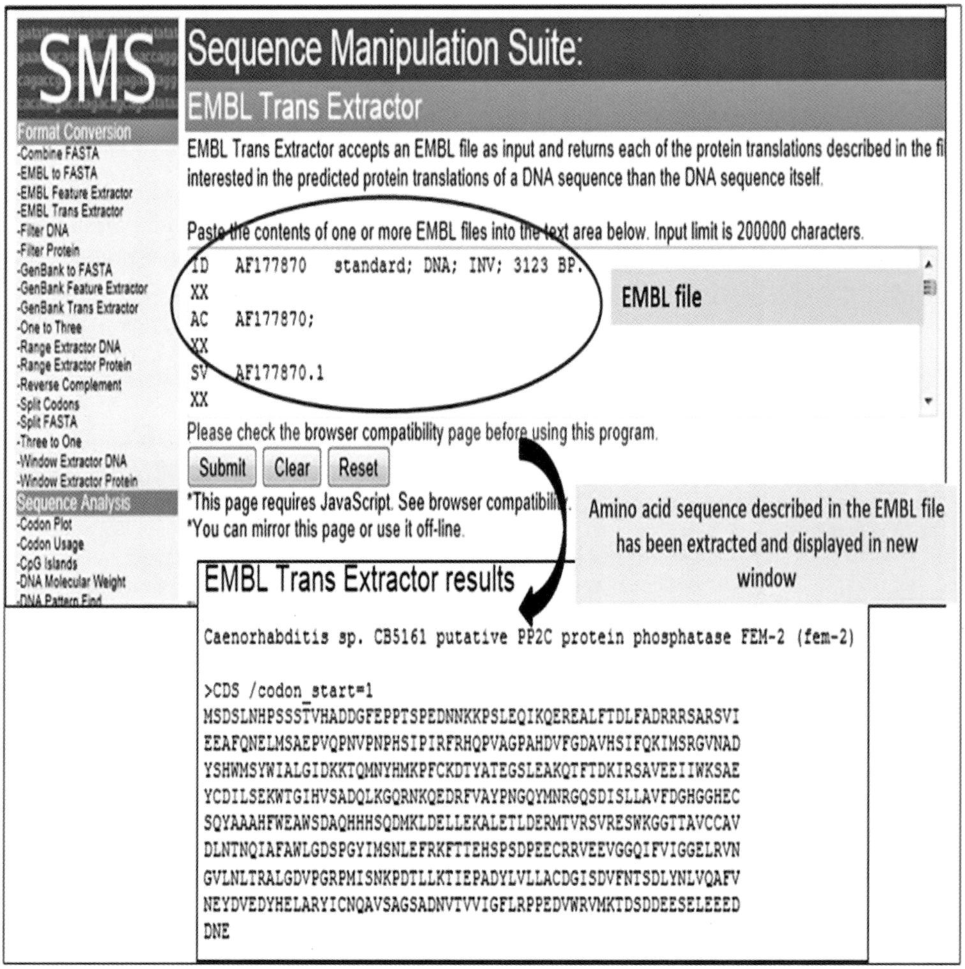

FIGURE 6.2 "EMBL Trans Extractor" input page, and the corresponding output page with extracted results.

6.4.4 Filter DNA

The input DNA sequence is filtered by eliminating the non-DNA characters (digits and blank spaces) from the whole sequence (input limit is 500 000) (Figure 6.3).

There are some options to modify filtration:

- What to replace: "Characters" and/or "white spaces".
- Replace with what: n/N/t/T/u/U/*/-/?
- Case conversion of characters.

6.4.5 Filter Protein

This is similar to the "Filter DNA" program. It filters out non-amino acid characters (digits, blank spaces, special characters) from an amino acid sequence. Some options

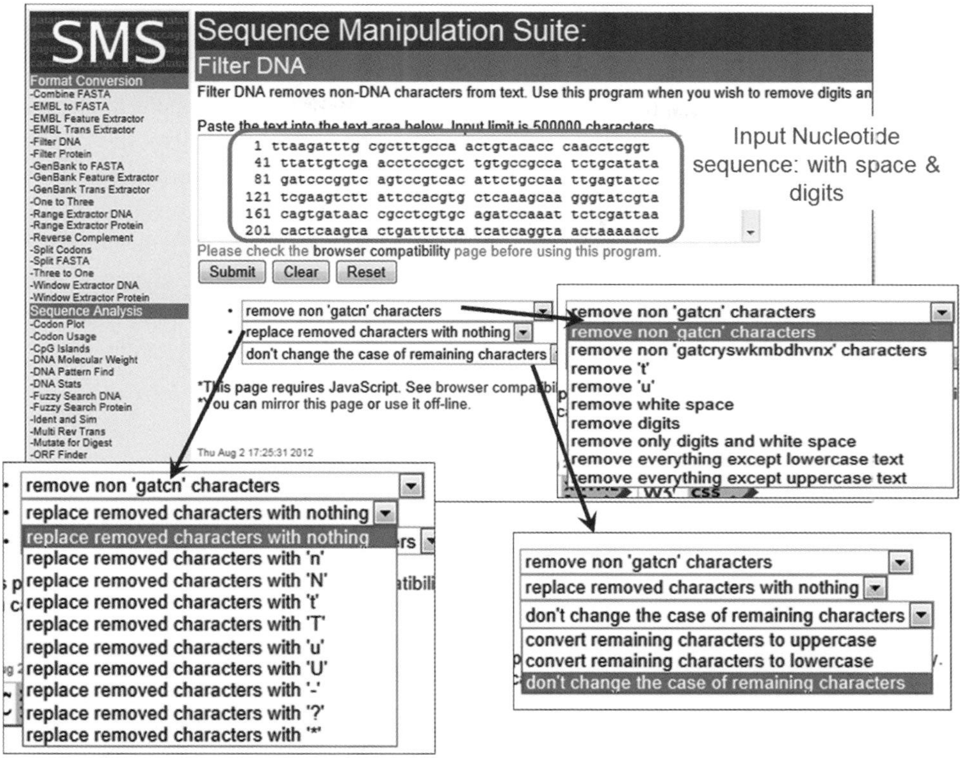

FIGURE 6.3 "Filter DNA" input page, along with various options as control parameters. (*See insert for colour representation of the figure.*)

are available on what to replace, replace with what and case conversion. The character limit of input is 500 000.

6.4.6 GenBank Feature Extractor

Similar to the "EMBL Feature Extractor" program. The input is nucleotide sequences in GenBank format.

6.4.7 GenBank Trans Extractor

Similar to the "EMBL Trans Extractor". The input is the nucleotide sequence in GenBank format.

6.4.8 One to Three

This program converts single-letter amino acid codes into three-character amino acid codes. Single or multiple amino acid sequence(s) in FASTA format (one letter code) is/are required and pasted into the sequence box. The input limit is 100 000 characters.

6.4.9 Range Extractor DNA

This returns the specific nucleotide sequence, based on the position and/or range(s) of nucleotide(s)/nucleotide sequence(s) specified in the input. The user needs to paste the DNA sequence in the sequence box. The specific position(s) (given by the position value(s) of the base(s)) and/or the ranges (two position values for the termini, separated by "..."), separated by comma(s), are then given. There are some options to output the results in FASTA format (either in upper or lower case) in one sequence, or in multiple sets of sequences (for multiple positions/ranges). The range(s) can be specified either in the original strand ("direct strand" option) or in the complementary strand to the input sequence ("complementary strand" option). The input limit is 500 000 characters.

6.4.10 Range Extractor Protein

This program is similar to "Range Extractor DNA", except that the input sequence is amino acid (in FASTA format) (Figure 6.4). Obviously, the drop-down options for "direct strand" and "complementary strand" are not there for amino acids in this program.

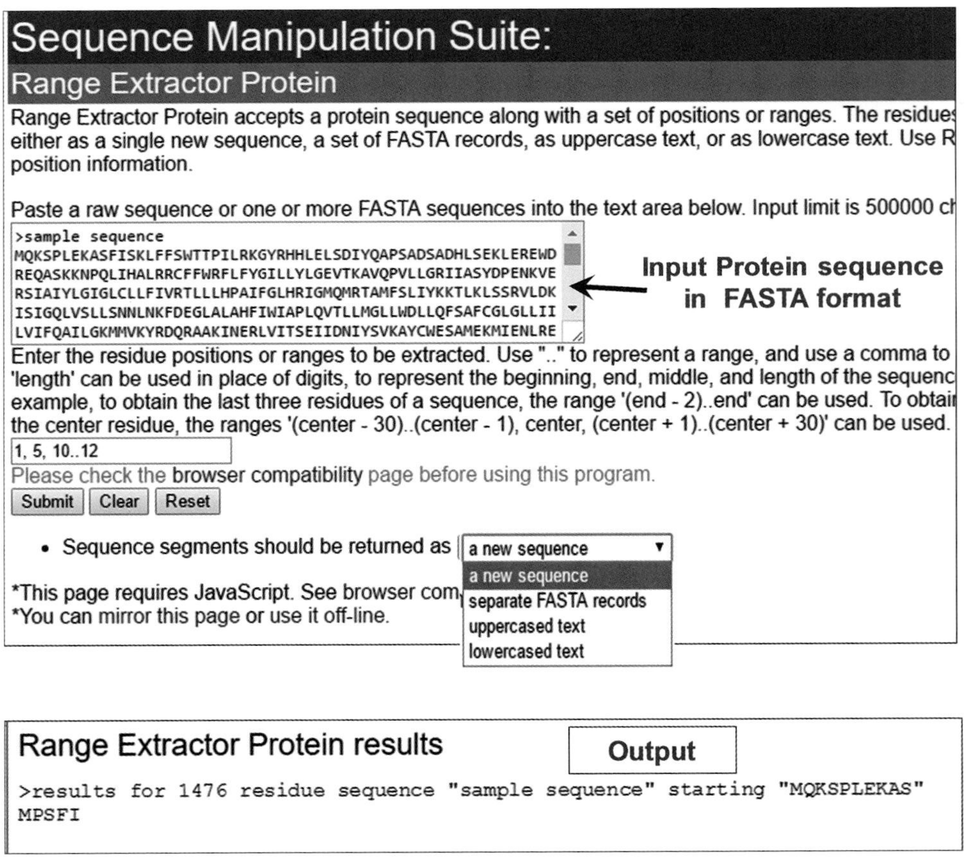

FIGURE 6.4 "Range Extractor Protein" input page and the corresponding output page with extracted sequences. (*See insert for colour representation of the figure.*)

6.4.11 Reverse Complement

This program is used to fetch the reverse-complement of the input sequence, or obtaining the reverse sequence(s), or only the complement of given nucleotide sequence(s). It can work with single or multiple DNA sequence(s) as input. It supports all the IUPAC DNA alphabets (Figure 6.5). The input limit is 100 000 characters.

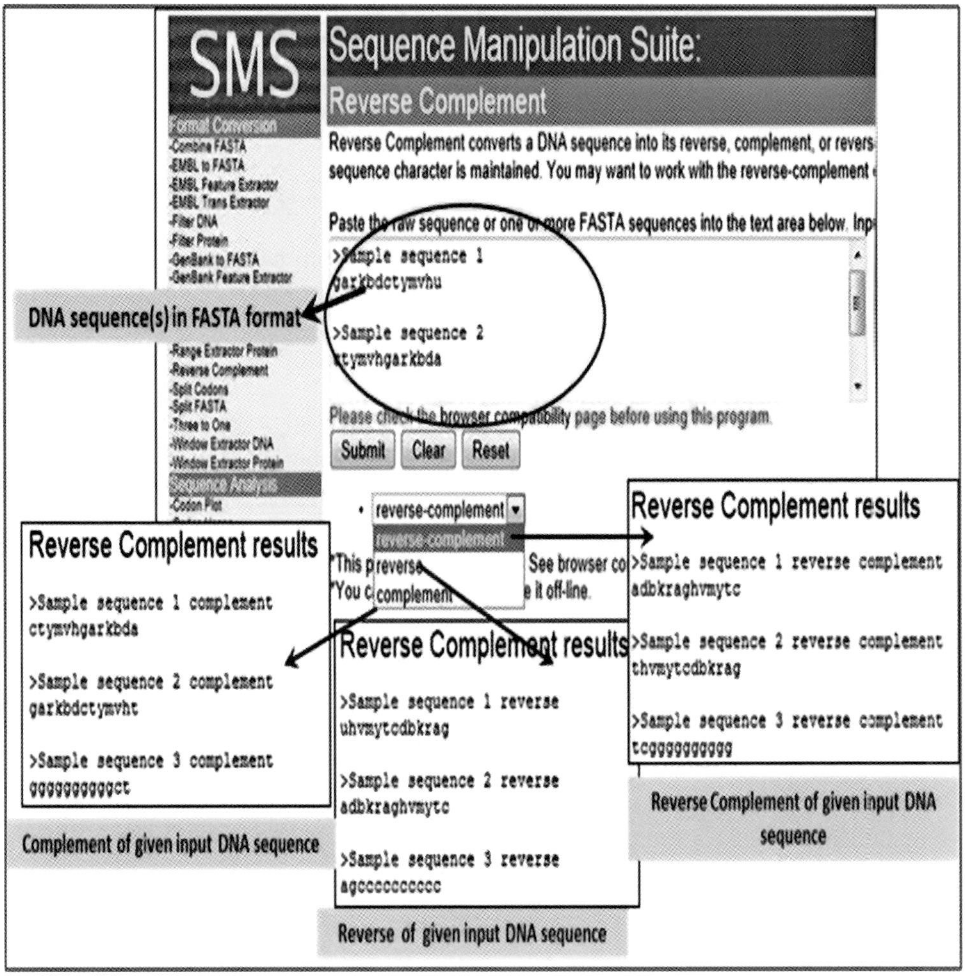

FIGURE 6.5 "Reverse Complement" input page and the corresponding output pages for "Complement", "Reverse" and "Reverse Complement", respectively (from left to right), of the input sequences.

6.5 SEQUENCE ANALYSIS

6.5.1 Codon Usage

This accepts single/multiple DNA sequence(s) in FASTA format, and estimates the number and frequency of usage of each of the 64 codons available in the specific genome (eukaryotic/prokaryotic, nuclear/mitochondrial, etc.). The output file presents the frequencies of occurrence of each of the codons in the given input sequence(s).The preference of a given sequence for a specific synonymous codon can be determined by this program. Input limit is 500 000 characters.

6.5.2 CpG Islands

This program estimates the Observed/Expected values for G/C dinucleotide contents in a 200 bp window, within a given DNA sequence and G/C content (Gardiner-Garden and Frommer, 1987). CpG islands are like islets within a given DNA sequence (split in windows of a specific length) that are characterized by a higher Observed/Expected ratio (>0.6) of Cytosine-Phosphate-Guanosine (CpG) dimers and GC content greater than 50%.

$$\text{Expected no. of CpG} = \frac{\text{No. of 'C'} * \text{No. of 'G'}}{\text{Window length}} \qquad [6.1]$$

This program can also be used for identifying the 5′ regions of vertebrate genes, since these regions are often thronging with CpG dimers in vertebrates. The maximum input limit is 100 000 characters.

6.5.3 DNA Molecular Weight

This calculates the molecular weight of double/single-stranded, linear/circular DNA sequence(s) (drop-down options are there to select the types of DNA molecule(s)) in FASTA format. Standard IUPAC base symbols are accepted. The character limit is 200 000. This program is used for calculating molecule copy number.

6.5.4 DNA Pattern Find

This program scans one or more submitted DNA sequence(s) for a specific pattern instructed by the user. The default pattern is "ctt[ca]", which searches for occurrences of "cttc" and "ctta". The user can modify it. The output file mentions the base positions (start and end) of the match, along with the number of times that it has been identified in the direct (original) or reverse strand. "*DNA Pattern Find*" is used to screen the input sequence (as a raw sequence of FASTA formatted sequence(s)) and localize the pattern of interest. The input limit is 500 000 characters.

6.5.5 DNA Stats

A very useful program to obtain the number, as well as the percentage, of each of the bases from the input sequence(s) in terms of the kinds of bases (means, pyrimidine, purine, A/T, etc.). The limit of input is 500 000 characters. The sequence(s) are submitted as a raw sequence, or as one or more FASTA-format.

6.5.6 Mutate for Digest

This program is used to explore mutable regions in a DNA sequence (provided in FASTA format) to generate a restriction site to study the effect of mutation on restriction digestion. The output file also displays the translation of the DNA (according to the reading frame indicated by the user), to determine the alterations in various reading frames (RFs) due to the proposed mutation. Thus, experiments involving polymerase chain reactions (PCR) or site-directed mutagenesis can be studied *in silico* using this program. Four parameters can be set for alteration of output:

a. *Search for future __<Restriction Enzyme Name>__ sites*: Almost all commercially available restriction enzymes (REs) have been enlisted. The user needs to choose one RE, according to the requirement or proposal.
b. *Show _<Number>_ of bases per line*: Choose any one of 30, 45, 60, 75, 90 and 105.
c. *Show the translation for reading frame*: RFs can be 1, 2, 3, all 3, upper case or none.
d. *Use the __<Genetic Code Options>_ genetic code*: The options are prokaryotic, eukaryotic, nuclear, or mitochondrial codes. The input limit is 10 000 characters.

6.5.7 Protein Isoelectric Point

The theoretical isoelectric point (pI) is calculated for single or multiple amino acid sequence(s) (in FASTA format, with input limit of 200 000 characters), to estimate the probable location of a protein on a 2D gel. The user can add up to five copies of one of the 21 optional epitopes and fusion protein tags listed (e.g., His6, HSV, Glu-Glu, etc.) to modify the pH of the submitted amino acid sequences (Figure 6.6).

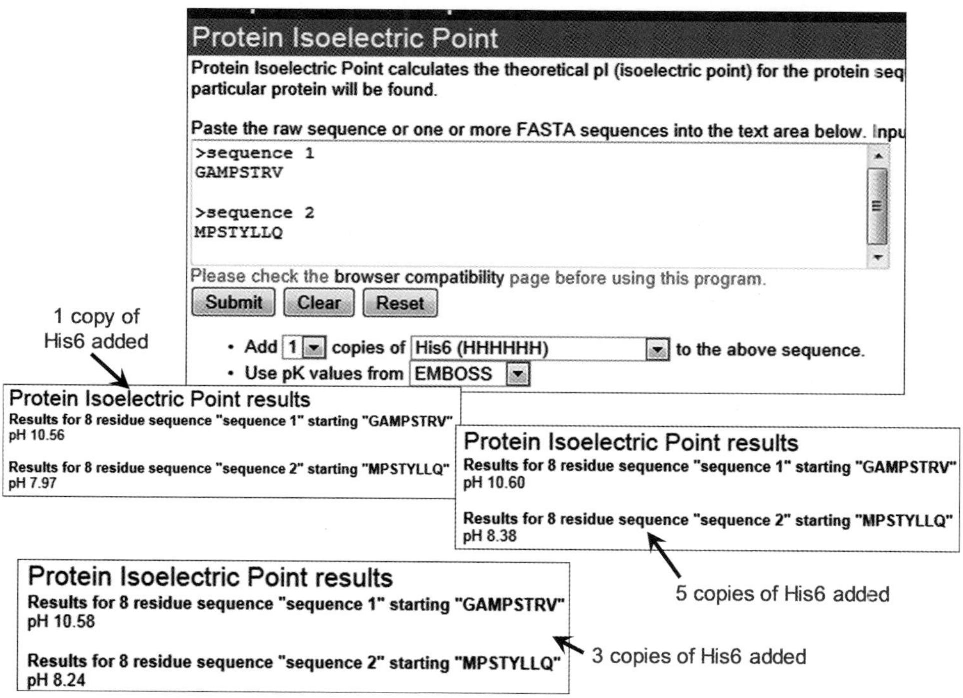

FIGURE 6.6 "Protein Isoelectric Point" input page and the corresponding output page with results, with respect to the parameters. (*See insert for colour representation of the figure.*)

6.5.8 Protein Molecular Weight

This calculates the molecular weight of one or more protein sequence(s), entered in FASTA format or as a raw (unformatted) sequence (character limit is 200 000). The user can append 1–5 copies of one out of the 21 enlisted epitopes and fusion proteins. This program is used to predict a recombinant or simple protein by determining the position of a particular protein on a gel, compared with a set of protein standards.

6.5.9 Protein Pattern Find

Similar to "DNA Pattern Find", this program is used to search a query (i.e., any consensus amino acid sequence) within one or more input sequence(s) (entered in FASTA format, and with character limit 500 000). The default search pattern is "X[^X]{0,5} X", which means that the user wants to search for the occurrence of two residues of the amino acid "X" which may be spanned by 0–5 amino acids (other than X) in between.

6.5.10 Protein Stats

Similar to the DNA Stats program, this is used to obtain data such as times of occurrences of each residue in one or more input sequence(s) (in FASTA format or raw sequence; input limit 500 000 characters).

6.5.11 Restriction Summary

Returns the positions of the restriction sites for all the enlisted regularly used REs against one or more linear or circular DNA sequence in FASTA format (100 000 base limit). This program is very useful to scan a DNA sequence for possible RE sites present.

6.5.12 Reverse Translate

This returns the reverse translated nucleotide sequence(s), along with a consensus sequence for each amino acid, from one or more input amino acid sequence(s) (in FASTA format with a limit of 20 000 characters), based on the codon usage table entered by the user (selected from http://www.kazusa.or.jp/codon/). This program is used to design oligos that target a (not yet sequenced) coding region belonging to a related species.

6.6 SEQUENCE FIGURES

6.6.1 Restriction Map

This displays a "textual map" for the *RE* sites in the template DNA (FASTA format input; input limit is 100 000 characters) which can be exploited for exploring the RE sites for cloning a sequence. It also returns the *in silico* translated amino acid sequence, according to the user-defined reading frame.

6.6.2 Translation Map

Depicts a textual map for displaying *in silico* translations of the input DNA sequence (in FASTA format; input limit is 500 000 characters), according to the first, second, third, or all three reading frames (RFs). This program understands IUPAC codes and different genetic codes being used.

6.7 RANDOM SEQUENCES

6.7.1 Mutate DNA

This introduces random mutation(s) in a coding sequence (presented in FASTA format as input sequence; input limit is 100 000 characters), which are studied to assess the effect of spontaneous mutation on the nature of the encoded peptide. The user can specify the number of mutation(s) and whether mutation is to occur in the start and stop codon of the mRNA.

6.7.2 Mutate Protein

Similar to the "Mutate DNA" program, this affects the mutation rate in an amino acid sequence. Multiple mutations can occur, just like in the "Mutate DNA" program, in the same amino acid position. This program is used to assess the effect of mutation on the chemical nature of the peptide, and the phenotypic effect on the trait.

6.7.3 Random Coding DNA

This produces a random coding sequence (ORF from start to stop codon), based on the user-specified genetic code and ORF length. Such ORFs are used to study the evolutionary perspectives and speciation.

6.7.4 Random DNA Sequence

Similar to "Random Coding Sequence", this generates random DNA instead of a coding sequence.

6.8 MISCELLANEOUS

a. **IUPAC codes**: The International Union of Pure and Applied Chemistry is a physical body that provides *codes for protein and DNA sequence*. The list has been given in the previous chapter (Chapter 5: Sequence Format Conversion).
b. Genetic codes, Browser compatibility, Reference, etc.

6.9 QUESTIONS

1. Find the reverse complement and reverse sequence of the following sequences:
   > Seq1_GenBank_Acc_No_ AB002707.1
   AGATAATACTTGAGACGTTCCAGTTTNTATTAGTACAAAATG
   NCCAATTCATTCAATGAATTGAGAAATGACATTCTAAGTGAG
   TTAGGAGCCACGACAATTGTAGAACACACAGTGTTTAACAAGT
   AACCAATGAGAATTNNTGATCTATCAATCAGTTGGTAGTATCG
   AGGACTACCAAGATTATAACGGAATAACGAGGAATT

> Seq2_GenBank_Acc_No_ KT779508.1
TGAGTAAATCAGTTATAGTTTGTTTGATGGTATCTACTA
CTCGGATAACCGTAGTAATTCTAGAGCTAATACGTGCAAC
AAACCCCGACTTCTGGAAGGGATGCATTTATTAGATAAAA
GGTCGACGCGGGCTCTGCCCGTTGCTGCGATGATTCATGA
TAACTCGACGGATCGCACGGCCATCGTGCCGGCGACGCAT
CATTCAAATTTCTGCCCTATCAACTTTCGATGGTAGGATA
GTGGCCTACCATGGTGGTGACGGGTGACGGAGAATTAGGG
TTCGATTCCGGAGAGGGAGCCTGAGAAACGGCTACCACAT
CCAAGGAAGGCAGCAGGCGCGCAAATTACCCAATCCTGAC
ACGGGGAGGTAGTGACAATAAATAACAATACCGGGCTCAAT
GAGTCTGGTAATTGGAATGAGTACAATCTAAATCCCTTAA

2. Determine whether the given sequence has a CpG island:
   > Seq1_GenBank_Acc_No_ XM_014823107.1
   ATGAGAAGCGGCATCATAGCGCAGTGCGCTTTCTGTGTAACTC
   GCGGCAACGTCGCTCAGGCAAGCTTTCGATTTCTGGCCCAGAAC
   TTCGGCCGCAAGATCTGTCCGCTAGCTTGGGCACACTCGTCGGA
   TCGGTGCCGCAGCTGCTTCTGGCGCGGCCGGATACCAGACATAC
   CAGAGCGTGATTACCTGCGTGTGTGGGCGCAAGAGGATCTCAAC
   GTCATCGTCATCGTCATGGCAACCCTTGGCAAGTTTGCCTTAAC
   GGTTACGTTCGCCGTCTGCTACCTGTACAGCGGTGAGATCTACC
   CGACTGCCATCCGGAATGTCGGACTTGGAAGCAATTCGGCTTGT
   GCGCGGGTCGGAGCGATGGTGGCGCCATATATCACCCTGCTGGC
   CAAGGACGTGGCGTGGCTGCCCATGGTACTGTTCGGCGCGCTGG
   CAGTGGTTGCTGCTCTGCTGGCAGCCATGTTGCCAGAGACGCGA
   AATTGCCATCTGCCAGAGACGATCGAAGACGGAGAGAATTT
   CAACAG
3. Enumerate the DNA statistics of the nucleotide sequence with NCBI Nucleotide accession number S78771.1
4. Open the sequence with NCBI GenBank acc. No. NM_001271282.2, and then extract the features of the sequence.
5. Provide the restriction summary of the sequence given below:
   > Seq1_GenBank_Acc_No_ XM_012883685.1
   ATGGTAGAGGACGAGGACGAAGACGAAGATACGTCTAACAACA
   GCAGCTCAGATGACAGCAGCAGCTCCGATGACGATGACGATGAC
   GTCCCAGACGATGACGAGTATGATGTTAAGAAAGTTAAGCACCG
   AGAGGAGGTGCCGCGCATTCAGATAGTTGGATCAAGGTCGCAAT
   GGTTGGAAGCAATCCGCAGAGACGGCACGGCAGGTGAGTCAGCT
   AGGATGAAGGCATTCTTAGAGGTATTTCGCGAAGCCCAACACCT
   TTATCCTGACCAGAGAGTTTCTGCTACCTCCGAGGAGACGAAGA
   CCCTTGATATCGTCGCCCTTATTCTAAAGGATGAAGGGAAAATC
   TGTGTGCAATATGATGGCATACTTCCGCCCCGCGATAGGGCAGC
   AGCGCTAAAGACATTCCAGGATGGGGCTCCAGCTACCTTTGTCTGA

Detection of Restriction Enzyme Sites

CHAPTER 7

CS Mukhopadhyay and RK Choudhary
School of Animal Biotechnology, GADVASU, Ludhiana

7.1 INTRODUCTION

The restriction enzyme (RE) sites present on a nucleotide sequence can be detected using a suitable *in silico* tool. A nucleotide sequence is subjected to detection of RE site(s) in some wet-lab experiments such as gene cloning, nucleotide sequencing, *in vitro* expression of a target protein, RFLP, AFLP, restriction mapping and restriction enzyme assays.

Several online tools for RE site detection are available. Some of the user-friendly and accessible web-based tools and their URLs (websites) have been tabulated below. Detailed procedures for determining RE sites using all these web tools will not be covered. This chapter will show how to use *NEBCutter* (New England Biolabs) as a tool for identifying RE sites.

a. *NEBCutter* (http://tools.neb.com/NEBcutter2/index): New England Biolabs hosts this RE site mapper software.
b. Webcutter 2.0 (http://rna.lundberg.gu.se/cutter2/): Another RE site detection software (online, free) for linear and circular DNA.
c. Mapper (http://arbl.cvmbs.colostate.edu/molkit/mapper/index.html): Java platform-based online software for mapping the RE sites on a target sequence.
d. Web Map (https://pga.mgh.harvard.edu/web_apps/web_map/start): This tool maps the RE sites for circular or linear nucleotide sequences. The reverse complement of the given sequence can also be checked for mapping of RE sites.
e. Restriction-Mapper (http://restrictionmapper.org/): Online, freely available tool for mapping restriction endonuclease sites on a DNA sequence.

7.2 OBJECTIVE

To identify the RE site(s) present in a given nucleotide sequence.

Basic Applied Bioinformatics, First Edition. Chandra Sekhar Mukhopadhyay, Ratan Kumar Choudhary and Mir Asif Iquebal.

7.3 PROCEDURE (USING NEBCUTTER)

7.3.1 Select the nucleotide sequence

This could be a nucleotide sequence of any length, depending on the purpose. Insert for gene cloning, or cDNA expression, an amplicon for detection of mutation (RFLP), or a DNA sequence which is to be screened for the presence of RE sites (RE assay).

7.3.2 Open the online tool for RE site detection, *NEBCutter* v2.0

This is hosted by New England BioLabs (Vincze *et al.*, 2003); the URL is http://tools.neb.com/NEBcutter2/index.

7.3.3 Paste or upload the sequence

There are several options to enter the sequence (one at a time) under study:

a. **Local sequence file**: Prepare a notepad (*.txt) file and maintain individual the DNA sequence in FASTA format. The maximum size of the file can be 1 Mb.
b. **GenBank Accession Number**: Paste/type the NCBI, GenBank accession number in the text box provided. You can browse GenBank online to select or view the input sequence flat file.
c. **Paste the DNA sequence**: You can directly paste the input DNA sequence (in FASTA format) in the sequence box provided. The maximum input limit is 300 kilobases.
d. **Standard sequences- # Plasmid vectors**: This is a drop-down menu from which you need to select your plasmid vector for RE analysis. Keep the parameters (e.g., 'Local sequence file', 'GenBank Accession Number' and 'Paste in your DNA sequence') blank to screen the Plasmid vector.
e. **Standard sequences- # Viral + phage**: Similar to the above, but enlist the sequences of phage and viruses.

7.3.4 Selection of options

- **Linear or circular**: Select the appropriate radio button to indicate whether your input sequence is a linear or circular sequence.
- **Enzymes to use**: There are five alternatives, of which you can select any one. The first three options enable you to select enzymes from New England Biolabs (NEB)

DEFINING YOUR OWN OLIGOS (FIGURE 7.1)

a. The user can define their own oligos (short stretch of nucleotide sequences) that will be searched in the given input sequence and then reported by the program.
b. Up to 40 oligos can be defined with a specific identifier/name (naming is optional).
c. The symbols used are: caret (^) to indicate a site on the forward strand; underscore (_) for a site on the complementary strand. Oligo showing such symbols signifies a sticky end cut, while a vertical bar or pipe symbol (|) means the presence of the site on both strands and, hence, refers to blunt cuts.

FIGURE 7.1 A short nucleotide sequence (oligo) can be searched in the input sequence for determining specific RE sites present in the oligos. (*See insert for colour representation of the figure.*)

or other commercially available ones. The last two options allow you to choose some defined oligonucleotide sequence(s). There is a link "[define oligos]" which allows the user to define his/her own oligos.

Users can also use the following options to make the search for RE sites more stringent by clicking "More Options" (Figure 7.2).

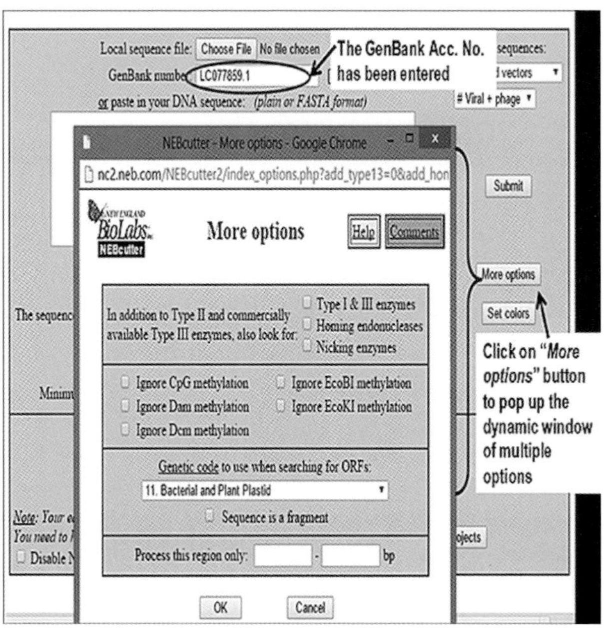

FIGURE 7.2 More options enable the user to make stringent selection of RE sites.

7.3.5 Following options are available under "more options"

7.3.5.1 "Type I and III enzymes" check box

NEBCutter, by default, screens for type II REs (the cleaving location is adjacent to or within the recognition site, independent of methylase, and the REs are magnesium-dependent). Checking this box will also enable *NEBCutter* to search for the Type I (cleaving location remote from recognition site; exerts both restriction and methylase activity, and are ATP-dependent) and Type III (cleaving location similar to Type II; complexed with methylase and are ATP-dependent) REs.

7.3.5.2 Homing endonucleases

The endonucleases that catalyze the hydrolysis of genomic DNA within the cells synthesizing them. Check this box to include homing endonucleases.

7.3.5.3 Nicking enzymes

These endonucleases cut only one strand of a double-stranded DNA at a specific recognition site. Checking will include these enzyme sites for screening the input sequence.

7.3.5.4 Check boxes to ignore some of the specific sites in the input sequences

The checked methylase(s) will be ignored while screening for overlapping methylation sensitivity of the enzymes. The methylation-sensitive restriction enzymes (MSREs) cannot cleave methylated cytosine and, thus, are used to analyze methylated DNA and the methylation status of cytosine residues in CpG sequences:

a. Ignore CpG methylation
b. Ignore EcoBI methylation
c. Ignore Dam methylation
d. Ignore EcoKI methylation
e. Ignore Dcm methylation

7.3.5.5 Genetic code to use when searching for ORFs

A drop-down list of open reading frames (ORFs) is available. The user needs to select the ORF, depending on the input sequence.

7.3.5.6 "Sequence is a fragment" check-box

Check this when a partial (i.e., an in-between fragment of a larger string) nucleotide sequence (missing Start or Stop codon) is submitted.

7.3.5.7 "Process this region only" check-box

Only the specified region is prepared for screening.

7.3.5.8 Set Colors

The user can set colors for different portions of the graphical output, such as Scale, Cut-Size (blunt, 5′ and 3′ extensions, highlighted ORF and so on).

7.3.5.9 Minimum ORF length to display

Applicable for coding sequence. Minimize the size of the ORF if the coding sequence is shorter.

7.3.5.10 Name of the sequence (optional)

The user needs to provide a name to the sequence being analyzed, as an identifier.

7.3.5.11 Other options

These include a check box for "Disable NEBcutter cookies" and a button to "Delete projects".

7.3.5.12 "Submit" button

To initiate screening of the input sequence for RE site(s).

7.3.6 Inferring the output

The output page displays the whole input sequence as a single line (if linear sequence option has been selected) with points of RE sites highlighted (Figure 7.3). Each of the RE shown on the RE-site is hyperlinked (appearing as blue-colored text), with the page containing detail for the RE. The color schema is displayed at the top-right part of the output page. Note that the meaning of orange-colored hash (#) indicates susceptibility of the enzyme to methylation caused by common methylases of *E. coli* origin. The asterisk (*) symbol indicates susceptibility of the enzyme to CpG methylation.

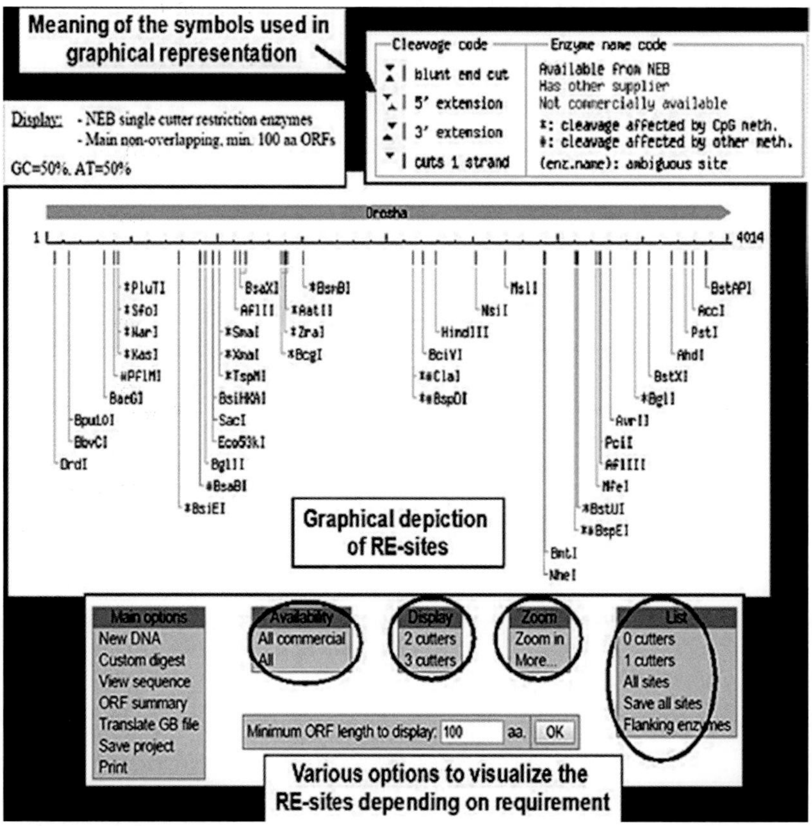

FIGURE 7.3 Result output window of *NEBCutter*. Details are discussed in the text under the sub-heading "Inferring the output".

The page also contains five small panes with hyperlinked words. These have been tabulated and explained in Table 7.1. The explanations have been adopted from the help provided by the *NEBCutter* tool (http://tools.neb.com/NEBcutter2/help/main_display.html), so sometimes the lines may be verbatim.

TABLE 7.1 Meaning of different terminologies used in *NEBCutter* (Vince *et al.*, 2003).

Term	Meaning
Main Options	
New DNA	This button, when clicked, opens the initial page
Custom digest	To check digestion of input DNA sequence using a set of REs. These enzymes can be further categorized based on the type of the restriction end (blunt or 3′-overhang or 5′-overhang) or position of the restriction sites in the target sequence, e.g., REs sites within the input DNA sequence
View sequence	To get the input sequence
ORF summary	Tabulates following information about the genes that are displayed: coordinates of the genes; length of polypeptide; GenBank protein IDs of the respective gene sequences; single-cutter REs
Translate GB file	The ORF-finder program of this tool predicts all the non-overlapping, large open reading frames
Save project	Saves the current project in the user's local disk as a compressed file which can be again uploaded to the site later.
Print	To produce a printable file of the current project in PDF, EPS or GIF format.
Availability	
All Commercial	Displays the REs commercially available from any agencies. The default is NEB-produced REs.
All	Displays all commercial appropriate literature cited, but not commercially available REs.
Display	
1, 2 or 3 cutters	The default is one cutter REs. The user can also specify displaying two or three cutters, separately.
Alternative/ Normal	"Alternative" will switch to alternative linear display for two and three cutter REs. Normal will display in the default fashion for all REs together on the scale.
Zoom	
Zoom in or Unzoom	This enlarges a selected region to a higher resolution (up to base level). More will pop up a window to specify the coordinates to be displayed.
List	
0, 1 or 2 cutters	The page contains a list of REs as specified by the users (non-cutter/single-cutter/double-cutter). The table contains the name of the enzyme and the RE site (specificity) which can be saved as a text file. The user can also modify the search on some cutters in this page.
All sites	Enlists all the RE sites according to their location along the input sequence
Save all sites	To save the list of all sites in computer, in *.txt format. The name of the file will be the same as the name of the sequence given by the user.
Flanking enzymes	This is very useful for some genetic studies. The user can identify the REs for the regions flanking a target region on the input sequence.

7.4 QUESTIONS

1. Let us say we are selecting the partial, intergenic spacer sequence of *Theileria annulata* (NCBI, GenBank Acc. no. AJ538184.1) and *Babesia bigemina* (AJ538183.1) to detect the presence of the RE site for Cfr131. The objective is to differentiate these two species, based on the presence of the reported RE site (*Current Science* (2007), **93**(12), pp. 1840–1843). Check which of the following sequences harbors the RE site.
2. Given the vector pBR322 (NCBI GenBank Acc. No. J01749.1), and the insert sequence AJ812216.1, select at least two suitable restriction enzymes which can be used for inserting the insert into the vector sequence at one position.
3. Diagrammatically depict (using the *NEBCutter* online tool) the RE sites on the mRNA sequence AY762972.1 for the rare-cutters SmaI and NotI.
4. Sickle cell anemia in human occurs due to a missense mutation in the 6th amino acid (Glutamate to Valine: dbSNP Id rs334) in the beta chain of hemoglobin. The mutation is GAG to GTG. Determine which RE can be used to detect the SNP in an individual if the sequence of the amplicon is 5′-GACACCATGGTGCATCTGACTCCTG[G/T] GGAGAAGTCTGCCGTTACTGCCCTG-3′.
5. Beta-Casein is an important constituent of milk. There are two types of allelic variants, namely, Type A1 and A2 ("CCT" and "CAT", respectively, at position 350 of both the sequences). Given the sequences of these types of beta casein of bovine milk, determine the specific restriction enzyme that can be used for RFLP study for discerning the two allele-types:

   > XM_010797953|CSNA2
   ATGCCATTAAATACTATATATAAACAACCACAAAATCAGA
   TCATTATCCATTCAGCTCCTCCTTCACTTCTTGTCCTCTA
   CTTTGGAAAAAAGGAATTGAGAGCCATGAAGGTCCTCATCC
   TTGCCTGCCTGGTGGCTCTGGCCCTTGCAAGAGAGCTGGAA
   GAACTCAATGTACCTGGTGAGATTGTGGAAAGCCTTTCAAG
   CAGTGAGGAATCTATTACACGCATCAATAAGAAAATTGAGA
   AGTTTCAGAGTGAGGAACAGCAGCAAACAGAGGATGAACTC
   CAGGATAAAATCCACCCCTTTGCCCAGACACAGTCTCTAGT
   CTATCCCTTCCCTGGGCCCATCCATAACAGCCTCCCACAAA
   ACATCCCTCCTCTTACTCAAACCCCTGTGGTGGTGCCGCCT
   TTCCTTCAGCCTGAAGTAATGGGAGTCTCCAAAGTGAAGGA
   GGCTATGGCTCCTAAGCACAAAGAAATGCCCTTCCCTAAAT
   ATCCAGTTGAGCCCTTTACTGAAAGGCAGAGCCTGACTCTC
   ACTGATGTTGAAAATCTGCACCTTCCTCTGCCTCTGCTCCA
   GTCTTGGATGCACCAGCCTCACCAGCCTCTTCCTCCAACTG
   TCATGTTTCCTCCTCAGTCCGTGCTGTCCCTTTCTCAGTCC
   AAAGTCCTGCCTGTTCCCCAGAAAGCAGTGCCCTATCCCCA
   GAGAGATATGCCCATTCAGGCCTTTCTGCTGTACCAGGAGC
   CTGTACTCGGTCCTGTCCGGGGACCCTTCCCTATTATTGTC
   TAAGAGGATTTCAAAGTGAATGCCCCCTCCTCACTTTTGAA
   TTGACTGCGACTGGAAATATGGCAACTTTTCAATCCTTGCA

TCATGTTACTAAGATAATTTTTAAATGAGTATACATGGAAC
AAAAAATGAAACTTTATTCCTTTATTTATTTTATGCTTTTT
CATCTTAATTTGAATTTGAGTCATAAACTATATATTTCAAA
ATTTTAATTCAACATTAGCATAAAAGTTCAATTTTAACTTG
GAAATATCATGAACATATCAAAATATGTATAAAAATAATTT
CTGGAATTGTGATTATTATTTCTTTAAGAATCTATTTCCTA
ACCAGTCATTTCAATAAATTAATCCTTAGGCATA

> XM_010806178|CSNA1
ATGCCATTAAATACTATATATAAACAACCACAAAATCAGATCAT
TATCCATTCAGCTCCTCCTTCACTTCTTGTCCTCTACTTTGG
AAAAAAGGAATTGAGAGCCATGAAGGTCCTCATCCTTGCCTGC
CTGGTGGCTCTGGCCCTTGCAAGAGAGCTGGAAGAACTCAATG
TACCTGGTGAGATTGTGGAAAGCCTTTCAAGCAGTGAGGAATC
TATTACACGCATCAATAAGAAAATTGAGAAGTTTCAGAGTGAG
GAACAGCAGCAAACAGAGGATGAACTCCAGGATAAAATCCACC
CCTTTGCCCAGACACAGTCTCTAGTCTATCCCTTCCCTGGGCC
CATCCCTAACAGCCTCCCACAAAACATCCCTCCTCTTACTCAA
ACCCCTGTGGTGGTGCCGCCTTTCCTTCAGCCTGAAGTAATGG
GAGTCTCCAAAGTGAAGGAGGCTATGGCTCCTAAGCACAAAGA
AATGCCCTTCCCTAAATATCCAGTTGAGCCCTTTACTGAAAGC
CAGAGCCTGACTCTCACTGATGTTGAAAATCTGCACCTTCCTC
TGCCTCTGCTCCAGTCTTGGATGCACCAGCCTCACCAGCCTCT
TCCTCCAACTGTCATGTTTCCTCCTCAGTCCGTGCTGTCCCTT
TCTCAGTCCAAAGTCCTGCCTGTTCCCCAGAAAGCAGTGCCCT
ATCCCCAGAGAGATATGCCCATTCAGGCCTTTCTGCTGTACCA
GGAGCCTGTACTCGGTCCTGTCCGGGGACCCTTCCCTATTATT
GTCTAAGAGGATTTCAAAGTGAATGCCCCCTCCTCACTTTTGA
ATTGACTGCGACTGGAAATATGGCAACTTTTCAATCCTTGCAT
CATGTTACTAAGATAATTTTTAAATGAGTATACATGGAACAAA
AAATGAAACTTTATTCCTTTATTTATTTTATGCTTTTTCATCT
TAATTTGAATTTGAGTCATAAACTATATATTTCAAAATTTTAA
TTCAACATTAGCATAAAAGTTCAATTTTAACTTGGAAATATCA
TGAACATATCAAAATATGTATAAAAATAATTTCTGGAATTGTG
ATTATTATTTCTTTAAGAATCTATTTCCTAACCAGTCATTTCA
ATAAATTAATCCTTAGGCATA

Sequence Alignment

SECTION

II

Dot Plot Analysis

CHAPTER

8

CS Mukhopadhyay and RK Choudhary
School of Animal Biotechnology, GADVASU, Ludhiana

8.1 INTRODUCTION

A two-dimensional (2D) plot depicting one or more of the various sequence features (sequence similarities, direct and/or inverted repeats, motifs, gaps, sequence inversions, etc.) is called a dot plot. A single sequence, or two different sequences (with the same type of residues), can be studied to *reveal the hidden sequence features*. Dot plot has been used for local (not global) alignment, and was identified as a very powerful tool for molecular sequence analysis as early as during the late 1960s (Fitch, 1969).

8.2 OBJECTIVE

To compare two homologous molecular sequences using a dot plot.

8.3 PROCEDURE

Molecular sequences can be subjected to dot plot analysis using online tools like Dotlet, Dotter, and so on.

a. Dotlet (http://myhits.isb-sib.ch/util/dotlet/doc/dotlet_help.html). This is freely available online and is used as a tool for diagonal plotting of sequences.
b. Dot plot(+) (http://www.hku.hk/bruhk/gcgdoc/dot plot.html). Dot plot(+) software can identify the overlapping portions of two sequences, and also any repeats and inverted repeats of a particular sequence.
c. Dotter (http://sonnhammer.sbc.su.se/Dotter.html). A graphical dot plot program for thorough comparison of two molecular sequences. Dotter can be run on any of the following operating systems: MAC, Linux, Sun Solaris and Windows OS.

Two different sequences, or a single sequence, can be placed along the vertical and the horizontal axes of a matrix for analysis using a dot plot. The query and the subject

Basic Applied Bioinformatics, First Edition. Chandra Sekhar Mukhopadhyay, Ratan Kumar Choudhary and Mir Asif Iquebal.

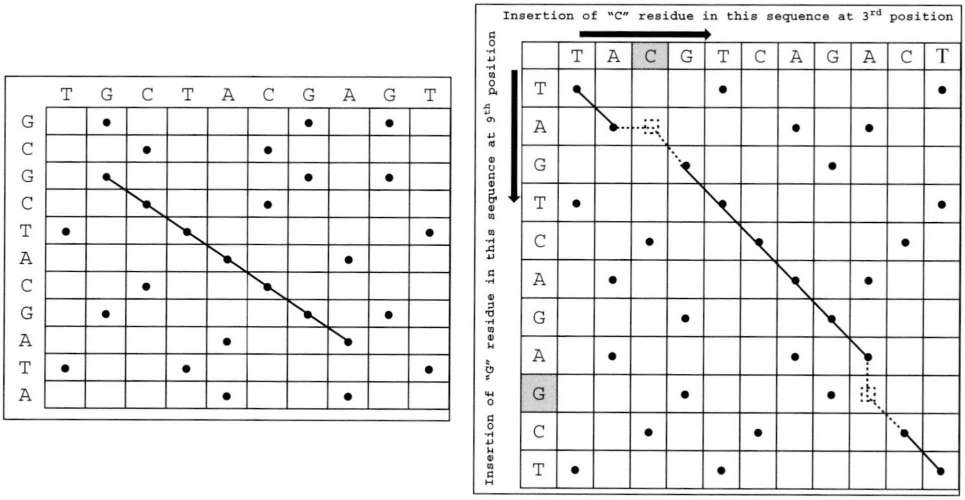

FIGURE 8.1 Depiction of plotting the straight line based on the runs of dots obtained from matches between residues along the *X*- and *Y*-axes. Insertion in any of the sequences will distort the run of the straight line.

sequences are placed along rows (*Y*-axis) and columns (*X*-axis), respectively. Next, a dot is placed in the cells, where the two axes have the same residue. Thereby, a subset of the sequence which has a run of identical residues will form a straight line (Figure 8.1).

8.4 PARAMETERS OF DOT PLOT ANALYSIS

There are two main parameters optimized during a dot plot analysis: window size and mismatch limit.

8.4.1 Window size

This determines the run of residues that must match in both sequences. If the specified number of residues at a stretch is matching, the graph will not indicate any mark of dot(s). Window size, thus, monitors the *background noise*. The smaller the window size, the more background noise there will be. Again, a very high window size will produce a clean plot, devoid of any indication of sequence similarity.

8.4.2 Mismatch limit

This parameter allows one to tolerate a specified number of mismatches, thereby indicating the stretches of residues with sequence similarity. The limit specified by different software ranges from 1 to 3.

Please note that these dot plot analyses have been done using http://www.vivo.colostate.edu/molkit/dnadot/ and https://wssp.rutgers.edu/StudentScholars/WSSP08/Dot plotter/Dot plotPractice.html?destination=StudentScholars/WSSP08/Dot plotter/Dot plotPractice.html, online tools which are no longer available.

8.5 INTERPRETATION

Dot plot analysis reveals several sequence features at a glance. Some examples have been given below:

8.5.1 Insertion(s)/deletion(s) in a pair of sequences

The sequence-pair being compared using dot plot may differ due to insertion(s)/deletion(s) at one or more positions. These InDels are reflected by a break in the straight line (Figure 8.1). Insertion in the horizontal sequence (or deletion in the vertical sequence) will necessitate horizontal movement, and a break in the straight line and insertion in the vertical sequence (or deletion in the horizontal sequence) will be indicated by vertical movement and discontinuity in the straight line. The third base (i.e., "C") of the horizontal sequence and the ninth base (i.e. "G") of the vertical sequence are the insertions (highlighted yellow in the second diagram, Figure 8.1) in those respective sequences.

8.5.2 Identifying repeat sequences

The presence of repeat sequence(s) can be detected by a dot plot (Figure 8.2). The same sequence is placed along the horizontal and vertical axes. There is four fold

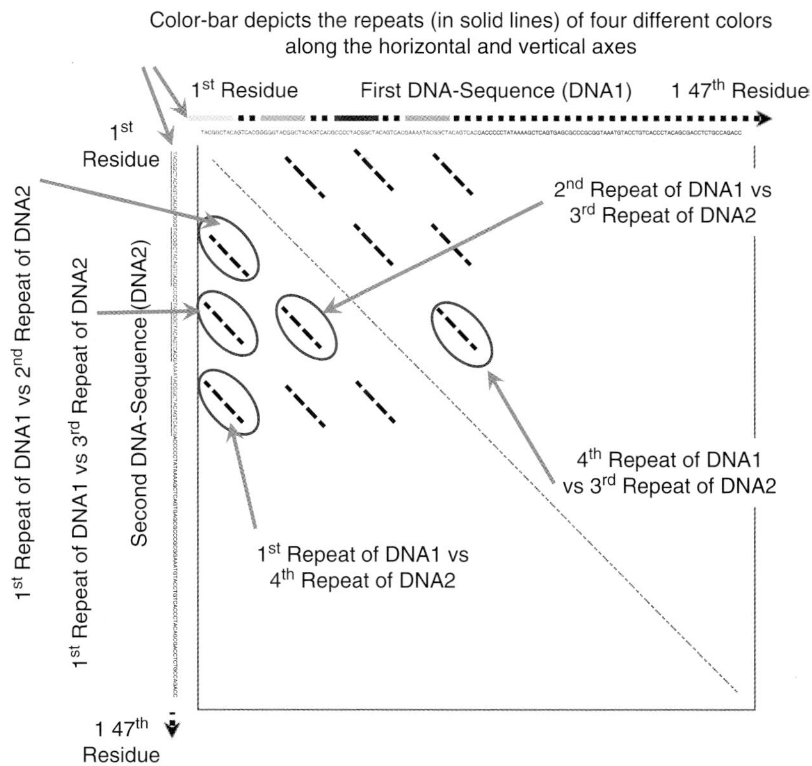

FIGURE 8.2 Interpretation of dot plot based on the same repeat sequence (shown above) which has been placed along both axes. The four different colors (yellow, green, blue and gray) have been shown to indicate the 1st, 2nd, 3rd and 4th repeat of "TACGGCTACAGTCACG". (*See insert for colour representation of the figure.*)

repetition of the same sequence "TACGGCTACAGTCACG", intervened by short tetramers of different sequences:

TACGGCTACAGTCACGGGGGTACGGCTACAGTCACG CCCCTACGGCTACAGTCACGAAAATACGGCTACAGTCACG ACCCCCTATAAAAGCTCAGTGAGCGCCCGCGGTAAATGTACC TGTCACCCTACAGCGACCTCTGCCAGACC

In the dot plot result, we find one diagonal line representing the full sequence, and some short fragmented lines parallel to the diagonal. These short lines represent the repeat sequences. Every fragment stands for alignment of the repeats with each other.

- An unbroken main diagonal signifies that both sequences are the same.
- If the insertion is in the first sequence (along the *X*-axis), the graph will progress along the *X*-axis.
- If the insertion is in the second sequence (*Y*-axis), the graph will proceed along the *Y*-axis.

Nucleic acid sequences have usually a poor signal/noise ratio, because there are only four different bases: decreasing the window size increases the noise.

8.5.3 Unraveling other sequence features

A nucleotide sequence may produce a *stem-loop secondary structure* when it has a palindromic sequence intervened by a short sequence. Similarly, there may be an inversion in the other half of a given sequence. Dot plot analysis can reveal such features. Inverted sequences will produce a main diagonal line between the other two corners (the corners adjacent to the end terminals of the sequences) of the matrix. Smaller diagonal lines are symmetrically parallel to the main diagonal, which indicates that the same repeat is there in the sequence in tandem.

8.6 QUESTIONS

1. Briefly describe how a dot plot will look like in the following conditions:
   a. Plotting a single sequence with itself.
   b. Plotting a sequence against its reverse sequence.
   c. Plotting a sequence with internal repeats with itself.
   d. Plotting a sequence against the same, but after inserting a short sequence somewhere within.
2. Prepare a dot plot with the following sequences, with window sizes 5, 7 and 10:
   >Seq1
   MMNRVQPENVHSTIFTPREYQVELVDACLKGNTLSVLASRST
   RTFLITMVTREMAHLVDACLKGNTLSVLASRSTRTLTRSKEQ
   GGKGQLVDACLKGNTLSVLASRSTRTRTLLTGWSGPGLVRAG
   EAIQQNTNLAVTTYTRLEQVDGWLPSRWSHTFTEAQVIIMTV
   DVLEKGLETGLLQLDMLNLLVITDAHRVATMMNRVQPENVHS
   TIFTPREYQVELVDACLKGNTLSVLASRSTRTFLITMVTREM
   AHLVDACLKGNTLSVLASRSTRTLTRSKEQGGKGQLVDACLK

GNTLSVLASRSTRTRTLLTGWSGPGLVRAGEAIQQNTNLAVT
TYTRLEQVDGWLPSRWSHTFTEAQVIIMTVDVLEKGLETGLL
QLDMLNLLVITDAHRVAT

>Seq2
TAVRHADTVNMDGTGKVDVTMVATTHSWRSWGDVRTYTTVAN
TNAGARVGGSWGTTRTRTSRSAVSTNGKCADVGKGGKSRTTRT
SRSAVSTNGKCADVHAMRTVMTTRTSRSAVSTNGKCADVVYRT
TSHVNVRNMMTAVRHADTVNMDGTGKVDVTMVATTHSWRSWGD
VRTYTTVANTNAGARVGGSWGTTRTRTSRSAVSTNGKCADVGK
GGKSRTTRTSRSAVSTNGKCADVHAMRTVMTTRTSRSAVSTN
GKCADVVYRTTSHVNVRNMM

3. Suppose a sequence has tandem repeats (like VNTRs). Explain how the dot plot will look when the sequence is plotted against itself.

> Hint: The greater the number of diagonals, the greater the number of tandem repeats.

4. The consensus palindrome sequence TGTGAGCGCTCACA is given (from *Proceedings of the National Academy of Sciences of the USA* **81**, 1624–1628). Describe the dot plot if you impose no restriction to window-size and word match.
5. Under what circumstances should we use dot plot before multiple or pairwise sequence alignment?

Needleman–Wunsch Algorithm (Global Alignment)

CHAPTER 9

CS Mukhopadhyay and RK Choudhary
School of Animal Biotechnology, GADVASU, Ludhiana

9.1 INTRODUCTION

The Needleman–Wunsch algorithm (NWA; Needleman and Wunsch, 1970) is used for global alignment. We compare homologous molecular sequences character by character to achieve sequence alignment. Global alignment is the end-to-end alignment between two sequences; hence, it introduces gaps that represent insertions/deletions. This is useful for identifying "InDels" (**In**sertion and **Del**etions), and for overall comparison of two or more comparable (i.e., similar) sequences. Phylogenetically close sequences of the same length are the most suited for global alignment.

The best alignment can be identified by quantifying or scoring the possible alternative alignments. Scoring matrices are used to award the match(es) and penalize the mismatch(es) and gap(s), so that the best alignment with the highest score can be identified. The scores in the matrix are integer values (e.g., +1, 0, –1).

> **Dynamic Programming (DP)** means that the scores of the subsequent cells can only be determined if the initial cells (towards the top left) have been scored. The DP methodology requires computation of the initial values (at the left and top side of the matrix) to obtain the later values of the cells (located towards the right and bottom side) in the matrix.

9.2 OBJECTIVE

To align two comparable sequences (nucleotide or amino acid) for obtaining a global alignment using the NWA.

Basic Applied Bioinformatics, First Edition. Chandra Sekhar Mukhopadhyay, Ratan Kumar Choudhary and Mir Asif Iquebal.

9.3 PROCEDURE

Let us start with two short sequences which are to be globally aligned using an NWA:

- Seq1: CTAGTAG
- Seq2: CAGGTAGTG

If the sequences are of length "*m*" and "*n*", respectively, we will obtain one scoring matrix of dimensions $m \times n$.

9.3.1 Step 1: define a scoring scheme

A scoring scheme is first defined, and then the dynamic scoring is done, starting from the top left to the bottom right of the matrix:

Match score (S_{ij}) = 2
Mismatch score (S_{ij}) = –1
Gap penalty (d) = –2

Here, "*i*" (varying from 1, 2, …, *m*) and "*j*" (varying from 1, 2, …, *n*) refer to the indexes for the row and column numbers, respectively, of the cell of the scoring matrix (described in Step 2). Please note that the scoring scheme can differ according to the *evolutionary relatedness* of the input sequences. A gap extension penalty is also introduced in advanced methods of such dynamic algorithms, to direct proper alignment and select the best one, based on the scores.

9.3.2 Step 2: initiation of matrix construction

This step starts with creating a matrix that represents the score for each pair of residues belonging to the pair of sequences.

a. Seq1: written along the vertical axis (or *Y*-axis) of the matrix (i.e., along the rows).
b. Seq2: written along the horizontal axis (or *X*-axis) of the matrix (i.e., along the columns).

TABLE 9.1

			Sequence 2 →								
			"j" varying from 1, 2, 3 …, n								
			C	A	G	G	T	A	G	T	G
← Sequence 1	"i" varying from 1, 2, 3 …, m	C									
		T									
		A									
		G									
		T									
		A									
		G									

Finally, we will get a matrix with each of its cells filled up with three scores obtained from three different types of movements, as shown below:

- **Horizontal movement**: represents a gap in the sequence written vertically (along the *Y*-axis, row numbers have been denoted by "*i*")
- **Vertical movement**: represents a gap in the sequence written horizontally (along the *X*-axis, column numbers are indicated by "*j*")
- **Diagonal movement**: representing either match or mismatch between the two sequences (cell positions indexed by "*i*" and "*j*", respectively) (Figure 9.1).

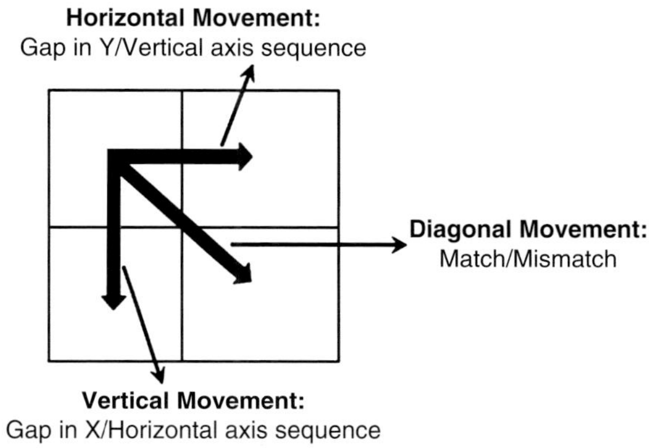

FIGURE 9.1 Three types of movement along the matrix in dynamic programming.

Annex one row at the top and one column at the left of the original $m \times n$ dimensional matrix (to make it an $(m+1) \times (n+1)$ dimensional matrix). These are termed the 0th row and 0th column, respectively. Next, fill the first cell (i.e., the cell located at the top left-most corner of the 0th row and 0th column) with a zero.

9.3.2.1 Scoring method: dynamic programming

The score of each cell can be determined from three different movements towards the cell. The formula used to calculate the scores for each cell is shown in Figure 9.2.

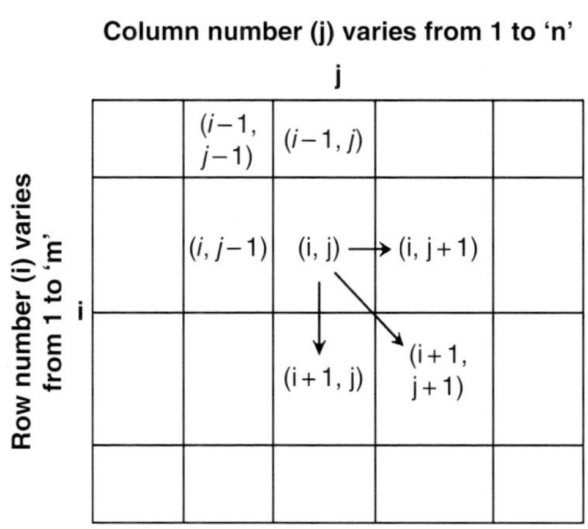

FIGURE 9.2 Increment in the respective indexes of the cells (denoting row and column numbers, respectively) of the matrix, to indicate the movement along the cells.

$$F(i,j) = \max \begin{Bmatrix} Fi-1, j+d\left(\text{Gap in horizontal}-\text{axis sequence; Vertical movement}\right) \\ Fi-1, j-1+s(i,j)\left(\text{match/mismatch; Diagonal movement}\right) \\ Fi, j-1+d\left(\text{Gap in vertical}-\text{axis sequence; Horizontal movement}\right) \end{Bmatrix}$$

[9.1]

Select the highest score(s): Now, after obtaining these three scores in the cell being studied, the highest value is selected, which is to be used as the score of that cell. This highest score is used for calculating the respective scores of the cells located just right (hence, pertaining to horizontal movement), just bottom (vertical movement) and just at the bottom-right (diagonal movement) positions of the current cell. If more than one score shows the highest value, then that highest value will be considered for calculating the score of the adjacent cells. However, the backtracking will be through all these cells contributing to the same highest values (see Figures 9.3 and 9.4).

Let us clarify this with the first few cells of the matrix shown in Step 3:

a. Put "0" value in the first cell (utmost top left cell).
b. The value of the next cell (right side) will be 0 + (–2) = –2; since it is a horizontal movement, so the gap penalty of –2 will be given. The next cell at the right side will have the score of –2 + (–2) = –4, due to the gap penalty for the same horizontal movement. Likewise, the subsequent cells on the right side will be awarded –6, –8, –10, –12, –14, –16, and the last one –18.
c. Now, for the downward movement in the very first column, the gap penalty will be consecutively awarded to each cell. The cells will have the scores of –2, –4, … –14.
d. The diagonal movement starts from 0 to the bottom right cell; we will check whether there is a match or mismatch. In this case, there is a match of "C" to "C". Hence, a score of +2 will be awarded.
e. Please note that, except for the first row and first column (which were annexed to the original matrix), every cell of the matrix can have three values, due to three possible movements towards that cell:
   - vertical movement from the cell just above,
   - horizontal movement from the cell just at left, and
   - diagonal movement from the adjacent cell just at the top-left position).

   In the following matrix, three colors have been used: Blue for vertical movement (Gap), Yellow highlight for diagonal movement (Match or Mismatch), and Red for horizontal movement (Gap).
f. Now, out of these three values, the highest one will be selected as the score for that cell. The chosen score(s) has/have been highlighted in yellow in the matrix below.

9.3.3 Step 3: trace-back step

Finally, the trace-back step starts from the last cell at the right side of the bottom row in a way such that arrow(s) will be drawn to the cell(s) from which the current score (highest of three scores of the present cell) has been obtained. If there are two or three cells which contribute to the highest value, then two (or three, respectively) arrows will be drawn to indicate the previous cells.

The arrows are drawn from the bottom right cell towards the top left cell. If certain cell(s) have more than one arrow, this will lead to more than one path at every such junction.

		C	A	G	G	T	A	G	T	G
	0	−2	−4	−6	−8	−10	−12	−14	−16	−18
C	−2	[−4, 2, −4]	[−6, −3, 0]	[−8, −5, −2]	[−10, −7, −4]	[−12, −9, −6]	[−14, −11, −8]	[−16, −13, −10]	[−18, −15, −12]	[−20, −17, −14]
T	−4	[0, −3, −6]	[−2, 1, −2]	[−4, −1, −1]	[−6, −3, −3]	[−8, −2, −5]	[−10, −7, −4]	[−12, −9, −6]	[−14, −8, −8]	[−16, −13, −10]
A	−6	[−2, −5, −8]	[−1, 2, −4]	[−3, 0, 0]	[−5, −2, −2]	[−4, −4, −4]	[−6, 0, −6]	[−8, −5, −2]	[−10, −7, −4]	[−12, −9, −6]
G	−8	[−4, −7, −10]	[0, −3, −6]	[−2, 4, −2]	[−4, 2, 2]	[−6, −3, 0]	[−2, −5, −2]	[−4, 2, −4]	[−6, −3, 0]	[−8, −2, −2]
T	−10	[−6, −9, −12]	[−2, −5, −8]	[2, −4, −7]	[0, 3, 0]	[−2, 4, 1]	[−4, −1, 2]	[0, −3, 0]	[−2, 4, −2]	[−4, −1, 2]
A	−12	[−8, −11, −14]	[−4, −4, −10]	[0, −3, −6]	[1, 1, −2]	[2, 2, −1]	[0, 6, 0]	[−2, 1, 4]	[2, −1, 2]	[0, 3, 0]
G	−14	[−10, −13, −16]	[−6, −9, −12]	[−2, −2, −8]	[−1, 2, −4]	[0, 0, 0]	[4, 1, −2]	[2, 8, 2]	[0, 3, 6]	[1, 4, 4]

FIGURE 9.3 Each cell is assigned three scores obtained from three possible movements – namely, horizontal, diagonal and vertical. The arrows indicate back-tracing based on the highest score out of the three scores.

Tracing back is done from the bottom right cell towards the top left cell. The path is determined based on the source of the cell which contributes to the highest score in the present cell.

The dynamic programming thus proceeds from one cell to the other (in the defined directions), until the whole matrix obtains the score for each of the cells.

		C	A	G	G	T	A	G	T	G
	0	−2	−4	−6	−8	−10	−12	−14	−16	−18
C	−2	2	0	−2	−4	−6	−8	−10	−12	−14
T	−4	0	1	−1	−3	−2	−4	−6	−8	−10
A	−6	−2	2	0	−2	−4	0	−2	−4	−6
G	−8	−4	0	4	2	0	−2	2	0	−2
T	−10	−6	−2	2	3	4	2	0	4	2
A	−12	−8	−4	0	1	2	6	4	2	3
G	−14	−10	−6	−2	2	0	4	8	6	4

FIGURE 9.4 Trace-back starts from the bottom right cell towards the top left cell, according to the highest score(s) obtained in the previous step. There could be more than one path at a point (i.e., cell), if that cell has been awarded more than one highest score, due to two or three movements in the previous step.

All possible paths are obtained from the scoring matrix during the process of global alignment.

9.3.4 Step 4: calculating the scores for each alignment

Points to remember during scoring:

When only the column number (but not the row number) is increased by 1 (i.e., "j" is increased to "$j+1$"):

a. Indicates no change in row position, but there is one cell movement to the right.
b. Horizontal sequence (written along the *X*-axis) has one residue which is missing in the vertically written sequence (i.e., along the *Y*-axis).
c. A gap is introduced in the vertically written sequence.

When only the row number (but not the column number) is increased by 1 (i.e., "i" is increased to "$i+1$"):

a. Indicates no change in column position, but there is one cell movement to the bottom.
b. The vertical sequence (written along the *Y*-axis) has a residue which is missing in the horizontally written sequence (i.e., along the *X*-axis).
c. A gap is introduced in the horizontally written sequence

When both row and column numbers are increased by 1 (i.e., "i" and "j" are increased to "$i+1$" and "$j+1$", respectively):

a. Indicates one-cell diagonal movement towards the bottom left.
b. The vertical sequence (written along the *Y*-axis) and the horizontally written sequence (i.e., along the *X*-axis) have one residue each, which may be matching (award match score) or mismatching (penalize with the mismatch score).
c. No gap is introduced in either vertical or horizontally written sequences.

The score for each of these alignments is calculated according to the scoring scheme set at the beginning of scoring. The alignment score with the highest value is considered as the best global alignment. One could get more than one highest score (same value) when there are multiple (more than one) pair-wise alignments. In such a situation, all those alignments with the highest score are equally good, and any one of these can be accepted as the global alignment.

In the example shown above, we get seven possible global alignments, all of which have the highest score (i.e., equal to 4). Here, we can select any one of these alignments as the best alignment (Figure 9.5).

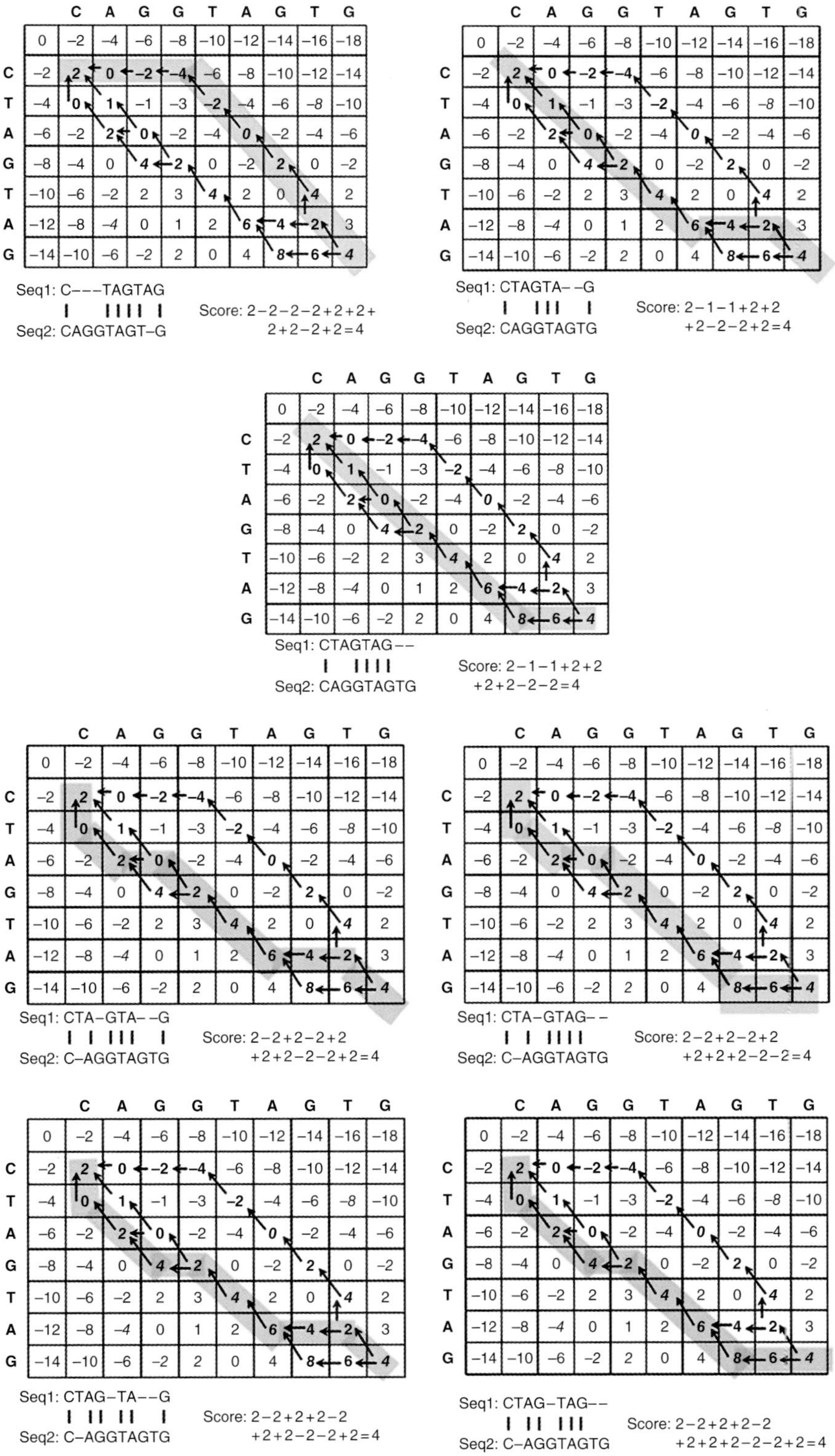

FIGURE 9.5 Global alignment (by NWA) has yielded seven equally good (same alignment score of 4) alignments.

9.4 QUESTIONS

1. Briefly describe the steps of the NWA.
2. What are the features of a dynamic programming? Why is the NWA considered to be dynamic programming?
3. Align the following pairs of sequences manually, using the Needleman–Wunsch dynamic algorithm (assuming scores and penalties to be the same as in the given example):
   a. Seq1: ACTGTGCGT
   b. Seq2: GACGCGTG
   c. Seq1: GTCACACATGT
   d. Seq2: GACCGTATTCGAGT
4. Let the Match score (S_{ij}) = 1, Mismatch score (S_{ij}) = −1 and Gap penalty (d) = −3. Align the following pairs of sequences globally,
   a. Seq1: GTACG
   b. Seq2: AGTATGCGA
   c. Seq1: GCTGTAGTGG
   d. Seq2: CGTGCA
5. Assume three different sets of weights for the match, mismatch and gap:

TABLE 9.2

Set	Match	Mismatch	Gap
1	1	−1	−3
2	2	0	−2
3	1	0	−1

   Now, globally align the following sequences and determine the best alignments for each of the sets of scores and compare:
   a. Seq1: GACTTAC
   b. Seq2: CGTTGAATTCAC

Smith–Waterman Algorithm (Local Alignment)

CHAPTER 10

CS Mukhopadhyay and RK Choudhary
School of Animal Biotechnology, GADVASU, Ludhiana

10.1 INTRODUCTION

The Smith–Waterman algorithm (Smith and Waterman, 1981) is a dynamic programming tool that is used for local alignment, to compare molecular sequences of any length with an aim to identify the conserved region(s). It is a modified form of the Needleman–Wunsch algorithm, where tracing back is stopped as soon as a score of zero (0) is encountered in the path. The scoring is done by replacing the negative values with zeroes in a cell.

10.2 OBJECTIVE

To align two sequences (nucleotide or amino acid) to find out the local region(s) of similarity, using the Smith–Waterman algorithm.

10.3 PROCEDURE

Two sequences will be utilized for local alignment using the Smith–Waterman algorithm:

Seq1: CTAGTAG
Seq2: CAGGTAGTG

The steps are similar to that of global alignment (Needleman–Wunsch algorithm):

1. Define a scoring scheme, as shown above.
2. Initiation of scoring matrix construction: Just like the NW algorithm, here we need to calculate three scores and select the highest when we move to the next cell (horizontally to right, vertically downward and diagonally towards bottom right).
3. Trace-back step: to find out the local alignment.

In the matrix shown in Figure 10.1, three scores are shown in each cell (vertical, diagonal and horizontal, respectively). The highest numerical value is chosen to calculate the scores for the next cells (hence, dynamic scoring scheme).

Basic Applied Bioinformatics, First Edition. Chandra Sekhar Mukhopadhyay, Ratan Kumar Choudhary and Mir Asif Iquebal.

10.3.1 Step 1: scoring scheme

The award for match and penalty for mismatch and gaps (horizontal and vertical) may be modified.

Match score (S_{ij}) = 2
Mismatch penalty (S_{ij}) = −1
Gap penalty (d) = −2

Here, "i" (varying from 1, 2, …, m) and "j" (varying from 1, 2, …, n) refer to the indexes for the row and column numbers of the cell of the scoring matrix, respectively.

Like the NW algorithm, dynamic scoring starts from the top left to the bottom right.

> **Note**: The assigned weights for scoring are arbitrary. The aforementioned values (for scoring) can be changed, and non-integer real numbers can also be used. However, the mismatch penalty is a negative number. The extent of homology, conservedness (hence, the need for introducing gaps during alignment), and so on, determines the weights for award or penalty during sequence alignment.

10.3.2 Step 2: matrix construction

a. Write the sequences (Seq1 and Seq2) along the Y-axis (along the rows) and the X-axis (along the columns), respectively.
   Scoring Method: Dynamic Programming
b. An award or penalty is determined by the movement along the cells, starting from the top left cell:

$$F(i,j) = \max \begin{Bmatrix} Fi-1, j+d\left(\text{Gap in horizontal} - \text{axis sequence; Vertical movement}\right) \\ Fi-1, j-1+s(i,j)\left(\text{match/mismatch; Diagonal movement}\right) \\ Fi, j-1+d\left(\text{Gap in vertical} - \text{axis sequence; Horizontal movement}\right) \end{Bmatrix} \quad [10.1]$$

The highest of these three scores is used for calculating the respective scores of the adjacent cells towards the right, bottom and diagonally bottom right (Figure 10.1).

10.3.3 Step 3: trace-back step

a. Round off all the negative values to zeroes.
b. Identify the cell having the highest score in the entire matrix.
c. Start tracing back from the cell (not necessarily at the bottom left of the matrix) with the highest score. Move to the cell located at the left, or top or top left (diagonal), based on the highest score among these three cells.

		C	A	G	G	T	A	G	T	G
	0	0	0	0	0	0	0	0	0	0
C	0	[0, 2, 0]	[0, 0, 0]	[0, 0, 0]	[0, 0, 0]	[0, 0, 0]	[0, 0, 0]	[0, 0, 0]	[0, 0, 0]	[0, 0, 0]
T	0	[0, 0, 0]	[0, 1, 0]	[0, 0, 0]	[0, 0, 0]	[0, 0, 0]	[0, 0, 0]	[0, 0, 0]	[0, 0, 0]	[0, 0, 0]
A	0	[0, 0, 0]	[0, 2, 0]	[0, 0, 0]	[0, 0, 0]	[0, 0, 0]	[0, 0, 0]	[0, 0, 0]	[0, 0, 0]	[0, 0, 0]
G	0	[0, 0, 0]	[0, 0, 0]	[0, 4, 0]	[0, 2, 2]	[0, 0, 0]	[0, 0, 0]	[0, 2, 0]	[0, 0, 0]	[0, 0, 0]
T	0	[0, 0, 0]	[0, 0, 0]	[2, 0, 0]	[0, 3, 0]	[0, 4, 1]	[0, 0, 2]	[0, 0, 0]	[0, 4, 0]	[0, 0, 2]
A	0	[0, 0, 0]	[0, 0, 0]	[0, 0, 0]	[1, 1, 0]	[2, 2, 0]	[0, 6, 0]	[0, 1, 4]	[2, 0, 2]	[0, 3, 0]
G	0	[0, 0, 0]	[0, 0, 0]	[0, 0, 0]	[0, 2, 0]	[0, 0, 0]	[4, 1, 0]	[2, 8, 2]	[0, 3, 6]	[1, 4, 4]

FIGURE 10.1 The scores in each cell are obtained from the movements from three directions – namely, horizontal, diagonal and vertical. The arrows indicate back-tracing based on the highest score out of the three scores.

		C	A	G	G	T	A	G	T	G
	0	0	0	0	0	0	0	0	0	0
C	0	**2**	**0**	**0**	**0**	0	0	0	0	0
T	0	**0**	**1**	0	0	**0**	0	0	0	0
A	0	0	**2**	**0**	0	0	**0**	0	0	0
G	0	0	0	**4**	**2**	0	0	**2**	0	0
T	0	0	0	2	3	**4**	2	0	**4**	2
A	0	0	0	0	1	2	**6**	**4**	**2**	3
G	0	0	0	0	2	0	4	**8**	**6**	**4**

FIGURE 10.2 Trace-back step: starting with the highest score in the matrix, moving towards the top left and stopping at the last positive score.

d. Proceed backward (towards top left) in a similar manner to that described for the Needleman–Wunsch algorithm.
e. *Stop as soon as the highest score is encountered in the path is 0 (zero).*
f. Draw the path of tracing using arrows from the starting cell and finally to the stopping cell.
g. There may be more than one highest score (of the same value). If a cell has initiated more than one arrow, then it will lead to more than one path at such a junction.

10.3.4 Step 4: calculating the scores for local alignment

The score for each of the local alignment is computed in a similar manner, as shown in global alignment.

```
Seq1: GTAG
      | | | |        Score: 2+2+2+2=8
Seq2: GTAG
```

TABLE 10.1 Similarities and differences between NW and SW algorithms.

SN	Particulars	NWA	SWA
1	Application	Global alignment: screens for the region(s) of high similarity, considering the whole sequences (from start to end).	Local alignment: identifies the discrete region(s) of high similarity (only similar fragment(s) of the sequences).
2	Scoring system	Award for 'match'. Penalty for 'mismatch' and 'gap'. The scores may vary depending on the sequence similarities	The scoring system is the same as NWA. The values may change accordingly
3	Algorithm type	Dynamic algorithm to construct the similarity (or scoring) matrix in an iterative manner.	Same as NWA.
4	Tracing Back	No modification of scoring matrix obtained before starting trace-back step from the bottom right corner.	All the negative scores are set to zero, and trace-back step may start from some cells with higher/highest positive values.
5	Number of trace back paths	Multiple (but linked as if branched with each other) paths can be obtained, due to same scores in a cell obtained from those three movements (horizontal, vertical and diagonal) during dynamic scoring of the similarity matrix	Multiple discrete paths can be obtained, due to the presence of higher positive scores in more than one cell. Each of the higher values gives a fresh start. However, in some cases, branch-like multi-paths could be found, like NWA.
6	Multiple results	More than one global alignment could be found if two or more alignments have the highest scores	There could be more than one local alignment.

10.4 QUESTIONS

1. Construct the scoring matrix using the Smith–Waterman algorithm and find out the local alignment based on the score, using the following pairs of sequences:
   a. Seq1: CAGTCAGT; Seq2: AGTTGCA
   b. Seq1: AGGCATGAA; Seq2: GGTTAA
   c. Seq1: CAGTCC; Seq2: AGTCCGCTAC
2. What are the applications of local alignment of nucleotide as well as amino acid sequences?
3. Given the pair of nucleotide sequences: Seq1: GATCGTCATG and Seq2: GACGTCACTG, for the following set of awards and penalty values, determine the local alignments:
   a. Match score (Sij) = 2; Mismatch score (Sij) = –1; Gap penalty (d) = –2
   b. Match score (Sij) = 1; Mismatch score (Sij) = –1; Gap penalty (d) = –2
   c. Match score (Sij) = 4; Mismatch score (Sij) = –3; Gap penalty (d) = –6

   Is there any change in the local alignment? If yes, why is this change observed? If no, what could be the reasons for no change in the best alignment?
4. Enumerate the differences between NW-algorithm and SW-algorithm. Let two similar sequences (with a couple of mismatches only) be aligned by both the methods; how far the results will differ? Please take the following example and work out:
   - Seq1: MACTAPRD
   - Seq2: MACCAPRE
5. Justify the following comment: "The SW algorithm is a database search algorithm".

Sequence Alignment Using Online Tools

CHAPTER 11

CS Mukhopadhyay and RK Choudhary
School of Animal Biotechnology, GADVASU, Ludhiana

11.1 INTRODUCTION

Algorithms used to do pairwise or multiple sequence alignments vary with the sequence alignment tool available online. Links to some useful sites for sequence alignment are given below:

1. **Color INteractive Editor for Multiple Alignments** (CINEMA 2.1). CINEMA (http://www.bioinf.man.ac.uk/dbbrowser/CINEMA2.1/) is freely available online. The sequence alignment is supported with a color editor.
2. **Multiple Alignment Construction and Analysis Workbench** (MACAW) (http://en.bio-soft.net/format/MACAW.html). This is downloadable software that is used to identify localized sequence similarities and edits blocks of multiple sequences.
3. **Java ALignment VIEWer** (JALVIEW) (http://www.jalview.org/). JALVIEW has freely accessible multiple alignment editors. Alignment tools like "EBI ClustalW" and protein domain database "Pfam" use this Java-based platform.
4. **Clustal W** (http://www.ebi.ac.uk/Tools/clustalw2/index.html). Used for pairwise and multiple sequence alignment.
5. **Clustal Omega** (http://www.ebi.ac.uk/Tools/msa/clustalo/). Clustal Omega is capable of handling several thousands of medium-to-large sized sequences simultaneously (Sievers and Higgins, 2014).
6. **Multiple Alignment using Fast Fourier Transform** (MAFFT version 6) (http://mafft.cbrc.jp/alignment/software/index.html). The multiple sequence alignment program MAFFT is available in both online and downloadable forms (Katoh *et al.*, 2002). Several multiple alignment methods are available in MAFFT: L-INS-i (accurate; for alignment of ≤ 200 sequences), FFT-NS-2 (fast; for alignment of ≤ 10 000 sequences).

Basic Applied Bioinformatics, First Edition. Chandra Sekhar Mukhopadhyay, Ratan Kumar Choudhary and Mir Asif Iquebal.

7. **Multiple Sequence Comparison by Log-Expectation** (MUSCLE) (http://www.ebi.ac.uk/tools/msa/muscle/). The quality of MSA yielded by MUSCLE is better than Clustal, and the algorithm is faster for larger alignments (Edgar, 2004). The user guide for MUSCLE is available at http://www.genebee.msu.su/muscle/help.html.
8. **Tree-based Consistency objective function for alignment evaluation** (T-Coffee) (http://www.ebi.ac.uk/Tools/msa/tcoffee/). This multiple sequence alignment program has been developed by Cedric Notredame of CRG Centro de Regulacio Genomica (Barcelona) (Notredame *et al.*, 2000). T-Coffee has been recommended as a very efficient multiple sequence aligner that outputs extra information on structural and evolutionary perspectives (Magis *et al.*, 2014). T-Coffee accepts sequences in PIR and FASTA format, and the default output format is Clustal. Being progressive alignment software, it generates a library of pair-wise alignments to direct the MSA. It can identify motifs and can also evaluate the alignment quality.

11.2 OBJECTIVE

To align multiple amino acid sequences of a given protein (SRY) using the online Clustal alignment program.

11.3 PROCEDURE

1. Open the Clustal Omega home page (http://www.ebi.ac.uk/Tools/msa/clustalo/). If one browser is not working due to incompatibility, the users may switch to a suitable browser.
2. **Input sequences**: The input sequences (let us use the nucleotide sequences: NCBI accession numbers AFG33955, ABV44686, AAW23363, ABS82755, AAG34436, AAG34440, AAG34393, AAB58342, AAL09287, AAL09284) can be either pasted in FASTA format in the sequence box or saved in one of the specific input formats (e.g., FASTA) in a text (*.txt) file. This file can be uploaded by clicking the "Upload a file" button. The input sequences for MSA can be of one of the following formats: NBRF/PIR, FASTA, EMBL/Swiss-Prot, Clustal, GCC/MSF, GCG, RSF, or GDE. The program yields output in any one of the following formats: PHYLIP, Clustal, GCG/MSF, NBRF/PIR, GDE, or NEXUS.
3. **Specify the type of input sequences**: Select the sequence type – "Protein", "DNA" or "RNA" – from the drop-down list in "Step 1 – Enter your input sequences".

4. **Parameters**: Clustal Omega makes use of seeded guide-trees and HMM profile-profile progressive alignments for multiple sequence alignment. Parameters are available in the section "Step 2 – Set your parameters". These enable the user to modify the MSA according to requirements:
   a. *Dealign Input Sequences*: Select "Yes" from the drop-down options to remove gaps present in the input sequences if these have been entered as "already aligned". The default value is "no".
   b. *Output Alignment Format*: The user can select any one of the six output formats (PHYLIP, Clustal, GCG/MSF, NBRF/PIR, GDE, or NEXUS). The default format is "Clustal".
   c. *mBed-like Clustering Guide-tree*: the mBed is a sampling method to accelerate the calculations for constructing a guide tree. The default option "yes" instructs the program to generate guide trees from the input sequences. It converts the sequences into vectors of distances, and then clusters the vectors using *k*-means (a clustering method for partitioning "*n*" number of observations into "*k*" number of clusters). Each of the "*k*" clusters is further clustered using a simple hierarchical clustering method UPGMA (**U**nweighted **P**air **G**roup **M**ethod with **A**rithmetic mean) to construct phenograms (diagrammatic representation of taxonomic relationship among organisms). Finally, the sub-clusters are joined to create a tree.
   d. *mBed-like Clustering Iteration*: Select "Yes" (default is "No") to imply mBed-like clustering during subsequent iterations.
   e. *Number of Combined Iterations*: Total number of iterations that include a guide tree (for constructing phenograms) using a hidden Markov model (for aligning multiple sequences). The user can increase the "default (0)" up to five combined iterations if the software fails to generate logically acceptable alignment or to construct the guide tree.
   f. *Max Guide Tree Iterations*: This refers to the number of iterations for the guide tree (generating phenogram) only, after the user has set the number of combined iterations. The default value is "default".
   g. *Max HMM Iterations*: Similar to the above, this restricts the number of iterations for the hidden Markov model for alignment of sequences.

> Please consult the FAQs page for Clustal Omega at: http://www.ebi.ac.uk/Tools/msa/clustalo/help/faq.html

5. **Job submission**: Click the "Submit" button to get the alignment and associated results (e.g., guide tree, the distance between sequences). Alternatively, MSA results can be obtained through email as specified by the user, after checking the box "be notified by email" (Figure 11.1).

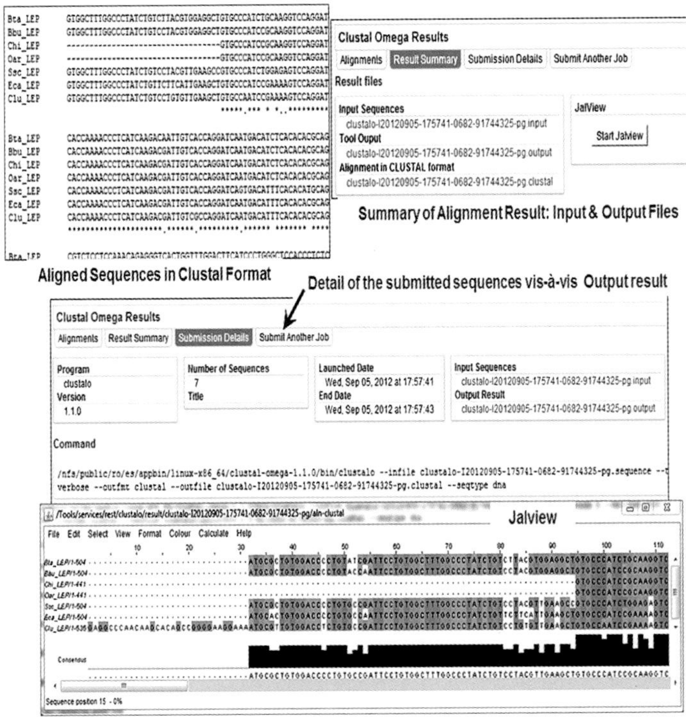

FIGURE 11.1 The output of multiple sequence alignment using Clustal Omega is obtained in different tabs – "Alignments", "Result Summary", "Submission Details". *Jalview* is the Java alignment viewer that displays the alignment, along with the consensus sequence.

11.4 INTERPRETATION OF RESULTS

The alignment results are displayed in Clustal format as rows of interleaved sequences. Gaps are introduced to show insertion-deletion (InDel). At the bottom of each block of sequences, one line of symbols indicates the matches and conserved residues:

a. "*" indicates match for all the residues in the same column;
b. ":" indicates conserved substitution observed;
c. "." indicates semi-conserved substitution observed;
d. No symbol, blank space: indicates a mismatch.

11.5 COLOR SCHEME FOR AMINO ACID RESIDUES

Amino acids with similar physicochemical properties are shown in the same color (http://www.hhmi.umbc.edu/toolkit/ClustalWGuide.html):

a. **Red**: small, hydrophobic, aromatic, not Y (A, V, F, P, M, I, L, W).
b. **Blue**: acidic (D, E).
c. **Magenta**: basic (R, H, K).
d. **Green**: hydroxyl, amine, amide, basic (S, T, Y, H, C, N, G,Q).
e. **Gray**: others.

11.6 QUESTIONS

1. Align the following nucleotide sequences and find the conserved domain(s): NM_005217.3, NM_004084.3, X52053.1, M21130.1, BC119706.2
2. Align the given sequences and show the overall alignment as graphical view (overview window) using online tool MAFFT (http://mafft.cbrc.jp/alignment/server/index.html): ABQ72077.1, AET17647.1, AEM98800.1, CCC62950.1, AEJ49160.1.
3. Align the given sequences using ClustalW and T-Coffee, and logically justify which program has given the more reliable results: KF469208.1, D73408.1, AM933377.1, XM003473564.2, XM004440021.2, AY970684.1, XM004680883.1, AF227738.1, DQ372924.1, AF231714.1, XM004049428.1, AY826184.1, NM001105535.2, NM010776.1
4. Align the given sequences and comment on the conserved patterns found: XM004087588.2, XM004286285.1, NM_001009005.2, NM001161885.1, NM_001009005.2, KF469209.1, NM001141497.1, XM004607626.1, KF469210.1
5. Specify how you will modify the Clustal Omega parameters to obtain an optimum alignment for distantly related sequences.

SECTION III

Basic Local Alignment Search Tools

Basic Local Alignment Search Tool for Nucleotide (BLASTn)

CHAPTER 12

CS Mukhopadhyay and RK Choudhary
School of Animal Biotechnology, GADVASU, Ludhiana

12.1 INTRODUCTION

The Basic Local Alignment Search Tool (BLAST) is a collection of programs for searching homologous sequences for a given query sequence or a set of sequences against selected database (called "Subject sequence"). Thus, BLAST finds regions of *local similarities* between these query and subject sequences. BLAST is a heuristic program, developed by Altschul and coworkers (Altschul *et al.*, 1990), that can yield results in a reasonable time. The term "heuristic" means that the developed algorithm is faster than the classical method but may not be the optimum method. Default parameters of BLAST can be modified according to need.

BLASTn (BLAST with suffix n) is one of the BLAST programs (Table 12.1) that is used to compare a nucleotide query sequence against a nucleotide database. Functional and evolutionary relationships between sequences can be deciphered using BLAST. In addition, it is used to identify member(s) of gene families.

12.2 OBJECTIVE

To search a homologous nucleotide sequence(s) from the nucleotide database, using query nucleotide sequence.

12.3 PROCEDURE

The BLASTn program is run by feeding the input sequence and setting the BLASTn parameters. General steps for setting BLASTn search are given below:

a. Feed the query sequence(s) of interest.
b. Selection of specific subject database.
c. Select the BLASTn program (MegaBLAST, Discontinuous MegaBLAST, blastn).
d. Selection of optional parameters, if required.

These steps are discussed in detail below to elucidate the operation of BLASTn.

Basic Applied Bioinformatics, First Edition. Chandra Sekhar Mukhopadhyay, Ratan Kumar Choudhary and Mir Asif Iquebal.

TABLE 12.1 Overview of various types of BLAST algorithms available at the National Center for Biotechnology Information (NCBI) website, with their applications.

BLAST type	Query	Database	Alignment level	Application
BLASTn	Nucleotide	Nucleotide	Nucleotide	Oligo-mapping, cross-species sequence study, cDNA, EST study, screening repetitive elements, gDNA annotation.
BLASTp	Protein	Protein	Protein	Protein homology, motif search, phylogeny study, characterize novel transcripts.
BLASTx	Nucleotide	Protein	Protein	Explore protein coding genes in cDNA/gDNA, characterize a novel transcript.
tBLASTn	Protein	Nucleotide	Protein	Mapping protein to genomic DNA, compare unknown proteins from multiple organisms to gDNA.
tBLASTx	Nucleotide	Nucleotide	Protein	Search for protein coding genes whose products are not in protein database, cross-species gene prediction at transcript level.

12.3.1 Open BLASTn homepage

Open the NCBI home page by typing http://www.ncbi.nlm.nih.gov/ and click "BLAST". Alternatively, it can also be opened by typing http://blast.ncbi.nlm.nih.gov/Blast.cgi. Then click "nucleotide blast" and the BLASTn window will appear.

The inputs required to specify the BLASTn parameters are broadly categorized into three sections:

12.3.1.1 Enter query sequence

The user may enter an accession number or nucleotide sequence (raw or FASTA formatted) in the specified sequence box. If the sequence of entry is in raw sequence format (i.e., without header line of FASTA), the output page will show "None" (instead of the header line of the FASTA sequence) under the heading "Description".

a. Multiple input sequences: one can provide more than one input query sequences in FASTA-sequence format (or NCBI accession numbers separated by Return or Enter) in the specified sequence box. Alternatively, a text file containing the query sequences (in FASTA format) could be uploaded by clicking the "Choose File" button. The results page will display one drop-down option under the "Results for:" heading, at the top left of the page. This allows the user to select the BLAST result for any one of the multiple input sequences. Once the user specifies the required option, the other parameters of the search result, including "Query ID", "Description", "Molecule type" and "Query Length", change accordingly.
b. Give a job title: It is a good practice to provide a job title to identify the results of BLASTn in saved searches. Please note that pasting your sequence in FASTA format will automatically pick up the descriptive line of FASTA as "Job Title".
c. Checking "Align two or more sequences" option: Checking this box will create another sequence box where your own subject sequence(s) is/are pasted. This is done to align a query sequence with a specific subject sequence. Please provide

the input sequences (query and subject) in FASTA format, so that BLAST can assign an identification tag to the subject sequence name. BLASTn also gives you one Dot plot ("Dot Matrix View") of the pairwise local sequence alignment when a single input is given as subject sequence.

d. Provide query sub-range (optional): To specify a particular range of a single input sequence (applicable for single query sequence) that is to be searched against the database. This is especially useful when a GenBank accession number is used instead of the whole sequence itself.

12.3.1.2 Choose search set

a. Select database: Three options are available: "Human Genomic+Transcript", "Mouse genomic+Transcript" and "Others (nr)". For a DNA database (for BLASTn), the default database is either human or mouse genomic, plus transcript database. Other commonly used databases include the nucleotide "nr" database or EST database. Choose the nucleotide nr database if the databases for microbes/plants/animals are to be searched. The drop-down menu enables the user to specify the required database against which the query sequence is to be searched. "Others (nr) etc" will provide the databases shown in Figure 12.1. Table 12.2 enlists the databases against which a query sequence can be searched in BLASTn.

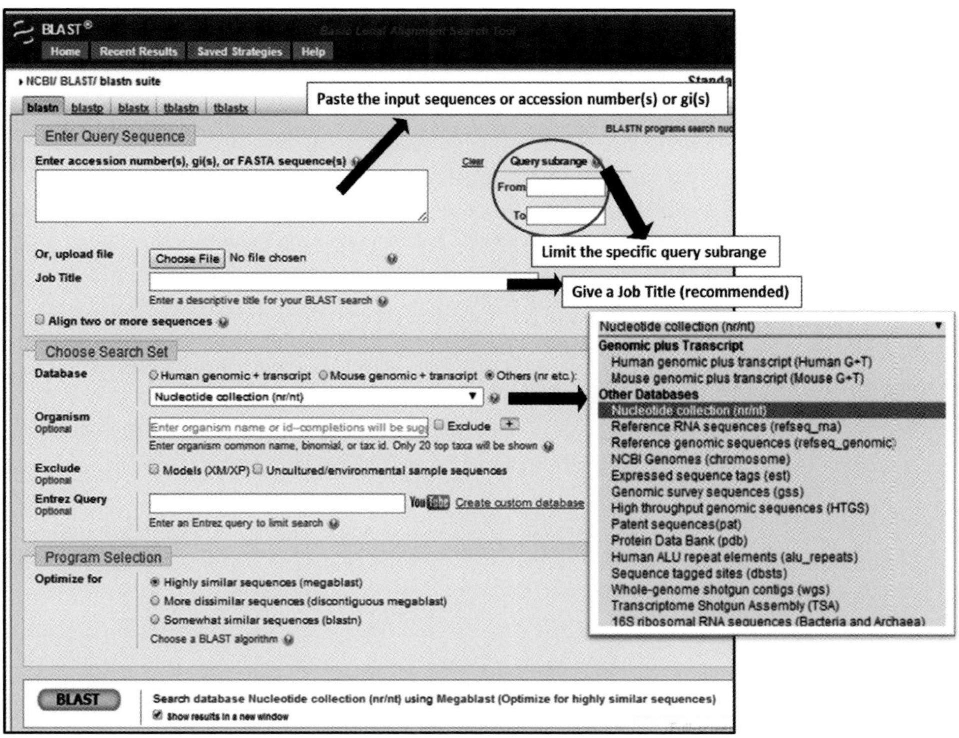

FIGURE 12.1 Main page for BLASTn search at NCBI. The sequence can be entered into the box as query sequences with either accession number or sequence in FASTA format. The gene identity number (i.e., the gi mentioned in this figure) is not currently used as sequence identifier in the NCBI nucleotide database.

TABLE 12.2 Optional BLASTn parameters. Numbered arrows refer to the serial number (SN) of discussion in Table 12.3.

SN	Terms	Explanation
1	Maximum target sequences	An user can opt (from the drop-down list) for the maximum number of aligned sequences to be displayed in the BLAST result. You can select a range of nucleotide searches.
2	Sort queries	Check this box if BLASTn needs to adjust for short queries (i.e., input "word size" or seed) or related parameters to improve results.
3	Expect threshold	BLAST alignment may result in chance hits to non-homologous sequences. This threshold value (E-value) should be lower to minimize the random matches in the databases – for example, if the match score (S-score) is 32.7, and E-value is 0.025, meaning that a score of 32.7 or better would be expected by chance 2.5 times in 100 times (i.e., one time in 40). E-value $\leq$ 0.005 is considered to be statistically significant.
4	Word size	BLAST follows a heuristic algorithm, where a seed word of specific length starts finding its match, and then gets extended in both directions. BLASTn needs an exact match for the seed word between both query and subject sequences. A drop-down menu of the word size has been provided. Taking a larger word size may end up in fewer results, while a shorter word size may lead to more random hits.
5	Max matches in a query range	This is a very useful parameter with practical implications. Sometimes a particular portion of a given query sequence gets a very large number of matches, due to strong similarity, while the other portion does not get a chance to display the result. This option sets a balance by limiting the occurrence of a strong match and offers an opportunity for the portions with weak matches within the same query sequence.
6	Match/mismatch scores	The user can set the ratio between award and penalty for match and mismatch, respectively. Selection can be made from the pull-down menu specifying a positive value for the match and negative value for a mismatch. A wider ratio should be used for identifying divergent sequences through BLASTn.
7	Gap costs	A drop-down menu displays a given range of gap costs. Linear costs are available for MegaBLAST only, while increasing the gap costs will minimize the occurrence of gaps in the aligned sequences.
8	Filter: low complexity region	Low complexity regions are repeat sequences which could introduce spurious results in BLASTn matching. Check this box.
	Filter: species-specific repeats for	If checked, this will mask the repeat elements for that particular selected species.
9	Mask: masks for lookup table only	BLASTn selects the seed word from the look-up table and then proceeds for an extension. If repeated filter is checked, then no seed is obtained from the low complexity region.
	Mask: mask lower case letter	Lower case characters (i.e., bases), indicating low complexity regions, are masked and not considered for BLAST.

TABLE 12.3 Databases against which a query can be searched in BLASTn (http://www.ncbi.nlm.nih.gov/books/NBK153387/).

SN	Databases	Description of database
1	Human genomics plus transcript	Genomic DNA sequences (from all assemblies and chromosomes) and RefSeq RNA sequences of human.
2	Mouse genomics plus transcript	Genomic DNA sequences (from all assemblies and chromosomes) and RefSeq RNA sequences of mouse.
3	Nucleotide collection	Non-redundant sequences from GenBank, EMBL, DDBJ, PDB and RefSeq; however, this excludes very specific databases like EST, STS, GSS, WGS, TSA, patent sequences and HTGS (phases 0–2) sequences.
4	Reference RNA sequence	Reference sequences for various transcripts at NCBI db.
5	Reference genomic sequences	Reference sequences for various genomic sequences at NCBI db.
6	NCBI genomes	NCBI chromosomal DNA sequences of all species in db.
7	Expressed sequence tags (EST)	EST sequences from GenBank, EMBL and DDBJ db.
8	Genomic survey sequences (GSS)	GSS, namely single-pass genomic data (a sequence that has been analyzed in sequencer machine only once), exon-trapped sequences (that are used to identify genes in cloned DNA, by recognizing and trapping carrier containing the exon sequence), and *Alu* PCR sequences.
9	HT genomic sequences (HTGS)	HTGS of Phases 0, 1 and 2; the unfinished HTGS.
10	Patent sequences	DNA sequences available at the patent division of GenBank.
11	Protein data bank (PDB)	Nucleotide sequence database maintained at PDB.
12	Human *Alu* repeat elements	Abundant *Alu* elements present in human genome.
13	Sequence tagged sites (STS)	STS sequences from GenBank, EMBL and DDBJ db.
14	Whole genome shotgun contigs (WGSC)	Database harboring the WGS contigs, except for the WGS data from Chromosome db.
15	Transcriptome shotgun assembly (TSA)	Database containing the computationally assembled mRNA sequences from primary data.
16	16s rRNA sequences (bacteria and archaea)	16s Ribosomal RNA data belonging to bacteria and archaea

b. Select organism (optional): specify if the database search is to be restricted to a specific organism, or if any specific organism is to be excluded from the search.
c. Entrez query (optional): the BLAST search can be refined by limiting the search to specific databases by restricting the sequences as per the Entrez query. Some examples of Entrez query are: 3000 : 5000 [mlwt], 100 : 450 [slen], protease NOT Bos [organism]. One can use AND, OR, NOT operators to refine the search.
   i. 3000 : 5000 [mlwt] means this search will limit protein sequences with a molecular weight of 3–5 kD.
   ii. 100 : 450 [slen] means the length of nucleotide or protein search will be limited to 100–450 residue/bases.
   iii. Protease NOT Bos [organism] means the search will be for all proteases except those in bovine (*Bos taurus* or *Bos indicus*).

12.3.1.3 Program selection

This is a very important parameter to be chosen:

a. MegaBLAST for highly similar sequences: this is very fast, but the target should have 95% or more identification with the query – for example, two nucleotides or protein sequences of same species (*B. taurus* vs. *B. indicus*).
b. Discontiguous MegaBLAST for more dissimilar sequences: this allows mismatches and is more suitable for cross-species comparison – for example, two nucleotide sequences between two different species (dog vs. bovine or more divergent species).
c. BLASTn for somewhat similar BLAST: this is somewhat slower than the other two options. It allows the user to search the database with a shorter word size that ultimately searches for a similar type of sequence that has a smaller degree of similarity – for example, two nucleotides or protein sequences of unrelated organisms (e.g., searching homologous sequences of yeast in mice).

12.3.2 Algorithm parameters

The default parameters of BLAST are fine to use on some occasions. The user, nevertheless, would need to optimize the parameters under certain conditions, such as if the query size is short, or the query is to be searched against divergent homologs, by expanding algorithm parameters (Figure 12.2). The meanings of the parameters are explained in Table 12.2.

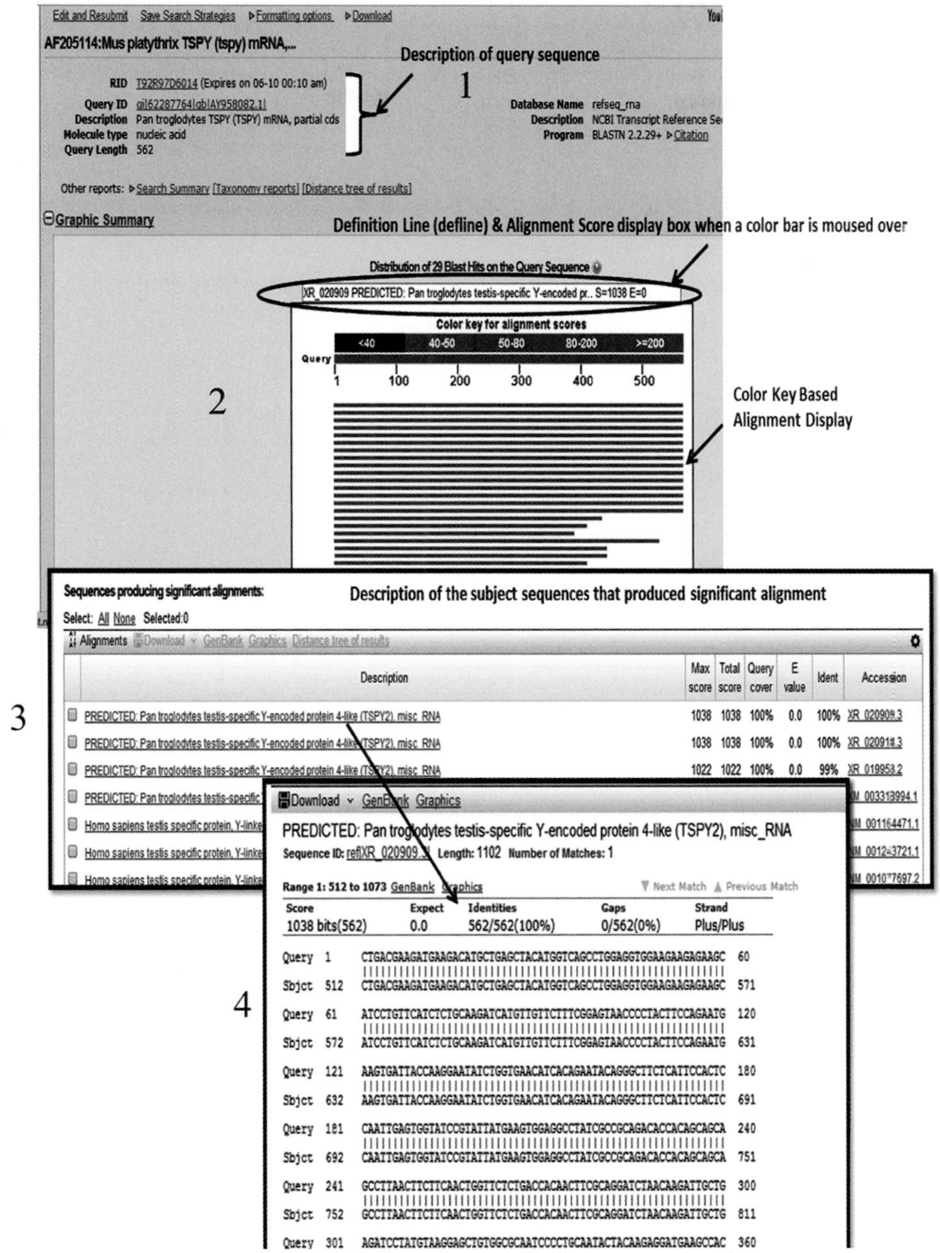

FIGURE 12.2 Optional BLASTn parameters. Numbered arrows refer to the serial number of discussion in Table 12.3.

12.3.3 Click on BLAST button

The Blast results can be obtained on a new Window (or tab) if the "Show results in a new window" box is checked.

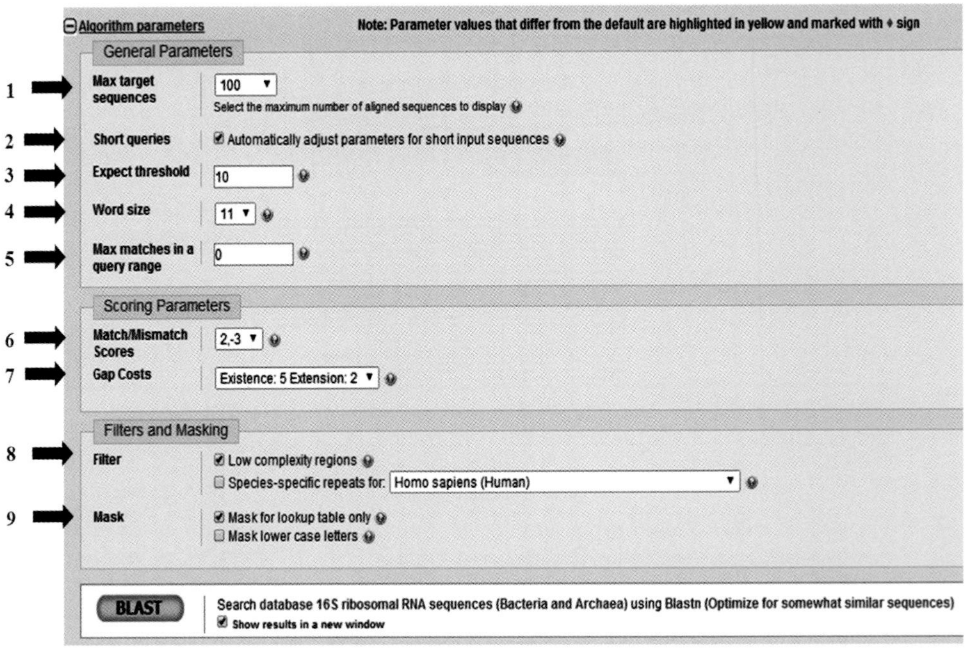

FIGURE 12.3 The result page of BLASTn contains the color key-based alignment display, followed by a tabular description of sequence alignments and, finally, alignments of each of the sequence pairs (query vs. database sequence).

12.3.4 Interpretation of BLAST results

a. **Query**: This refers to the input sequence (or accession number of a sequence given as input) that is to be compared against the entries (i.e., subject) in a database.
b. **Raw alignment score (*S*)**: The score of an alignment, calculated on the basis of match, mismatch/substitution and gap in the alignment. The BLAST program awards the substitution score according to PAM or BLOSUM matrices, while the gaps are penalized with gap-open penalty (higher value) and gap-extension penalty (lower value than gap-open penalty).
c. **Bit score (*S*′)**: The raw alignment score is normalized for the scoring system to determine the bit score, in order to compare alignment scores from different searches. The higher the bit score, the better the alignment.
d. **High-scoring Segment Pair (HSP)**: A local alignment (without gaps) with maximal (or near the highest) alignment scores in a given search. A single query may reveal more than one HSP with a single subject of the database sequence. These are presented as the ranges of the subject sequence. Two situations can

arise when aligning the 'Query' and 'Subject' sequences, due to the occurrence of considerably large gaps in any one sequence (i.e., gaps arising due to intron).

   i. When the query-sequence is a gene sequence with intron, but the subject is a coding sequence (or mRNA) without gap, the color key for alignment score will show a black pipe symbol on the colored line.
   ii. When the subject sequence is a gene sequence with intron, but the query is a coding sequence (or mRNA) without gap, the color key for alignment score will show a blank space on the colored line.
   iii. Note that there could also be several ranges for a single pair of query and subject alignments, which can be overlapping over the subject sequence.

e. **Max Score**: This is inversely proportional to the E-value. The Max Score is the highest bits value out of more than one HSP for a single pair of alignments between query and a subject.
f. **Total Score**: The sum of the bit scores from all HSPs obtained in an alignment between query and subject sequences.
g. **Maximum Identity**: The highest percentage of matches for a set of HSPs with respect to the subject sequence.
h. **E-value or Expectation value or Expect value**: The statistical likelihood that the alignment between the query and subject sequences has *occurred by chance*. The hit obtained is not due to homology, but is due to mere random matches between the two sequences. Thus, it "describes the chance of randomly achieving the same alignment in a database of a particular size". The E-value is the number of alignments with scores superseding *S*, however, which occur due to any random cause but not homology between the sequences. Hence, the lower the probability (i.e., E-value), the better the alignment is. The E-value is calculated by relating the observed alignment score, *S*, to the expected distribution of HSP scores from comparisons of random sequences of the same length and composition as the query to the database. The E-value is calculated as: $E = (\text{query length}) * (\text{length of database}) * 2^{-(S)}$
i. **Query Coverage**: The proportion (expressed in %) of the query sequence that has a homologous counterpart in the subject sequence (i.e., the percentage of the query sequence that has been included in the alignments over all the HSPs).
j. **Maximum Identity**: The highest percentage of matches for a set of HSPs with respect to the subject sequence.

12.4 QUESTIONS

1. Download the sequence EF432553.2 (Partial mRNA sequence of taurine) from the NCBI GenBank and then BLAST it to find the bubaline mRNA sequence of the same gene. Give reasons for selecting the particular bubaline sequence.
2. Given the same sequence (EF432553.2), how will you obtain the transcript variants of the bovine TSPY gene?
3. Suppose BLASTn of a given nucleotide sequence (200 bases length) shows an E-value of >0.05 for a set of sequences. Will you consider these sequences to be worth further study?

4. Explain the following terms:
    a. E-value
    b. HSP
    c. Bit-score
    d. Megablast
    e. Discontiguous megablast
5. Interpret the given BLASTn output in your own language. Explain each of the terms given in the output:

Sequences producing significant alignments:

Select: All None Selected:0

Alignments Download GenBank Graphics Distance tree of results

Description	Max score	Total score	Query cover	E value	Ident	Accession
Bubalus bubalis Drosha mRNA, partial cds, clone: Jn1-1	1229	1229	100%	0.0	100%	LC065664.1
Bubalus bubalis drosha ribonuclease III (DROSHA), mRNA	1205	1205	98%	0.0	99%	NM_001319795.1
PREDICTED: Bubalus bubalis drosha, ribonuclease type III (DROSHA), transcript variant X2, mRNA	1205	1205	98%	0.0	99%	XM_006067803.1
PREDICTED: Bos indicus drosha ribonuclease III (DROSHA), transcript variant X2, mRNA	1186	1186	97%	0.0	99%	XM_019982967.1
PREDICTED: Bos indicus drosha ribonuclease III (DROSHA), transcript variant X1, mRNA	1186	1186	97%	0.0	99%	XM_019982966.1
PREDICTED: Bos taurus drosha ribonuclease III (DROSHA), transcript variant X5, mRNA	1186	1186	97%	0.0	99%	XM_005196187.3
PREDICTED: Bos taurus drosha ribonuclease III (DROSHA), transcript variant X4, mRNA	1186	1186	97%	0.0	99%	XM_015468377.1

FIGURE 12.4

Download GenBank Graphics

PREDICTED: Bubalus bubalis drosha, ribonuclease type III (DROSHA), transcript variant X1, mRNA

Sequence ID: ref|XM_006067802.1| **Length:** 4191 **Number of Matches:** 1

Range 1: 2112 to 2463 GenBank Graphics ▼ Next Match ▲ Previous Match

Score	Expect	Identities	Gaps	Strand
645 bits(349)	0.0	351/352(99%)	0/352(0%)	Plus/Plus

```
Query  42    TGCTTCAGTGGGAGGAACTTGAGTGGCAGAAATACGCAGAAGAATGCAAAGGCATGATCG  101
             ||||||||||||| ||||||||||||||||||||||||||||||||||||||||||||||
Sbjct  2112  TGCTTCAGTGGGAAGAACTTGAGTGGCAGAAATACGCAGAAGAATGCAAAGGCATGATCG  2171

Query  102   TCACCAACCCTGGAGCGAAACCAAGCTCTGTCCGCATCGATCAACTGGATCGTGAACAAT  161
             ||||||||||||||||||||||||||||||||||||||||||||||||||||||||||||
Sbjct  2172  TCACCAACCCTGGAGCGAAACCAAGCTCTGTCCGCATCGATCAACTGGATCGTGAACAAT  2231

Query  162   TCAACCCTGATGTGATTACTTTTCCGATTATCGTTCACTTTGGGATACGCCCTGCACAGT  221
             ||||||||||||||||||||||||||||||||||||||||||||||||||||||||||||
Sbjct  2232  TCAACCCTGATGTGATTACTTTTCCGATTATCGTTCACTTTGGGATACGCCCTGCACAGT  2291

Query  222   TGAGTTATGCTGGAGACCCACAGTACCAGAAACTGTGGAAGAGTTATGTTAAGCTTCGCC  281
             ||||||||||||||||||||||||||||||||||||||||||||||||||||||||||||
Sbjct  2292  TGAGTTATGCTGGAGACCCACAGTACCAGAAACTGTGGAAGAGTTATGTTAAGCTTCGCC  2351

Query  282   ACCTCCTAGCAAATAGTCCAAAAGTCAAACAGACTGACAAGCAGAAGCTGGCACAGAGGG  341
             ||||||||||||||||||||||||||||||||||||||||||||||||||||||||||||
Sbjct  2352  ACCTCCTAGCAAATAGTCCAAAAGTCAAACAGACTGACAAGCAGAAGCTGGCACAGAGGG  2411

Query  342   AGGAAGCACTCCAGAAAATACGACAGAAGAATACCATGAGGCGAGAAGTAAC  393
             ||||||||||||||||||||||||||||||||||||||||||||||||||||
Sbjct  2412  AGGAAGCACTCCAGAAAATACGACAGAAGAATACCATGAGGCGAGAAGTAAC  2463
```

FIGURE 12.5

6. When can you infer that you have obtained a unique sequence in the output? Does E-value play any role in finding the unique match?
7 From 5a output obtained, can we find out or reach the page indicating its cytogenetic location? If yes, how can we do so?

Basic Local Alignment Search Tool for Amino Acid Sequences (BLASTp)

CHAPTER 13

CS Mukhopadhyay and RK Choudhary
School of Animal Biotechnology, GADVASU, Ludhiana

13.1 INTRODUCTION

BLASTp is a set of programs that searches the protein databases using an amino acid sequence as the query. There are four different algorithms, with well-defined applications in BLASTp: blastp, psi-blast, phi-blast and delta-blast.

13.2 OBJECTIVE

To search a homologous protein sequence from the protein database, using the given amino acid sequence as query.

13.3 PROCEDURE

13.3.1 Protein-protein BLAST (BLASTp)

The necessary steps are the same for BLASTp and BLASTn (see Chapter 12: "Basic Local Alignment Search Tool for nucleotide (BLASTn)"), regarding:

- selection of sequence of interests (query sequences);
- specifying the BLAST program;
- selecting the sequence database;
- adjusting the optional parameters.

13.3.1.1 Open the BLASTp homepage

Open the URL: http://www.ncbi.nlm.nih.gov/blast/Blast.cgi? PROGRAM=blastp&PAGE_TYPE=BlastSearch&LINK_LOC=blasthome to get the homepage for BLASTp.

Basic Applied Bioinformatics, First Edition. Chandra Sekhar Mukhopadhyay, Ratan Kumar Choudhary and Mir Asif Iquebal.

BASIC TERMINOLOGIES

a. **Position-Specific Iterative BLAST (PSI-BLAST)**: An iterative (repeating the same set of instructions or steps) BLASTp search used for constructing a scoring matrix based on the frequencies of each amino acid in each position (hence, "position-specific") of protein sequence alignment (i.e., a profile). The profile is refined with the resulting sequences obtained after each cycle of sequence search. The first round of search is identical to BLASTp, but the same procedure is repeated to extend the search to distantly related proteins. PSI-BLAST is the underlying tool of *PsiPred*, a secondary structure (protein) prediction tool (see Chapter 28: Prediction of Secondary Structure of Protein).
b. **Position Hit Initiated BLAST (PHI-BLAST)**: This is a motif (a pattern in protein sequence defined in PROSITE format) specific iterative BLASTp that constructs a position-specific scoring matrix (PSSM) around the motif. PHI-BLAST follows the PSI-BLAST steps in order to use the pattern (within the query) for one or more rounds of PSI-BLAST searching. PHI-BLAST is preferred when a short query (probably harboring a pattern) is used and we need to minimize random matches due to the short nature of the input sequence.
c. **Delta BLAST**: The domain enhanced look-up time accelerated (DELTA) BLAST algorithm searches the conserved domain database (CDD: contains annotated multiple sequence alignment models for domains and full-length proteins) to construct the PSSM. The PSSM is further used to search the BLAST databases for the given query. This combined information on database search using the CDD library improves the homology detection.
d. **Position-Specific Scoring Matrix (PSSM)**: A scoring matrix (i.e., profile) derived from alignment of functionally related proteins. The matrix gives weights to each position of the sequence, according to obtained diversity specific to the family. The PSSM is created by PSI-BLAST after multiple sequence alignment (seeded by the query sequence) from the database matches.

13.3.1.2 Enter query sequences

a. Enter accession number(s) or FASTA sequence(s): Paste one or more query sequence(s) in FASTA format, or the respective NCBI Protein accession number(s), into the specified sequence box. Alternatively, a text file containing the query sequences (in FASTA format) could be uploaded by clicking the "Choose File" button.
b. Give a job title to identify the BLAST results from saved searches.
c. Uncheck "Align two or more sequences": When this checkbox is ticked, the page will be refreshed to provide the user with another sequence box where the subject sequence(s) is/are to be pasted. Such alignment of query and specific subject sequences is done to study sequence homology, according to the requirements of the user. However, when this option is checked, the "Database", "Organism", Exclude" and "Entrez Query" parameters are not required and so do not remain available on the page.
d. Provide Query Sub-range (optional) to specify a particular range of the input sequence which is to be searched against the database. It is used when the NCBI Protein Accession number is used instead of the whole sequence itself (Figure 13.1).

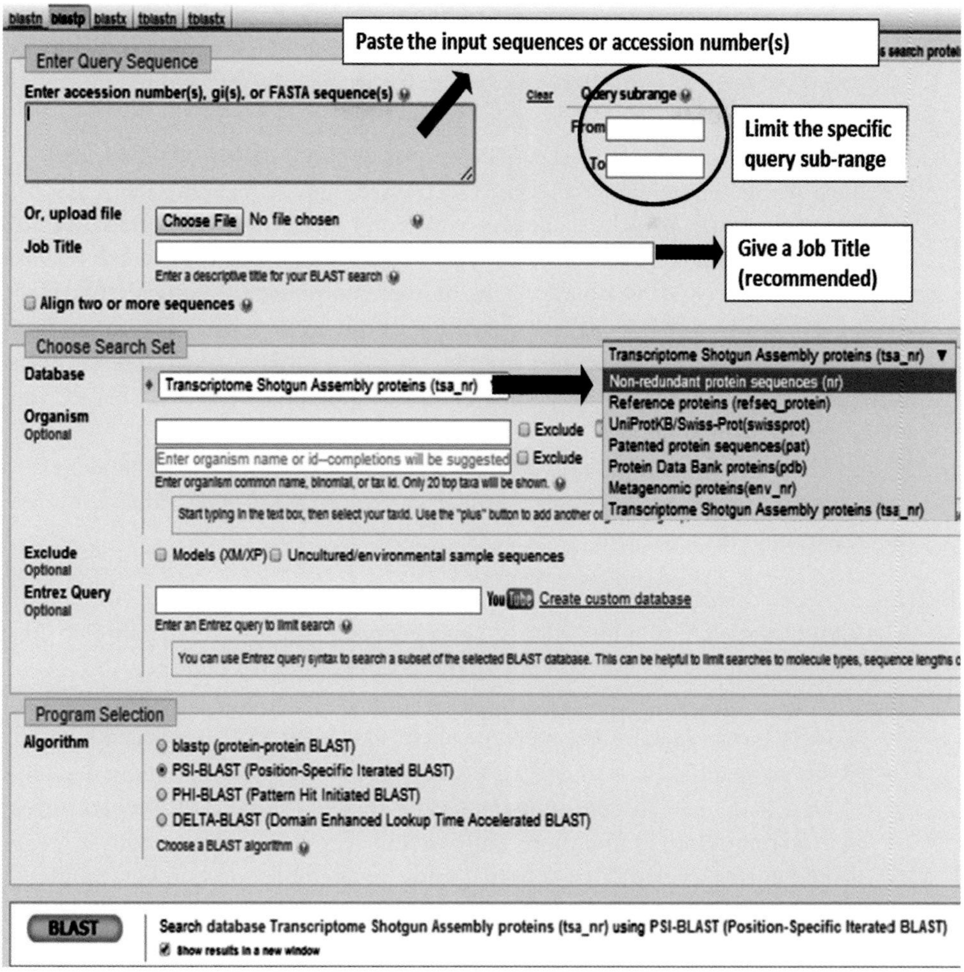

FIGURE 13.1 Setting the parameters for BLASTp search at NCBI. The sequence(s) can be entered into the box as query sequence(s), with either NCBI Protein accession number or sequence(s) in FASTA format.

13.3.1.3 Choose search set

a. Database: You need to choose one of the following databases:
   i. *Non-redundant protein sequences (nr)*: this contains translated non-redundant protein sequences, PIR, Swiss-Prot, PDB, PRF, excluding those in env_nr.
   ii. *Reference proteins (refseq_protein)*: contains amino acid sequences from the NCBI Reference Sequence project.
   iii. *UniProtKB/Swiss-Prot (swissprot)*: includes the latest major release of the Swiss-Prot database.
   iv. *Patented protein sequences (pat)*: consists of the proteins maintained by the Patent Division of NCBI GenBank.
   v. *Protein Data Bank proteins (pdb)*: the amino acid sequences derived from the reported 3D structure records of PDB.

   vi. *Metagenomic proteins (env_nr)*: *in silico* translated, non-redundant coding sequence entries from env_nt.
   vii. *Transcriptome Shotgun Assembly proteins (tsa_nr)*: the non-redundant coding sequences are translated from the TSA archive.

b. Organism (optional): Specify the organism, by common name, binomial name or taxonomical ID, to do the search against its protein sequences. Conversely, you can also check the small check box adjacent to the entry box to exclude any one or more organisms (click on the "+" sign to add more organisms) from your search results.
c. Exclude Models (XM/XP) and/or Uncultured/environmental sample sequences (optional): You can check one or both of the check boxes to exclude one or both options. Models (XM/XP) stands for the "model reference sequences". This is determined and annotated from the Genome Annotation Project of NCBI and, hence, could be incomplete.
d. Entrez Query (Optional): Same as BLASTn, and used to restrict the search to specified Entrez query. It allows Boolean operators AND, OR, NOT to define the database to be searched.

13.3.1.4 Program selection

a. Algorithm: There are four algorithms; choose any one of these, depending on your sequence and the end results you are interested in getting from the BLAST.
   i. BLASTp: searches protein database using protein query. Recently, NCBI protein BLAST has included a new method called "Quick BLAST" or faster BLASTp.
   ii. Position-Specific Iterative BLAST (PSI-BLAST): used to find more distantly related matches. The preliminary search results, by default, present information on permitted mutations; subsequent searches use these data to create a substitution matrix. That is how it finds the members of a protein family.
   iii. Position Hit Initiated BLAST (PHI-BLAST): a variation of the earlier PSI-BLAST and used when the protein family has a known signature pattern (e.g., structural domain, active site, evolutionarily conserved sequence, etc.), with the aim of eliminating false positives. A pattern (protein domain or motif) is specified in the sequence box, which must be matched during the database search.
   iv. Domain enhanced lookup time accelerated (DELTA) BLAST: faster and more accurate than BLASTp, as it uses the Reversed Position Specific BLAST(RPSBLAST) search to construct the PSSM. DELTA-BLAST results are used to initiate a PSI-BLAST search for better accuracy.
b. Click "BLAST": Click on the button to begin the BLASTp search. Click the adjacent checkbox (before executing "BLAST" command) to open the search result in a new window.

13.3.2 Algorithm parameters

These are of the following subtypes. Details of each have been provided in Table 13.1.

a. General parameters
b. Scoring parameters
c. Filters and masking

TABLE 13.1 Algorithm parameters of BLASTp: Numbered arrows in Figure 13.2 refer to the serial number (SN) of discussion in this table.

SN	Terms	Explanation
1	Maximum target sequences	Just like BLASTn, the user can opt (from the drop-down list) for the highest number of aligned sequences to be displayed in the BLAST result.
2	Sort queries	Check this box if BLASTp needs to adjust for short queries (i.e., input "word size" or seed) regarding related parameters to improve results.
3	Expect threshold	BLAST alignment may occur due to chance hits to non-homologous sequences. This threshold value (E-value) should be lower, to minimize the random matches in the databases. The default value of 10 means that, out of the search results, ten matches could be due to chance. Reducing the Expect Threshold value will reduce the search output
4	Word size	You can opt for either 2 or 3. Since BLAST follows a heuristic algorithm, a seed word of particular length starts finding its match, and then extends in both directions. Taking a larger word size may end up in fewer results, while a shorter word size can lead to more random hits.
5	Max matches in a query range	This is a very useful parameter with practical implications. Sometimes, a particular portion of a given query sequence gets a very large number of matches due to high similarity, while the other portion does not get a chance to display the result. This option sets a balance by limiting the occurrence of strong match, and offers an opportunity for the portions with weak matches within the same query sequence.
6	Matrix	The user can select between the Percent Accepted Mutations (PAM: used to score alignment between closely related sequences) or Blocks Substitution Matrix (BLOSUM: for evolutionarily divergent sequences), where higher values for matrix indicate greater evolutionary distance in PAM and vice versa in BLOSUM. The BLOSUM62 scoring matrix is a useful all-square matrix.
7	Gap costs	A drop-down menu displays a given range of gap costs. Increasing gap cost will reduce the number of gaps. It is better to use the default value, unless the results obtained are very irrelevant regarding false positives.
8	Compositional adjustments	This matrix is used to adjust or compensate the compositional differences between the sequences being compared. The adjustment thus improves the E-value of the search. "Conditional compositional score matrix adjustment" is more sophisticated than "Composition-based statistics". One can use the default option.
9	Filter: low complexity region	Low complexity regions are repeat sequences which could introduce spurious results in BLASTp matching.
10	Mask: masks for lookup table only	BLASTp selects the seed word from the look-up table and then proceeds for an extension. If repeated filter is checked, no seed is found from the low complexity region.
	Mask: mask lower case letter	The lower-case characters (i.e., bases) indicating low complexity regions are masked and are not considered for BLAST.
11	Upload PSSM	This is a very useful, advanced, but optional tool. One can download a PSSM from PSI-BLAST or DELTA-BLAST search. That PSSM can then be uploaded for searching a different database to find out a required homology.
12	PSI-BLAST threshold	The threshold for statistical significance is set for including a protein sequence in the PSSM in the following iteration. The default is 0.005.
13	Pseudo count	The default is "0", which enables BLASTp to determine the pseudo-count value, based on minimum length description principle.

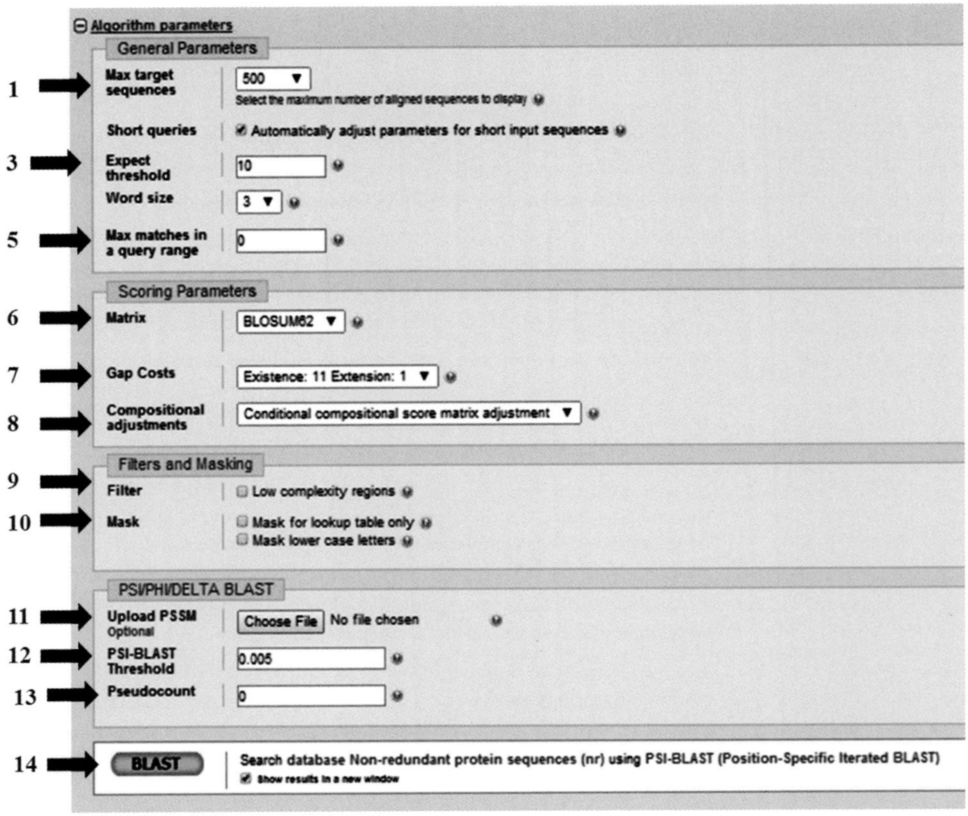

FIGURE 13.2 Optional BLASTp parameters. The numbered arrows refer to the serial number of discussion in Table 13.1.

13.3.3 Interpretation of BLASTp results

13.3.3.1 Results of Protein-Protein BLAST (BLASTp)

a. This is similar to BLASTn results. The final result compares the query and the database sequence intervened by the line of characters containing the matched residues (indicating identity), "+" symbol (indicating a positive substitution, but not an identity), and gap (indicating mismatch).
b. The "Method" indicates the compositional adjustment selected during BLASTp parameter selection.
c. GenPept: Clicking this hyperlink will open the flat file containing the sequence with annotation in NCBI GenBank format.
d. Graphics: Clicking on this hyperlink will open the Graphics window for the protein under consideration.

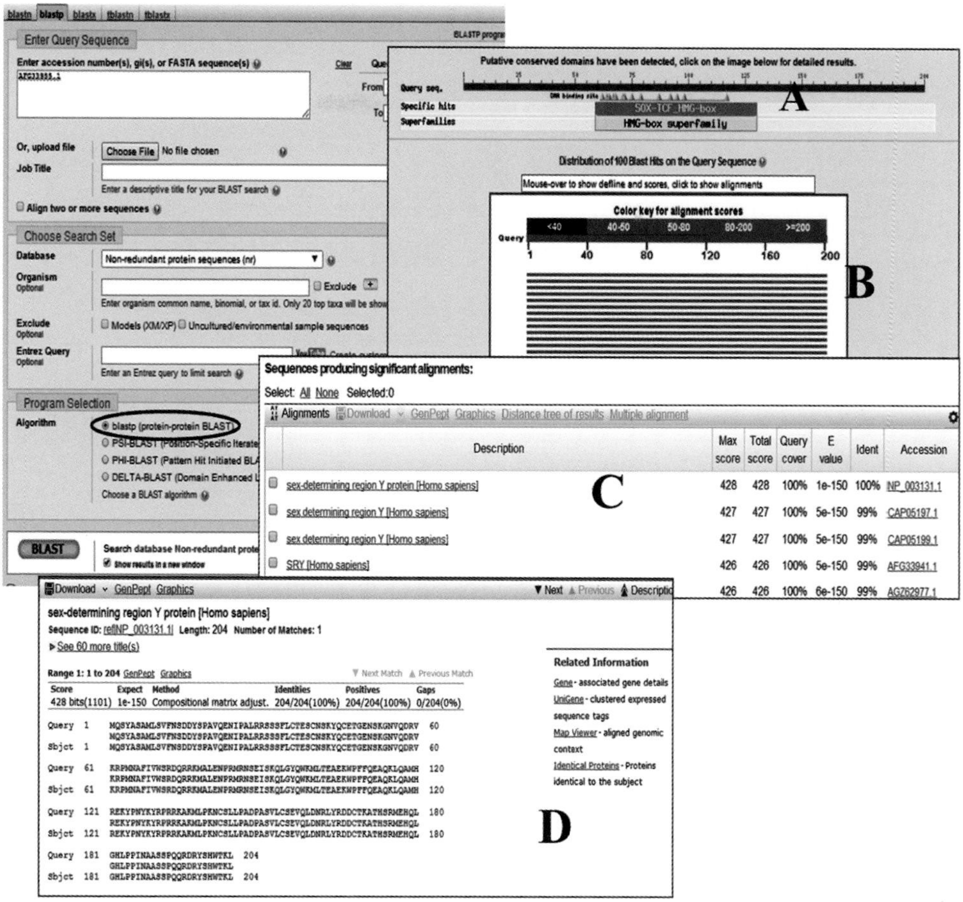

FIGURE 13.3 Different sections of the result page of BLASTp.
"A" indicates the putative conserved domain(s) detected by BLASTp search. Clicking on this image will open the graphical summary of the conserved domain(s) of that protein.
"B" indicates the alignment and the scores in terms of color key, for each of the alignments.
"C" indicates the table of alignment detail (Description, Max score, Total score, Query coverage, E-value, Identity, and Accession).
"D" shows the detail of the alignment residue-wise.

13.3.3.2 Results of Position-Specific Iterative BLAST (PSI-BLAST)

The results window is almost the same as previous ones, except for a new column added: "Select for PSI-blast" (indicated by "E" in Figure 13.4). This column contains checkboxes against each of the PSI-BLAST results. All the checkboxes are, by default, ticked; however, they can be reversed (unchecked) by unchecking them individually. PSI-BLAST can then be run for a second iteration by indicating the number of sequences (default is 500) and then clicking on the "Go" button.

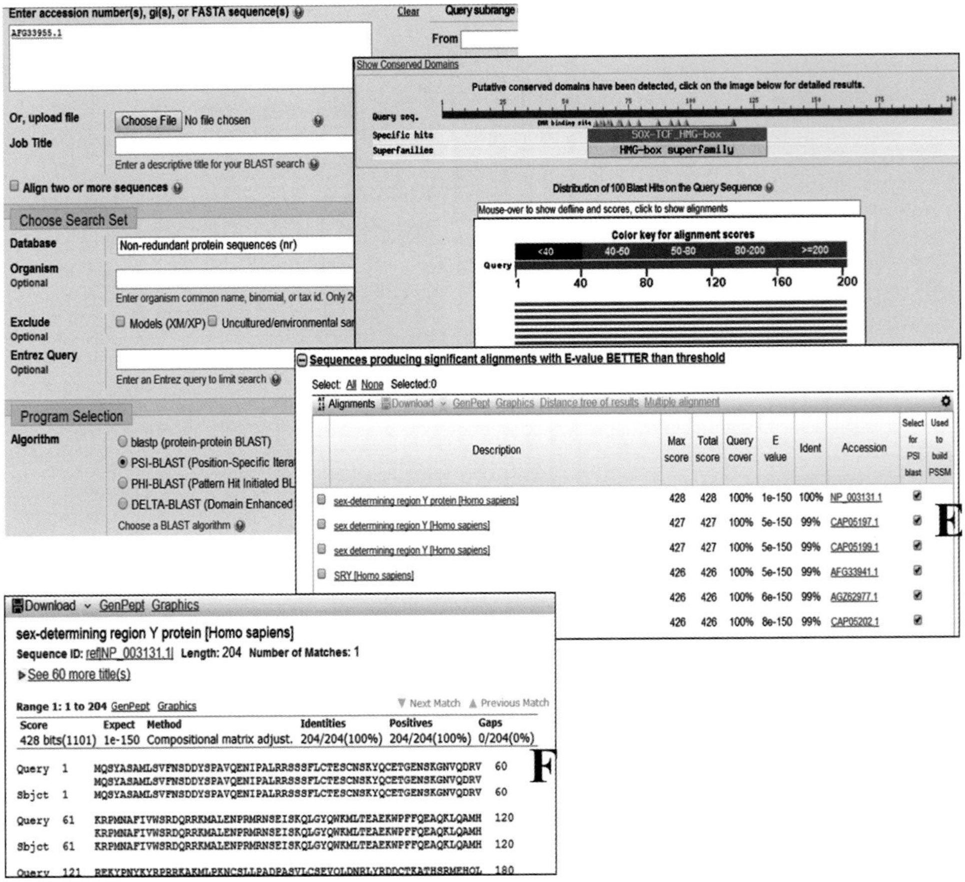

FIGURE 13.4 Results of PSI-BLAST. 'E' indicates the "Select for PSI blast" column, and "F" indicates the detailed result for each alignment.

The last column, "Used to build PSSM" is checked to indicate the sequences which have been used in the second iteration. The rows harboring the sequences which have not been used in this iteration to build PSSM are highlighted in yellow.

13.3.3.3 Results of Position Hit Initiated BLAST (PHI-BLAST)

In this example, the amino acid sequence "krpmnafivw srdqrrkmal" has been used as a PHI pattern to run PHI-BLAST. The specified pattern is indicated by asterisk marks above the alignment.

13.3.3.4 Results of domain enhanced lookup time accelerated (DELTA) BLAST

The output of DELTA-BLAST (Figure 13.6) is more sensitive and accurate than PSI-BLAST.

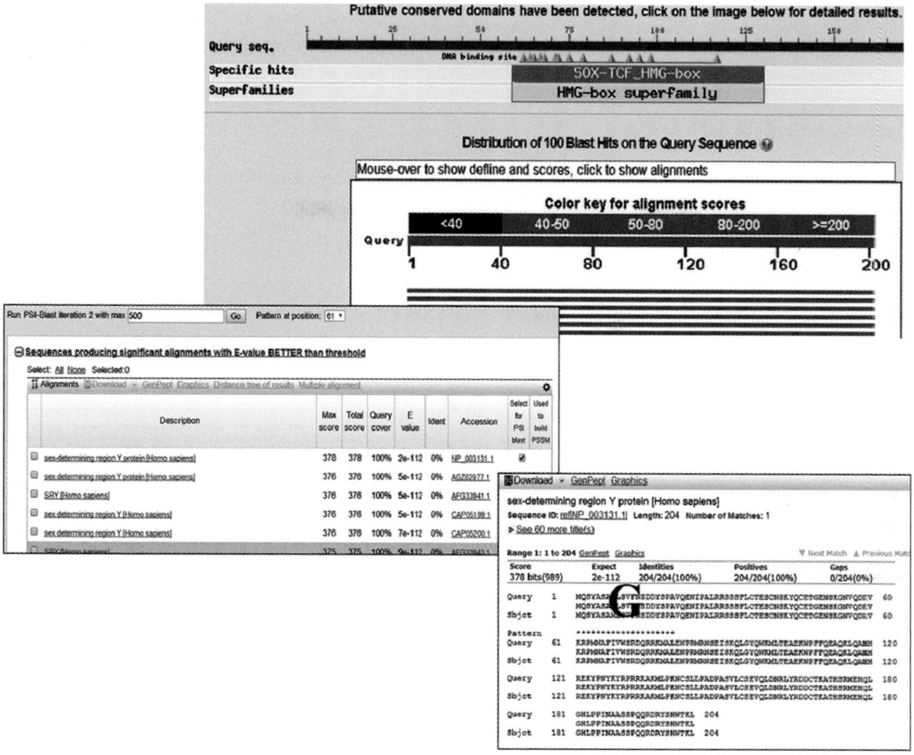

FIGURE 13.5 Result of PHI-BLAST. 'G' indicates the detailed result of each alignment. The asterisks in the second row of alignment indicate the pattern which has been given for PHI-BLAST analysis.

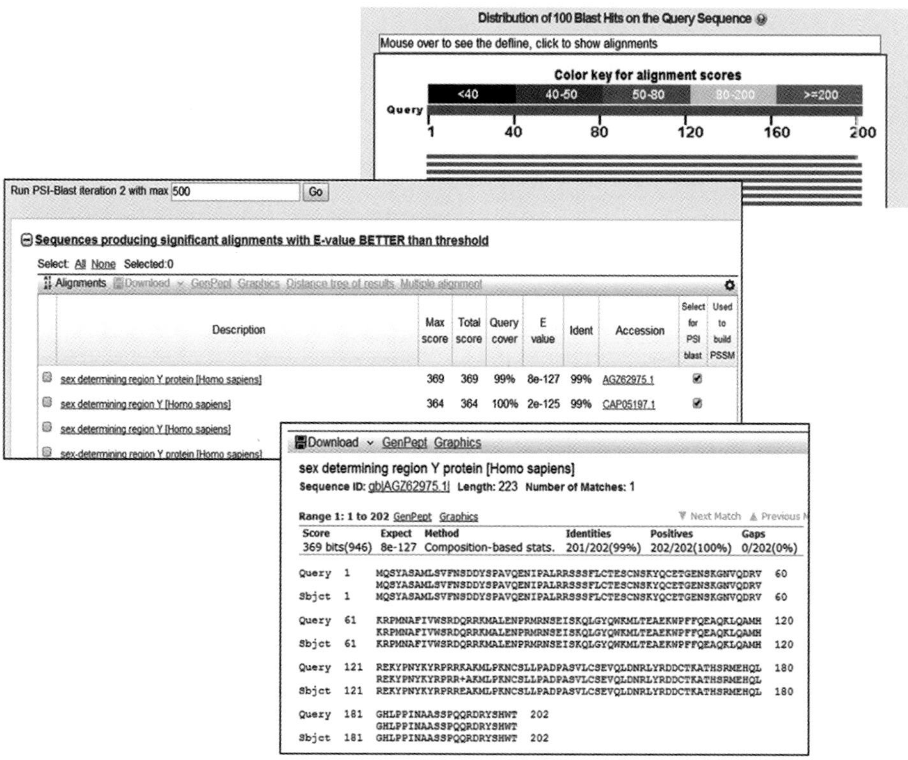

FIGURE 13.6 The result page of DELTA-BLAST. The components and parameters are similar to PSI-BLAST.

13.4 QUESTIONS

1. Enumerate in brief the principle and use of the following types of protein BLASTs: BLASTP, PSI-BLAST, PHI-BLAST, DELTA-BLAST
2. Download a set of divergent peptide sequences (at least ten different species) using the given query: ACC61291.
3. Interpret various parameters obtained from the BLASTp result.
4. Use the following sequence as a PHI pattern and identify the pattern: gkqesmdskl in the sequence **NP_005517.1**.
5. Explain the results for each of these parameters:

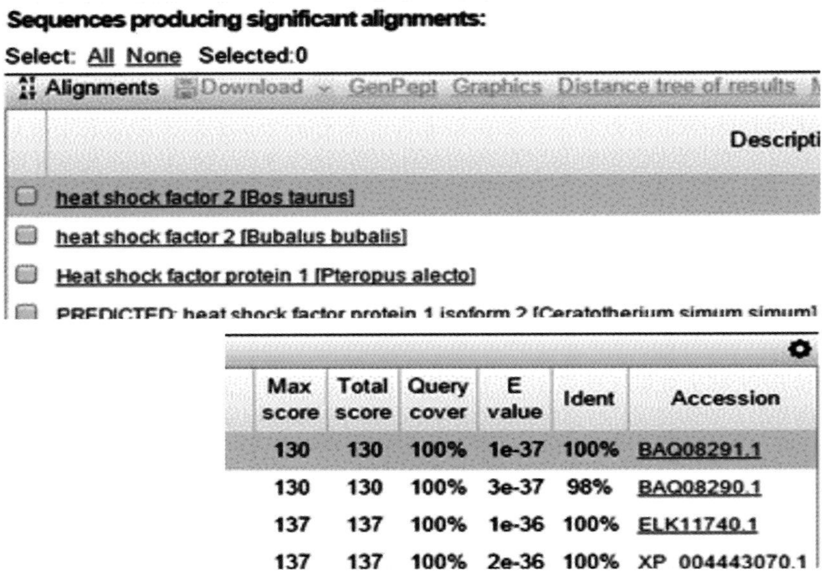

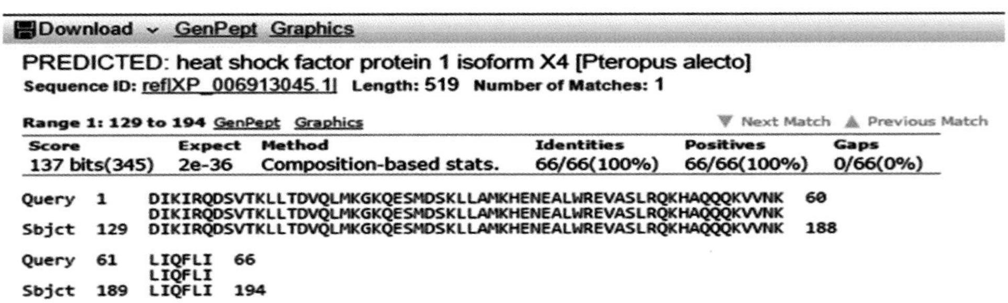

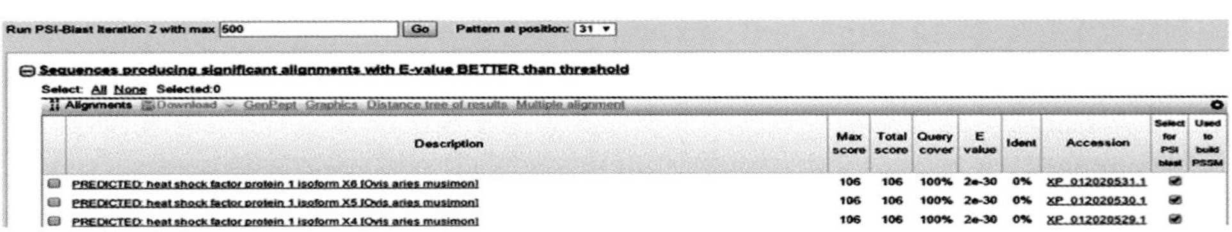

Download ⌄ GenPept Graphics

PREDICTED: heat shock factor protein 1 isoform X2 [Ovis aries musimon]

Sequence ID: ref|XP_012020527.1| **Length:** 564 **Number of Matches:** 1

Range 1: 140 to 205 GenPept Graphics ▼ Next Match ▲ Previous Match

Score	Expect	Identities	Positives	Gaps
106 bits(287)	2e-30	66/66(100%)	66/66(100%)	0/66(0%)

```
Pattern                                       *******
Query    1    DIKIRQDSVTKLLTDVQLMKGKQESMDSKLLAMKHENEALWREVASLRQKHAQQQKVVNK   60
              DIKIRQDSVTKLLTDVQLMKGKQESMDSKLLAMKHENEALWREVASLRQKHAQQQKVVNK
Sbjct    140  DIKIRQDSVTKLLTDVQLMKGKQESMDSKLLAMKHENEALWREVASLRQKHAQQQKVVNK   199

Query    61   LIQFLI   66
              LIQFLI
Sbjct    200  LIQFLI   205
```

BLASTx

CHAPTER 14

CS Mukhopadhyay and RK Choudhary
School of Animal Biotechnology, GADVASU, Ludhiana

14.1 INTRODUCTION

BLASTx is one of the three translated BLAST algorithms – namely, BLASTx, tBLASTn and tBLASTx. In BLASTx, a nucleotide sequence is used as a query, which is first translated in all six reading frames, and then each of the translated amino acid sequences is compared to the protein sequences in protein databases. Thus, the comparison occurs at the level of amino acid and, so, the result is the aligned amino acid sequences (i.e., the translated query versus homologous sequence in protein database), although the query is a nucleotide sequence. BLASTx runs at a slower pace, due to matching all the six reading frames to the protein databases. The result ultimately gives the open reading frame as a match with its homologous sequence.

BLASTx is a powerful gene-finding or gene-predicting tool. It is recommended for identifying the protein-coding genes in genomic DNA/cDNA. It is also used to detect whether a novel nucleotide sequence is a protein-coding gene or not, and it can be used to identify proteins encoded by transcripts or transcript variants.

14.2 OBJECTIVE

To determine the open reading frame and the name of the gene from the given coding sequence (cds).

14.3 PROCEDURE

The basic steps are same as for BLASTn. However, parameters like "Genetic Code", "Organism", and "Database" may be required to be modified. Open the NCBI home page with the URL http://www.ncbi.nlm.nih.gov/ and click "BLASTx". It can also be opened by entering http://blast.ncbi.nlm.nih.gov/Blast.cgi?PROGRAM=blastx&PAGE_TYPE=BlastSearch&LINK_LOC=blasthome in the space for URL. The BLASTx main web page is now open (Figure 14.1).

Basic Applied Bioinformatics, First Edition. Chandra Sekhar Mukhopadhyay, Ratan Kumar Choudhary and Mir Asif Iquebal.

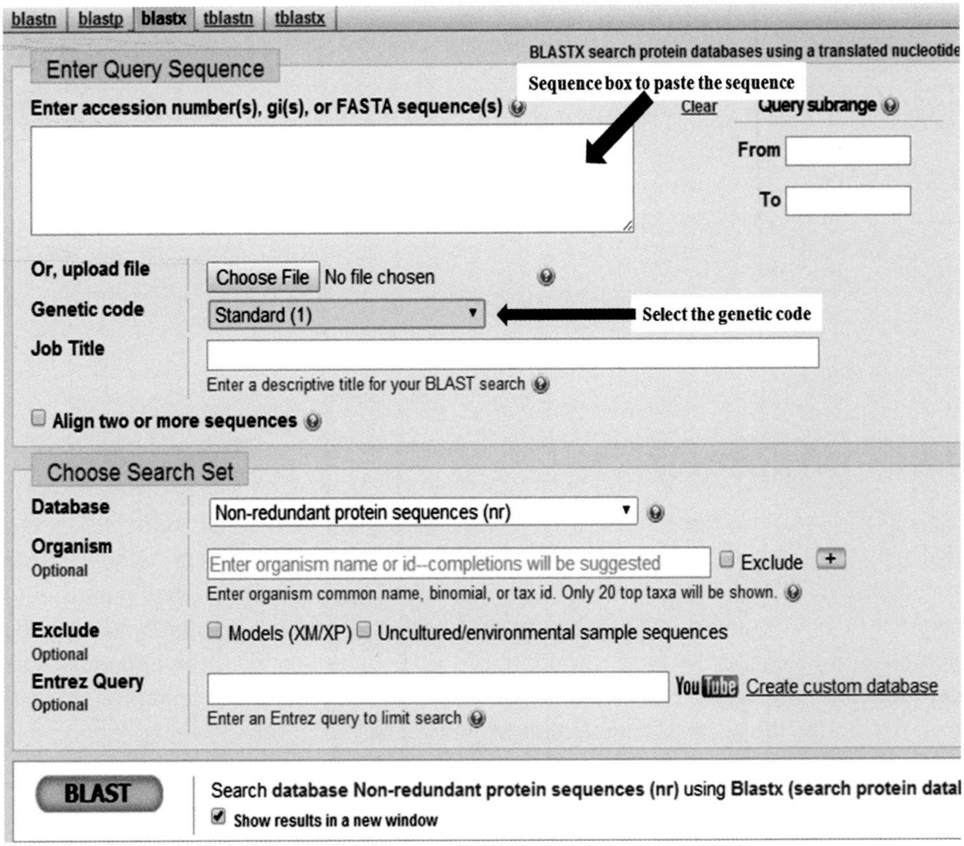

FIGURE 14.1 Homepage of BLASTx at NCBI. The sequence can be entered into the box (angled arrow) as query sequences, either with accession number(s) or as sequence(s) in FASTA format.

14.3.1 Enter query sequences

Enter accession number(s) or FASTA sequence(s), pasting one or more nucleotide query sequence(s) in FASTA format, or the respective NCBI accession number(s) in the specified sequence box. Alternatively, a text file containing the query sequences (in FASTA format) could also be uploaded by clicking the "Choose File" button.

- Provide Query Sub-range (optional): This specifies a particular range of the input sequence which is to be searched against the database. It is especially useful when the GenBank accession number is used instead of the sequence itself.
- Genetic Code: The default is "Standard", which can be used for eukaryotic genomic DNA-derived sequences. Other options, for prokaryotic DNA or mold or yeast or vertebrate or invertebrate mitochondrial DNA, are also available.
- Give a Job Title: To identify the BLAST results from saved searches.
- Checking "Align two or more sequences": This checkbox, if checked, will refresh the page to provide the user with another sequence box, where the subject sequence(s) is/are to be pasted. The application is the same as that discussed in the previous BLASTn or BLASTp chapters.

14.3.2 Choose search set

- Database: Choose any one of the protein databases against which the search is to be made. The list of databases is same as that of BLASTp.
- Organism (Optional): Specify the organism by common name or binomial name or taxonomical ID. You can also check the small checkbox adjacent to the entry box to exclude one or more organisms (click on the "+" sign to add more) from the search results.
- Exclude Models(XM/XP) and/or Uncultured/environmental sample sequences (optional): Check one or both of the check boxes to exclude one or both of the options. Models(XM/XP) stands for "model reference sequences", determined and annotated from the Genome Annotation Project of NCBI and, thus, could be incomplete.
- Entrez Query (Optional): Same as for BLASTn. This is used to restrict the search to specified Entrez query. It allows the Boolean operators AND, OR, NOT to define the database to be searched.
- BLAST: Click on the button to initiate the BLASTx search. Click the checkbox to open the search result in a new window (Figure 14.2).

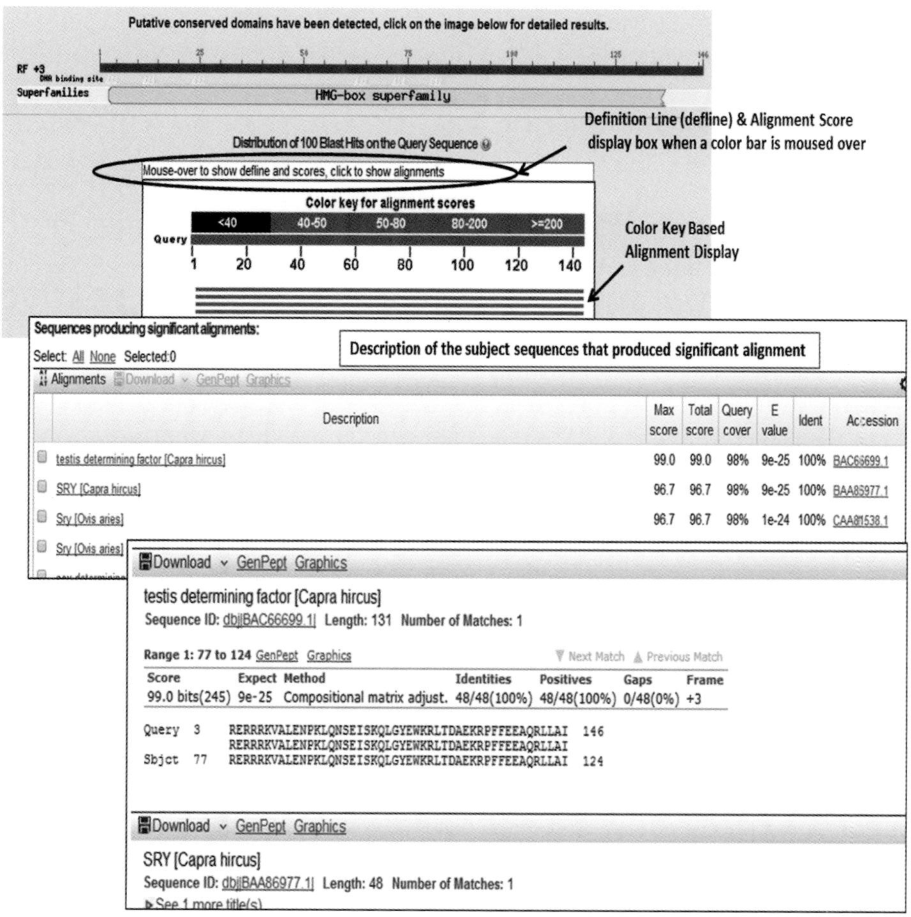

FIGURE 14.2 The results page of BLASTx contains a color key-based alignment display, followed by a tabular description of sequence alignments and, finally, alignment of each of the sequence pairs (a query versus database sequence, called a subject sequence).

14.3.3 Program selection

a. Algorithm parameters: These are very much the same as those for BLASTp, except for the parameter "Short Queries", which has been dropped in BLASTx.
b. Optional parameters: These are of the subtypes shown in Table 12.2 (Chapter 12 of this book).

14.4 INTERPRETATION OF BLASTx RESULTS

a. The output of BLASTx is similar to that of BLASTp.
b. The color key-based alignment depiction and the table indicating the BLASTx output for various homologous sequences are the same as BLASTx.
c. Individual pairwise alignment is also the same as BLASTp. However, the open reading frame out of all the possible six reading frames is indicated by "Frame".

14.5 QUESTIONS

1. Explain how BLAST can be used as a gene prediction tool.
2. Suppose the following partial cDNA amplicon (JQ911700.1 of NCBI GenBank) has been custom sequenced in yak (*Bosgrunniens*):
CCGAAGAAGAAAATGGCCATAACCAGGTCCCAAATATTAGGA
CTTTTCATCACTGTCCTGATCGGCCTACAGGAATCGTGGGCT
ATTAAAGAGAATCATGTGATCATCCAAGCTGAGTTCTATCTG
AAACCTGAGGAATCAGCCGAGTTTATGTTTGACTTTGATGGT
GATGAGATTTTCCACGTGGATATGGGGAAGAAGGAGACGGTG
TGGCGGCTTCCAGAATTTGGACATTTTGCCAGCTTTGAGGCT
CAGGGTGCCCTGGCCAATATGGCTGTGATGAAAGCCAACCTG
GACATCATGATAAAGCGCTCCAACAACACCCCAAACACCAAT
GTTCCTCCAGAAGTGACTCTGCTCCCAAACAAGCCTGTGGAA
CTGGGAGAGCCCAACACACTCATCTGCTTCATTGACAAGTTC
TCCCCACCCGTGATCAGTGTCACATGGCTTCGAAATGGCAAA
CCTGTCACTGATGGAGTGTCACAGACGGTCTTCATGCCCAGG
AATGACCACCTTTTCCGCAAGTTCCACTACCTCCCCTTCCTG
CCCACAACAGAGGATGTCTATGACTGCAAGGTGGAGCACTTG
GGTTTGAATGAGCCTCTTCTCAAGCACTGGGAGTATGAAGCT
CCAGCCCCCCTCCCAGAGACCACAGAGAATGCAGTGTGTGCC
CTGGGCCTGATTGTGGCTCTGGTGGGCATCATTGCAGGGACC
ATCTTCATCATCAAGGGCGTGCGCAAAGCCAACACCGTTGAA
CGCCGAGGGCCTCTGTGAGGCGCCTGCAGGTAATGGACTTTG
TTACAGAGAAGATCAATGAAGATATTTCTGCCTTAATAGCTT
TACAAACCTGGCAATTCTCCAATTGTTCACCTCACTGAAGAC
CACCATGCTTCAGCACTTCCCAGTCCTTTACTTACCCTAAGA
GTAAGATGCCTTCCACAATCTCC

Determine whether it belongs to some protein-coding gene, along with the reading frame.

3. Identify the nucleotide sequence (AY095312.1) and comment on whether it is a part of a coding sequence.
4. What is the principle and what are the applications of BLASTx?
5. Examine and interpret the following output:

Download ˅ GenPept Graphics

sex determining factor [Bubalus bubalis]

Sequence ID: gb|AFX62419.1| **Length:** 229 **Number of Matches:** 1

Range 1: 1 to 116 GenPept Graphics ▼ Next Match ▲ Previous Match

Score	Expect	Method	Identities	Positives	Gaps	Frame
227 bits(579)	2e-72	Compositional matrix adjust.	109/116(94%)	112/116(96%)	0/116(0%)	+1

```
Query  1    MFRVLNDDVYSPAGVQHQNILAFRKDSSSCTDSHSANDQCERGENVRESSQDHIKRPMNA  180
            MFRVLNDDVYSPA VQ QNILAFRKDSSSCTDSHSANDQCERGENVRESSQDHIKRPMNA
Sbjct  1    MFRVLNDDVYSPAVVQQQNILAFRKDSSSCTDSHSANDQCERGENVRESSQDHIKRPMNA  60

Query  181  FILGSREQKQKLALENPKMKNSKISKQLGYEGKRLTDAEKRPFFEEAQRLLSIHRD  348
            FIL SREQ++KLALENPKMKNS+ISKQLGYE KRLTDAEKRPFFEEAQRLLSIHRD
Sbjct  61   FILWSREQRRKLALENPKMKNSEISKQLGYEWKRLTDAEKRPFFEEAQRLLSIHRD  116
```

tBLASTn

CHAPTER 15

CS Mukhopadhyay and RK Choudhary
School of Animal Biotechnology, GADVASU, Ludhiana

15.1 INTRODUCTION

tBLASTn is another type of translated BLAST algorithm, in which an amino acid sequence is used as a query to compare with the **t**ranslated **n**ucleotide (coding sequence) database. The amino acid sequence is compared at the protein level with each subject nucleotide sequence translated in all six reading frames. Thus, tBLASTn is very useful for searching protein homolog(s) in unannotated nucleotide data such as expressed sequence tags (maintained in BLAST database "est") and draft genome records (located in the BLAST database "htgs"), which remain unannotated in the respective databases.

15.2 OBJECTIVE

To search for the homologous protein sequences of a pair of given protein sequences (NP_001028007, NP_001028008).

15.3 PROCEDURE

The basic steps of tBLASTn are the same as for BLASTx:

15.3.1 Open the tBLASTn page

Open the NCBI home page by typing http://www.ncbi.nlm.nih.gov/ and click "tBLASTn". Alternatively, it can also be opened by entering http://blast.ncbi.nlm.nih.gov/Blast.cgi?PROGRAM =tblastn&PAGE_TYPE=BlastSearch&LINK_LOC=blasthome in the URL.

The main page of tBLASTn will be displayed (Figure 15.1).

Basic Applied Bioinformatics, First Edition. Chandra Sekhar Mukhopadhyay, Ratan Kumar Choudhary and Mir Asif Iquebal.

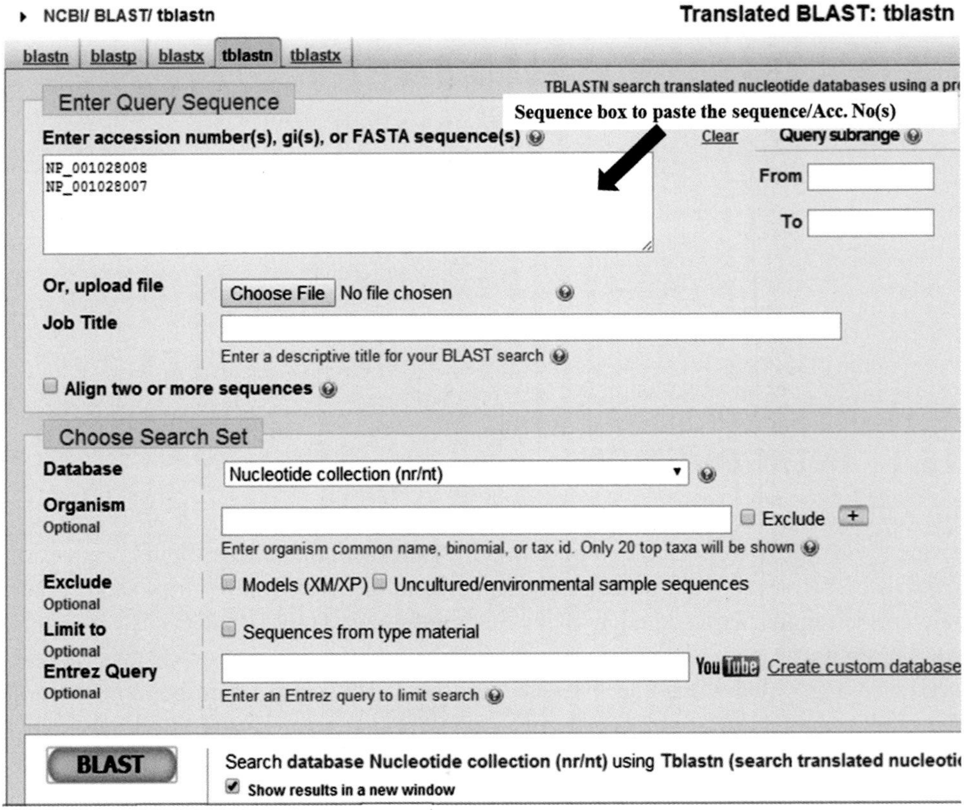

FIGURE 15.1 Homepage for tBLASTn at NCBI. The query sequence(s) can be entered with either accession numbers or sequence(s) in FASTA format.

15.3.2 Enter query sequences

- Enter accession number(s) or FASTA sequence(s): Paste one or more protein query sequence(s) in FASTA format, or the respective NCBI accession number(s) (separated by Enter or Return key) for protein in the specified sequence box. Alternatively, a text file containing the amino acid query sequences (in FASTA format) could be uploaded by clicking the "Choose File" button.
- Give a Job Title to identify the tBLASTn results from saved searches.
- Checking "Align two or more sequences": If this check box is checked, the page will be refreshed to provide the user with another sequence box, where the subject nucleotide sequence(s) is/are pasted.
- Provide Query Sub-range (optional): To specify a range of the input sequence that is to be searched against the database. This is especially useful when the GenBank accession number is used instead of the whole sequence itself.

15.3.3 Choose search set

- Database: Choose any one of the nucleotide databases against which the search is to be made. The list of databases is almost the same as that for BLASTn, except for two options: "Human Genomics plus Transcript" and "Mouse Genomics plus Transcript" are absent.
- Organism (Optional): Specify the organism (by common name or binomial name or taxonomical ID), if required. You can also check the small check box adjacent to the entry box to exclude any one or more (click on the "+" sign to add more organisms to be excluded) organisms from your search results.
- Exclude Models (XM/XP) and/or Uncultured/environmental sample sequences (optional): Check one or both of the check boxes to exclude one or both of the options. Models (XM/XP) stands for the "model reference sequences", determined and annotated from the Genome Annotation Project of NCBI and, thus, could be incomplete.
- Entrez Query (optional): As with BLASTn, this is used to restrict the search to the specified Entrez query. It allows the Boolean operators, AND, OR, NOT, to define the database to be searched.
- BLAST: Click on the button to initiate the tBLASTn search. Click the check box to open the search result in a new window.

15.4 ALGORITHM PARAMETERS

These are the same as those for BLASTx:

a. General parameters
b. Scoring parameters
c. Filters and masking

15.5 INTERPRETATION OF tBLASTn RESULTS

a. The output of tBLASTn is similar to that of BLASTp or BLASTx.
b. The color key-based alignment depiction and the table indicating the tBLASTn output for various homologous sequences are also the same as that for BLASTx (Figure 15.2).
c. Individual pairwise alignment is also the same as that for BLASTp. However, the open reading frame out of all the possible six reading frames is indicated by "Frame".
d. Variants of a protein can also be identified from the tBLASTn results.

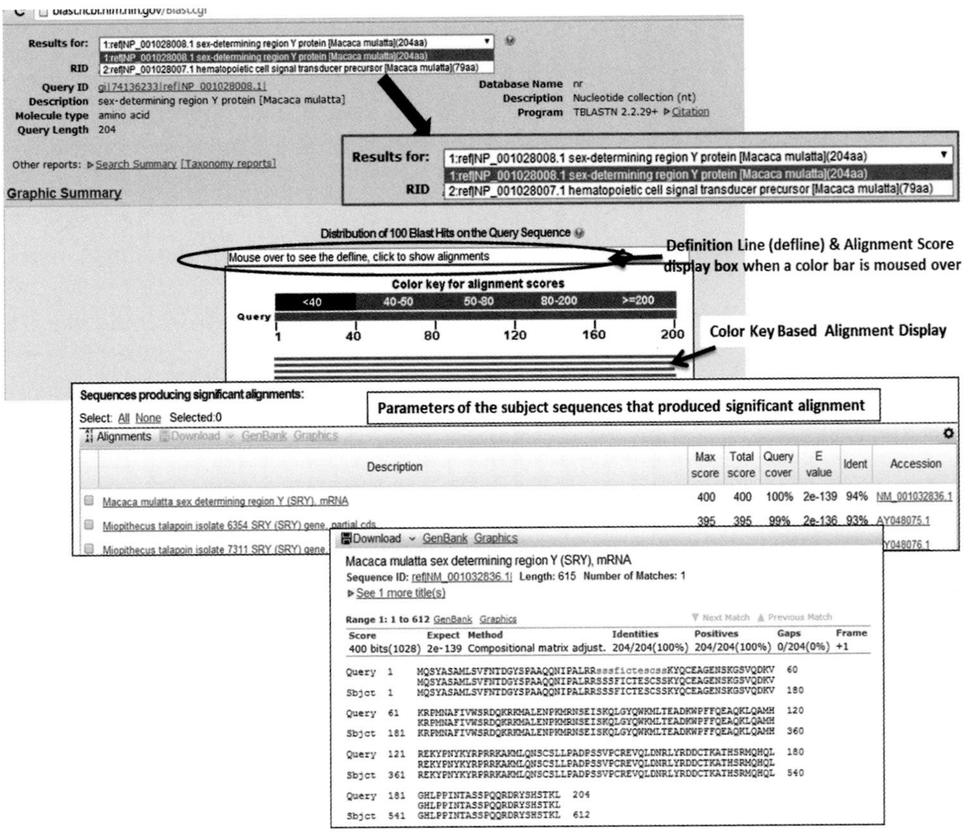

FIGURE 15.2 The results page of tBLASTn contains the color key-based alignment display, followed by a tabular description of sequence alignments and, finally, alignment of each of the sequence pairs (query versus database sequences).

15.6 QUESTIONS

1. The given amino acid sequence is to be checked for possible transcript variants (transcripts of the same gene with varying length and encoded protein sequences) in non-humped cattle:
   DPLKLATEVGNTENQQGSASKSKVEMSCEGSAEPSDTTTTLCVQESIYG ISEIPLVSSGDGAKDPNDECEVNSGNSMPDLEAEEELSEDHSQIHGNSVV LTNSTEPASEDPFVADENSTE
2. Discover the protein homologs in the equine genome for the following genes, using taurine amino acid sequences as the query sequence: TSPY (Testis-specific protein, Y-encoded), Cathelicidin, TLR4.
3. Discuss the applications of tBLASTn.
4. Explain the result of tBLASTn given in Figure 15.2, systematically.
5. Assume that the tBLASTn tool is not working for some days (or is not available). How will you proceed to analyze a given novel amino acid sequence to annotate its encoding gene-specific features?

tBLASTx

CHAPTER 16

CS Mukhopadhyay and RK Choudhary
School of Animal Biotechnology, GADVASU, Ludhiana

16.1 INTRODUCTION

tBLASTx, or translated BLASTx, accepts nucleotide query sequence(s) as well as database subject sequences, translates both to 6-frame amino acid sequences and, finally, compares them at the amino acid level. tBLASTx is a valuable tool for discovering novel genes in the nucleotide sequences, such as single pass expressed sequence tags and draft genome records which are unannotated and riddled with errors (e.g., wrong bases and frame shifts). These errors often make one coding sequence difficult to be detected. However, the limitations associated with tBLASTx are:

i. tBLASTx is very much resource-intensive and time-consuming;
ii. large queries are not recommended, due to the inherent limitation of the time required.

16.2 OBJECTIVE

To determine the homology of a given nucleotide query sequence against the database of draft genome records, as well as expressed sequence tag data to identify the sequence.

16.3 PROCEDURE

The necessary steps are the same as for BLASTx. Open the NCBI home page by typing http://www.ncbi.nlm.nih.gov/and click "tBLASTx"; alternatively, it can also be opened by entering the URL in the address bar: http://blast.ncbi.nlm.nih.gov/Blast.cgi?PROGRAM=tblastx&PAGE_TYPE=BlastSearch&LINK_LOC=blasthome.

Basic Applied Bioinformatics, First Edition. Chandra Sekhar Mukhopadhyay, Ratan Kumar Choudhary and Mir Asif Iquebal.

16.3.1 Enter query sequences

- Enter accession number(s) or FASTA sequence(s): Paste one or more nucleotide query sequence(s) in FASTA format, or the respective NCBI accession number(s), into the specified sequence box. A text file containing the nucleotide query sequence(s) (in FASTA format) can be uploaded by clicking the "Choose File" button.
- Provide Query Sub-range (optional) to specify a particular range of the input sequence that is to be searched against the database. This is especially useful when the GenBank accession number is used instead of the actual whole sequence.
- Genetic Code: The default is "Standard", which can be used for eukaryotic genomic DNA-derived sequences. Other options, for prokaryotic DNA, or mold or yeast or vertebrate or invertebrate mitochondrial DNA, are also available.
- Give a Job Title to identify the tBLASTx results from saved searches.
- Checking "Align two or more sequences": This check box, if checked, will refresh the page to provide the user with another sequence box, where the subject sequence(s) is/are to be pasted.

16.3.2 Choose search set

- Database: Choose any one of the nucleotide databases against which the search is to be made. The list of databases is almost the same as that of BLASTn, except for these two options: "Human Genomics plus Transcript", "Mouse Genomics plus Transcript".
- Organism (optional): Specify the organism by common name, binomial name or taxonomical ID. You can also check the small check box adjacent to the entry box to exclude any one or more organisms (click on the "+" sign to add) from your search results.
- Exclude Models(XM/XP) and/or Uncultured/environmental sample sequences (optional): Check one or both of the check boxes to exclude one or both of the options. Models(XM/XP) stands for "model reference sequences", determined and annotated from the Genome Annotation Project of NCBI and, thus, could be incomplete.
- Limit to: Check the small check box if you want to restrict the search to type materials only. Type material refers to any preserved specimen of an organism or bacterial strains as cultures like culture collection.
- Entrez Query (optional): The same as for BLASTn. This option restricts the search to the specified Entrez query only. It allows the Boolean operators, AND, OR, NOT, to define the database to be searched.
- **BLAST**: Click on the button to initiate the tBLASTx search. Click the check box to open the search result in a new window.

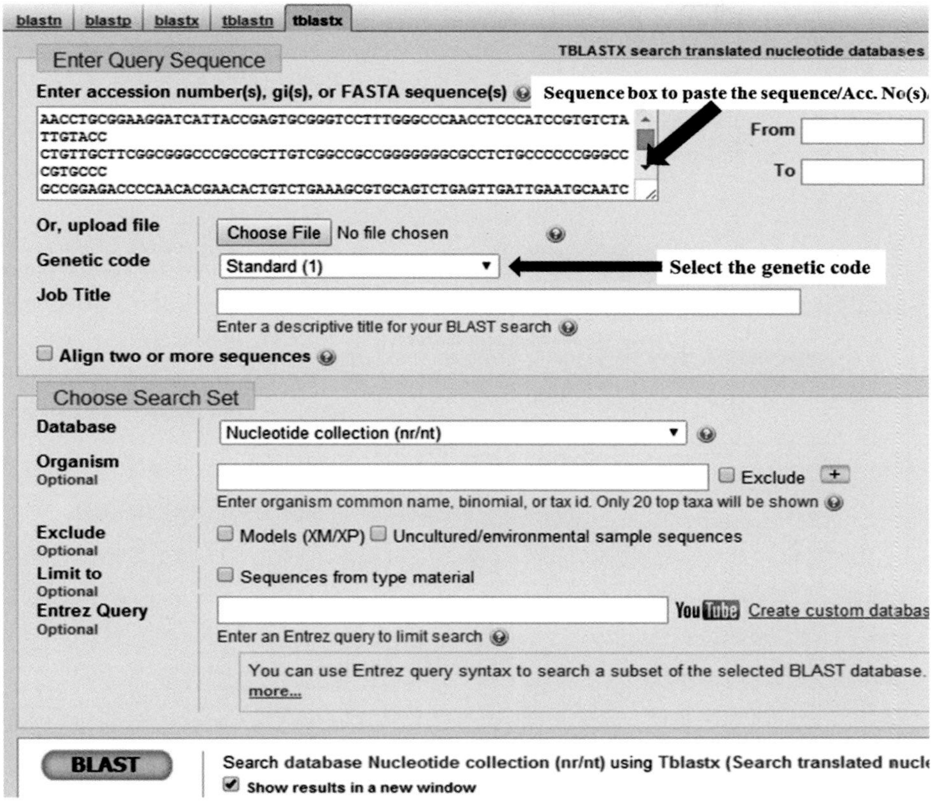

FIGURE 16.1 Main page for tBLASTx search at NCBI. The sequence can be entered into the box as query sequences, with either accession no. or sequence, in FASTA format.

16.4 ALGORITHM PARAMETERS

These are the same as those for BLASTx.

a. General parameters
b. Scoring parameters
c. Filters and Masking

16.5 INTERPRETATION OF tBLASTx RESULTS

a. The output of tBLASTx is similar to that of BLASTp or BLASTx.
b. The color key-based alignment depiction and the table indicating the tBLASTx output for various homologous sequences are the same as those for BLASTx.
c. There is an extra column "N" in the tabular display of the tBLASTx parameters description. "N" represents the number of different segment pairs that have been used to produce the ungapped alignment.

d. Individual pairwise alignment is also the same as that for BLASTp. However, the open reading frames for both the sequences (query vs. database entries) are indicated by "Frame". The asterisk (*) symbol indicates translation stop. The amino acid matches (+/+, +/–, –/+ and –/– strands) are shown as range 1, range 2 and so on, in descending order of score.

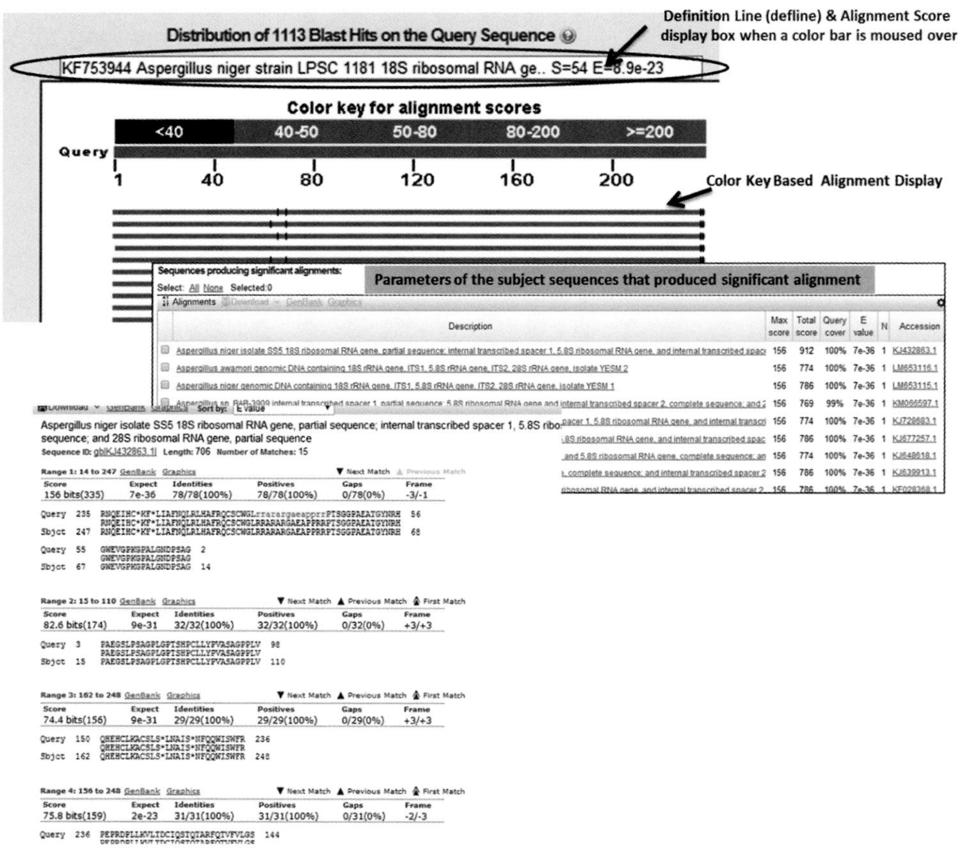

FIGURE 16.2 The result page of tBLASTx contains the color key-based alignment display, followed by tabular description of sequence alignments and, finally, alignment of each of the sequence pairs (query versus database subject sequences).

16.6 QUESTIONS

1. What are the applications of BLASTx?
2. Explain the various parameters obtained in a tBLASTx result.
3. Interpret the given tBLASTx output:

Sequences producing significant alignments:

Select: All None Selected:0

Alignments Download GenBank Graphics

Description	Max score	Total score	Query cover	E value	N	Accession
PREDICTED: Ovis aries musimon drosha, ribonuclease type III (DROSHA), transcript variant X6, mRNA	285	1542	96%	3e-74	1	XM_012159286.1
PREDICTED: Ovis aries musimon drosha, ribonuclease type III (DROSHA), transcript variant X5, mRNA	285	1542	96%	3e-74	1	XM_012159285.1
PREDICTED: Ovis aries musimon drosha, ribonuclease type III (DROSHA), transcript variant X4, mRNA	285	1542	96%	3e-74	1	XM_012159284.1
PREDICTED: Ovis aries musimon drosha, ribonuclease type III (DROSHA), transcript variant X3, mRNA	285	1542	96%	3e-74	1	XM_012159283.1
PREDICTED: Ovis aries drosha, ribonuclease type III (DROSHA), transcript variant X2, mRNA	285	1542	96%	3e-74	1	XM_012097133.1

Download GenBank Graphics Sort by: E value

PREDICTED: Ovis aries musimon drosha, ribonuclease type III (DROSHA), transcript variant X6, mRNA

Sequence ID: ref|XM_012159286.1| **Length:** 4650 **Number of Matches:** 7

Range 1: 2788 to 3120 GenBank Graphics ▼ Next Match ▲ Previous Match

Score	Expect	Identities	Positives	Gaps	Frame
285 bits(617)	3e-74	111/111(100%)	111/111(100%)	0/111(0%)	+3/+1

```
Query  3     KTGIRSDVCQHAMMLPVLTHHIRYHQCLMHLDKLIGYTFQDRCLLQLAMTHPSHHLNFGM  182
             KTGIRSDVCQHAMMLPVLTHHIRYHQCLMHLDKLIGYTFQDRCLLQLAMTHPSHHLNFGM
Sbjct  2788  KTGIRSDVCQHAMMLPVLTHHIRYHQCLMHLDKLIGYTFQDRCLLQLAMTHPSHHLNFGM  2967

Query  183   NPDHARNSLSNCGIRQPKYGDRKVHHMHMRKKGINTLINIMSRLGQDDPTP  335
             NPDHARNSLSNCGIRQPKYGDRKVHHMHMRKKGINTLINIMSRLGQDDPTP
Sbjct  2968  NPDHARNSLSNCGIRQPKYGDRKVHHMHMRKKGINTLINIMSRLGQDDPTP  3120
```

Range 2: 2786 to 3121 GenBank Graphics ▼ Next Match ▲ Previous Match ⩓ First Match

Score	Expect	Identities	Positives	Gaps	Frame
267 bits(577)	1e-68	106/112(95%)	107/112(95%)	0/112(0%)	-1/-3

```
Query  336   RELGHLGQGVT*YLLRC*FLFSACACDELFYLHIWAAESHS*TENSWHDQDSFQNLSDDL  157
             RELGHLGQGVT*YLLRC*FLF ACACDELFYLHIWAAE  S*T+NSWHDQDSFQNLSD L
Sbjct  3121  RELGHLGQGVT*YLLRC*FLFCACACDELFYLHIWAAEFRS*TKNSWHDQDSFQNLSDGL  2942

Query  156   GGSWPAVTDNDLGKYILSTCPDALGTGGSEYDGSELAASWHADRHQNGCQFS  1
             GGSWPAVT NDLGKYILSTCPDALGTGGSEYDGSELAASWHADRHQNGCQFS
Sbjct  2941  GGSWPAVTGNDLGKYILSTCPDALGTGGSEYDGSELAASWHADRHQNGCQFS  2786
```

Range 3: 2787 to 3125 GenBank Graphics ▼ Next Match ▲ Previous Match ⩓ First Match

Score	Expect	Identities	Positives	Gaps	Frame
261 bits(565)	5e-67	106/113(94%)	109/113(96%)	0/113(0%)	+2/+3

```
Query  2     ENWHPF*CLSACHDAASSDPSYSLPPVPNASGQVDRIYFPRSLSVTAGHDPPKSSLKFWN  181
             ENWHPF*CLSACHDAASSDPSYSLPPVPNASGQVDRIYFPRSL VTAGHDPPK SLKFWN
```

4. Let us say you have obtained the following sequence after custom-sequencing one cloned DNA fragment. Run tBLASTx and comment on the DNA fragment:
AGCGGCCGCCAGTGTGATGGATATCTGCAGAATTCGCCCTTT
CCAGCTTCAAGAGCCAAATACTGATCGACAACTTATTGAGAC
TTCTCCAGTTCTACAGAAACTTACTGAATTTGAAGATGCAAT
TGGAGTAATTTTTACACATGTTCGACTTTTGGCAAGAGCATT
CACATTGAGAACTGTGGGATTTAACCATCTGACCCTAGGCCA
CAATCAGAGGATGGAACTCCTTGGTGACTCCATAATGCAGCT
GGTAGCAACGGAGTACTTATTCATTCATTTTCCAGATCATCA
CGAAGGACACTTAACTCTGTTGCGAAGCTCTTTGGTGAATAA
CAGAACTCAGGCCAAGGTGGCGGAGGAACTGGGCATGCAGGA
ATACGCCATCACCAACGACAAGACCGAAAGACCTGTCGCCCT
GAGAACCAAGACCTTGGCTGACCTTTTGGAATCATTTATTGC
AGCACTATACATTGATAAAGACTTGGAATACGTTCATACTTT
CATGAATGTTTGTTTCTTTCCACGATTAAAAGAATTCATTTT
GAATCAGGATTGGAACGACCCCAAGTCCCAGCTTCAGCAGTG
CTGCCTGACTCTTAGGACAGAAGGAAAAGAACCAGACATTCC
GCTATACAAGACTCTGCAGACGGTGGGGCCATCCCATGCAAG
GACCTACACTGTGGCTGTCTACTTCAAGGGAGAAAGAATTGG
CTGTGGGAAAGGACCAAGTATTCAGCAAGCAGAAATGGGA
GCAGCAA

5. Explain the following tBLASTn output:

Select: All None Selected:0

Alignments Download GenBank Graphics

Description	Max score	Total score	Query cover	E value	N	Accession
Mus musculus ribonuclease III, nuclear, mRNA (cDNA clone IMAGE:5698108), partial c	147	1266	95%	5e-33	2	BC060265.1
Mus musculus 2 days neonate sympathetic ganglion cDNA, RIKEN full-length enriched	147	1266	95%	6e-33	2	AK148640.1
Mus musculus nullipotent stem cell CRL-2070 NE cDNA, RIKEN full-length enriched lib	147	1256	95%	6e-33	2	AK144147.1
Mus musculus ribonuclease III, nuclear, mRNA (cDNA clone MGC:115770 IMAGE:641	147	1266	95%	6e-33	2	BC088999.1
Mus musculus drosha, ribonuclease type III (Drosha), transcript variant 2, mRNA	147	1266	95%	6e-33	2	NM_026799.3
Mus musculus drosha, ribonuclease type III (Drosha), transcript variant 1, mRNA	147	1266	95%	6e-33	2	NM_001130149.1

Primer Designing and Quality Checking

SECTION IV

Primer Designing – Basics

CHAPTER 17

CS Mukhopadhyay and RK Choudhary
School of Animal Biotechnology, GADVASU, Ludhiana

17.1 INTRODUCTION

A primer is a short synthetic oligonucleotide, which is used to initiate amplification of DNA/RNA in a polymerase chain reaction (PCR). Literally, "to prime" means to "initiate" or "start". *In vivo*, a short oligo-sequence (i.e., the primer) is required, because the enzyme "DNA polymerase" has no capacity to initiate DNA replication without any primer. During this process of molecular photocopying, *in vitro* amplification of the target nucleotide sequence is initiated by a short complementary oligo.

Specificity and efficiency are two important factors for designing primers. Specificity of a primer pair is the ability of a PCR primer pair to amplify a specific product (i.e., no spurious amplification). The length and sequence of the oligo-sequence pattern (repetitive or single copy, part of a multi-gene family) of the template are factors that affect the specificity of primers. The efficiency of a primer pair refers to the fold increase of amplicon in each cycle, which should be ideally two folds in each cycle (practically, between 1.8 and 1.95).

17.2 OTHER IMPORTANT FEATURES FOR DESIGNING "GOOD" PRIMERS

17.2.1 Adding RE sites to primers

Restriction endonuclease (RE) site (4–6 nucleotides) is added at the 5′-terminus of the oligo to use the amplicon in cloning and genetic engineering. Add 2–3 more bases before this RE site to facilitate the RE to attach to the sequence. Two different set of T_ms – namely, for the core primer (18 nt long) and whole primer (18 + 4 or 6 bases) – are to be considered. The T_m should not differ much for these two sets.

Basic Applied Bioinformatics, First Edition. Chandra Sekhar Mukhopadhyay, Ratan Kumar Choudhary and Mir Asif Iquebal.

TABLE 17.1 Important parameters to be considered for designing "good" primers (http://www.premierbiosoft.com/tech_notes/PCR_Primer_Design.html).

SN	Feature	Ideal value or range	Pros	Cons
1	Primer length	18–25 nucleotides (nt).	Too short primers show low specificity, while very long primers reduce template-binding efficiency due to formation of secondary structures, and also require more time to anneal and denature.	Primers for multiplexing may be as long as 30–35 bp, while primers used for random priming (e.g., RAPD) are kept short, e.g., 8 (octamers) to 12-mer, to promote random priming.
2	Melting temperature (T_m)	Mean annealing temperature is 5 °C less than the average T_m of the primer pair.	T_m refers to the temperature at which half proportion (or 50%) of the primer and its complementary nucleotides of the template are hybridized.	A primer should anneal to the template before the template strands re-nature.
3	Optimal T_m	52–62 °C.	T_m below 45 °C could encourage secondary annealing and spurious amplification. If T_m > 65–70 °C of the primers for automated sequencing, secondary priming artifacts, and noisy amplicons are evident.	Higher T_m (75–80 °C) is recommended for amplifying high G/C content of targets
4	Primer pair T_m mismatch	Ideal 2 °C; at most, 5 °C.	Out of the primer pairs, the primer which has more T_m will misprime at lower temperatures, while the other primer may not anneal to template at higher temperatures.	–
5	Cross-homology	No cross-homology within the same species.	Mispriming will result.	Primer-BLAST checks possible spurious amplifications. Exon-exon junctions of complementary DNA strand (CDS) should be targeted to amplify mRNA/cDNA (genomic DNA will not be amplified).
6	Primer G/C content	Optimum is 45–55%.	Can vary between 40 and 60%.	This determines the annealing temperature.
7	G/C Clamp	2 G/C within the last four bases at 3′ end of primer.	Three or more G/C clamps could make the primer "sticky", due to higher T_m at 3′ terminus	3′-terminus of the primer is crucial. G/C at 3′-end increases the efficiency of the primers.
8	Stretches of nucleotides	Max. of four dinucleotides or four mono-nucleotide repeats in a primer.	Runs of same bases increase the probability of primer-dimer and hairpin loop formation.	Could lead to annealing of primer(s) to an unintended template, due to chance similarity between complementary sequences that leads to low Gibb's free energy at the 3′-end.
9	Optimal amplicon size	SSCP: < 400 nucleotide; cloning: 200- several kilobases (kb); Real time- qPCR: 80–200 bp; RFLP studies: variable.	In real-time qPCR, amplicon size < 200 increases amplification efficiency close to 100%.	The length of the product is determined by the purpose of the experiment, rather than the rules for designing "good" primers.

17.2.2 Secondary structures in primers

Secondary structures are the various combinations of the primers formed among themselves (self and heterologous) via a loop-like structure produced by the same primer.

a. *Hairpins*: these are formed due to intra-molecular interactions among the nucleotides of the same primer.
b. *Self-dimer (homodimer)*: these dimers are formed by intermolecular interactions between two primers that are the same.
c. *Cross-dimer (hetero-dimer)*: produced by inter-molecular interactions between the forward and reverse primers.

17.2.3 Max 3′-end stability

Higher 3′-end stability improves priming efficiency; however, increased 3′-stability can inversely influence the specificity of primer, because 3′-terminal partial hybridization induces non-specific extension. It is rendered by the maximum ΔG of the 3′-end.

17.2.4 Analyzing Gibbs free energy (Delta G)

The "Gibbs Free Energy" (ΔG) measures the quantity of work yielded by a process or system that is operational at a steady pressure. A higher absolute value, however, with a negative symbol, indicates the spontaneity of the reaction. In other words, a ΔG value of –9 signifies that the oligo and the template will anneal more spontaneously than another oligo which has ΔG = –5 kcal/mol. Thus, the former primer will require more energy to be dissociated or denatured than the second one. Therefore, ΔG is the amount of energy that can break a particular secondary structure, which is why the ΔG value is recommended to be less (absolute value) for the secondary structures than the primer-template duplex. The ΔG values of the primers can be estimated using software such as "IDT Oligo Analyzer" online (http://eu.idtdna.com/analyzer/applications/oligoanalyzer/) for the 3′ ends of each of the oligos of the homo- and hetero-dimers.

TABLE 17.2 The acceptable values of Gibb's free energy for various secondary structures of primer (http://ls23l.lscore.ucla.edu/Primer3/primer3web_help.htm).

SN	Secondary structures	Acceptable ΔG
1	3′-end hairpin	> –2 kcal/mol
2	Internal hairpin	> –3 kcal/mol
3	3′-end self-dimer	> –5 kcal/mol
4	Internal self-dimers	> –6 kcal/mol
5	3′-end cross-dimer	> –5 kcal/mol
6	Internal cross-dimer	> –6 kcal/mol
7	5 bases from the 3′-end of primers	> = –9 kcal/mol

17.3 QUESTIONS

1. Suppose you have designed a primer of 21 bases, which is as follows: 5′ ACGTAG CTATGGCAATGTAAT 3′. Enumerate what are the parameters that are not optimum in this primer.
2. The ΔG value of the 3′-end dimer of a given primer is –8 kcal/mol. Explain your logic as to the acceptability of this primer.
3. Enlist the main parameters, along with the optimum values, that are used for evaluation of a designed primer.
4. What do you mean by secondary structure? How are these formed? What are the implications of secondary structure formation during PCR amplification?

Designing PCR Primers Using the *Primer3* Online Tool

CHAPTER 18

CS Mukhopadhyay and RK Choudhary
School of Animal Biotechnology, GADVASU, Ludhiana

18.1 INTRODUCTION

The previous chapter (Chapter 17) discussed in brief the salient features of PCR-primers. Here we will learn how to use Primer3, an online software tool, for designing PCR primers.

18.2 OBJECTIVE

To design primers for bubaline Dicer I cds, using the online primer designing tool Primer3.

18.3 PROCEDURE

18.3.1 Downloading a nucleotide sequence

Type the URL (www.ncbi.nlm.nih.gov/) to open the NCBI home page. Search the required nucleotide sequence, select the target sequence from NCBI nucleotide and save it in FASTA format in a text file.

18.3.2 Open Primer3 online tool

Open the Primer3 (version 4) software by using the URL http://primer3.ut.ee/.

18.3.3 Obtaining nucleotide sequence of interest

Paste the nucleotide sequence (in FASTA format) in the box for source sequence in the Primer3 page.

Basic Applied Bioinformatics, First Edition. Chandra Sekhar Mukhopadhyay, Ratan Kumar Choudhary and Mir Asif Iquebal.

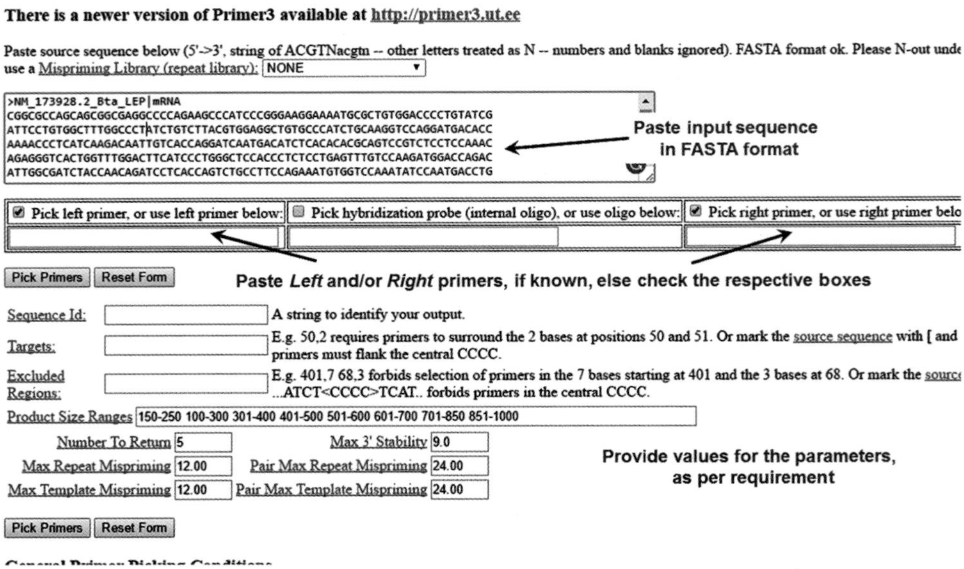

FIGURE 18.1 Setting the parameters of the Primer3 online tool for primer designing. (*See insert for colour representation of the figure.*)

18.3.4 Set the required parameters

All the parameters mentioned on the Primer3 page have been explained, along with the optimal values, in Table 18.1.

All the parameters are not necessarily equally important. The experience of the user will cultivate a discretionary ability to select the parameters and set the optimal values. However, the parameters mentioned in the "General Primer Picking Conditions" section (Primer Size, Primer T_m, Maximum T_m difference, Primer GC%, Maximum selfcomplementary and Max 3′ Self-Complementary) must be set, or else the software will design a set of primers on its default values for the parameters.

18.3.5 Get the primers

Finally, click on the "Pick Primers" option to get the primers. If any mistake has been made while setting the parameters, the "Reset Form" button should be clicked to initiate the parameter-setting afresh.

18.4 OUTPUT

a. Primer3 gives more than one set of primers (based on the number set for the "Number to Return" parameter).
b. The primers are to be critically evaluated for various parameters, including product size and the target covered, as well as other parameters:
    i. Start: There are two start values – one for each of the primers in the pair. These refer to the index value of the starting base of the amplicon (for Forward primer) and the last base (for Reverse primer) nucleotide on the input (i.e., target) sequence on which the forward primer or the reverse primers bind.
    ii. Length (given as "len"): Length of each primer.

TABLE 18.1 Primer3 parameters, description and their optimal values/options (http://ls23l.lscore.ucla.edu/Primer3/primer3web_help.htm).

SN	Parameter	Description	Optimal value
1	Library (repeat library)	Allows the user to screen the databases (called libraries) for reported repeat sequences in human (microsatellite), rodents (VNTRs) and Drosophila (VNTRs) to skip the interspersed repeats as the location of primer(s).	If the nucleotide sequence belongs to human, or rodents, Drosophila, select that species from the drop-down menu; else select "None".
2	Source sequence	The nucleotide sequence from which the primers are to be designed should be in FASTA or EMBL format.	Avoid repeat sequences, unless primer is being designed for microsatellites.
3	Pick left primer, or use left primer below	Check the small box to get the left primer designed by Primer3. One can also paste the left (i.e., forward) primer in the rectangular sequence box, if the user wants to use the specified primer only.	If a specific left primer is to be used, insert the sequence in the given space; else tick the check box on the left.
4	Pick hybridization probe (internal oligo), or use oligo below	The software will choose a probe for the sequence given; else, we can put our own probe in the space below. If we do not need any probe, the check box is left unchecked.	Same as "Pick left primer" given above. This option is left unchecked if probe is not required.
5	Pick right primer, or use right primer below	Same as "Pick left primer"; this option is used for the right primer.	–
6	Sequence ID	This is an identifier or tag for the sequence given as input that is specified in the Primer3 output.	Specify one name for the sequence as the ID of the sequence.
7	Targets	Target means a specific nucleotide sequence (e.g., simple repeat sequence/base pair polymorphism) within the input/source sequence that is to be flanked by the designed primer-pairs. More than one target(s) can also be specified. The syntax of mentioning the targets is using a space delimited list of START, LENGTH. START indicates the start of the target, while LENGTH specifies the length of the target.	Mention the target(s) in the space separated START, LENGTH syntax if any specific target is to be flanked by the primers; else leave this parameter blank. The target regions are specified by asterisks (*) in the Primer3 output. Multiple targets can be specified as START1, LENGTH1 START2, LENGTH2.

(*Continued*)

TABLE 18.1 (Continued)

SN	Parameter	Description	Optimal value
8	Excluded regions	The portion(s) of nucleotide in the source sequence that must be excluded while designing primers by the software. The Primer3 will just skip those excluded regions and primers will not flank those specified regions. The syntax is same as for Targets, i.e., mention space separated list of START, LENGTH of the excluded regions. Do not put comma between the values of different sets of START and LENGTH.	To exclude regions of "low sequence quality" or repeat sequences, specify the base locations as space-separated list of START, LENGTH of those excluded regions. Otherwise, the parameter should be left blank. The excluded regions are specified by "X" markings in the Primer3 output.
9	Product size ranges	This is the desired amplicon length or the product size we need in our wet lab. A list of ranges has been specified in the software, but any other set of ranges can be specified. User needs to specify one or more ranges, which will be considered during primer designing, starting from lower ranges to higher ranges specified (if the lower range does not succeed in picking primers, the software shifts to the next higher range).	Specify one or more ranges as comma separated ranges, e.g., 80–100, 201–300. If the user skips this option, Primer3 will automatically screen ranges from the lower to higher side to pick primers.
10	Number to return	It specifies the number of primer pairs to be returned by Primer3. The time required to design the primers is proportional to the number of primers returned.	By default, the number of return is 5. This may be changed if good primers are not obtained in the first round, or we need to screen for another set of alternative primers.
11	Max 3′ stability	The "Max 3′ stability" parameter estimates the maximum stability for the five terminal bases at the 3′-end of the primer. More value (ignoring sign) means more stable 3′-ends. The ΔG ranges between 6.86 (i.e., the highest for "GCGCG") and 0.86 kcal/mol (the lowest for "TATAT") for the 3′-pentamers (SantaLucia, 1998). The value is calculated using the Nearest-Neighbor parameter for the maximum ΔG to denature the 3′-pentamers.	Default value is 9.0 (it is not the ΔG), which is the maximum recommended value. The value can be kept unchanged; however, if the software fails to design the primers, the values may be reduced. Reducing the value of "Max 3′ Stability" simply compromises the efficiency of the primers.
12	Max repeat mispriming	This parameter limits the maximum extent of mispriming due to the repeat sequence, based on the weighted similarity in the mispriming library. The higher the value set for the parameter, the lower the probability of mispriming – hence, the less the weighted similarity with any repeat sequence in the database.	Default value: 12. Keep the value as default value. The libraries mentioned in the drop-down list of "Library (repeat library)" (i.e., human, rodents and Drosophila) can only be screened for similarity with repeat sequences for mispriming.

13	Pair max mispriming	Principally, this parameter is the same as the previous one; however, it calculates the weighted similarity for both the primers (sense and antisense) jointly.	Default value: 24. Set the value to 24. Applicable for primers to be designed for human, rodents and Drosophila.
14	Max template mispriming	This parameter estimates the probability of mispriming on a different location of the same template for each of the primers individually.	Default value: 12 Keep the value as default value.
15	Pair Max Template Mispriming	The scores of "Max Template Mispriming" for both the primers (forward and reverse) are summed up. It estimates the likelihood that both the primers will also anneal to the template provided on a different location.	Default value: 24. Use the default value.
General primer picking conditions			
16	Primer size	The primer length optimally ranges between 18 and 25 bp; however, for multiplexing the primer length may increase up to 30–35 bp. Allowed range of values are 1 (minimum) and 36 (maximum) for the number of bases in a primer.	Enter the number of bases (bp) as given: Minimum size: 20 Optimum: 23 Maximum: 25
17	Primer T_m	The melting temperature (T_m) value can ideally range between 52–62 °C. The software follows the primer-T_m formula and the thermodynamic parameters are obtained from the table given by Breslauer *et al.* (1986).	Enter the T_m as given (for normal primer): Minimum T_m: 50 °C Optimum T_m: Between 55 and 60 °C Maximum: 65 °C.
18	Maximum T_m difference	Wider T_m values for the primers lead to less efficiency of primers and mispriming.	Set the value to 2 °C. If required to increase the difference, the highest allowed T_m difference is normally 5 °C.
19	Table of thermodynamic parameters	The Nearest-Neighbor thermodynamic parameters, as well as the method to calculate the T_m, can be obtained from any one of the two following sources: Breslauer *et al.* (1986) and Santa Lucia (1998). Click on the drop-down menu and select "Santa Lucia 1998".	The recommended option is "Santa Lucia 1998".
20	Product T_m	Product T_m refers to the temperature at which 50% of the amplified product remains single-stranded. Lower product T_m will denature the template too early; it also signifies high A/T content. Again, too high product T_m will complicate the PCR, due to the requirement of more time to denature the template which, in turn, may affect the functionality of *Taq* polymerase. It is recommended to select the product T_m to 50 °C.	Enter the product T_m as: Minimum: 45 °C Optimum: 50 °C Maximum: 55 °C.

(Continued)

TABLE 18.1 (Continued)

SN	Parameter	Description	Optimal value
21	Primer GC%	The G/C content of primer determines its T_m. Ideally, 50–65% G/C works well to maintain the efficiency of the designed primers.	Enter the G/C% as: Minimum: 35 Optimum: 65 Maximum: 80.
22	Max self-complementary	Primers with self-complementarity produces homodimer and hairpin loop structures that hinder the annealing of primers with the complementary sequence. Non-negative scores are assigned to the primers, based on the extent of local alignment between the sequences of the same primers.	Initially set the value to 3.00. If it does not yield primer sequences, increase by 1 for each iteration of Primer3 run.
23	Max 3′ self-complementary	The 3′ terminus of the primer is very critical. Self-complementarity at the 3′ end of any of the primer will thwart it from extending the template DNA during PCR. The terminal 3 bases should be strictly non-complementary either to self or to the other primer.	Initially set the value to 3.00. Gradually increase the score by 1 in each run of the Primer3 software, if required.
24	Max #N	It is the number of maximal allowable unknown nucleotides in the flanked region to be primed. The "N" stands as a place-holder for any of the bases. The Ns may increase the possibility of mispriming.	Set the value to 0. However, sometimes it becomes challenging to exclude the Ns, especially in a novel cDNA sequence. In such cases, it may be increased to 1 or 2.
25	Max poly-X:	This option mentions the maximum number of allowable mononucleotide repeats in the primer. Long single-base repeats can promote mispriming. Runs of 3 or more G/Cs or A/Ts at the 3′ terminus of primer should be avoided.	Default is 5. Set the value to 3.
26	Inside target penalty	This parameter is very useful if a specific target (e.g., a gap junction) is to be overlapped. Thus, this option allows the software to select a primer that overlaps with the target. The default value is "BLANK". The only two non-default values, (i.e., 0 and 1) are allowed to penalize the primer that spans or overlaps the single target (dual targets are not possible to be overlapped simultaneously or by two different primers, according to the code written for Primer3), by multiplying the value assigned (0 or 1) with base-counts of the primer overlapping the target.	Default value is "blank". Set the value to "0" (if no target is there), or "1" (if a single target is to be overlapped).

27	Outside target penalty	It is just the opposite of the above, where the software awards penalty to the primer that does not overlap the target. The chosen value (0 or 1) is multiplied by the number of bases from the 3′ terminus of oligo to the target for obtaining the 'position penalty'. Thus this parameter enables us to reach to the target.	Default value is "0". Set the value to "0" (if no target is there), or "1" (if proximity to a single target is to be considered).
28	First base index	This refers to the key or index of the initial base in the source (i.e., input) sequence. This option informs Primer3 about the type of indexing of the initial base in the source sequence. GenBank (NCBI) uses one-based indexing.	Default index is "1". Set to "1".
29	GC clamp	This refers to the number of the G/Cs at the 3′ end of both the primers. 1 or 2 G/C clamp(s) (no more than that) is/are recommended in the 3′-pentamers for each primer. This parameter does not introduce G/C clamp in the hybridization probe.	Default: 0. Set: Default works fine; however, it can be increased to 1 or 2, if required.
30	Concentration of monovalent cations	The concentration of monovalent salt (KCl) in mM in PCR master-mix for calculating the T_m of the primers.	Default: 50 mM. Set to default.
31	Concentration of divalent cations	The concentration of $MgCl_2$ is required, as it has an immense effect on primer annealing and the net output of the PCR. Primer3 converts this to monovalent cation concentration for calculating the primer T_m. The concentration of deoxynucleotide triphosphate (dNTP) must be less than the concentration of divalent cations; otherwise, excess of dNTPs may chelate with $MgCl_2$ and, thereby, hinder the progress of PCR.	Default is 0. Set to default, unless it is required to increase, as in the case of SYBR green primer designing, which requires approximately 3 mM $MgCl_2$.
32	Concentration of dNTPs	This refers to concentration of dNTPs in the PCR master mix in mM.	Default: 0. When the concentration of divalent cations (e.g., Mg^{2+}) is provided, then only this argument is to be used.
33	Annealing oligo concentration	Stands for the concentration of primers (nM) in the polymerase chain reaction. The software uses this argument to calculate primer T_m.	Default: 50 nM Set to default.

(*Continued*)

TABLE 18.1 (Continued)

SN	Parameter	Description	Optimal value
34	Liberal base	This enables the Primer3 to accept any unrecognized bases (characters other than A/T/G/C), even symbols like "*" and "–", by changing the base to "N". To make this parameter work, the parameter "Max #N's" must be set to a non-0 value.	It is better to click on the check box. However, this will only work if the parameter "Max #N's" is set to a positive value. Default is "checked".
35	Show debugging information	Checking this parameter will include the input to Primer3-core as a part of the output.	Default: Unchecked. Depends on user's requirement.
36	Do not treat ambiguity codes in libraries as consensus	This is an instruction (self-explanatory) to the software, and applicable to the species mentioned in the list of the mispriming libraries.	Default: Checked.
37	Lower-case masking	Lower-case letters signify the low-complexity regions, i.e., the repeat region which, when present in the 3′ terminus of the primer, may negatively impact on the primer efficiency and annealing ability. Hence, when this parameter is checked, lower-case letters are deleted from the 3′ end of the primers.	Default: Unchecked Check the box, if it is required.
Other per-sequence inputs			
38	Included region	We may need to exclude some sequences, such as repeat sequence, vector sequence, signaling region, or the untranslated regions. This parameter specifies the region in the source-sequence to include for designing primers. The syntax is START LENGTH. The START is the base from which the Primer3 software will look for suitable primers, and the LENGTH signifies the end of the region of choice.	Set the value as per requirement; else leave blank.
39	Start codon position	The parameter is very useful, but requires expertise to select an in-frame amplicon. Any positive value will mark the start of "ATG" (i.e., the default start codon) in the sequence at that specified base onwards. An error will be pinged if there is no "ATG" at the specified location of the source sequence. A negative value indicates that the "ATG" sequence is upstream to the start of the source code, while a value less than or equal to 10^{-6} commands Primer3 to ignore this parameter.	Set the value according to the position of the start codon; else leave the parameter blank.

40	Sequence quality	• "Min Sequence Quality" • "Min 3′ Sequence Quality" • "Sequence Quality Range Min" • "Sequence Quality Range Max" The parameter indicates the quality of the source sequence. Each base should represent only one integer. A higher value of integer indicates higher confidence for base-calling of the source-sequence. The parameters under "Sequence Quality" are considered only when quality values are assigned to each of the base.	Default: Blank (for Sequence Quality). It is better to leave this blank unless we are confident enough about the quality of the source sequence, and that primers will work on the template, having exactly the same sequence as the source.
41	Objective function penalty weights for primers and primer-pairs	In a nutshell, this argument assigns weights to each of the parameters required to qualify a primer, as per the user's requirement. The software takes care of the weights while designing primers. The users can change the weights as per their own discretion (based on the use of primers).	The default values are set in the software. It is better to change the weights after gaining adequate experience in primer designing. This is a very powerful parameter to select the best primers. If no weight is given, Primer3 will use the values as fed by the user at the beginning (i.e., "General primer picking conditions").
42	HybOligo (internal oligo) per-sequence inputs	This parameter is very important for selecting the hybridization oligos (i.e., the probe). These parameters work similarly to the parameters for primers. However, the parameter "Max 3′ Complementarity" is an exception, since it cannot be applied for internal probes which do not produce primer-dimer.	Specify the parameters as per requirement. Set the value of "Max 3′ Complementarity" to 24.

iii. T_m(given as "tm"): T_m differences and GC%, for each oligo of the primer-pair,

iv. Self-complementarity (cited as "any" in the output): A value less than or equaling 3 indicates that the homodimer and hairpin structures produced by the particular primer can be tolerated during amplification. Higher values of self-complementarity suggest that the secondary structure formed by the primer will require more energy (dissociation temperature) to dissociate. This could hinder the efficiency of the PCR.

v. 3′ self-complementary (cited as "3" in the output): This parameter is critical, as the 3′ end of the primer should be able to dissociate during denaturation and, again, the clamping of the 3′ end should be correct to effect efficient amplification. The permitted value of this parameter is 3 or less.

c. Next, the primers are to be tested for the secondary structure formation using "IDT Oligo Analyzer" online software (http://eu.idtdna.com; click "Tool", and then click "OligoAnalyzer").

```
No mispriming library specified
Using 1-based sequence positions
OLIGO              start  len      tm     gc%   any    3' seq
LEFT PRIMER          129   20   60.02   55.00  4.00  0.00 CAGCTTCGTTACTGCCCTTC
RIGHT PRIMER         187   24   60.01   41.67  8.00  2.00 GGGGAAGTAATTTAAAGTCCTGGT
SEQUENCE SIZE: 1739
INCLUDED REGION SIZE: 1739

PRODUCT SIZE: 59, PAIR ANY COMPL: 7.00, PAIR 3' COMPL: 1.00
TARGETS (start, len)*: 150,10 300,12
EXCLUDED REGIONS (start, len)*: 600,5

    1 CATCAGAGCCCTCCAGAGAGGCTTCAACAGCAGAGAAGCTGGATGGCTCCAGAGAGGAGT

   61 GAATAGTGAGGATTTGCCTTGGAGGAACTTGCGCTGTGGTGATGATGAACGACATCCCAC

  121 TGCAGAGGCAGCTTCGTTACTGCCCTTCAAAGGACTTATCACAACCAGGACTTTAAATTA
              >>>>>>>>>>>>>>>>>>>> **********     <<<<<<<<<<<<<<<<<

  181 CTTCCCCTGAGATCAAGTAAAACGAAATGCTTTTGGAAGTGCTGAACCCAAGGCATTACA
      <<<<<<<

  241 ATGTCACCAGCATGGTGTCCGAAGTTGTGCCTATTGCTAGCATTGCAGTCCTGCTGCTCA
                                                                 *

  301 CTGGATTTCTTCTCTTGGTTTGGTATTATGAGGACACATCCTCAATACCAGGTCCCAGCT
      ***********

  361 ACTTTCTGGGAATTGGGCCCCTCATTTCCCACTGCAGGTTCCTCTGGATGGGGATCGGCA

  421 GTGCCTGCAACTACTACAACAAGATGTATGGAGAATTCATGAGAGTCTGGGTATGTGGAG

  481 AGGAAACCCTTATTATTAGCAAGTCCTCAAGTATGTTCCATGTAATGAAGCACAGTCACT

  541 ACATATCCCGATTTGGCAGTAAACTTGGGTTGCAATTCATCGGCATGCACGAGAAAGGCA
                                                                 X

  601 TCATATTTAACAATAATCCAGCACTCTGGAAAGTTGTTCGACCTTTCTTTACAAAAGCTT
      XXXX

  661 TGTCCGGCCCTGGCCTGGTGCGCATGGTGACCATCTGTGCTGATTCCATCACCAAGCATC

  721 TGGACAGGCTGGAGGAGGTCTGCAATGACTTGGGCTATGTGGACGTGTTGACCCTCATGC
```

FIGURE 18.2 Output page of Primer3 online tool, displaying one pair of primers and their position in the input target sequence (asterisks below the bases).

TABLE 18.2 Important parameters based on which primer is selected.

SN	Feature	Normal PCR	Multiplexing
1	Primer length	18–25 nucleotide (nt)	30–35 nt
2	Optimal T_m	50–60	50–65: No much difference between various primer pairs
3	Primer G/C content	Optimum is 45–55%	45–55: The primer pairs should not vary much among themselves for the average G/C content
4	3′ stability of primers	ΔG > = –9 kcal/mol Primer3 value = 9	ΔG > = –8 kcal/mol Primer3 value > 7
5	Self complementarity	Primer3 value < = 3	Primer3 value < =3
6	Permitted ΔG for homo- or heterodimers	>–5 for 3′ end dimmers >–6 for internal homo- or heterodimers	>–5 for 3′ end dimmers >–6 for internal homo- or heterodimers
7	ΔG for hairpins	> –2 for 3′ end hairpin > –3 for internal hairpins	> –2 for 3′ end hairpin > –3 for internal hairpins
8	Max template mispriming	Primer3 value = 12 for each primer	Primer3 value = 12 for each primer

18.5 SELECTION OF THE BEST PRIMER-PAIRS BY COMPARATIVE EVALUATION OF THE DESIGNED PRIMERS

The Primer3 software will give some (by default five pairs of oligos) primer pairs, from which the user needs to select the best pair for custom synthesizing (Table 18.2).

18.5.1 Primer3

a. Length of primer
b. T_m (given as "tm") and T_m difference
c. GC%
d. Self-complementarity (cited as "any" in the output) and 3′ self-Complementarity

18.5.2 Oligo analysis

a. Mispriming
b. GC clamp
c. Self-complementarity, etc.

18.6 QUESTIONS

1. Design a primer pair that will flank the region between 100 and 150 bases of the given sequence: >Test1(NCBI Acc. No. AB969677.1)
   TGGATGAATGAAAAGCCCTGCTTTGCAACCCCTCAGCATGGC
   AGGCCTGCAGCTCATGACCCCTGCTTCCTCACCAATGGGTCC
   TTTCTTTGGACTTCCATGGCAACAAGAAGCAATTCATGATAA

CATTTATACGCCAAGAAAATATCAGGTTGAACTGCTTGAAGC
AGCTCTGGATCATAATACCATAGTCTGTTTAAACACTGGCTC
AGGGAAGACGTTTATTGCAGTACTACTCACTAAAGAGCTGTC
CTATCAGATCAGGGGAGACTTCAACAGAAATGGCAAAAGGAC
GGTGTTCTTGGTCAACTCTGCAAACCAGGTTGCTCAACAAGT
GTCAGCTGTCAGAACTCACTCAGATCTCAAGGTCGGGGAATA
CTCAAACTTAGAAGTAAGTGCATCTTGGACAAAAGAGAAATG
GAACCTAGAGTTTACTAAACATCAGGTTCTCGTTATGACTTG
CTATGTCGCCTTGAATGTTTTGAAAAATGGTTACTTATCACT
GTCAGACATTAACCTTTTGGTGTTTGATGAGTGTCATCTTGC
AATCCTAGACCACCCCTACCGAGAAATTATGAAGCTTTGTGA
AAATTGTCCATCATGTCCTCGTATTTTGGGACTAACTGCTTC
CATTTTAAATGGGAAATGTGATCCAGAGGAATTGGAAGAAAA
GATTCAGAAACTGGAGAAAATTCTTAAGAGTAATGCTGAAAC
TGCAACTGACTTGGTGGTCTTAGACAGATATACTTCTCAGCC
ATGTGAGATTGTGGTAGACTGTGGACCATTTACTGACAGAAG
TGGGCTTTATGAAAGACTGCTGATGGAGTTAGAAGAAGCACT
TAATTTTATCAATGACTGTAACATATCTGTACATTCAAAAGA
AAGAGATTCTACTTTAATTTCTAAACAGATACTCTCAGACTG
CCGTGCGGTCCTGGTTGTCCTGGGACCCTGGTGTGCCGATAA
AGTAGCTGGAATGATGGTCAGAGAGCTGCAGAAACACATCAA
ACATGAGCAAGAGGAGCTGCACCGGAAGTTTCTGTTGTTCAC
AGACACTTTCCTACGGAAAATCCACGCCCTGTGTGAAGAGCA
CTTCTCCCCTGCCTCGCTTGACCTGAAGTTTGTCACTCCTAA
AGTAATAAAGCTGCTCGAGATCTTACGCAAATACAAACCGTA
TGAGCGGCAGCAGTTTGAAAGCGTGGAGTGGTATAATAATAG
GAACCAGGATAATTACGTGTCCTGGAGCGATTCTGAGGATGA
CGAGGAAGATGAAGAGATTGAAGAGAAAGAAAAGCCAGAGAC
AAATTTTCCTTCTCCATTTACCAATATTTTGTGTGGAATTAT
TTTTGTGGAAAGAAGATACACAGCCGTGGTCTTAAACAGATT
GATAAAGGAAGCTGGCAAACAAGATCCAGAGCTGGCTTACAT
CAGCAGCAATTTTATAACTGGACATGGCATCGGAAAGAATCA
GCCTCGGAACAAACAGATGGAAGCAGAATTCAGAAAGCAGGA
AGAGGTACTTAGGAAATTTCGAGCACATGAAACCAACCTGCT
TATTGCAACAAGTATTGTGGAGGAGGGTGTCGACATACCGAA
ATGCAACTTGGTGGTTCGTTTCGATCTGCCCACAGAGTATCG
ATCCTACGTTCAGTCGAAGGGAAGAGCGAGGGCACCCATCTC
TAATTATGTAATGTTAGCAGACACAGATAAAATAAAGAGTTT
TGAAGAAGACCTTAAAACATACAAAGCTATTGAAAAGATCTT
GCGAAACAAATGTTCCAAGTCGGTTGATACCGGGGAGGCCGA
CACGGAGCCCGTGGTGGATGACGACGATGTTTTCCCACCGTA
CGTGCTGAGGCCTGAGGACGGTCCCCGTGTCACGATCAACAC
AGCCATTGGGCATGTCAACAGATACTGTGCTAGATTACCAAG
TGATCCATTTACTCATCTGGCTCCTAAATGTAGAACCCGAGA
GTTGCCTGATGGTACATTTTATTCAACTCTTTATCTGCCAAT
TAACTCACCTCTTC

2. Design a primer pair that amplifies the conserved domain of SRY-HMG box of *Bos taurus*.
3. Suppose you need to design a primer for a novel gene in buffalo. The nucleotide sequence in buffalo is not available, but it is available in taurine cattle. Write the steps you will take to design a set of primers for the target gene in buffalo.

> **Hint**: Download the sequence from a phylogenetically closely related species (here, taurine cattle). If the gene is reported to be conserved, then the downloaded sequence can be used as a template for primer designing. Otherwise, download the homologous sequence (BLASTn) and compare which regions are conserved. Design the primer so that the primers anneal the conserved locations only.

4. For some reason, you do not have any access to the internet, and you are asked to determine the quality of a given four sets of primers (listed below) and select the best pair out of them. How will you proceed?

TABLE 18.3

SN	Primer	5′<-----Sequence----->3′
1	Prmr1-F	ACGGTCTTGGAGGCTACTCT
2	Prmr1-R	GCTCGTGATCGGACCTGTAG
3	Prmr2-F	GAATGTCGTTCCACCCAGGA
4	Prmr2-R	CCTGTGGTCGTCGTAATGCT
5	Prmr3-F	AGGAGTTATGGTTTGCCGGT
6	Prmr3-R	GCCTGGGTCGTTGTACCAAA
7	Prmr4-F	TACGTCACGATGCAAGGCAG
8	Prmr4-R	CTGTTACTAGCACTGGGCGG

5. Which parameters are the most important to select the best primer out of a set of designed primers? Elaborate with reasoning.

Quality Checking of the Designed Primers

CHAPTER 19

CS Mukhopadhyay and RK Choudhary
School of Animal Biotechnology, GADVASU, Ludhiana

19.1 INTRODUCTION

Biocomputationally designed primers may produce secondary structures such as homo- and hetero-dimers and hairpin loop structures. These structures interfere with the efficiency and specificity of the primers, and produce noise in SYBR green chemistry-based real-time PCR assays. The online OligoAnalyzer Version 3.1 tool (Integrated DNA Technology: http://eu.idtdna.com/site) can be used to detect possible secondary structures produced by primer(s).

19.2 OBJECTIVE

To determine the quality of the designed primers for specificity to the template and possibility of secondary structure formation.

19.3 PROCEDURE

Open the IDT Oligo Analyzer Version 3.1 online tool using the URL: https://eu.idtdna.com/analyzer/Applications/OligoAnalyzer/

19.3.1 Sequence box

This is located at the left top of the page. Each of the primer sequences (i.e., Forward and Reverse primers in 5′ to 3′ direction) is provided one at a time in this box (Figure 19.1). The software can understand a wide array of modified bases, including "standard bases" (symbols of the bases are not case-sensitive), "mixed bases" (i.e., degenerate or wobble bases, in uppercase only), RNA (e.g., rA), and so on.

Basic Applied Bioinformatics, First Edition. Chandra Sekhar Mukhopadhyay, Ratan Kumar Choudhary and Mir Asif Iquebal.

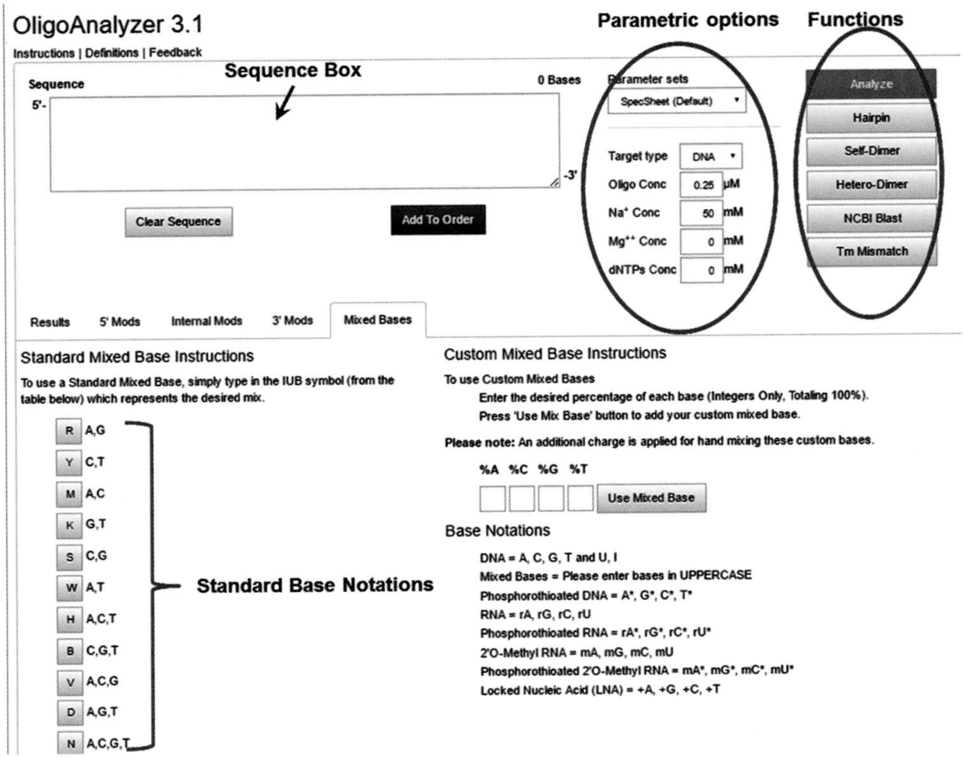

FIGURE 19.1 Homepage of Oligoanalyzer 3.1, indicating different parameters and functions for the output of the function "Analyze". (*See insert for colour representation of the figure.*)

19.3.2 Parameters

The software uses the following parameters:

- Target Type: either "DNA" or "RNA". Select "DNA" for standard oligos.
- Enter oligo concentration: Primer concentration is at least six times higher than that of the complementary target (using Nearest-Neighbor Thermodynamics)
  a. Default value: 0.25 µM.
  b. Acceptable range: 100 pM to 100 mM for conventional PCR. SYBR green assay may need to increase the oligo concentration.
  c. Primer3 software assumes that the standard oligo concentration is 50 nM.
- Enter Na^+ concentration:
  a. Default value: 50.0 mM,
  b. Acceptable range: 5 mM to 1.5 M.
  c. The monovalent cation can be K^+ (as in Primer3), instead of Na^+.
- Enter Mg^{++} concentration:
  a. Default concentration: 0.0 mM.
  b. Range: 0.01 mM to 0.60 M.
  c. The lowest range of Mg^{++} is calculated by the oligo concentration weighted by the primer length.
  d. Set the value to 0.2 mM for standard PCR primers and to 3 mM for SYBR green primers.

- Enter dNTP concentration:
  a. Default: 0.0 mM. Set the concentration to "0", since it is not required for primer-dimer and hairpin formation.
  b. Range: 0.0 mM to 1.0 M – with the imposed restriction that the upper bound may not exceed 120% of the Mg^{++} concentration.

19.3.3 Functions

- *Analyze*: This functional parameter is used to analyze the physical properties of the given primer.
  a. Complementary sequence.
  b. Primer length.
  c. GC content.
  d. Melting temperature.
  e. Extinction coefficient (the strength to which a substance absorbs light at a given wavelength per molar concentration; this differs for each base), calculated for a primer from the table of extinction coefficients, using the formula: 125.9*A260 units/μmol
  f. The molecular weight (expressed as μg/OD and nmoles/OD, respectively).
- *Hairpin*: This command predicts the hairpin formation by the input/primer sequence using the mFold algorithm (Zuker, 2003). It gives the detailed results for ΔG, T_m (in °C), ΔH (enthalpy of primer) and ΔS (entropy of primer) (Figure 19.2). Interpretation of results: lower ΔG (with symbol) (i.e., Gibbs free energy < –6 kcal/mol)

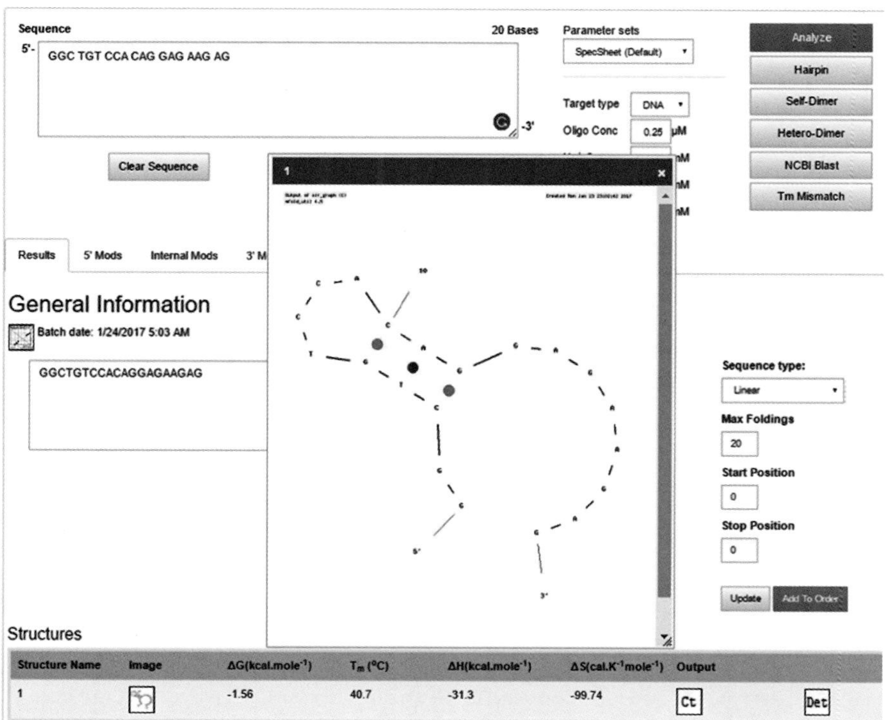

FIGURE 19.2 Output of the function "Hairpin" of the Oligoanalyzer 3.1 tool, displaying the possible hairpins and the related thermodynamic values. (*See insert for colour representation of the figure.*)

values indicate stronger loop formation that will hinder the progress of PCR. The permitted limit of ΔG is more than –3 kcal/mol (for internal loops) and –2 kcal/mol (at the 3′ end). If several hairpin loops are formed, it will be better to reject the primer.

- *Self-dimer*: This reveals the possible duplexes and their stabilities when a primer hybridizes to itself. Interpretation of results: Check whether the ΔG values of the self-dimers are more than the permitted limit of ΔG values, which are more than –5 kcal/mole (for 3′-end) and –6 kcal/mole (for internal self-dimers).
- *Hetero-dimer*: This command shows the formation of possible duplexes between the primer pairs. The other primer sequence is to be entered in the sequence box provided in the interface after clicking on the "HETERO-DIMER" button. The "Create Complement" button will give the complementary sequence of the primer. Interpretation of results: the ΔG of hetero-dimer should exceed the permitted limit of ΔG (i.e., more than –5 kcal/mole (for the 3′-end) and –6 kcal/mole (for internal self-dimers). There should be a minimum number (<5 or 6) of heterodimers formed.
- *NCBI BLAST*: The primer sequences can be BLAST-ed using this "NCBI BLAST" button for searching short nearly exact matches. Interpretation of results: The specificity of the primer (for the intended target only) and a BLAST score more than 40 (for an oligo of 18 or more bases) should be obtained. The color key for alignment scores will show blue line(s).
- *T_m mismatch*: New base(s) can be included by using the drop-down boxes at 5′ and 3′ termini flanking the duplex region. Click on the "red target base" to get the drop-down box from which you can select one mismatch to get inserted into the target sequence.

19.4 IDT UNAFOLD – CHECKING THE SECONDARY STRUCTURE FORMATION OF THE AMPLICON

It is possible to determine the exact amplicon size and detect the secondary structure formation of the amplicon using *in silico* analysis of target sequence. First, the amplicon size and sequence are determined using Primer3 software (discussed in Chapter 18) with available primer sequences and target sequences. Later, these amplicons are checked for secondary structure formation, as discussed below. It is important to consider the secondary structure of the amplicons, because the amplicon, when in the denatured state, may prevent primer annealing with the target and/or amplification, provided the ΔG is high (more than –3.0 kcal/mol). This is important while designing primers for SYBR Green assay in real-time PCR.

19.4.1 Procedure

a. Open the free online tool "UnaFold" of IDT (http://eu.idtdna.com/UNAFold?).
b. Type an input name in the box provided (to identify your results).
c. Paste the amplicon sequence in the blank sequence-box (Figure 19.3).
d. Change the Mg^{++} ion concentration to 0.2 or 0.25 mM (for conventional PCR, else 3.0 mM for SYBR green assay using Q-PCR), and the melting temperature according to the average melting temperature (T_m) of oligos.

e. Specify the start and end of the amplicon within the input sequence in the given text boxes.
f. The dNTP concentration has no role in the secondary structure formation of the amplicon, so it is omitted from the list of the parameters.
g. Click on the "Submit" button.

19.4.2 Interpreting the results

The inference is same as the interpretation of the results for hairpin structure formation in an oligo sequence. If the ΔG values are significantly high (more than –3.0 kcal/mol) and the T_m values are less than 60 °C, the amplicon will be suitable for use with PCR.

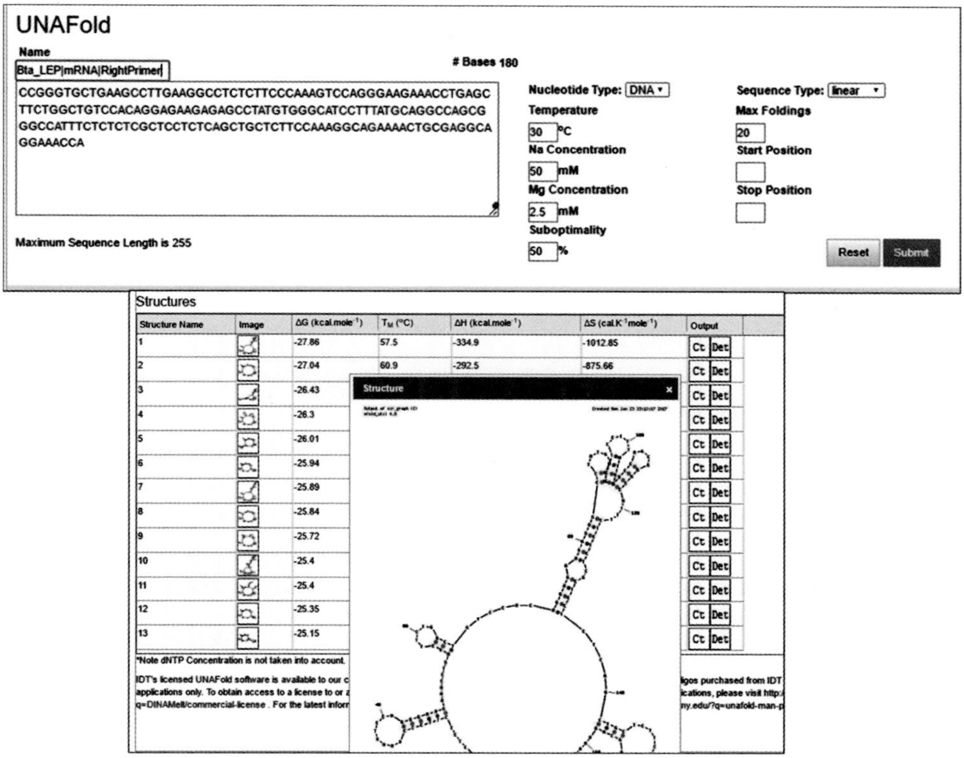

FIGURE 19.3 Prediction of secondary structure in the amplicon using the UNAFold tool of IDT. (*See insert for colour representation of the figure.*)

a. If you are using SYBR green chemistry for real-time PCR, it could be a very serious concern when the ΔG is low (–4.0 kcal/mol or less, which means the secondary structure needs more energy to dissociate) and/or the T_m of the secondary structure is higher than 60 °C (which means, at lower temperatures, most of the secondary structure will not dissociate).
b. In such cases, discard the primers, since the secondary structures formed within the single stranded amplicon may impede the progress of PCR.

19.5 PRIMER-BLAST – TO DETECT POSSIBLE SPURIOUS AMPLIFICATION

A number of factors could have contributed to mispriming:

a. Random chance of sequence match.
b. Conserved sequence (certain portions of the promoters or genes belonging to a gene family).
c. The presence of repeat sequences in the primers. This is not a property of a good primer; however, sometimes it cannot be avoided while targeting a specific template.

The specificity of a primer can be checked by using the freely available online Primer-BLAST from NCBI (Ye *et al.*, 2012). Primer-BLAST searches the primer(s) against a nucleotide database (specified by users). The primer(s) that show(s) matches, particularly at the 3′-end, with unintended target sequences will mislead the SYBR green assay of real-time PCR with a false positive signal.

The steps for using the software have been shown here. Interested readers can go through the protocol given in detail in the following link: ftp://ftp.ncbi.nih.gov/pub/factsheets/HowTo_PrimerBLAST.pdf/

i. Open the Primer-BLAST homepage: http://www.ncbi.nlm.nih.gov/tools/primer-blast/index.cgi?LINK_LOC=BlastHome
ii. Paste the target DNA sequence (in FASTA format) or accession number or gi ID in the sequence box if we need to design our primers through primer-BLAST. Otherwise, this step is to be skipped.
iii. Next, paste the forward and reverse primer sequences (5′–3′ direction) in the spaces provided for the forward and reverse primers, respectively.
iv. Primer Parameters, such as "PCR product size", "# of primer return", and "Primer melting temperature", are to be set only if we seek to design new primers using Primer-BLAST. Otherwise, we do not need to use these parameters for checking the specificity of our primers (which have already been designed using software such as "Primer3").
v. The parameters under "Exon/Intron selection" portion are required for designing new parameters; hence, these are not to be set for checking the specificity of already designed primers.
vi. The critical parameters that are to be set for detecting the specificity of our primers to the intended target sequence are those under the section "Primer Pair Specificity Checking Parameters". These parameters are discussed here:
    a. Specificity check: Tick the checkbox for instructing the software to search the specified database for a possible match with all unintended sequence.
    b. Database: The drop-down menu has five options for database selection, namely:
        - RefSeq mRNA;
        - genome (reference assembly from selected organisms);
        - genome (chromosomes from all organisms);
        - RefSeq_RNA, non-redundant (nr);
        - custom;
        - the user needs to select RefSeq mRNA if the primers have been designed using mRNA or cDNA as template. Otherwise, the genome-based options are to be chosen.

c. Organism: Select the name of the target organism (English or scientific name or taxonomy ID).
d. Exclusion (optional): The user can exclude two categories of sequences from the database search – namely, predicted RefSeq transcripts (accession with XM, XR prefix) and/or uncultured/environmental sample sequences.
e. Entrez query (optional): This option enables the user to put a limit to the database search to a "regular Entrez query". It will search only the given nucleotide sequence for primer specificity and, thereby, excludes all other sequences from the search, no matter what option has been selected in the database parameter by the user. It is better to leave this option blank.
f. Primer specificity stringency: This parameter imposes a restriction on the number of mismatches in total within the last five base-pairs of at least one primer.
g. Misprimed product size deviation: This option limits the size of the unintended product by ignoring the off-target products of unusual size.
h. Splice variant handling: Applicable only for the "Refseq mRNA" option. Checking this option enables the program to include all mRNA splice variants of the same gene that could be flanked by the primer sequences. It makes the primers gene-specific rather than transcript-specific.

vii. Click on "Get Primers" to obtain the specificity.
viii. *Output*: The output will show us possible unintended amplicons. The primers should be discarded if these show annealing at the last seven bases at the 3′-end.

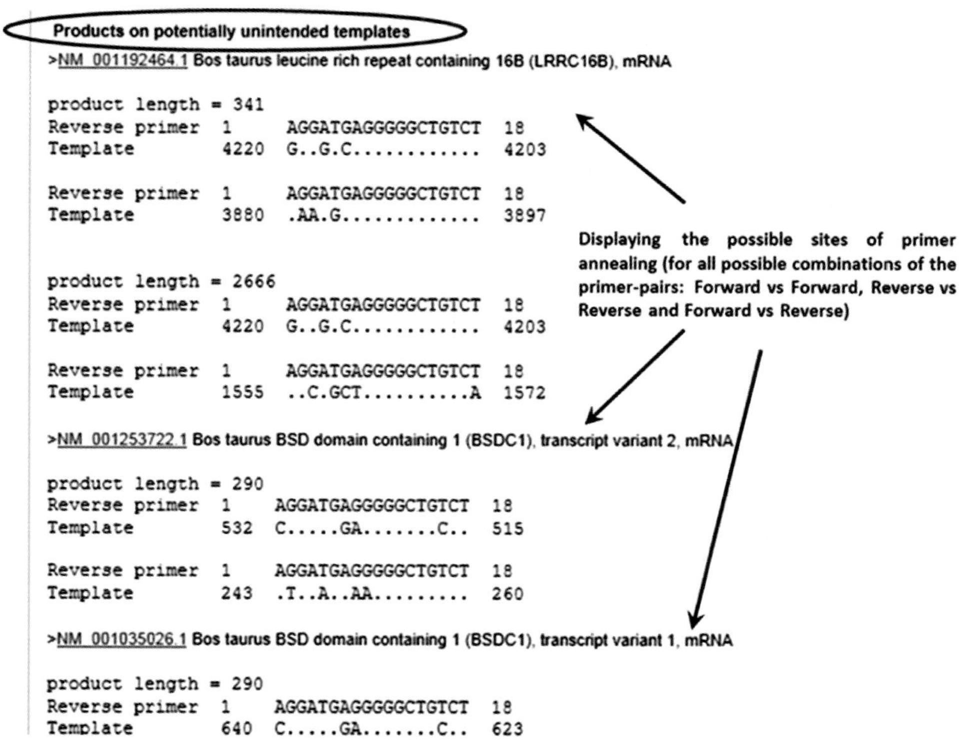

FIGURE 19.4 Output result of Primer-BLAST and selection of primers from the list displayed.

19.6 QUESTIONS

1. Check the quality and comment on the given set of primers:

TABLE 19.1

SN	Primer	5'<-----Sequence----->3'
1	Prmr1-F	GCCTCATCGATTTTATGTCGCT
2	Prmr1-R	CCAGGTGACTCGCTGAACAA

2. You have designed a pair of primers for BuLA (bubaline leukocytic antigen) cds in buffalo. While using Primer-BLAST, you are not getting any match. What could be the possible reasons? How will you modify your primer-BLAST parameters?
3. Design a set of primers (using suitable tool) for the cds for the beta chain of hemoglobin, covering the 15th to 20th bases. Check the quality of designed primers using suitable tools.
4. What is Gibbs free energy? What are the optimal permitted values of ΔG for different types of secondary structures in primers and the target sequence?
5. Given the following sets of primers for siRNA amplification, select the best one based on possible secondary structures formation:

TABLE 19.2

SN	Oligo	5'<-----Sequence----->3'
1	RN1siR2-F	GCGTAATACGACTCACTATAGGGAGACCCTGGGCTCATTCTTCAGG
2	RN1siR2-R	GCGTAATACGACTCACTATAGGGAGAGCATCCCAGGAGCTGTAGTC
3	RN2siR3-F	GCGTAATACGACTCACTATAGGGAGAACACTGCGTCACCTGATCTC
4	RN2siR3-R	GCGTAATACGACTCACTATAGGGAGAGACTCTGACCTTGCCATCGT

Primer Designing for SYBR Green Chemistry of qPCR

CHAPTER 20

CS Mukhopadhyay and RK Choudhary
School of Animal Biotechnology, GADVASU, Ludhiana

20.1 INTRODUCTION

SYBR is a fluorophore that illuminates green after binding to double-stranded DNA at the minor grooves. It is used in quantitative PCR (qPCR) to determine the amount of amplicon generated following each cycle of amplification. As real-time PCR is very sensitive, and the presence of secondary structures and spurious amplicon would inflate the quantified amplicon, adequate care should be exercised while designing very specific and high-quality primers for the SYBR Green chemistry of qPCR.

TABLE 20.1 Optimal and permissible ranges of parameters of qPCR primers (SYBR green chemistry).

SN	Feature	Optimal	Limits	Pros and cons
1	Length of each primer	21–23 nt	18–25 nt	Shorter primers could increase mispriming, and longer primers would reduce efficiency.
2	Amplicon	120–200 bp	100–220 bp	Primers for qPCR using chemistry other than SYBR Green use shorter amplicon (60–150 bp).
3	Primer T_m	64 °C	58–67 °C	Difference between the T_m of two primers should not be more than 2 °C.
4	Annealing temperature	60 °C	57–64 °C	The annealing and extension temperature of the Taq is recommended to be same.
5	GC% of primer	55–65%	35–80%	Higher or lower GC% would affect T_m of primers.

(*Continued*)

Basic Applied Bioinformatics, First Edition. Chandra Sekhar Mukhopadhyay, Ratan Kumar Choudhary and Mir Asif Iquebal.

TABLE 20.1 (Continued)

SN	Feature	Optimal	Limits	Pros and cons
6	GC-clamp at 3′-end	1 clamp	Maximum 2 clamps	More than 2 G/C clamp with the last five bases at 3′-end will make the primers sticky.
7	Run of identical bases	2 bases	Maximum 4	This will increase the possibility of secondary structure formation.
8	Proportion of different bases	Equal percentage of each base	Minimum 20–30% of one base	—
9	3′-end stability of primers	Delta G >= −9 kcal/mol	−7 to −11 kcal/mol	Delta G of last five bases at 3′-end should be higher (ignoring the symbol) than that of the internal region of the primer.
10	Concentration of Mg^{++}	3–4 mM of $MgCl_2$	3–6 mM	SYBR green buffer itself contains appropriate amount of $MgCl_2$.
11	Max repeat mispriming	12 or more	11–15	Larger value inversely proportionate to the probability of repeat mispriming of the primers.
12	Check for template mispriming	No mispriming	Do not select primers with template mispriming	Run Primer-Blast of NCBI to identify the possible mispriming in the same species.
13	Number of unknown bases	0	0	No non-specific bases should be present, to prevent non-specific amplification.
14	Run of the same base	2	2–4	Run of the same base will increase the mispriming.
15	Checking the primer quality	Delta G values should be higher for primer-template and lower for primer secondary structures		Identify potential hairpins and primer-dimer/self-dimer formation.

20.2 QUESTIONS

1. Design a pair of primers for SYBR Green chemistry-based real-time PCR of the TLR4 mRNA in *Catla catla*. Check the quality of the primers and evaluate them.
2. List the key parameters that determine the efficacy of qPCR primers (SYBR Green Chemistry).

Molecular Phylogenetics

SECTION V

Construction of Phylogenetic Tree: Unweighted-Pair Group Method with Arithmetic Mean (UPGMA)

CHAPTER 21

CS Mukhopadhyay and RK Choudhary
School of Animal Biotechnology, GADVASU, Ludhiana

21.1 INTRODUCTION

UPGMA is a clustering algorithm that works by joining the branches of a tree on the basis of maximum similarity criteria among pairs of sequences, and by calculating the means of joined pairs. UPGMA is "*ultrametric*", so all the terminal nodes are equally distanced from the root. Hence, at the end, when a root is added, the rooted tree is produced.

- *Unweighted*: It indicates equal contribution of all the pair-wise distances. There is no weighting of any specific taxa-pairs to indicate a different evolutionary rate compared with another pair(s). This is the opposite of the Weighted-Pair Group Method with Arithmetic mean (WPGMA).
- *Pair-groups*: Any two taxa or any two clusters (clade) or one taxon and a cluster are always combined in pairs (that is, interpreted as dichotomies).
- *Arithmetic mean*: Pair-wise distance of each group is the mean distance to all members of that group.

21.2 ASSUMPTIONS

a. Constant rate of evolution (i.e., mutation-rate) amongst all the sequences.
b. Distance data are ultrametric: This enables clustering by satisfying the "three point condition" to generate the tree.

Basic Applied Bioinformatics, First Edition. Chandra Sekhar Mukhopadhyay, Ratan Kumar Choudhary and Mir Asif Iquebal.

TERMINOLOGIES AND POINTS

a. *Neighbors*: the external nodes with the smallest number of mismatches.
b. *Ultrametric*: in distance methods, an ultrametric tree is characterized by a "Three point condition", where three points (a, b, c) satisfy the condition for the distance (d): $d(a,b) \leq \max\{d(a,c), d(b,c)\}$. Unless the data are ultrametric (i.e., the OTUs of any branch have different lengths), this condition will not be satisfied. The ultrametric tree assumes a molecular clock of consistent evolutionary rate.
c. *Rate of evolution*: UPGMA is very sensitive to the rate of evolution. If the assumption of equal evolutionary rate is not met, it will generate a rooted tree with the wrong topology.
d. When there is weak similarity between sequences, the distance methods (UPGMA, neighbor joining, minimum evolution, Fitch Margoliash) are the best options.
e. Note that the UPGMA will produce a rooted tree at the final step if a root is added; else the tree itself will be unrooted.

21.3 OBJECTIVE

To construct a phylogenetic tree (dendrogram), using the UPGMA method, from a set of molecular sequences.

21.4 PROCEDURE

a. Calculate the raw pair-wise distance data from a set of sequences and construct a distance matrix:

TABLE 21.1

	A	B	C	D
B	3			
C	7	7		
D	10	10	10	
E	10	10	10	8

Note: while constructing the tree from the distance values, one needs to select the closest pair (with minimum distance among all possible pairs) from the distance matrix, and then merge these two objects to yield one.

b. Identify the *least-distant pair*: Here, the minimum distance is d(AB) = 3 (i.e., between the taxa "A" and "B").
c. Place these two taxa in a single group as a cluster and consider the duo as a single external node.
d. The distance is d(AB) = AB = 3. Hence, the depth of divergence (for each branch) of this sub-tree will be 3/2 = 1.5 units.

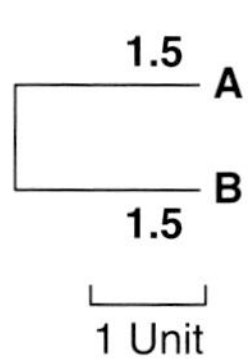

FIGURE 21.1

e. Now consider "AB" as a single taxon and repeat the same steps, as above. Find the shortest distance and make the respective taxa a single cluster.

$$d(AB)C = [d(AC) + d(BC)]/2 = (7 + 7)/2 = 7$$
$$d(AB)D = [d(AD) + d(BD)]/2 = (10 + 10)/2 = 10$$
$$d(AB)E = [d(AE) + d(BE)]/2 = (10 + 10)/2 = 10$$

TABLE 21.2

	AB	C	D
C	7		
D	10	10	
E	10	10	8

f. Add the new taxa (C) with "AB" cluster, since "AB" and "C" have the least distance (i.e., 7), to produce the sub-tree of ABC, in which the AB sub-tree (drawn just in the last step) is attached to the point M. The length of AK + AN = OC.

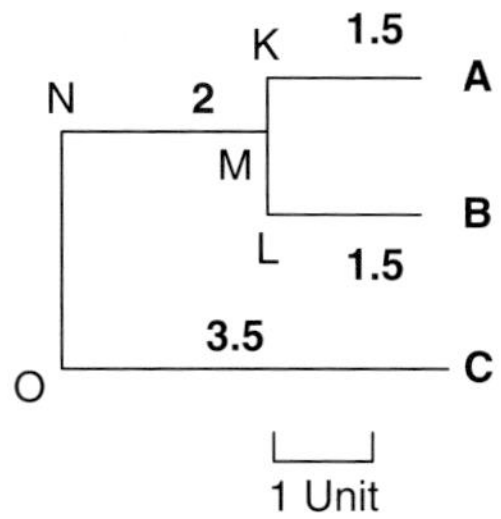

FIGURE 21.2

For the sake of understanding, the internal nodes of the branches have been marked (as K, L, N, etc.). These are not required for tree construction in general.

g. Now, repeat the last two steps, – that is, calculate the mean distance between (AB) C cluster and Sequence "D" and then draw the phylogenetic tree:

$$d(ABC)D = [d(AD) + d(BD) + d(DC)]/3 = (10 + 10 + 10)/3 = 10$$
$$d(ABC)E = [d(AE) + d(BE) + d(CE)]/3 = (10 + 10 + 10)/3 = 10$$

TABLE 21.3

	ABC	D
D	10	
E	10	8

h. Now, the least distance is DE = 8. We will repeat the same steps as before: The branch length will be 8/2 = 4 units.

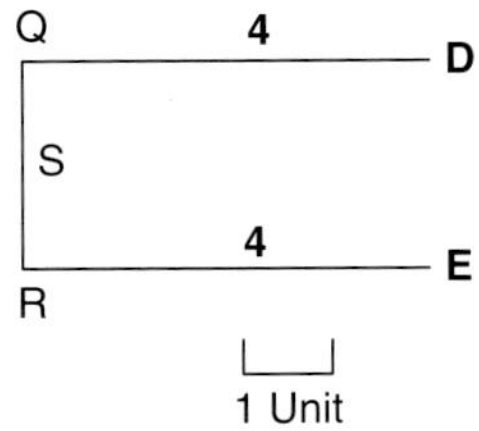

FIGURE 21.3

i. Now, calculate the distance between these two clusters ABC and DE:

$$d(ABC)(DE)$$
$$=\left[d(AD)+d(BD)+d(DC)+d(AE)+d(BE)+d(CE)\right]/6$$
$$=(10+10+10+10+10+10)/6=10$$

DIFFERENTIATING BETWEEN ULTRAMETRIC AND NON-ULTRA-METRIC DATA

i. The example given above is of an ultrametric data-set that follows the principle of "three point condition", i.e., d(AC)<=max{d(AB, BC)}
ii. Now let us take this example of non-ultrametric data where the three point condition is not fulfilled, hence d(AC)>max{d(AB, BC)}
The actual tree should look like this, and the non-ultrametric dataset for this tree is as follows:

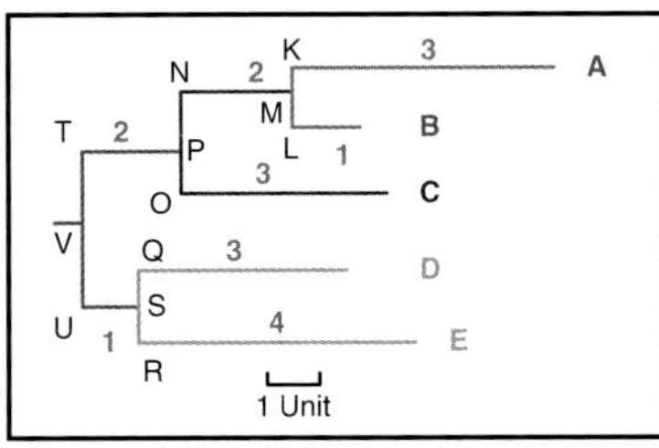

	A	B	C	D
B	4			
C	8	6		
D	11	9	9	
E	12	10	10	7

FIGURE 21.B1

Please note that d(AC) > max{d(AB), (BC)}, i.e., 8 > max{4,6}, so the dataset is non-ultrametric.
iii. Now, if this distance matrix is solved as earlier, we will ultimately get the following tree: In text iterations, as earlier, we get:

	A B	C	D
C	6		
D	9	8	
E	11	10	7

→

	A BC	D
D	13	
E	16	7

→

	ABC	DE
D E	9.67	

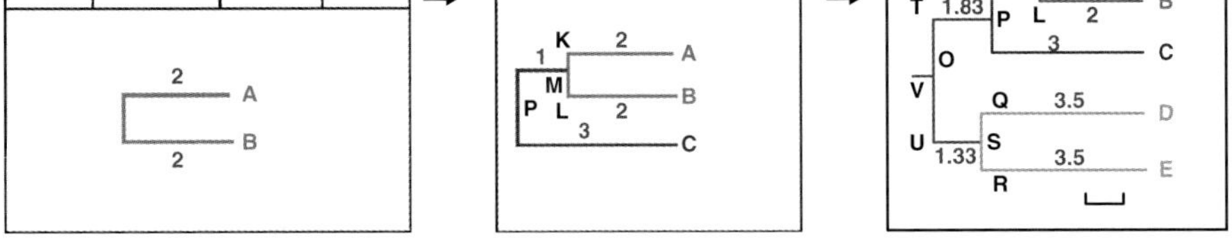

FIGURE 21.B2

iv. Now we can compare how the actual tree topology has been totally changed due to the non-ultrametric nature of the data.

v. The ultrametric dataset will give the same result in UPGMA and WPGMA; however, it is not necessary for non-ultrametric data.

vi. *Branch length* represents the evolutionary distance, not the time lapse of evolution. Thus, two sister taxa, though having the same evolutionary time, may have different distances or branch lengths (view NJ method). In distance-based methods, the distances between the nodes are additive – that is, to get the distance between two taxa, add up the distances of all the branches (main and internal branches) connecting these two taxa.

vii. The vertical width (perpendicular to the branches) does not carry any meaning in terms of evolutionary distance, time-scale or mutation rate.

viii. The terms "cladogram" and "phylogram" differ in that the latter represents true evolutionary history through the evolutionary time scale, while the former does not represent evolutionary time.

ix. In the above diagram, the scale bar of 1 unit measures evolutionary distance by 1 residue.

x. The scale bar often represents values of 0.01, 0.02, etc., which indicates 1% or 2% divergence (or 99% or 98% homology, respectively) among the taxa. The percentage homology is calculated through multiple sequence alignment.

xi. Base or amino acid substitution takes place after divergence of two homologous sequences. Such substitutions occur at different rates in different species; hence, this happens irrespective of the direction of time. The number of substitution of residue (nucleotide or amino acid) is the indicator of evolutionary distance.

21.5 INTERPRETATION OF UPGMA TREE

The distance from the root to the OUT of each cluster = 10/2 = 5 units.

Hence, the distance of TP = 5 – OC = 5 – 3.5 = 1.5 units.

The distance of US = 5 – 4 = 1.0 unit.

The UPGMA tree obtained in our example depicts evolutionary distances between the taxa. We need to add up the distances connecting these two taxa to calculate the distance between any two taxa ("A" and "D"): 1.5 + 2 + 1.5 + 1 + 4 = 10. This is exactly the value given in the distance table. Assuming equal evolutionary rates, these values indicate the evolutionary distances between the taxa.

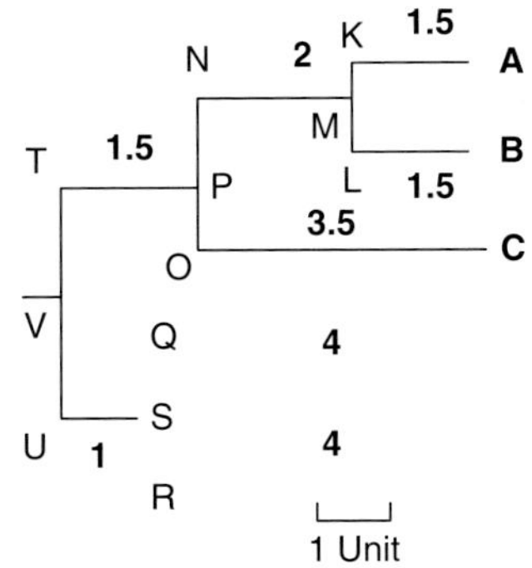

FIGURE 21.4

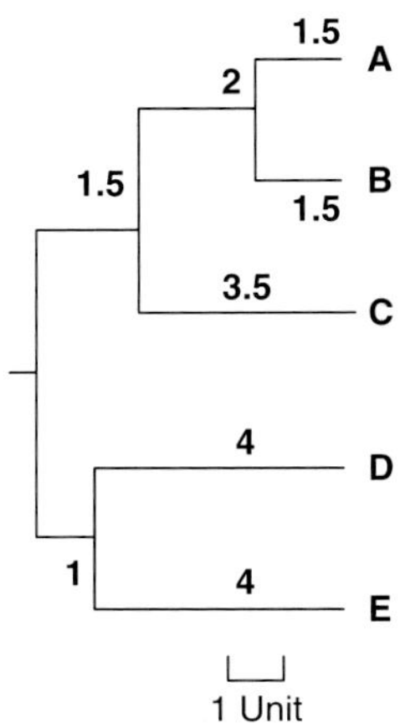

	A	B	C	D
B	3			
C	7	7		
D	10	10	10	
E	10	10	10	8

FIGURE 21.5

21.6 QUESTIONS

1. Draw a phylogenetic tree manually using the following distance matrices:

a. **TABLE 21.4**

	A	B	C	D	E
B	8				
C	18	18			
D	18	18	10		
E	18	18	10	4	
F	20	20	20	20	20

b. **TABLE 21.5**

	A	B	C	D	E
A	0				
B	4	0			
C	8	8	0		
D	8	8	6	0	
E	8	8	6	2	0

2. What are the merits and demerits of the UPGMA method of phylogenetic tree construction?
3. Explain in detail why ultrametric data are needed for UPGMA tree construction.
4. Under what circumstances do we prefer the UPGMA tree? How do you interpret the results of the UPGMA tree?
5. Suppose we have morphological data from which similarity and distance matrices can be constructed. Can we use such a distance matrix for the construction of a UPGMA tree? Justify your answer.

Construction of Phylogenetic Tree: Fitch Margoliash (FM) Algorithm

CHAPTER 22

CS Mukhopadhyay and RK Choudhary
School of Animal Biotechnology, GADVASU, Ludhiana

22.1 INTRODUCTION

This is the first algorithm based on least squares principle for phylogenetic tree reconstruction. It was developed by Walter Fitch and Emanuel Margoliash in 1967 (Fitch and Margoliash, 1967; Fitch, 1970, 1971). The evolutionary distances between the taxa are determined by the Jukes–Cantor model when DNA sequences (instead of distances) of the same length are entered.

22.1.1 Principle

The algorithm is based on optimality criteria that select the tree with a minimum amount of residual (difference between actual and expected summed evolutionary distance). The algorithm estimates the total branch length (distance) and clusters in accordance to taxa pair in order to determine the unrooted tree with minimum distance.

a. The FM algorithm does not assume a constant rate of evolution, which is quite realistic.
b. Optimized tree can be selected out of a number of possible trees.
c. Its demerit is underestimation of very long evolutionary distances, because it ignore homoplasies (absence of a character in the common ancestor, though it is being shared by a group of related species originating from the common ancestor).
d. It ignores the role of intermediate ancestor(s); hence, consistency of evolution is not the basic assumption.
e. Outgroup is added to the sequences in order to generate a rooted tree using the FM method.

Basic Applied Bioinformatics, First Edition. Chandra Sekhar Mukhopadhyay, Ratan Kumar Choudhary and Mir Asif Iquebal.

22.1.2 Assumption

a. The algorithm does not assume a constant mutation rate.
b. It assumes *additivity of distances* – that is, additivity of the branch length of the trees to yield the total branch length or distance.

22.2 OBJECTIVE

To construct a phylogenetic tree using the Fitch Margoliash (FM) method, given the distances among a set of molecular sequences.

22.3 PROCEDURE

Let us start with four sequences: "A", "B", "C" and "D", and consider that the given distances (d) between the four sequences are as follows:

TABLE 22.1

	A	B	C	D
A	0			
B	20	0		
C	26	22	0	
D	34	30	16	0

The iterative steps in this algorithm are as follows:

a. Consider two of the taxa (say, "A" and "B") for determining the distance from the third composite taxa (denoted by "X"). The composite taxa ("X") are a combination of the rest of the taxa (here, "D" and "C").
b. The distance between "A" and "X" (d_{AX}) is calculated by averaging both the distances from "A" to "C" and "D" taxa (all the component OTUs of the composite taxa "X").
c. Similarly, the distance between "B" and "X" (d_{BX}) is also calculated:

TABLE 22.2

	A	B	C	D	
A	0				$d_{AB} = 20$
B	20	0			$d_{AX} = (d_{AC} + d_{AD})/2 = (26 + 34)/2 = 30$
C	26	22	0		$d_{BX} = (d_{BC} + d_{BD})/2 = (22 + 30)/2 = 26$
D	34	30	16	0	

d. In the next step, the distance (P1) between the terminal node (taxa "A") and its intermediate ancestor ("P") is calculated using the formula: $P1 = (d_{AB} + d_{AX} - d_{BX})/2$.
e. Similarly, the distances between taxon "B" and intermediate ancestor "Q", as well as taxon "X" and its intermediate ancestor "R", are calculated.

TABLE 22.3

	A	B	X	$P_1 = (d_{AB} + d_{AX} - d_{BX})/2 = (20 + 30 - 26)/2 = 12$
A	0			$Q1 = (d_{AB} + d_{BX} - d_{AX})/2 = (20 + 26 - 30)/2 = 8$
B	20	0		$R1 = (d_{AX} + d_{BX} - d_{AB})/2 = (30 + 26 - 20)/2 = 18$
X	30	26	0	

The obtained distances (in the P1, Q1 and R1) are put in a tree:

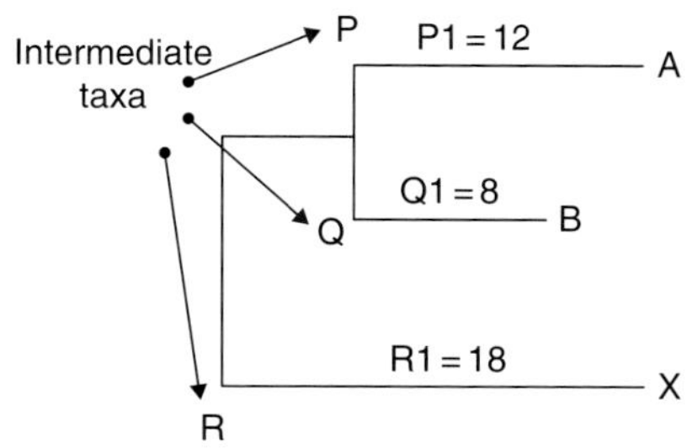

FIGURE 22.1

f. Now "A" and "B" are combined as "AB" (as we have obtained the distances of the taxa from the respective intermediate ancestors).
g. The combined taxon "X" is expanded into its component taxa (here, "D" and "C").
h. The immediate taxon ("C") is considered a second taxon to estimate the distance with its intermediate ancestor, and the other taxa are again combined into taxon "X", so that the same steps can be iterated.
i. The same notations are used as for the previous iteration. However, the subscripts are changed to "2", – that is, P2 (distance between "AB" node with its intermediate ancestor, designated by "P" again), Q2 (distance between "C" node with its intermediate ancestor, designated as "Q" again) and R2 (distance between "X" node with its intermediate ancestor "R").

TABLE 22.4

	AB	C	D	
AB	0			$(AB)C = (d_{AC} + d_{BC})/2 = (26 + 22)/2 = 24$
C	24	0		$(AB)D = (d_{AD} + d_{BD})/2 = (34 + 30)/2 = 32$
D (or X)	32	16	0	

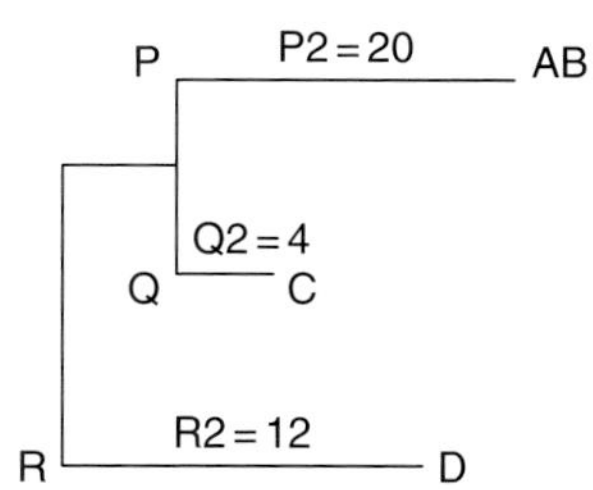

FIGURE 22.2

j. At this point, one additional parameter, internal branch length (IBL), is calculated for the combined taxa "AB":

IBL Calculation

$\alpha 1 = \text{Distance between "AB" and C} = (dAC + dBC)/2 = (26 + 22)/2 = 24$

$\beta 1 = \text{Avg. of branch lengths between "A" and "B"} = (P1 + Q1)/2 = (12 + 8)/2 = 10$

$\gamma 1 = \text{Distance between "C" and intermediate ancestor} = (d(AB)C + dCX - d(AB)X)/2$

$= (24 + 16 - 32)/2 = 4$

$\alpha = \beta + \gamma + \text{Internal Branch Length(IBL)}$

$IBL1 = \alpha 1 - (\beta 1 + \gamma 1) = (24 - (10 + 4)) = 10$

k. In the last step, no more additional information for calculating IBL between "D" and "ABC" is available. In this situation, the length of the internal branch (designated as "IBL2" for the second time calculation) is determined by the following formula:

$$IBL_2 = \left(\left(R_2\right) - \left(Q_2\right)\right)/2$$
$$= (12 - 4)/2 = 4$$

The final tree constructed by the FM algorithm is as follows:

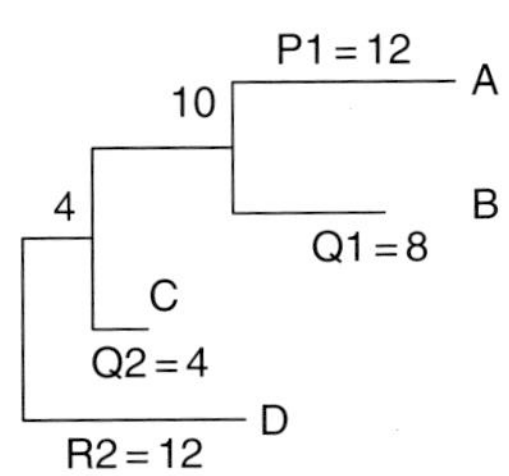

FIGURE 22.3

22.4 INTERPRETATION OF THE FM TREE

The phylogenetic tree has been constructed assuming a different rate of evolution among different branches (or taxa). The feature of additivity of the branches holds true to determine distances between any two OTUs.

22.5 QUESTIONS

1. Construct the phylogenetic tree using the FM method:

TABLE 22.5

	A	B	C	D
A	0			
B	10	0		
C	14	15	0	
D	24	18	11	0

2. In the last chapter, you constructed the phylogenetic tree using UPGMA (Q1a). Now construct the tree using the FM method and compare with the previous one.

TABLE 22.6

	A	B	C	D	E
B	8				
C	18	18			
D	18	18	10		
E	18	18	10	4	
F	20	20	20	20	20

3. What is the meaning of the term "internal branch length"? How is it important in calculating the phylogenetic tree using the FM method?
4. Differentiate between the principle and applications of the FM and UPGMA methods of phylogenetic tree construction.

Construction of Phylogenetic Tree: Neighbor-Joining Method

CHAPTER 23

CS Mukhopadhyay and RK Choudhary
School of Animal Biotechnology, GADVASU, Ludhiana

23.1 INTRODUCTION

The term "neighbors" refers to a node-pair which is separated by another node. The NJ method (Saitou and Nei, 1987) is a particular case of the "star decomposition method", where raw data are arranged in a distance matrix and nodes are created (see the example below) whereas, in the NJ method, the separation of nodes is adjusted by average divergence from all other nodes.

23.1.1 Principle

The principle of the neighbor-joining (NJ) technique is minimum evolution, which selects the tree with minimum branch-length. It is based on a very fast, greedy heuristic algorithm that generates sub-trees, and the closest sub-trees are joined to each other to yield the final tree, in a step-wise manner. The total branch length is the shortest for the true tree.

a. The NJ method can be applied for large datasets relating to the taxa with varying degrees of divergence (hence, the tree will show different lengths for different branches).
b. Multiple substitutions can be corrected.
c. Some of the sequence information is lost in the NJ method due to the nature of the algorithm.

23.1.2 Assumptions

a. Minimum mutational events explain the evolution of the molecular sequences.
b. The branch length of the tree with known topology represents the different rate of evolutionary changes.

Basic Applied Bioinformatics, First Edition. Chandra Sekhar Mukhopadhyay, Ratan Kumar Choudhary and Mir Asif Iquebal.

23.2 OBJECTIVE

To construct a phylogenetic tree, using the neighbor-joining method, employing a distance matrix obtained from a set of molecular sequences.

23.3 PROCEDURE

The essential points to remember are:

a. No preference is exercised in the pairing of sequences.
b. It searches to find the pair of sequences that minimizes the branch length.
c. The NJ method uses the Fitch–Margoliash Algorithm to create a rate corrected new distance table.

23.3.1 Start with a distance matrix

Let us consider a distance matrix of five sequences (A, B, C, D and E)

TABLE 23.1

Distance matrix	A	B	C	D
B	4.000			
C	7.000	8.000		
D	6.000	7.000	6.000	
E	8.000	9.000	10.000	9.000

23.3.2 Construction of a star tree

A star tree is first drawn, based on the number of input sequences (OTUs). This star tree is a random tree with a central hub that joins all the branches.

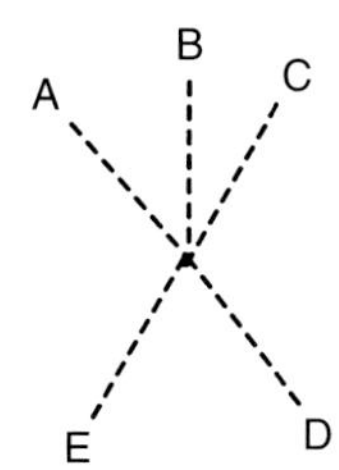

FIGURE 23.1

23.3.3 Calculation of net divergence

Net divergence (V_i) is calculated for each of the OTUs using the formula $V_i = \sum_{i=1}^{j} x_{ij}$, where, $i \neq j$ and $i, j = 1, 2, \ldots, 5$, i.e., the number of OTUs incorporated.

TABLE 23.2

Net divergence	Equations	Value
V_A =	$d_{AB} + d_{AC} + d_{AD} + d_{AE}$ =	25.000
V_B =	$d_{AB} + d_{BC} + d_{BD} + d_{BE}$ =	28.000
V_C =	$d_{AC} + d_{BC} + d_{CD} + d_{CE}$ =	31.000
V_D =	$d_{AD} + d_{BD} + d_{CD} + d_{DE}$ =	28.000
V_E =	$d_{AE} + d_{BE} + d_{CE} + d_{DE}$ =	36.000

23.3.4 Calculation of new distance values from the original distance and net divergence

The mean divergence (M_i) is calculated by dividing individual net divergence (V_i) by (N – 2), where N is the number of OTUs.

TABLE 23.3

M values	Equation	Calculation	Value
M_A =	V_A / (N – 2) =	25 / (5 – 2) =	8.333
M_B =	V_B / (N – 2) =	28 / (5 – 2) =	9.333
M_C =	V_C / (N – 2) =	31 / (5 – 2) =	10.333
M_D =	V_D / (N – 2) =	28 / (5 – 2) =	9.333
M_E =	V_E / (N – 2) =	36 / (5 – 2) =	12.000

23.3.5 Calculation of new distances

New distances (n_{ij}) are calculated by subtracting the mean divergence (M_i, M_j) of the two OTUs which are being studied from the distance (d_{ij}) between these two corresponding OTUs (i^{th} and j^{th} OTUs being studied).

TABLE 23.4

Distance	Equation	Calculation	Value
n_{AB} =	$d_{AB} - (M_A + M_B)$ =	4 – (8.333 + 9.333) =	**– 13.666**
n_{AC} =	$d_{AC} - (M_A + M_C)$ =	7 – (8.333 + 10.333) =	– 11.666
n_{AD} =	$d_{AD} - (M_A + M_D)$ =	6 – (8.333 + 9.333) =	– 11.666
n_{AE} =	$d_{AE} - (M_A + M_E)$ =	8 – (8.333 + 12.000) =	– 12.333
n_{BC} =	$d_{BC} - (M_B + M_C)$ =	8 – (9.333 + 10.333) =	– 11.666
n_{BD} =	$d_{BD} - (M_B + M_D)$ =	7 – (9.333 + 9.333) =	– 11.666
n_{BE} =	$d_{BE} - (M_B + M_E)$ =	9 – (9.333 + 12.000) =	– 12.333
n_{CD} =	$d_{CD} - (M_C + M_D)$ =	6 – (10.333 + 9.333) =	**– 13.666**
n_{CE} =	$d_{CE} - (M_C + M_E)$ =	10 – (10.333 + 12.000) =	– 12.333
n_{DE} =	$d_{DE} - (M_D + M_E)$ =	9 – (9.333 + 12.000) =	– 12.333

23.3.6 Construction of new distance matrix from the new distance values (n_{ij})

TABLE 23.5

	A	B	C	D
B	**– 13.666**			
C	– 11.666	– 11.666		
D	– 11.666	– 11.666	**– 13.666**	
E	– 12.333	– 12.333	– 12.333	– 12.333

Here, there are two pairs of OTUs – namely, "A", "B" and "C", "D" exhibiting least divergence with value $n_{ij} = -13.666$ (bold). We can select any one of these two pairs. In this example, n_{CD} (shown in bold) is selected.

23.3.7 Calculation of branch length of the internal node

a. First, the *least distance value* is identified from the new distance matrix.
b. The two taxa with the minimum n_{ij} distance value are taken as neighbors. In the present example, "C" and "D" are the neighbors, with the minimum n_{ij} value of –13.666.
c. Now, let us assume that "X" is the new node between the neighbors "C" and "D". The branch lengths from the internal node "X" and the external nodes "C" and "D" (denoted as L_{CX} and L_{DX}, respectively) are calculated.

TABLE 23.6

Branch length	Equation	Calculation	Value
L_{CX} =	$(d_{CD} / 2) + ((M_C - M_D) / (2))$ =	(6 / 2) + ((10.333 – 9.333) / (2))	=3.500
L_{DX} =	$d_{CD} - L_{CX}$ =	(6 – 3.500)	=2.500

These values of the branch lengths are now used to construct the tree:

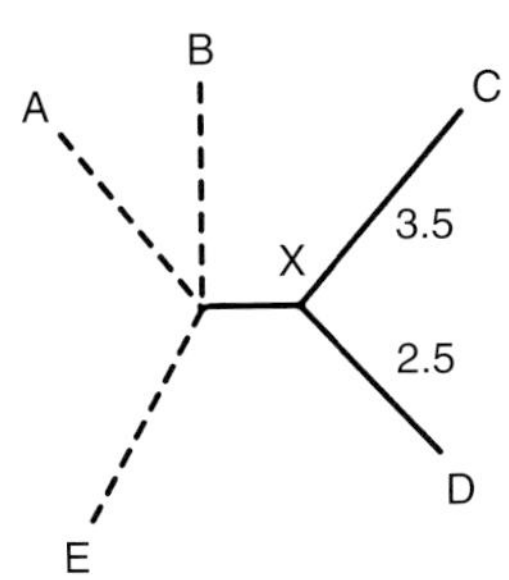

FIGURE 23.2

23.3.8 Distance of the other OTUs from internal node (X)

Next, the distances between rest of the terminal nodes (here, "A", "B" and "E") with the internal node ("X") are calculated:

TABLE 23.7

Equation	Calculation	Value
$m_{AX} = (d_{AC} + d_{AD} - d_{CD}) / 2$	= (7 + 6 – 6) / 2	= 3.500
$m_{BX} = (d_{BC} + d_{BD} - d_{CD}) / 2$	= (8 + 7 – 6) / 2	= 4.500
$m_{EX} = (d_{CE} + d_{DE} - d_{CD}) / 2$	= (10 + 9 – 6) / 2	= 6.500

Now the first iteration is over, the number of OTUs has been reduced to four ($N - 1$), where the OTUs "C" and "D" have been merged into "X". The second iteration will start with the same steps as the first iteration.

23.3.9 New distance matrix to start the second iteration

TABLE 23.8

Distance matrix	A	B	X
B	4.000		
X	3.500	4.500	
E	8.000	9.000	6.500

CROSS-CHECKING

The values of L can be alternatively determined by simply placing the distance between the nodes from the second OTU as the first one to be calculated. Hence, any one of the neighboring OTUs can be considered as the first OTU to be calculated.

Branch length	Equation	Calculation	Value
LDX =	(dCD / 2) + ((MD – MC) / (2)) =	(6 / 2) + ((9.333 – 10.333) / (2)) =	2.500
LCX =	dCD – LDX =	(6 – 2.500) =	3.500

23.3.10 Calculation of net divergence

TABLE 23.9

Net Div.	Equations	Value
$V_A =$	$d_{AB} + d_{AX} + d_{AE}$	=15.500
$V_B =$	$d_{AB} + d_{BX} + d_{BE}$	=17.500
$V_X =$	$d_{AX} + d_{BX} + d_{EX}$	=14.500
$V_E =$	$d_{AE} + d_{BE} + d_{EX}$	=23.500

23.3.11 Calculation of new distance values from the original distance and net divergence

TABLE 23.10

M values	Equation	Calculation	Value
M_A =	V_A / (N – 2) =	13.5 / (4 – 2) =	6.750
M_B =	V_B / (N – 2) =	17.5 / (4 – 2) =	8.750
M_X =	V_X / (N – 2) =	14.5 / (4 – 2) =	7.250
M_E =	V_E / (N – 2) =	21.5 / (4 – 2) =	10.750

23.3.12 Calculation of new distances

TABLE 23.11

Distance	Equation	Calculation	Value
n_{AB} =	d_{AB} – (M_A + M_B) =	4 – (6.75 + 8.75) =	– 11.500
n_{AX} =	d_{AX} – (M_A + M_X) =	3.5 – (6.75 + 7.25) =	– 10.500
n_{AE} =	d_{AE} – (M_A + M_E) =	8 – (6.75 + 10.75) =	– 9.500
n_{BX} =	d_{BX} – (M_B + M_X) =	4.5 – (8.75 + 7.25) =	– 11.500
n_{BE} =	d_{BE} – (M_B + M_E) =	9 – (8.75 + 10.75) =	– 10.500
n_{XE} =	d_{XE} – (M_X + M_E) =	6.5 – (7.25 + 10.75) =	– 11.500

23.3.13 Construction of new distance matrix from new distance values (n_{ij})

Now we select A and B as the neighbors out of the three pairs with minimum distances.

TABLE 23.12

	A	B	X
B	– 11.500		
X	– 10.500	– 11.500	
E	– 9.500	– 10.500	– 11.500

23.3.14 Calculation of the branch length of the internal node

a. The two taxa with the minimum n_{ij} distance value are taken as neighbors. In the present example, "A" and "B" are the neighbors with a minimum n_{ij} value of –11.500.

b. Now, let us assume that "Y" is the new node between the neighbors "A" and "B". The branch lengths from the internal node "Y" and the external nodes "A" and "B" (denoted as L_{AY} and L_{BY}, respectively) are calculated.

Distance with the internal node (Y):

TABLE 23.13

Branch Length	Equation	Value
L_{AY} =	$(d_{AB} / 2) + ((V_A - V_B) / (2))$	=1.000
L_{BY} =	$d_{AB} - L_{AY}$	=3.000

Next, the distances between the rest of the terminal nodes (here, "E", "X") with the internal node ("Y") is calculated:

TABLE 23.14

Equation	Calculation	Value
$m_{EY} = (d_{AE} + d_{BE} - d_{AB}) / 2 =$	(8 + 9 – 4) / 2 =	6.500
$m_{XY} = (d_{AX} + d_{BX} - d_{AB}) / 2 =$	(3.5 + 4.5 – 4) / 2 =	2.000

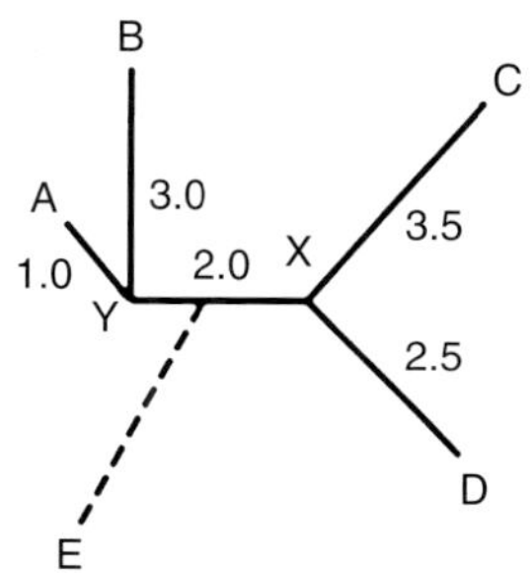

FIGURE 23.3

Now, a new distance matrix is created in the third iteration ($n = 3$):

TABLE 23.15

Distance matrix	Y	X
X	2.000	
E	6.500	6.500

23.3.15 Calculation of net divergence

TABLE 23.16

Net div.	Equations	Value
V_E	$= d_{EX} + d_{EY}$	13.000
V_X	$= d_{EX} + d_{XY}$	8.500
V_Y	$= d_{EY} + d_{XY}$	8.500

23.3.16 Calculation of new distance values from the original distance and net divergence

TABLE 23.17

M values	Equation	Calculation	Value
M_E	$= V_E / (N - 2)$	13 / (3 – 2) =	13.000
M_X	$= V_X / (N - 2)$	8.5 / (3 – 2) =	8.500
M_Y	$= V_Y / (N - 2)$	8.5 / (3 – 2) =	8.500

23.3.17 Calculation of new distances

TABLE 23.18

Distance	Equation	Calculation	Value
n_{EX} =	$d_{EX} - (M_E + M_X)$ =	6.5 – (13 + 8.5) =	–15.000
n_{EY} =	$d_{EY} - (M_E + M_Y)$ =	6.5 – (13 + 8.5) =	–15.000
n_{XY} =	$d_{XY} - (M_X + M_Y)$ =	2 – (8.5 + 8.5) =	–15.000

23.3.18 Construction of new distance matrix from the new distance values (n_{ij})

TABLE 23.19

New distance matrix	X	Y
Y	–15.000	
E	–15.000	–15.000

23.3.19 Calculation of branch length of the internal node

a. The two taxa with the minimum n_{ij} distance values are taken as neighbors. In the present example, "E" and "Y" are the neighbors with a minimum n_{ij} value of –15.000.
b. Now, let us assume that "Z" is the new node between the neighbors "E" and "Y". The branch lengths from the internal node "Z" and the external nodes "E" and "Y" (denoted as L_{EZ} and L_{YZ}, respectively) are calculated.

TABLE 23.20

Branch length	Equation	Calculation
L_{EZ} =	$(d_{EY} / 2) + ((V_E - V_Y) / (2))$ =	5.500
L_{YZ} =	$d_{EY} - L_{EZ}$ =	1.000

23.3.20 Distance with internal node (Z)

Next, the distances between the rest of the terminal nodes (here, "E", "Y") with the internal node ("Z") are calculated:

TABLE 23.21

Equation	Calculation	Value
$m_{XZ} = (d_{XE} + d_{YX} - d_{EY}) / 2 =$	((6.5) + (2) – (6.5)) / 2 =	1.000

The final neighbor-joining tree obtained is shown below:

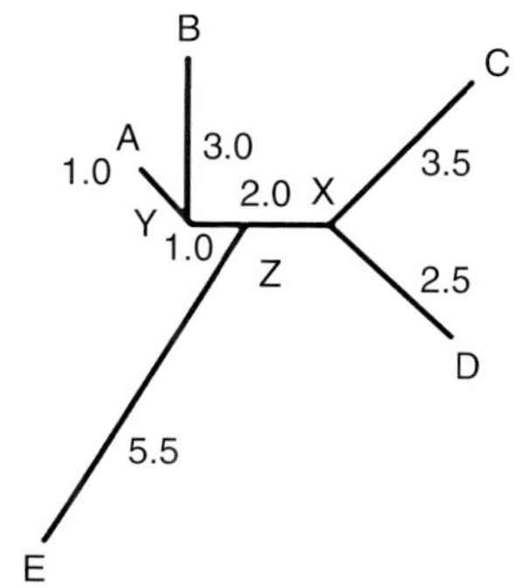

FIGURE 23.4

The rooted tree can be obtained by positioning the branches in a rectangular format of branches:

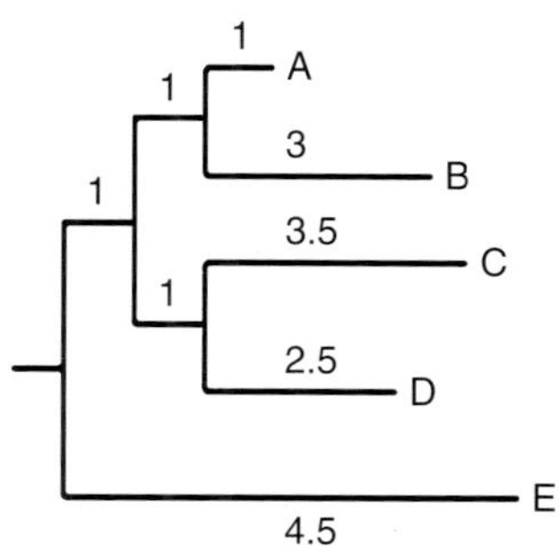

FIGURE 23.5

Thus, trees are generated for different topologies, and the tree with the shortest total length is to be selected, as a principle.

23.4 INTERPRETATION OF NJ TREE

The values shown in each branch denote the rate-corrected distances, which are not proportional to the time-scale. The inference is thus that the rate of evolution is not the same in different branches, and the distances between two taxa are the sum of the distances indicated in the branches. The distances are calculated using a suitable method (assuming a different rate of substitution).

Minimum evolution is a distance-based method to construct additive trees. Presently, the minimum evolution method utilizes two criteria: unweighted least squares (to identify the shortest branch length) and tree topology. However, the use of weighted least squares or generalized least squares can yield an erroneous tree (Gascuel *et al.*, 2001). Different topologies are tested using unweighted least squares, and the tree with the shorted sum of branch length is assumed to be the true tree. However, this is not necessary for all instances.

23.5 QUESTIONS

1. Construct a phylogenetic tree using the given distance matrix by applying the NJ method:

TABLE 23.22

Distance matrix	A	B	C	D
B	8.000			
C	7.000	8.000		
D	6.000	12.000	6.000	
E	4.000	7.000	5.000	7.000

2. Use the following distance matrix to construct an NJ method-based phylogenetic tree:

TABLE 23.23

Distance matrix	A	B	C	D	E
B	5.000				
C	4.000	8.000			
D	7.000	10.000	7.000		
E	6.000	9.000	6.000	6.000	
F	7.000	12.000	8.000	9.000	7.000

3. Given the following distances, construct a phylogenetic tree using the NJ method:

TABLE 23.24

Distance matrix	A	B	C	D
B	3.000			
C	5.000	8.000		
D	6.000	7.000	6.000	
E	7.000	10.000	12.000	8.000

4. How is the NJ method principally different from the UPGMA method of phylogeny construction?
5. Enumerate the conditions where the NJ method is the best suited for phylogenetic tree construction. Logically explain the reasons why.

Construction of Phylogenetic Tree: Maximum Parsimony Method

CHAPTER 24

CS Mukhopadhyay and RK Choudhary
School of Animal Biotechnology, GADVASU, Ludhiana

24.1 INTRODUCTION

"Parsimony" means penny-pinching or thriftiness. Here, the term "maximum parsimony" (MP) indicates extreme caution to opt for simpler hypotheses to select a tree (out of many) with minimum length (sounds like the minimum evolution (ME) method). This approach is applied when the ancestral relationship is required to be reconstructed. The MP method is not the same as the ME method. MP identifies position-specific differences between all pairs of sequences, while ME indicates the distance (in terms of the number of differences among any two sequences).

24.1.1 Principle

MP is a character-based method, not distance-based like ME. The MP method identifies the "most parsimonious tree" through optimality criteria (not by clustering methods like UPGMA or NJ). Thus, the tree that needs a minimum number of evolutionary events (characterized by residue substitutions) to explain the sequence variation is selected.

24.1.2 Assumption

A tree that needs fewer substitutions is better than a tree that requires more (Li, 1997).

24.2 OBJECTIVE

To construct a phylogenetic tree using the MP method from a set of molecular sequences.

Basic Applied Bioinformatics, First Edition. Chandra Sekhar Mukhopadhyay, Ratan Kumar Choudhary and Mir Asif Iquebal.

a. MP is applicable for small sequences with strong similarity (sequence of a gene/peptide belonging to different strains/breeds/cultivars of a species) (Mount, 2009).
b. MP is applicable to molecular or non-molecular (i.e., dummy or morphological) data.
c. MP is a cladistic method (not a phenetic method like UPGMA, NJ, ME, or FM), which is based on synapomorphies – namely, sharing of the derived characters due to common evolutionary relationships. Thus, it does not reduce sequence information to a single number.
d. If the number of substitution per site is small, MP does not need exact constancy for rates of change between the branches.
e. Assumptions like equal evolutionary rate are not made, so the obtained tree is not under questionable screening for practical acceptability.
f. It differs from ME in that the length of the tree in ME is inferred from the genetic distances, while MP counts individual residue substitutions to construct the tree and is not a distance-based method.

24.3 PROCEDURE

a. The tree is constructed following multiple sequence alignment (MSA). Each column of MSA is considered to determine the "most informative columns" to initiate the tree building (elaborated below). The most informative site is the column which has at least two different residues, and each residue is represented at least twice (thus, at least three input sequences). It is based on the smallest number of evolutionary changes. Analysis of evolutionary changes for each residue is scored vertically in MSA.
b. The score is assigned to each tree. Finally, the best one (with the *least score*) is selected.
c. "Most parsimonious tree" means the tree with a minimum number of evolutionary changes.

24.3.1 Multiple Sequence Alignment (MSA)

Align the input nucleotide sequences (given below) to determine the most informative column(s) to determine the weights of each of the type of substitution.

TABLE 24.1

Sequences	Residues					
Seq 1	A	G	**G**	A	C	A
Seq 2	C	G	**A**	A	A	T
Seq 3	G	G	**C**	A	G	C
Seq 4	T	C	**A**	A	T	G

The third column of the aligned sequences is the most informative site (shown in **bold**), as it contains at least two base variants present at least twice in the particular column.

24.3.2 Selection of algorithm

Two types of algorithms are broadly used to calculate the distance for maximum parsimony: Fitch (abbreviated as "F" in the following demonstration) and Transversion Parsimony (abbreviated as "T-P"). These two scoring schemes for base substitution differ in the weights given to the type of substitution.

24.3.3 Fitch (F)

The weight assigned to both transversion, as well as transition to different bases, is one (1), while transition to the same state (i.e., "G" to "G", or "A" to "A", etc.) is assigned no weight (Table 24.2).

24.3.4 Transversion Parsimony (T-P)

The weights assigned are 4 for transversion, 1 for transition to other bases (other than self) and 0 for transition to the same base (Table 24.3).

24.3.5 Construct a topology for the sequences

Now construct a topology of the unrooted tree to initiate the analysis. In our current example, four sequences (Seq 1, Seq 2, Seq 3 and Seq 4) are considered. One of the possible topologies is ((Seq 1, Seq 3), (Seq2, Seq 4)). The alternative topologies are

TABLE 24.2

Scoring scheme of nucleotide substitution for Fitch method

	G	A	C	A
G	0	1	1	1
A	1	0	1	0
C	1	1	0	1
A	1	0	1	0

TABLE 24.3

Scoring scheme of nucleotide substitution for T-P method

	G	A	C	A
G	0	1	4	1
A	1	0	4	0
C	4	4	0	4
A	1	0	4	0

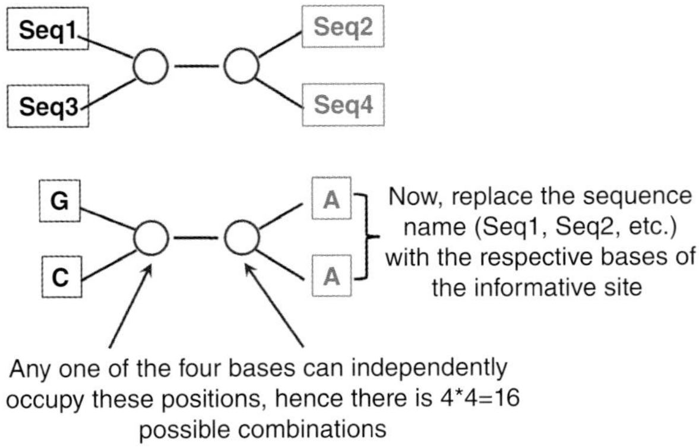

FIGURE 24.1

((Seq1, Seq2), (Seq3, Seq4)) and (Seq1, Seq4), (Seq2, Seq3)), which are also to be tested to find the trees yielding the minimum score(s).

The bases in the third column of the aligned sequences (i.e., the informative site) are "G", "A", "T" and "A" for Seq 1, Seq2, Seq3 and Seq 4, respectively.

24.3.6 Scoring for substitution of the bases

Now, these blank spaces will be filled with all possible combinations of the nucleotide bases.

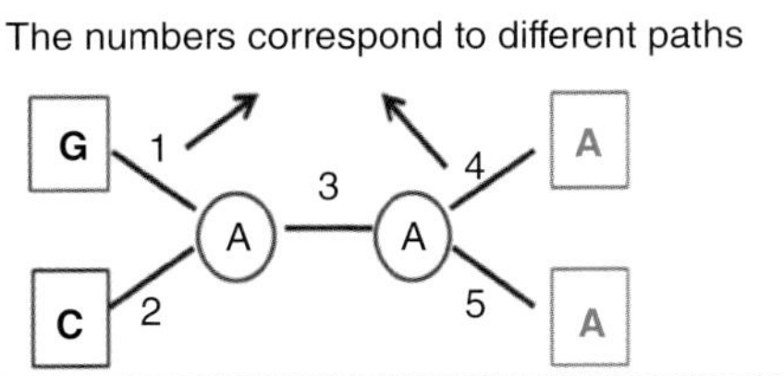

	Paths						
	1	2	3	4	5		Score
F	1	1	0	0	0	=	2
T-P	1	4	0	0	0	=	5

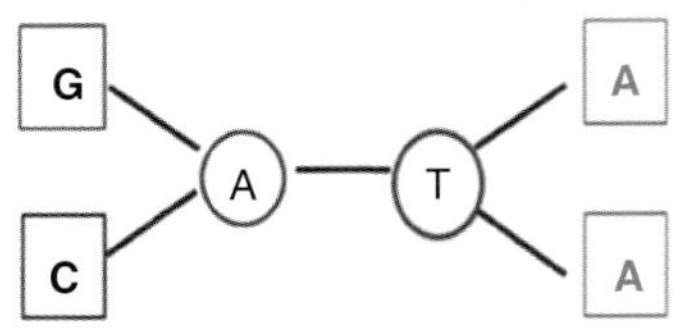

	Paths						
	1	2	3	4	5		Score
F	1	1	1	1	1	=	5
T-P	1	4	4	4	4	=	17

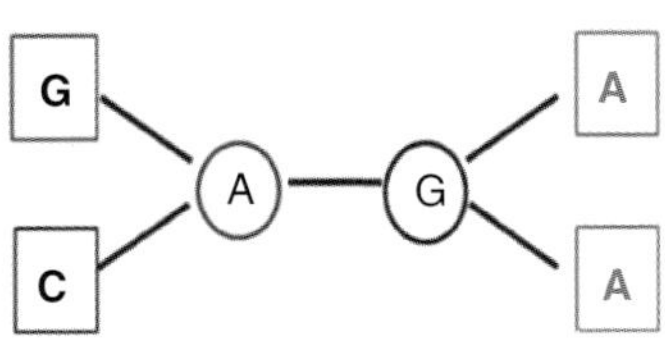

	Paths						
	1	2	3	4	5		Score
F	1	1	1	1	1	=	5
T-P	1	4	1	1	1	=	8

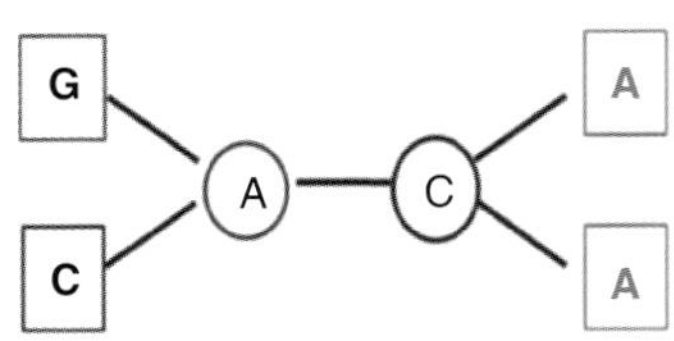

	Paths						
	1	2	3	4	5		Score
F	1	1	1	1	1	=	5
T-P	1	4	4	4	4	=	17

FIGURE 24.2

The scores of each of the combinations are calculated using both of the weighing schemes (F and T-P).

MP searches the optimal tree out of all the possible combinations. The paths with the least scores are to be determined individually for each of the weighing schemes.

The Fitch scheme shows that the minimum score is 2 for the four paths given below:

The T-P scheme shows a least score of 5 for two of these paths. Hence, these two paths are the best ones for the current topology.

It is necessary to iterate the same procedure for assessing the other topologies:

a. ((Seq1, Seq2), (Seq3, Seq4))
b. ((Seq1, Seq4), (Seq2, Seq3)).

The final optimal tree with the minimum score is selected after obtaining all the topologies, based on the *least score* in the Fitch or Transition Parsimony methods of scoring.

The number of paths exponentially increases with the number of sequences (OTUs). The branch and bound method is used to find the optimal tree without verifying all possible trees, in order to save computational resource requirements and time.

The number of rooted [obtained as (2n-3)*(2n-5)*(2n-7)*…*3*1=(2n-3)!/((2^(n-2))*(n-2)!)] and unrooted trees [obtained as (2n-5)*(2n-7)*…*3*1 = (2n-5)!/(2^(n-3)*(n-3)!)] required to be analyzed in this method is given in Table 24.4:

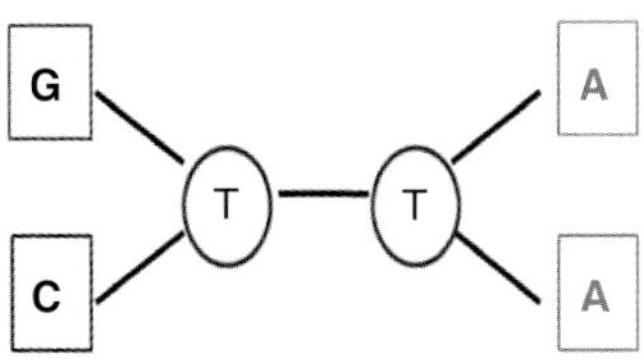

	Paths						
	1	2	3	4	5		Score
F	1	1	0	1	1	=	4
T-P	4	1	0	4	4	=	13

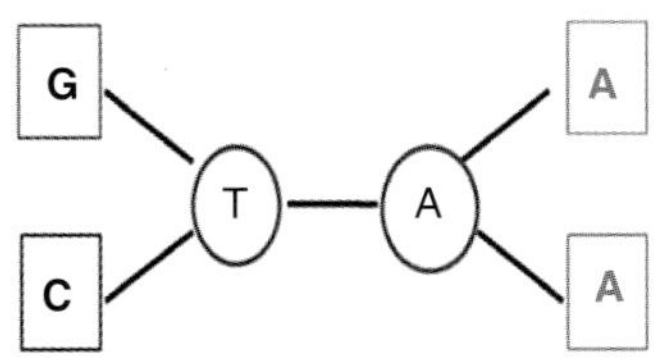

	Paths						
	1	2	3	4	5		Score
F	1	1	1	0	0	=	3
T-P	4	1	4	0	0	=	9

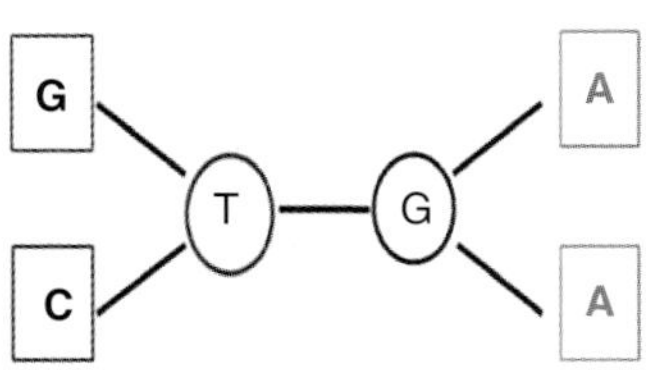

	Paths						
	1	2	3	4	5		Score
F	1	1	1	1	1	=	5
T-P	4	1	4	1	1	=	11

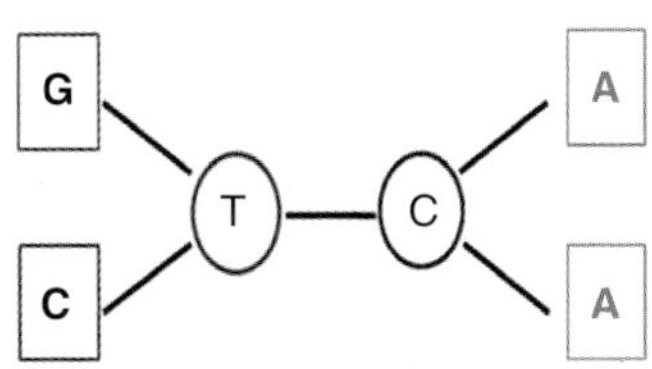

	Paths						
	1	2	3	4	5		Score
F	1	1	1	1	1	=	5
T-P	4	1	1	4	4	=	14

FIGURE 24.3

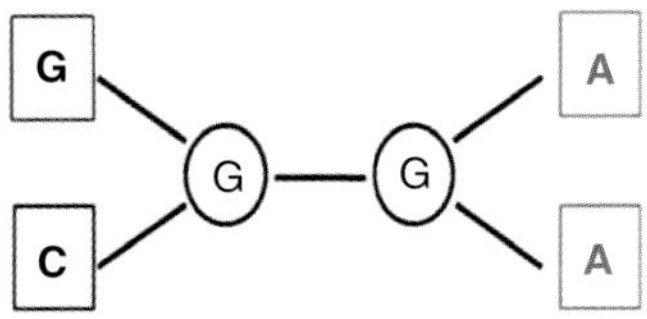

	Paths						
	1	2	3	4	5		Score
F	0	1	0	1	1	=	3
T-P	0	4	0	1	1	=	6

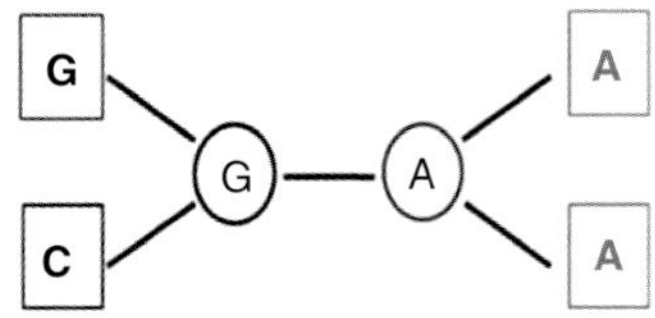

	Paths						
	1	2	3	4	5		Score
F	0	1	1	0	0	=	2
T-P	0	4	1	0	0	=	5

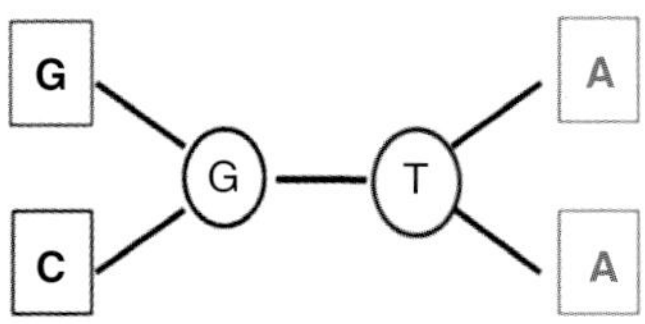

	Paths						
	1	2	3	4	5		Score
F	0	1	1	1	1	=	4
T-P	0	4	4	4	4	=	16

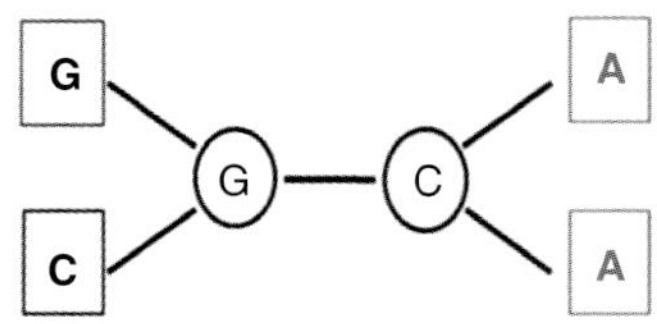

	Paths						
	1	2	3	4	5		Score
F	0	1	1	1	1	=	4
T-P	0	4	4	4	4	=	16

FIGURE 24.4

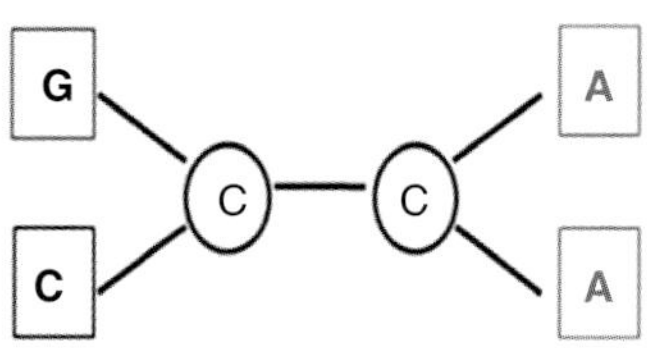

	Paths						
	1	2	3	4	5		Score
F	1	0	0	1	1	=	3
T-P	4	0	0	4	4	=	12

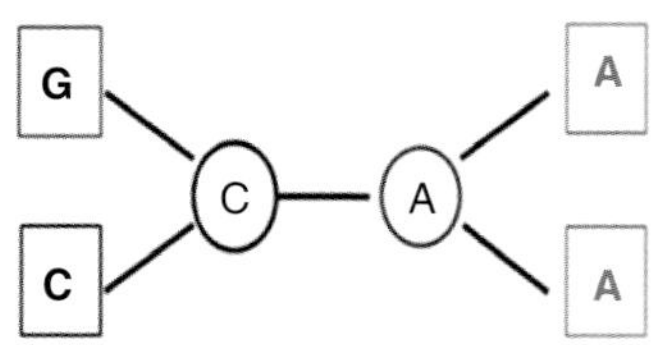

	Paths						
	1	2	3	4	5		Score
F	1	0	1	0	0	=	2
T-P	4	0	4	0	0	=	8

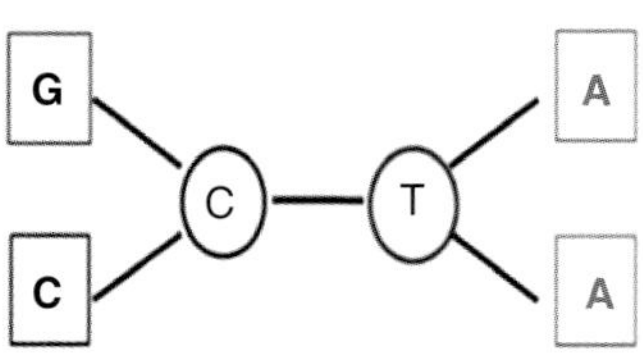

	Paths						
	1	2	3	4	5		Score
F	1	0	1	1	1	=	4
T-P	4	0	1	4	4	=	13

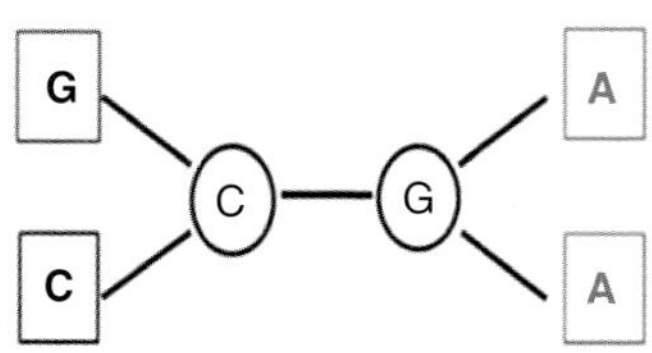

	Paths						
	1	2	3	4	5		Score
F	1	0	1	1	1	=	4
T-P	4	0	4	1	1	=	10

FIGURE 24.5

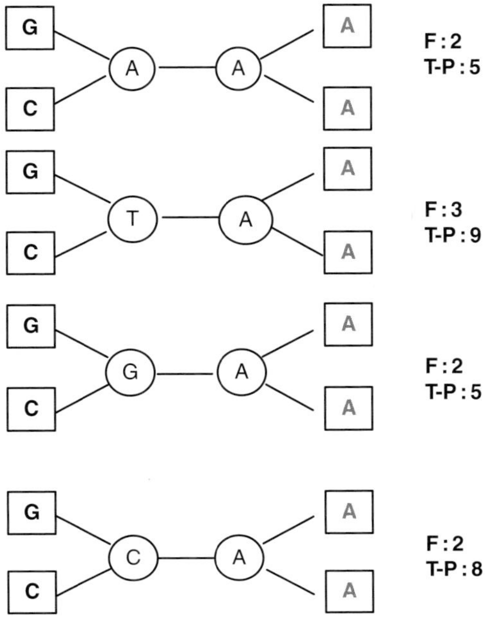

FIGURE 24.6

TABLE 24.4

OTU Count	Number of unrooted tree(s)	Number of rooted tree(s)
2	1	1
3	1	3
4	3	15
5	15	105
10	34 459 425	2.13E15
15	2.13E + 15	8.E + 21
50	2.84E + 74	2.75E + 76
n	$\frac{(2n-5)!}{[2^{n-3} * (n-3)!]}$	$\frac{(2n-3)!}{[2^{n-2} * (n-2)!]}$

24.4 INTERPRETATION OF MP TREE

The tree with a minimum number of steps has been selected as the optimal tree out of the finally obtained set of valid trees. Parsimony always chooses the tree that sought a minimum number of changes to explain the data. MP trees do not show branch length as an indicator of distance, but the number of substitutions is counted. The total number of steps required to reach the final step is known. The validity of branching tested through bootstrapping is denoted at the nodes of the consensus tree, which is defined as the tree obtained through an agreement for each node among several bootstrap trees.

A method such as "weighted parsimony" is used to identify the sites with the highest or the lowest rate of evolution.

24.5 QUESTIONS

1. Construct the most parsimonious tree using the following sequences:

TABLE 24.5

Seq.	Bases								
Seq 1	T	A	G	A	T	C	G	C	A
Seq 2	T	C	C	C	T	C	G	C	G
Seq 3	T	C	T	G	A	C	C	C	A
Seq 4	T	C	A	G	A	C	C	C	G

2. How does one determine the most informative sites in a given set of sequences for maximum parsimony analysis?
3. Differentiate between the maximum parsimony and minimum evolution methods of phylogenetic tree construction.
4. Why is there no scale given in the phylogenetic tree obtained from the MP method?

Construction of Phylogenetic Tree: Minimum Evolution Method

CHAPTER 25

CS Mukhopadhyay and RK Choudhary
School of Animal Biotechnology, GADVASU, Ludhiana

25.1 INTRODUCTION

The minimum evolution (ME) method is a distance-based method, and simple enough to avoid the least-squares approach to determine the optimal tree with minimum branch length. The ME method described here is different from the maximum parsimony (MP) method in its approach. ME identifies the number of sites differing among the input sequences to produce the distance matrix, and ME yields an unrooted tree from the given set of sequences.

25.1.1 Principle

The tree length is obtained by summing up the character differences, and finally the tree with minimum branch length is reconstructed.

25.1.2 Assumptions

a. The character states (residues) change independently along the lineages.
b. Constant rate of evolution over lapse of time.
c. Additivity of the branch length.
d. The tree with the smallest summed branch length is the true one (Rzhetsky and Nei, 1993).

25.2 OBJECTIVE

To construct a phylogenetic tree using the ME method from a given set of nucleotide sequences.

Basic Applied Bioinformatics, First Edition. Chandra Sekhar Mukhopadhyay, Ratan Kumar Choudhary and Mir Asif Iquebal.

a. ME is quick and is not a resource-intensive method.
b. It is applicable for a set of sequences derived from closely related organisms.
c. Results are not reliable for the sequences derived from distantly related species, due to generation of many differences in the sequences, which inflates distances.
d. The algorithm does not differentiate between transversion and transition, so the rate of evolution is assumed to be uniform for all types of substitution.
e. All sorts of sequence information are not considered for constructing the tree.

25.3 PROCEDURE

Start with four (an arbitrarily chosen number) nucleotide sequences:

a. First, the sequences (S1–S4) are to be aligned:

TABLE 25.1

	1	2	3	4	5	6	7	8	9	10	11	12	13	14	15	18	19	20
S1	C	G	T	A	G	G	A	T	A	G	A	C	A	T	A	G	G	C
S2	C	G	A	A	G	G	G	T	A	G	A	C	G	T	A	G	G	C
S3	C	A	A	A	G	G	A	G	A	G	C	T	A	T	G	G	G	C
S4	C	C	A	A	G	G	A	C	A	G	G	T	A	T	T	G	G	C

b. Next, determine the distance matrix by summing the number of different residues (designated by S) between each pair of sequences:

TABLE 25.2

	S1	S2	S3	S4
S1	0			
S2	3	0		
S3	6	7	0	
S4	6	7	4	0

c. The phylogenetic tree is constructed from the distances between sequence-pairs, indicating the difference between each pair of sequences:
   i. Just like UPGMA, the principle of ultrametric tree construction (assuming a molecular clock) is followed here. First, one needs to find out the sequence pairs (or taxa) with minimum distance (here, $d(S1 - S2) = 3$). Hence, S1 and S2 form a single cluster.
   ii. The ultrametric three-point condition is checked for the taxa: $dxy <= \max(dxz, dyz)$ when x, y and z are three taxa taken into consideration (Desper and Gascuel, 2005).

iii. Now, for four-point condition, note that the sum of $d(S1 - S2) + d(S3 - S4) <= d(S2 - S3)$ or $d(S2 - S4)$. This implies that the two largest sums are equal (i.e., 7): $d(S1 - S2) + d(S3 - S4) <= \max\{d(S1 - S3) + d(S2 - S4), d(S1 - S4) + d(S2 - S3)\}$

iv. The tree-metric is thus constructed, and this enables one to place S1 and S2 as a cluster and S3 and S4 as another cluster.

25.3.1 Difference between minimum evolution and maximum parsimony methods (Figure 25.1)

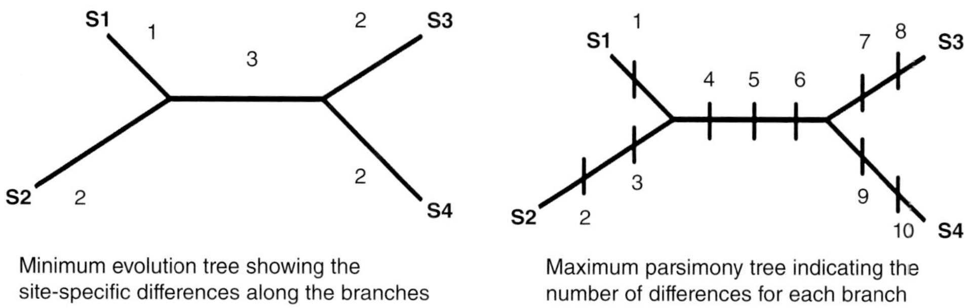

FIGURE 25.1 Comparative depiction of the phylogenetic tree constructed from the same data set, using the MP method.

25.4 INTERPRETATION OF THE ME TREE

Here, the ME tree is inferred, and the corresponding differences with the MP tree are shown.

ME is a distance-based method that selects the unrooted tree based on optimality criteria, whereas MP is a discrete method that determines the branch length from the fit of the individual residue of sequence-pairs to a tree.

The ME tree only specifies the number of differences (as distance) in a branch for each pair of sequence. However, the position-specific differences are not indicated. MP identifies each site of difference between any pair of sequences and shows this in the respective branches.

25.5 QUESTIONS

1. Construct a phylogenetic tree using the following homologous partial sequences by the ME method:

   > Bbu
   MASFRVKETVCPRTSQQPLEQCDFKENG

   > BtaTV4
   MTSFTVKETVCPRTSPQPPEQCDFKENG

> BtaTV2
MTSFTVKETVCPRTSPQPPEQCDFKENG

> BtaTV1
MVSFRVKETDCPRTSQQPLEQCDFKENG

> BtaTV3
MVSFRVKETDCPRTSQQPLEQCDFKENG

2. Construct a phylogenetic tree using this set of peptide sequences (ME method):
   > Bbu
   NELQSVRRFRPRRPRLPRPRPRPLPL

   > BtaTV4
   NELQSVR-FRP-PIRRPPIRP---PF

   > BtaTV2
   NELQSVRRIRPRPPRLPRPRPRPLPF

   > BtaTV1
   NELQSVRRIRPRPPRLPRPRPRPLPF

   > BtaTV3
   NELQSVR-FRP-PIRRPPIRP---PF
3. Now compare these two phylogenetic trees and explain what are the possible reasons for their differences, even though the source sequences are the same but from different portions of the peptide sequences.
4. Enumerate the differences between ME and MP trees. Also, explain the conditions when these two algorithms are best suited.
5. Discuss the assumptions and the practicality of the assumptions of the ME algorithm.

Construction of Phylogenetic Tree Using MEGA7

CHAPTER 26

CS Mukhopadhyay and RK Choudhary
School of Animal Biotechnology, GADVASU, Ludhiana

26.1 INTRODUCTION

The Molecular Evolution and Genetic Analysis (MEGA) is a freely downloadable (for research and education) integrated tool for analyzing molecular data (nucleotide and protein sequences) and construction of phylogenetic trees. The latest version, MEGA 7.0.14, is used for some bio-computational analyses, for example in: sequence alignment; determining the best evolutionary model; construction of phylogenetic trees as well as inferring ancestral sequences; mining online databases; estimation of divergence times; rates of molecular evolution; and testing evolutionary hypotheses. The software is freely available at http://www.megasoftware.net/ and can be run in Windows (both GUI and command-line-based), as well as in Linux and Mac operating systems (command-line-based).

In this chapter, we will see how a phylogenetic tree can be constructed using MEGA7 suit and inferred.

26.2 OBJECTIVE

To build a phylogenetic tree from a given set of molecular sequences.

26.3 PROCEDURE

26.3.1 Prepare the sequence file

Download and then arrange the molecular sequence data (nucleotide or amino acid sequences) in FASTA format, and save in a notepad (*.txt) file. It is not necessary that all the sequences should be of the same length, but the sequences should be homologous (depending on the hypothesis being tested in the experiment). The descriptive line of each FASTA formatted sequence may be shortened (Figure 26.1).

Basic Applied Bioinformatics, First Edition. Chandra Sekhar Mukhopadhyay, Ratan Kumar Choudhary and Mir Asif Iquebal.

FIGURE 26.1 Compile the unaligned, homologous molecular sequences in FASTA format in a text file.

26.3.2 Uploading data file/pasting the sequences

a. Open MEGA7 and click on Align → Click on Edit/Build Alignment (the first option) in the drop-down menu.
b. A small dialogue box will appear with the options in a radio button.
   i. *Create a new alignment*: Select this if you are starting afresh. Selecting the option will direct the user to another window to select the type of input sequence data, DNA or amino acid. Select the correct option and proceed. Copy all the sequences (in FASTA format) and paste in the Alignment Explorer.
   ii. *Open a saved alignment session*: Select this if you have already saved a previous alignment (on which someone has worked earlier). Select the file from the folder and proceed.
   iii. *Retrieve sequence from a file*: Click if you want to upload a sequence from a text file. The text (.txt) file containing molecular sequences (in FASTA, PAUP, MEGA, ALN, Phylip, GCG, PIR, NBRF, MSF or IG formats) is opened in the sequence editor for further analysis (MSA).

26.3.3 Align the sequences

Click on "Alignment" on the menu bar and select any one of the two options in the drop-down menu:

a. *Align by ClustalW*: Opt for ClustalW when the input sequences are of comparable length and homologous (Figure 26.2).
b. *Align by Muscle*: This option is preferred for sequences with considerably varying length, although belonging to the same super-family. Out of these two algorithms, namely, ClustalW (progressive algorithm) and Muscle (iterative algorithm), the performance of Muscle is considerably good when the input sequences vary in sequence lengths.

26.3.4 Save session

The alignment session can be saved as a *.mas file for future use (Figure 26.3).

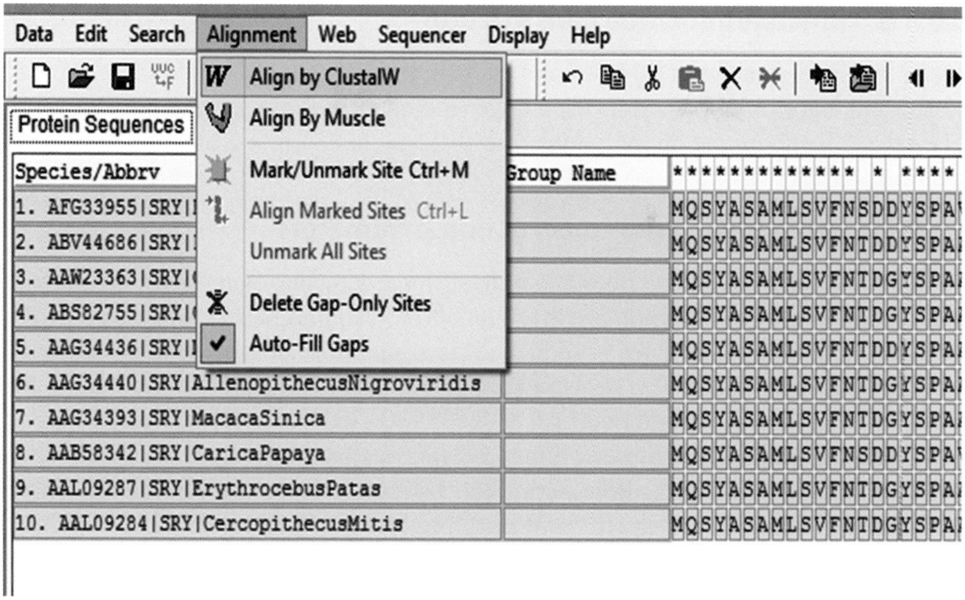

FIGURE 26.2 Aligning the input sequences using either ClustalW or Muscle available in MEGA7 interface. (*See insert for colour representation of the figure.*)

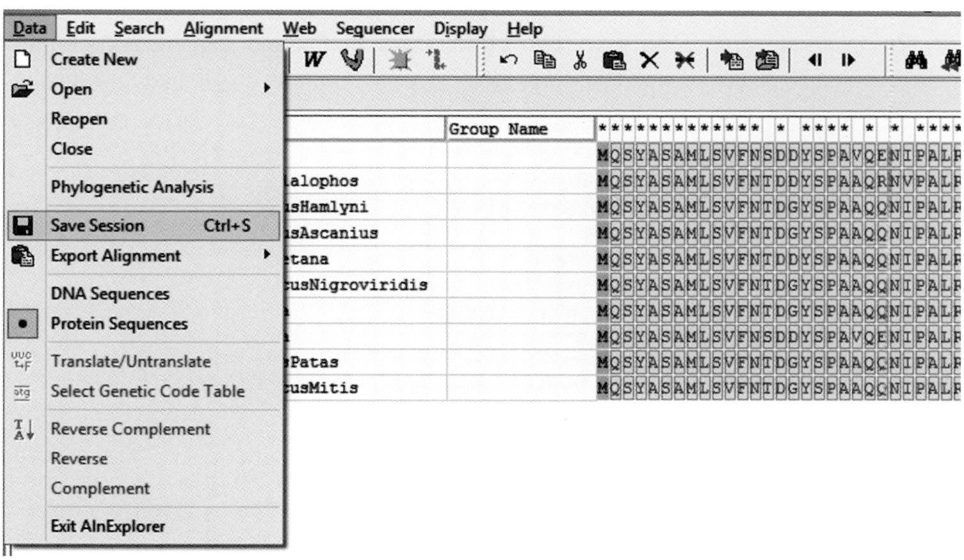

FIGURE 26.3 Exporting the alignment file and saving the alignment session for further use. (*See insert for colour representation of the figure.*)

26.3.5 Export alignment

The alignment data can be exported in any of the following file formats: MEGA, FASTA, PAUP.

Now, close the alignment explorer window to proceed for phylogenetic analysis.

26.3.6 Phylogenetic tree construction

Open the main window of MEGA7 and click on the "Phylogeny" tab in the menu bar. Select the algorithm you need for phylogenetic analysis from the drop-down menu. Here we will choose the option "Construct/Test Neighbor-Joining Tree".

26.3.7 Selection of tree construction parameters

a. *Test of Phylogeny*: Select "Bootstrap method" for re-sampling of the branching pattern.
b. *Number of Bootstrap replications*: Run 500 re-samplings if the sequence length is long and/or the number of sequences is higher; else consider 1000 bootstrap-replications. At least 100 bootstrap re-samplings are suggested for validating the branching of constructed tree.
c. *Model/Method*: The drop-down menu displays a list of models (i.e., Number of differences, p-distance, Poisson model, JTT, etc., depending on the algorithm chosen). It is better to run the program "Find Best DNA/Protein Models (ML)", available under the "Models" tab in the menu bar (Figure 26.4). However, selection of model is time-consuming, and is more applicable for the Maximum Likelihood-based algorithm. We can, in general, select an advanced model such as Jones–Taylor–Thornton (JTT). Please remember that the NJ method assumes different rates of evolutionary changes, while the ME method assumes the same rate of transversion and transition. Thus, accordingly, select the model based on the method you opt for phylogenetic tree construction (Figure 26.5).
d. *Rates among Sites*: There are two options (for nucleotide sequences as input): "Gamma Distributed" and "Uniform rates". Opt for Gamma distributed if sequences are divergent enough.

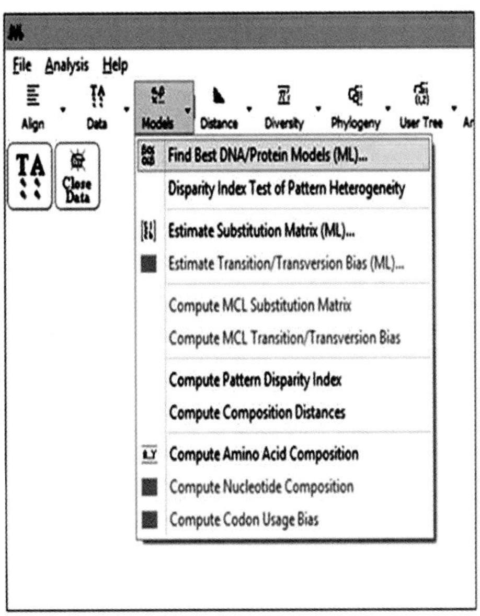

FIGURE 26.4 Selection of the best evolutionary model for further analyses.

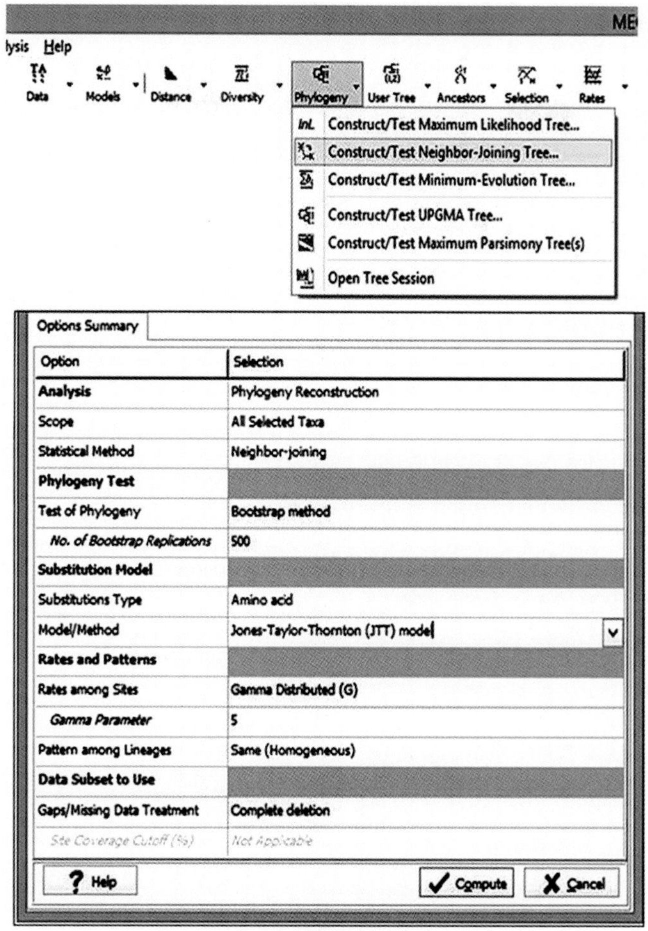

FIGURE 26.5 Setting the parameters for phylogenetic analysis.

e. *Gamma parameter*: Gamma distribution is specified with *Gamma parameter* (or shape parameter varying from 1 to 5) for modeling the evolutionary rates. Here it is assumed that the substitution rate varies from site to site.
f. Gaps/Missing Data Treatment: complete deletion.
g. Now, run the analysis for tree construction, by clicking on "Compute".

26.4 INTERPRETATION OF PHYLOGENETIC TREE

The phylogenetic tree displays the branch scale at the bottom.

a. *Node IDs*: Each of the internal nodes is given discrete and unique numerical IDs for specification.
b. *Branch length*: Each branch has a length (corresponding to the scale given at the bottom) that indicates the substitution of residues.
c. *Bootstrap value*: This indicates the stability of the branching pattern but bears no relationship to the accuracy of the tree.

26.4.1 Controlling the output of phylogenetic tree

The generated tree can be manipulated to suit the requirement of a presentation by changing its size, branch positions, toggling the bootstrap values, branch length, etc. (Figure 26.6). Since MEGA is very user-friendly software, everything can be controlled through the menu-bar options, or the buttons displayed in the left-hand side pane (Windows OS). Figure 26.7 clearly indicates the various buttons on the GUI for controlling the appearance of the tree.

26.4.2 Diagrams for each of the taxa

These can be inserted as follows:

a. Click on "Subtree" in Menu-bar → Select Use Subtree draw options → Click on the "Image" tab of Subtree Drawing options" and select the image from the saved image in the particular folder (Figure 26.8).

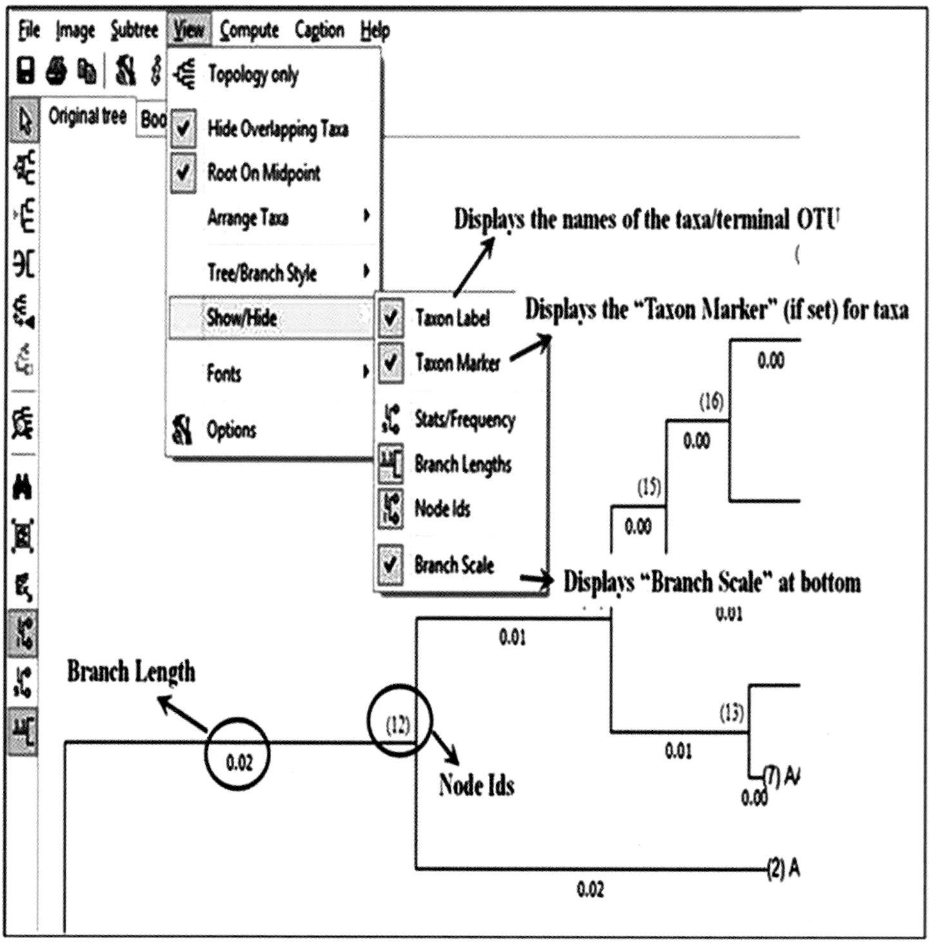

FIGURE 26.6 Controlling the display parameters using the menu bar parameters.

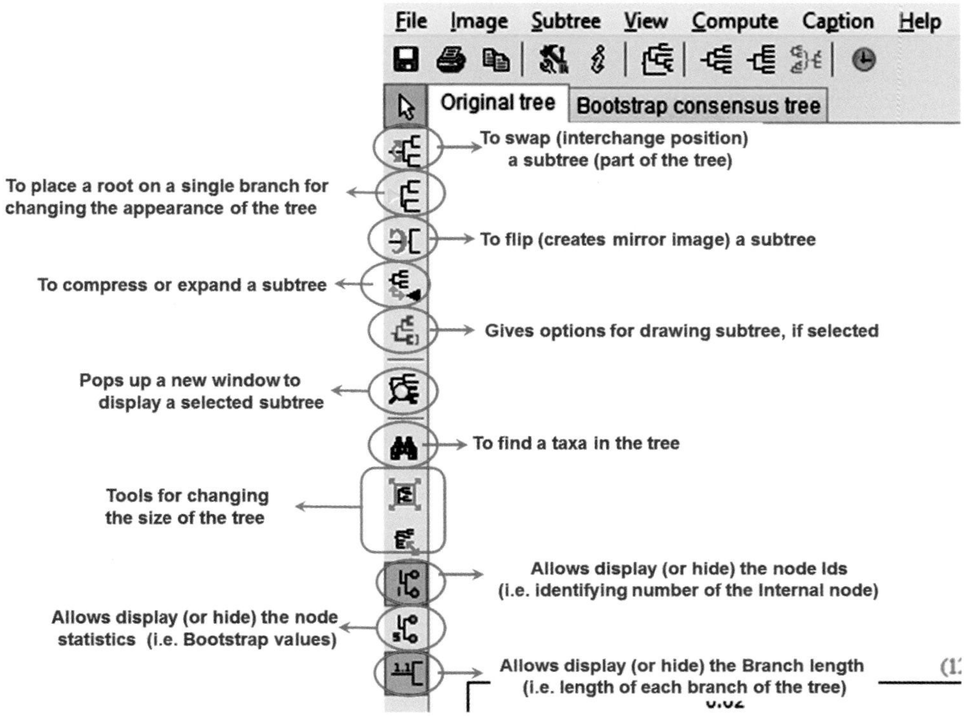

FIGURE 26.7 Controlling the tree display parameters using the left-hand-side buttons. (*See insert for colour representation of the figure.*)

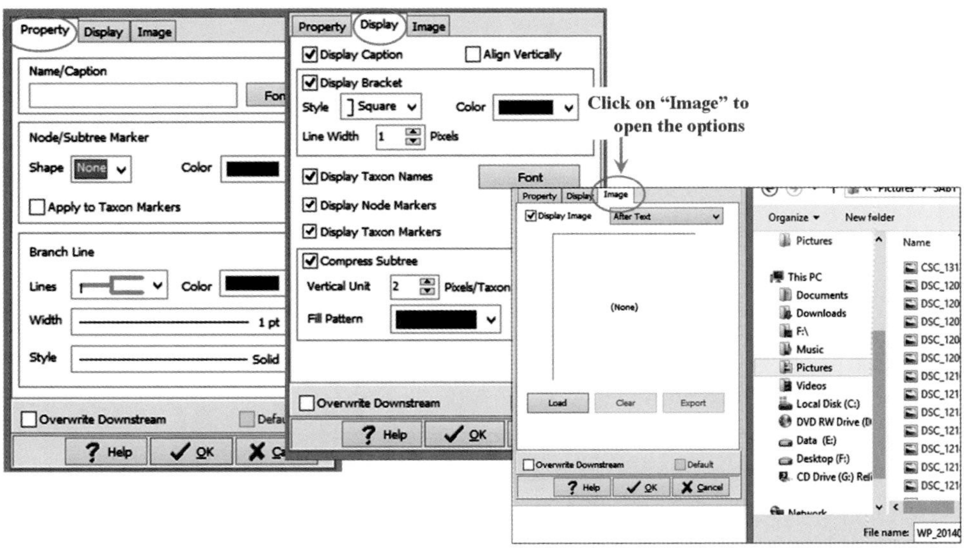

FIGURE 26.8 Insertion of figures for the external nodes (species name). (*See insert for colour representation of the figure.*)

b. Save the Phylogenetic Tree: Click on the "Image" option in the menu bar → Click on "Save as PNG file" (Figure 26.9).

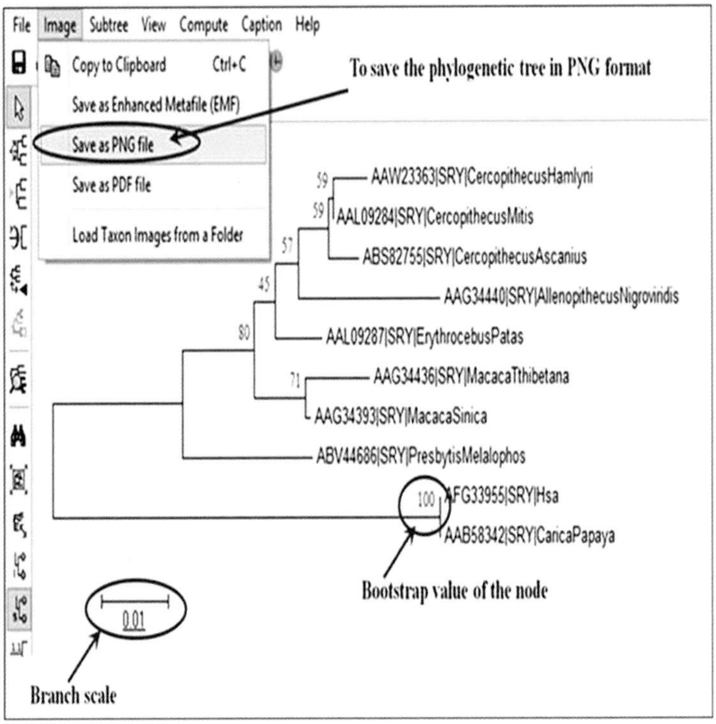

FIGURE 26.9 Saving the output phylogenetic tree as a PNG file.

26.5 QUESTIONS

1. Construct a phylogenetic tree using the neighbor-joining method, with bootstrap re-sampling of 500, using a set of homologous protein sequences.
2. Consider the previous example and increase the bootstrap re-sampling to 1000. Is there any change in the branching pattern reliability values (i.e., bootstrap values)? Display the tree so that only bootstrap values of more than 75 are shown in the nodes.
3. Construct a phylogenetic tree with the following algorithms: ME, NJ, UPGMA, Maximum Likelihood. Now, compare the trees using the protein sequences: NP001272506.1 AAI20478.1 CAH23217.1 XP005909397.1 XP005955229.1. The bootstrap re-sampling should be 500 for all the algorithms. Please determine the best evolutionary model before running the phylogeny analysis.
4. Determine the best model for phylogenetic tree construction using the following nucleotide sequences, and then construct a circular phylogenetic tree with bootstrap re-sampling and minimum evolution algorithm: AB974690.1 AB973433.1 NM001009772.1 NM001009406.1 NM001009787.1 NM001285577.1

5. Interpret the given output generated by MEGA using the NJ method:

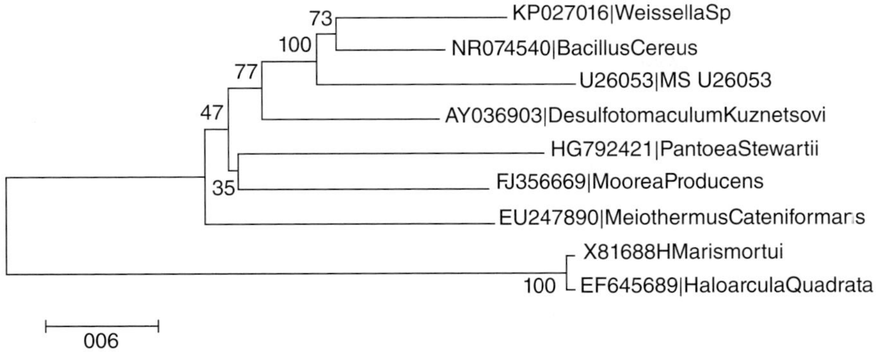

FIGURE 26.10

Interpretation of Phylogenetic Trees

CHAPTER 27

CS Mukhopadhyay and RK Choudhary
School of Animal Biotechnology, GADVASU, Ludhiana

27.1 INTRODUCTION

Phylogenetic trees are frequently encountered in research papers related to evolution, population diversity, microbial studies, and genetics. It is critical to infer the meaning of a given phylogenetic tree, as terms such as "cladogram", "phylogram", "phenogram", etc. sound quite confusing to a novice. This chapter starts with such terminologies, and then the phylogenetic tree is explained to decipher the meaning depicted in general.

SOME TERMINOLOGIES USED FOR PHYLOGENETIC TREES

a. **Dendrogram** is a broad term used to represent a phylogenetic tree. More precisely, "dendrogram" is a generic term applied to any type of phylogenetic tree (scaled or unscaled).
b. **Cladogram** is a representation of the ancestor-to-descendant relationship through a branching tree. The length of a branch denotes nothing (neither genetic change nor time-scale). A clade simply means a common ancestor and the descendants (here, species) of that common ancestor.
c. **Phylograms** are diagrammatic representations of molecular phylogeny, constructed by statistical analysis of molecular data. The tree shows the divergence of species from internal nodes, and the branch length signifies the degree of evolutionary change of the taxa. The common ancestor is shown only in rooted trees.
d. **Phenograms** are also statistically constructed trees which indicate only the degree of resemblance (or similarity) among the taxa. Phenograms are not phylograms, and the branch length does not indicate genetic change or time-scale. These are very rarely used, due to their limited informative nature.
e. **Chronograms** are cladograms that indicate evolutionary changes through time. The branch length of a chronogram is proportional to the time-scale (generally, in millions of years). The basic assumption is the same rate of genetic change among all the taxa.

Basic Applied Bioinformatics, First Edition. Chandra Sekhar Mukhopadhyay, Ratan Kumar Choudhary and Mir Asif Iquebal.

27.2 UNDERSTANDING PHYLOGENETIC TREES

A rectangular, horizontal phylogenetic tree is shown in Figure 27.1, which has been constructed using 18s rRNA sequences from nine divergent taxa. In general, a phylogenetic tree is two-dimensional, consisting of horizontal (analogous to the *X*-axis of a graph) and vertical (as *Y*-axis) axes.

The leaves or the terminal taxa are connected by internal nodes (solid circles) (Figure 27.1). The tree gradually shrinks towards the left and ends at the hypothetical common ancestor (solid square).

27.2.1 Horizontal dimension of a phylogenetic tree

This is the scale that signifies evolutionary distance (of a dendrogram) or time-scale (of a chronogram). The branch length of the current dendrogram (Figure 27.1) denotes the evolutionary distance between two taxa – the longer the branch, the more genetic change that taxon (or cluster of taxa) has experienced over the time of evolution.

A scale at the bottom of the tree acts as the unit of substitution of residue per site (base or amino acid, depending on the type of tree) and, thus, measures the substitution of residues. The following formula determines it:

$$Scale = \frac{\text{Number of mismatches (not gaps) in a pairwise sequence alignment}}{\text{Number of total aligned residues, excluding the gaps}} \qquad [27.1]$$

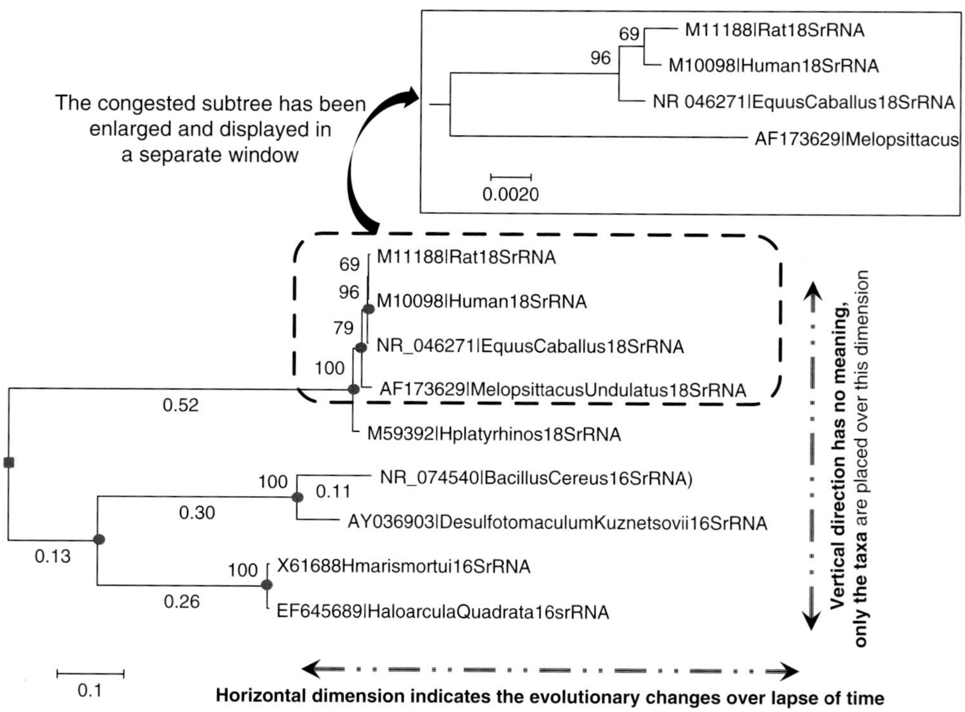

FIGURE 27.1 This dendrogram represents the evolutionary relationship among the taxa. The horizontal axis represents the evolutionary changes over time.

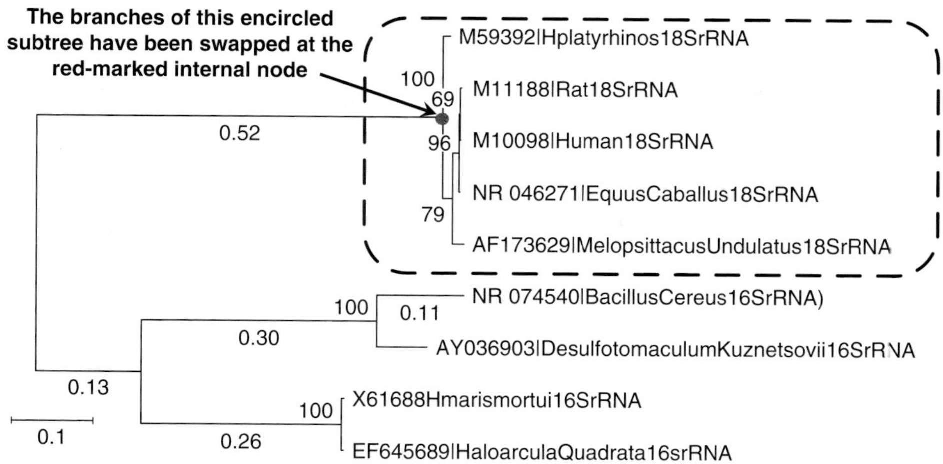

FIGURE 27.2 Swapping of the branches of the sub-tree of the main tree does not change any meaning represented by the tree. The evolutionary distances between the OTUs remain unchanged.

A scale of 0.1 means the amount of genetic change is 0.1 per unit length of the branch (indicated by the scale length). Thus, the total amount of genetic change will be

$$(0.1) \times (\text{branch length with respect to scale length}).$$

When the scale is represented as a percentage (here, it is 10%), this means that ten nucleotides have been substituted out of 100 residues. Please note that this does not necessarily mean that ten different nucleotides have been substituted, but that a single residue could have experienced substitution for multiple times. That is why a given value of 1.0 (or 100% in percent scale) does not mean all bases have been substituted but, rather, that 100 substitutions have taken place, some of which have occurred at the same residue position. Sometimes, the evolutionary scale is also represented as integer values, indicating the net number of base substitutions.

27.2.2 Vertical dimension of a phylogenetic tree

This direction has no meaning so far as evolutionary distance (or genetic changes) or the time-scale is concerned. This dimension is used only to place the taxa while building the phylogenetic tree. The branches of a sub-tree, or sub-sub-tree, or the whole tree, can be swapped without altering the meaning of the tree (depicted in terms of evolutionary relationship) (Figure 27.2). One can also increase the distance (width along the vertical axis) among the taxa, although it will have no impact on the meaning depicted by the tree.

27.3 REPRESENTATION OF PHYLOGENETIC TREES

A dendrogram can be drawn in several ways without distorting its meaning. The depictions are useful under different circumstances.

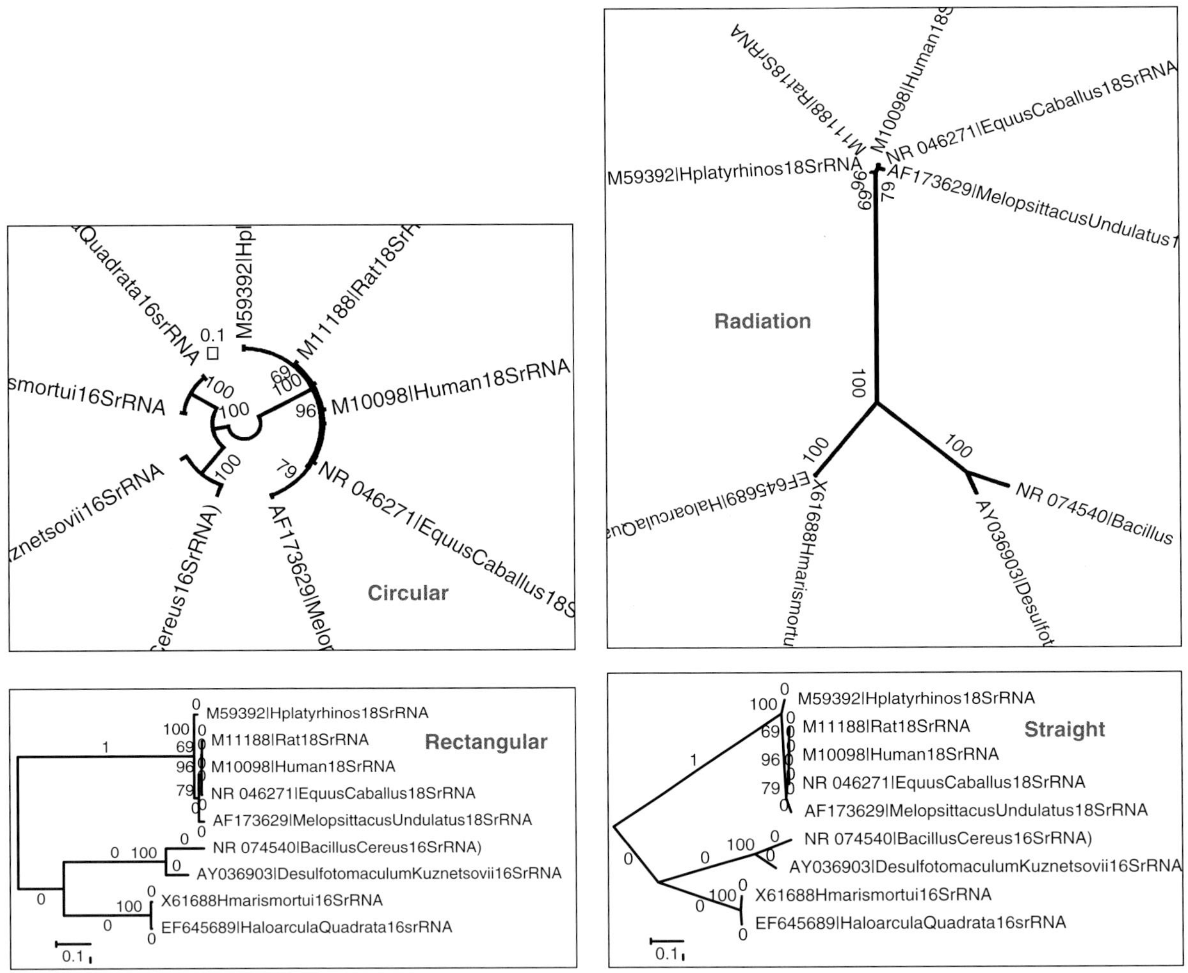

FIGURE 27.3 Representing the same phylogenetic tree as circular, radiation, rectangular and straight orientations.

27.3.1 Rectangular tree

This representation is well suited for both rooted and unrooted trees, and such trees are most easily understood. The branches connecting the taxa are separated by a vertical line (of an arbitrary length). The midpoint of the vertical line indicates the internal node (representing the hypothetical common ancestor of these taxa, which are not available at the present time) between two taxa being connected.

27.3.2 Straight tree

The rectangular trees are modified to straight tree by joining the taxa to the respective internal nodes directly (no vertical line is used), which makes the appearance of the tree more convergent towards the common ancestor. A straight tree depicts the same information as a rectangular tree.

27.3.3 Radiation tree

The typical tree-like appearance is substituted with a comparatively simple depiction. The divergence of the component taxa is not shown from a hypothetical ancestor (i.e., internal node). Figure 27.3 depicts how a straight tree can be converted to a radiation tree. The evolutionary scale may not be shown in this type of tree, though the node statistics (bootstrap values) and scale are present.

27.3.4 Circular tree

Both rooted and unrooted trees can be depicted by a circular tree. The distance from the center denotes the branch length. The distance at the periphery counts as nothing (like the vertical axis of rectangular or straight trees).

27.4 METHODS FOR CONSTRUCTING EVOLUTIONARY TREES FROM INFERENCES

There are two broad methods of phylogenetic tree construction: distance-based and character-based methods.

27.4.1 Distance-based methods

A distance matrix containing the pairwise distances between the input sequences is first generated through multiple sequence alignment (MSA). The number of substitutions of residues (spanned throughout the length) between each pair of multiple molecular sequences is calculated and is then converted into a single value (for each pair), using a suitable model. Examples of distance-based phylogenetic algorithms are UPGMA, Neighbor-joining (NJ), and Fitch–Margoliash. An appropriate evolutionary model is selected, based on the underlying evolutionary process in distance-based methods. Examples of such evolutionary models are: JC69 (Jukes and Cantor, 1969),

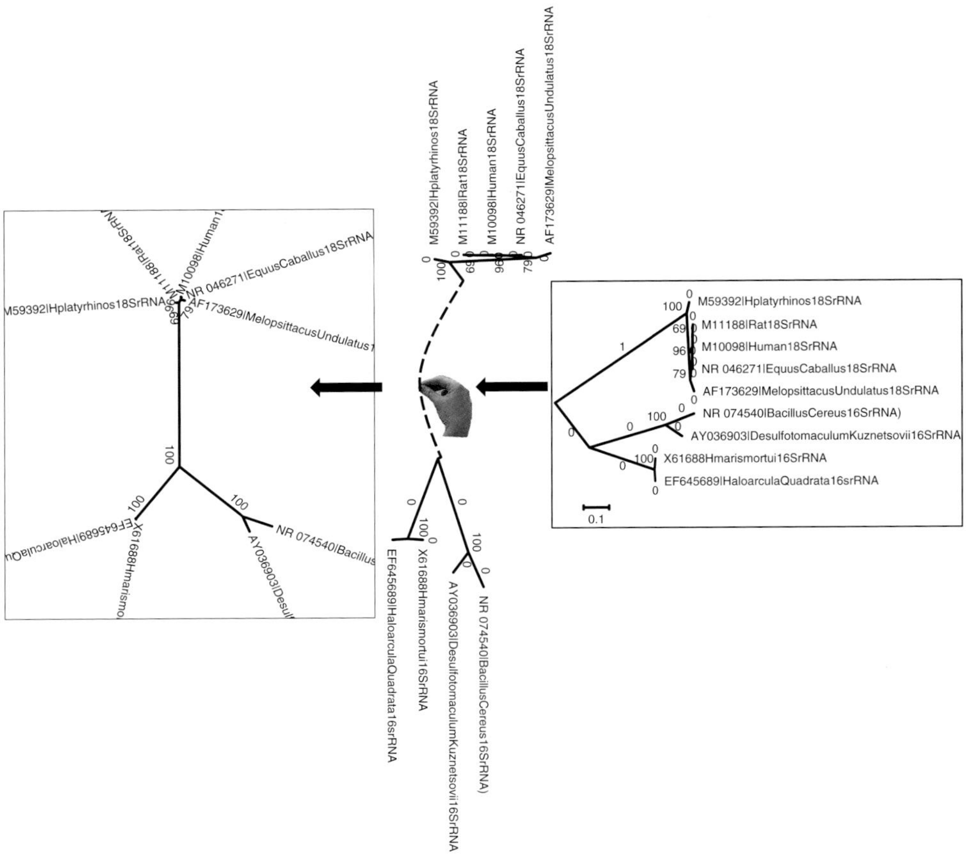

FIGURE 27.4 Converting a straight tree to a radiation tree by eliminating the depiction of divergence from common ancestor.

K80 (Kimura, 1980), F81 (Felsenstein, 1981), HKY85 (Hasegawa *et al.*, 1985), T92 (Tamura, 1992), TN93 (Tamura and Nei, 1993), and GTR (generalized time-reversible; Tavaré, 1986).

The evolutionary model is required to calculate the number of substitution, based on certain assumptions. Thus, selection of the evolutionary model is as critical as the selection of the appropriate phylogenetic algorithm. The later depends on the sequence type (amino acid or RNA or Coding DNA or non-coding DNA or intergenic DNA), sequence divergence, sequence length, and so on.

27.4.2 Character-based methods

Individual residues of the sequences are taken into account to construct the tree. Here, instead of calculating the distances among the taxa, the sequences are aligned, to find out the similarity and dissimilarity among characters in each of the columns of aligned sequences. The total number of different residues (over the length) is not calculated but, rather, some particular state (or location) of the aligned residues is identified to define the evolution of the sequences. Examples of a character-based method are maximum parsimony, maximum likelihood, and Bayesian inference. Maximum likelihood utilizes both approaches (distance- and character-based).

Again, phylogenetic trees can be constructed by any one of the following two methods:

27.4.3 Cladistic methods

This is a method that discovers the evolutionary relationship among taxa through intermediate, as well as common, ancestry. This approach yields a cladogram; for example, maximum parsimony.

27.4.4 Phenetic method

This studies the degree of similarity among a group of organisms to unveil the relationship through a tree-like network (called a phenogram), e.g. UPGMA, maximum likelihood method. The rate of divergence is assumed to be uniform among the taxa.

Table 27.1

		Types of data	
Tree construction methods		Distance	Character
	Clustering algorithm	UPGMA, NJ	
	Optimality criterion	ME	MP, ML

> Maximum parsimony adopts a strategy to search the tree with minimum number of mutations. Different methods are available for doing this: Subtree-Pruning-Regrafting (SPR), Heuristic search, Tree-Bisection-Reconnection (TBR) and Branch and Bound method (available in MEGA7 software). An heuristic search performs branch swapping through step-wise addition, followed by rearrangement of the OTUs to find the tree with minimum mutation.

27.5 INFERRING PHYLOGENETIC TREES

Now we will compare the outputs of different molecular phylogeny methods, using a set of nucleotide sequences (18s rRNA) belonging to nine organisms representing distant taxa.

At the outset, the best model (i.e., TN93 + G) was selected, based on the least Bayesian information criterion (BIC) score (which was 10145.019). Parameters selected for each of the methods have been specified along with the tree in Figure 27.5.

Parametric details for each of the algorithms used in constructing the phylogenetic trees are as follows:

- **MP Tree**: (parameters for tree construction are: subtree-pruning-regrafting (SPR); number of initial trees: 10; MP Search Level:1; no. of trees to retain: 100; bootstrap: 500 replicates)

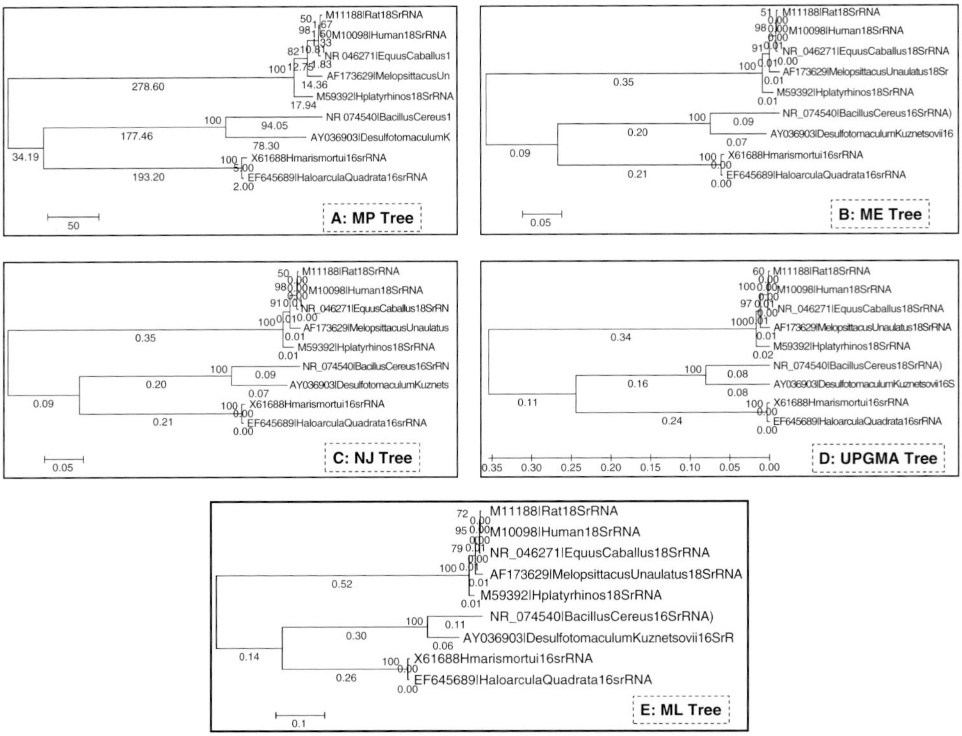

FIGURE 27.5 Phylogenetic trees constructed from nine sequences of 18s rRNA gene belonging to divergent species using various algorithms.

- **ME Tree:** (parameters for tree construction are: 'Tajima-Nei + Gamma' model; gamma distribution to determine rates among sites; gamma parameter: 5; same pattern of lineages; bootstrap: 500 replicates)
- **NJ Tree:** (parameters for tree construction are: 'Tajima-Nei + Gamma' Model; gamma Distribution to determine rates among sites; gamma parameter: 5; same pattern of lineages; bootstrap: 500 replicates)
- **UPGMA Tree:** (parameters for tree construction are: 'Tajima-Nei + Gamma' Model; gamma distribution to determine rates among sites; gamma parameter 5; same pattern of lineages; bootstrap: 500 replicates)
- **ML Tree:** Tajima-Nei Model; (Parameters for tree construction are: 'Tajima-Nei + Gamma' Model; gamma distribution to determine rates among sites; Gamma parameter: 5; ML heuristic method: nearest-neighbor-interchange (NNI); initial number of default tree: make initial tree automatically (Default – NJ/BioNJ); bootstrap: 500 replicates)

TABLE 27.2 Comparison between the features of the trees generated from the following important phylogenetic algorithms (Desper and Gascuel, 2005).

SN	Tree	Characteristic features
1	Maximum parsimony	• Scale-bar does not correspond to genetic distance but, rather, counting of substitution is done for sequence-pairs. • Consensus tree shows the agreement of branching based on bootstrap values. • The minimum number changes needed to explain the sequence data is available from the analysis. This is a character-based approach (not distance-based). Hence, those particular sites (or columns of MSA) of the aligned sequences are identified that reveal maximum information about the evolution. Such most-informative sites are only utilized to yield the phylogenetic tree. The scale bar indicates nucleotide substitution per site (value in decimal), or nucleotide substitution per unit length is given as the scale bar (value more than one).
2	UPGMA	• The scale bar at the bottom indicates the evolutionary distance, which is additive in nature. In the given example, the unit of the scale bar is 0.05, which means 95% homology and 5% divergence. Please note that an equal and constant evolution rate has been assumed for all the branches. Hence, an unrealistic assumption with UPGMA will yield an erroneous branch length and the wrong tree. • Being ultrametric, it is assumed that molecular clock and all terminal taxa are equally distant from the root. • The distance matrix shown below indicates the divergence between all pairs of sequences. The distance between the 18srRNA sequences of rat and horse (1st and 2nd sequences, respectively) is 0.004, which indicates that four bases out of every 1000 bases are different. Hence, these two sequences have 99.6% (i.e., 100–0.4) identity. • Sometimes the distance is also given as a raw number of difference in the number of residues between two sequences, and then the scale is an absolute value instead of a percentage.
3	NJ	• NJ is also a distance-based method, with additive nature of branch lengths. • The additive nature of a tree is characterized by the feature that the sum of each branches connecting two taxa makes up the distance between those two taxa. • NJ is fit for non-ultrametric distance data, where the additivity property is restored (and NJ tree becomes equivalent to ME tree). • NJ is based on clustering, and the molecular clock is not assumed (i.e., mutation or substitution rates are different for different OTUs). • The distances have been indicated in each arm, and the scale below is the indicator of the distance.
4	ME	• A distance-based method that ensures that the minimum total length of its branches indicates a minimum number of evolutionary events. • However, the tree with minimum total branch length is not necessarily the true tree, especially for short input sequences. • ME is comparable with MP, but the difference is that ME is inferred from a genetic distance, while MP is based on counting individual base substitutions over the tree.
5	ML	• This approach is more robust and utilizes information from a distance between sequences and the character information. • The distance scale indicates evolutionary distances between sequence pairs. However, the scale is not comparable to time-scale. • A scale bar of 0.1 indicates 0.1 substitutions per nucleotide.

TABLE 27.3 Pairwise distances (calculated by maximum composite likelihood model, using MEGA7) between the input sequences are shown in the lower triangular matrix.

	M11188	NR_046271	M59392	AF173629	M10098	NR_074540	X61688	EF645689	AY036903
M11188		0.002	0.009	0.007	0.002	1.721	1.463	1.472	1.706
NR_046271	0.004		0.009	0.007	0.002	1.723	1.464	1.473	1.708
M59392	0.035	0.034		0.010	0.009	1.699	1.461	1.470	1.717
AF173629	0.026	0.025	0.039		0.007	1.690	1.487	1.496	1.689
M10098	0.003	0.003	0.034	0.026		1.721	1.462	1.472	1.706
NR_074540	2.582	2.593	2.551	2.571	2.584		1.201	1.196	0.912
X61688	2.250	2.261	2.239	2.285	2.253	1.868		0.003	1.111
EF645689	2.266	2.276	2.253	2.299	2.268	1.859	0.006		1.112
AY036903	2.546	2.557	2.556	2.553	2.548	0.249	1.716	1.723	

27.6 QUESTIONS

1. How do you differentiate between distance-based and character-based methods of phylogeny?
2. What are the differences between a cladogram and phenogram? Under what circumstances will you use these terms?
3. What is the meaning of the scale bar given in a phylogenetic tree that has been constructed using the following methods?
   i. UPGMA
   ii. Maximum parsimony
   iii. Minimum evolution
4. Is there any difference between a circular tree and a vertical tree, so far as the meaning conveyed by the depictions is concerned?
5. Why do we first select the best evolutionary model before constructing a phylogenetic tree?

Protein Structure Prediction

SECTION VI

Prediction of Secondary Structure of Protein

CHAPTER 28

CS Mukhopadhyay and RK Choudhary
School of Animal Biotechnology, GADVASU, Ludhiana

28.1 INTRODUCTION

Secondary structures of a protein (e.g., α-helix, β-sheets, loops, and coils) are produced due to the patterns of hydrogen bonds between amino and carboxyl groups of the adjacent amino acids of the peptide backbone. The prediction of the secondary structure of proteins consists of a set of biocomputational approaches directed towards assigning the regions of local folding of protein (i.e., the secondary structure), based on the amino acid sequence (i.e., the primary structure).

28.2 OBJECTIVE

To predict the secondary structure of a given peptide molecule, using an online secondary structure prediction tool.

28.3 SECONDARY STRUCTURE PREDICTION USING ONLINE TOOL PSIPRED

28.3.1 Procedure

28.3.1.1 Download the amino acid sequence

Download a peptide sequence (e.g., taurine Keratin: NCBI protein Acc. No. Q148H4.1) from NCBI-Protein (http://www.ncbi.nlm.nih.gov/protein/) or UniProt (http://www.uniprot.org/uniprot/) database, in FASTA format.

> Q148H4.1|BovineKeratin
M T C G S G F R G R A F S C V S A C G P R P G R C C I T A A P Y R G I S C Y
R G L T G G F G S R S I C G G F R A G S F G R S F G Y R S G G V G G L N P P C
I T T V S V N E S L L T P L N L E I D P N A Q C V K Q E E K E Q I K C L N N R
F A A F I D K V R F L E Q Q N K L L E T K L Q F Y Q N R Q C C E S N L E P L F

Basic Applied Bioinformatics, First Edition. Chandra Sekhar Mukhopadhyay, Ratan Kumar Choudhary and Mir Asif Iquebal.

N G Y I E T L R R E A E C V E A D S G R L S S E L N S L Q E V L E G Y K K K Y
E E E V A L R A T A E N E F V A L K K D V D C A Y L R K S D L E A N V E A L I
Q E I D F L R R L Y E E E I R V L Q A H I S D T S V I V K M D N S R D L N M D
N I V A E I K A Q Y D D I A S R S R A E A E S W Y R S K C E E I K A T V I R H
G E T L R R T K E E I N E L N R V I Q R L T A E V E N A K C Q N S K L E A A V
T Q A E Q Q G E A A L N D A K C K L A G L E E A L Q K A K Q D M A C L L K E Y
Q E V M N S K L G L D I E I A T Y R R L L E G E E Q R L C E G V G S V N V C V
S S S R G G V V C G D L C V S G S R P V T G S V C S A P C S G N L A V S T G L
C A P C G P C N S V T S C G L G G I S S C G V G S C A S V C R K C

Secondary structures do not specify the atomic positions in three-dimensional space:

A. α-helices:

i. prevalent at the protein surface to provide an interface with the aqueous environment via free NH_2 groups
ii. The inner-facing side of the helix tends to have hydrophobic amino acids, and the outer-facing side hydrophilic amino acids: thus every third of four amino acid along the chain tends to be hydrophobic
iii. Alpha-helix regions:
    a. rich in alanine (A), glutamic acid (E), leucine (L), and methionine (M)
    b. Poor in proline (P), glycine (G), tyrosine (Y), and serine (S)
iv. Proline destabilizes or breaks α-helix, but can be present in longer helices, forming a bend.

B. β-sheets:

i. Hydrogen (H) bonding 5–10 consecutive amino acids in one portion of the peptide chain, with another 5–10 amino acids located downward (nearby or somewhat distant) along the chain
ii. Prediction of β-sheets is more challenging than prediction of α-helices:
    a. **Loops** are the regions located between α-helices and β-sheets, located on the surface of the structure. Thus, they interact with the surrounding aqueous environment and other proteins. Loops tend to have charged and polar amino acids.
    b. **Coils**: any region of a secondary structure that is not an α-helix, a β-sheet, or a recognizable turn is commonly referred to as a coil.

28.3.1.2 Open PSIPRED

Open PSIPRED using the URL http://bioinf.cs.ucl.ac.uk/psipred/:

a. Paste the amino acid sequence in the sequence box. A short sequence identifier may be entered.
b. Checkboxes offer options for multiple analyses in one go through Profile Based Fold Recognition (pGenTHREADER), Rapid Fold Recognition (GenTHREADER), etc.
c. Click on the "Predict" button.
d. Provide your email ID in the specified box to get the results mailed to your inbox. In general, analysis takes several hours to complete.

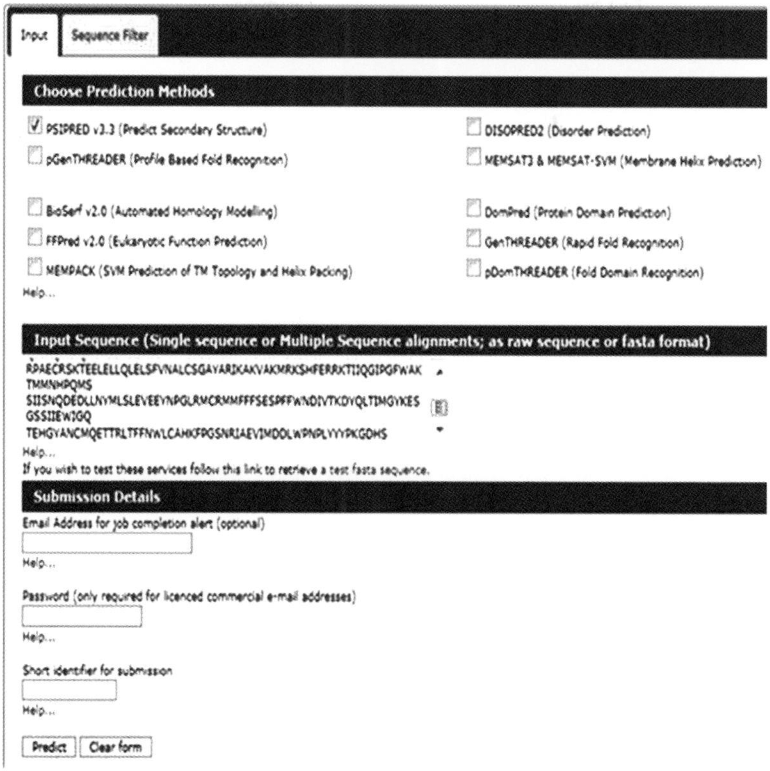

FIGURE 28.1 Graphical user interface (GUI) of PSIPRED and filling the inputs in the Input tab.

28.3.2 Output

The output is available in three tabs on the same screen:

- **Summary**: provides an easy-to-understand presentation of the secondary structure throughout the peptide sequence.
- **PSIPRED**: thumbnails hyperlinked with the detailed result. Click on the thumbnail and the output is opened in a new window, which can be downloaded as a figure in.png format.
- **Downloads**: The whole output or a part of it can be downloaded in different formats, viz.
  - the whole output in a zipped file,
  - only the results as plain text,
  - the raw scores in plain text format,
  - the results in Postscript or PDF format.

28.3.3 Interpretation of results

The output shows the type of secondary structure for each of the amino acids, using the following notations. Please note the proportions of the different types of secondary structure from the result.

The detailed result diagrammatically represents the different types of secondary structure predicted. Each of the blocks contains the following rows:

- **Confidence (Conf)**: The blue bars indicate the confidence (or probability of accuracy of prediction) for each of the amino acids. The taller the bar, the more confident we can be about the predicted secondary structure for that residue.
- **Prediction represented by the cartoon (Pred)**: The black, thin straight line represents coils (C), the yellow-colored arrow represents beta sheet (E), and the pink-colored barrel or pipe indicates helix (H).
- **Prediction coded by alphabetic notations (Pred)**: "H" for helix, "C" for coil, and "E" for beta sheet.
- **Amino acid sequence (AA)**: Each of the amino acid residues is placed in sequence.

Finally, the amino acid count is given as multiples of 10 (viz. 110, 20, 30, ….).

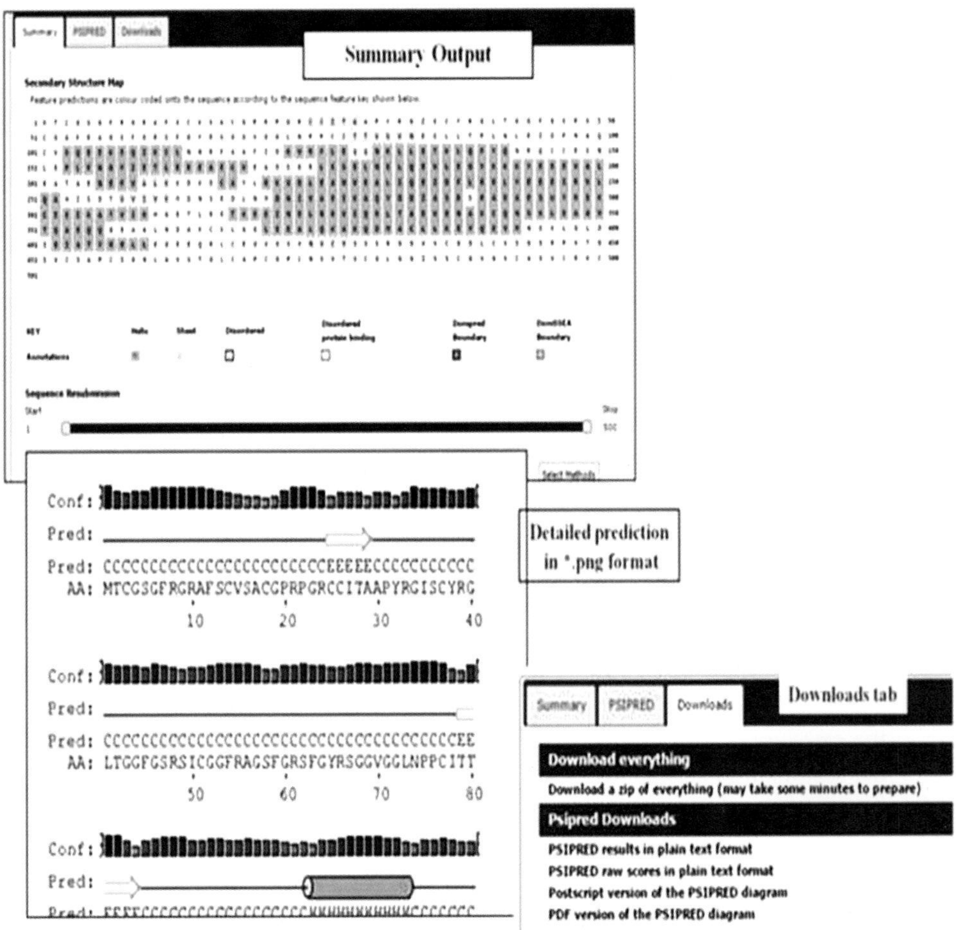

FIGURE 28.2 The output tabs of PSIPRED shown in three sections.

28.4 SECONDARY STRUCTURE PREDICTION USING THE ONLINE CDM TOOL

The CDM tool has evolved from GOR (Garnier, Osguthorpe, and Robson) methods of secondary protein structure prediction.

28.4.1 Procedure

a. *Download an amino acid sequence*: Download a peptide sequence in FASTA format.
b. *Open URL*: http://gor.bb.iastate.edu/
c. Open your workspace and enter your working mail ID.
d. *Input sequence*: One can either enter the PDB ID or paste the amino acid sequence (maximum 1000 amino acids) into the specified sequence box. Please ensure that the sequence is in raw sequence format (not in FASTA format, and without any other alphabets or symbols, other than single-letter alphabetic notations of 20 amino acids).
e. Click on "Submit".
f. The result will be sent to the given email ID.

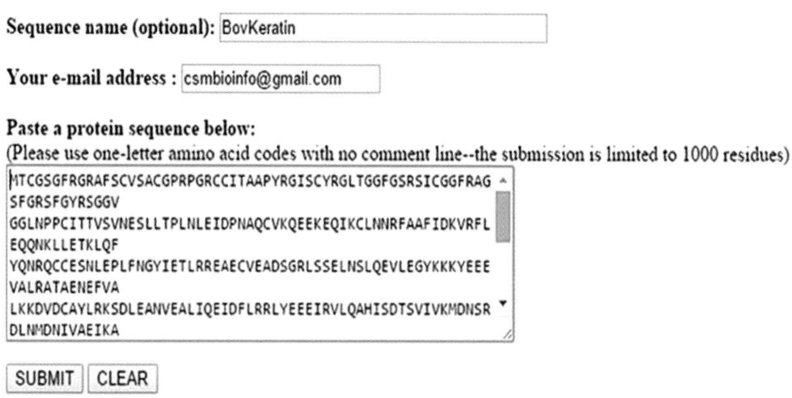

FIGURE 28.3 GUI of the online CDM tool for prediction of protein secondary structures.

28.5 QUESTIONS

1. Compare the secondary structures of caprine beta-defensin (GenBank: ABF71365.1) versus bubaline lingual antimicrobial peptide (ABE66309.1) using the PsiPred tool.
2. The NCBI Protein Accession number for bubaline Dicer peptide is given as BAP00765.1. Predict the secondary structure of the following domains: PAZ, RIBOc, DSRM and RNaseIII.
3. Compare the secondary structures of the antimicrobial domains of the following cathelicidin variants: NCBI Protein Id: AGA63736.2, XP006065246.1, NP001277882.1.
4. What do you mean by "secondary structure of protein"? What are the applications of secondary structure prediction?
5. What are the constraints in predicting beta-pleated sheets?

Prediction of Tertiary Structure of Protein: Sequence Homology

CHAPTER 29

CS Mukhopadhyay and RK Choudhary
School of Animal Biotechnology, GADVASU, Ludhiana

29.1 INTRODUCTION

Prediction of the tertiary structure of a peptide sequence is done by either comparative modeling (when the experimentally determined structure of a homologous protein is available) or *ab initio* approaches (when the structure of the homologous protein is not available).

COMPARATIVE PROTEIN MODELING

This approach uses previously reported structures as templates, as it assumes that two homologous proteins (descended from a common ancestral amino acid sequence) will share similar structures, and that the protein folds have become comparatively more conserved during the process of evolution than the respective amino acid sequence.

De novo or *ab initio* method

The principle behind this method is to predict the protein structure, based on physical principles and energy minimization of the best predicted protein structure. The salient features of all of the *ab initio* methods are:

1. Representing the target protein with regard to protein conformation space. This assumes that the conformations which minimize the energy function could be the native structures of the target protein.
2. Determining the compatibility of energy functions with the protein representation is crucial.
3. Suitable algorithms to search the conformational space to minimize the energy function are used.

29.2 OBJECTIVE

To predict the tertiary structure of a peptide using the homology modeling approach with the online tool SWISS-MODEL.

Basic Applied Bioinformatics, First Edition. Chandra Sekhar Mukhopadhyay, Ratan Kumar Choudhary and Mir Asif Iquebal.

29.3 PROCEDURE (SWISS-MODEL PROGRAM)

29.3.1 Query sequence

Let us say we are interested in determining (predicting) the structure of pancreatic ribonuclease of buffalo (*Bubalus bubalis*), which has not been experimentally determined and reported. Hence, we need to search for the reported structure from a related species (e.g., cattle), so that this can be used as a template for homology modeling.

The sequence of the 124-residue-long bubaline pancreatic ribonuclease (obtained from PDB) is:

>tr|Q95NE6|Q95NE6_BUBBU Pancreatic ribonuclease (Fragment) OS = Bubalusbubalis PE = 3 SV = 1

KETAAAKFQRQHMDSSTSSASSSNYCNQMMKSRSMTSDRCKP VNTFVHESLADVQAVCSQKNVACKNGQTNCYQSYSTMSITDC RETGSSKYPNCAYKTTQANKHIIVACEGNPYVPVHFDASV

29.3.2 Downloading the template structure

Download a reported structure of a related peptide (in a closely related species, with at least 70% sequence similarity) from a protein data bank (PDB). In the present example, we will download the .pdb file of the experimentally determined structure (X-ray diffraction, resolution of 1.95 Å) of bovine pancreatic ribonuclease (PDB ID: 4RTE).

29.3.3 Pairwise sequence alignment

Determining the percentage similarity between the sequences (template vs. query sequences of pancreatic ribonucleases):

```
CLUSTAL O(1.2.1) multiple sequence alignment


Q95NE6_Bbu_PancRNase      KETAAAKFQRQHMDSSTSSASSSNYCNQMMKSRSMTSDRCKPVNTFVHESLADVQAVCSQ  60
4RTE_Bta_PancRNase        KETAAAKFERQHMDSSTSAASSSNYCNQMMKSRNLTKDRCKPVNTFVHESLADVQAVCSQ  60
                          ********:*********:**************.:*.***********************

Q95NE6_Bbu_PancRNase      KNVACKNGQTNCYQSYSTMSITDCRETGSSKYPNCAYKTTQANKHIIVACEGNPYVPVHF  120
4RTE_Bta_PancRNase        KNVACKNGQTNCYQSYSTMSITDCRETGSSKYPNCAYKTTQANKHIIVACEGNPYVPVHF  120
                          ************************************************************

Q95NE6_Bbu_PancRNase      DASV  124
4RTE_Bta_PancRNase        DASV  124
                          ****
```

FIGURE 29.1 Pairwise sequence alignment to determine the extent of sequence identity between the query and template sequences.

The alignment result (percent identity matrix) clearly shows that there is a 95.97% match between the two sequences. Hence, the template of the target protein from bovine origin is fit to be used for homology modeling of the structure of bubaline pancreatic RNase.

29.3.4 Open SWISS-MODEL workspace

The URL is http://swissmodel.expasy.org/workspace/index.php?func=show_workspace.

The first-time user needs to create a new workspace and then start using the tool. A returning user needs to open the existing "workspace", and then enter the email ID and project title.

29.3.5 Working mode

Click on "Modelling" at the top of the page, then select "Automated Mode" from the drop-down menu.

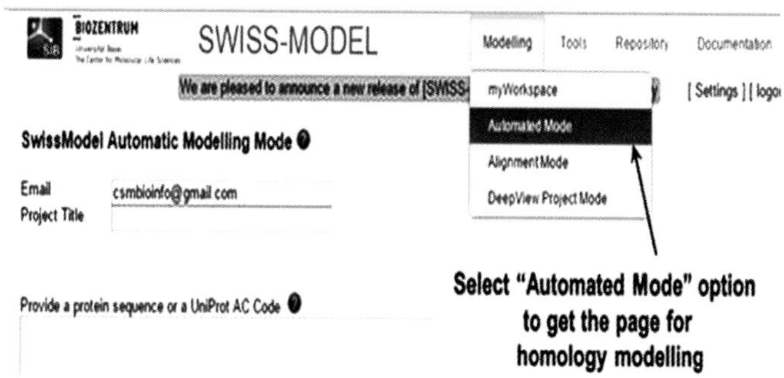

FIGURE 29.2 Open the page to initiate homology modeling using the SWISS-MODEL workspace

29.3.6 Input sequence

Paste the amino acid sequence (up to 1000 residues) into the sequence box, or enter the PDB ID of the peptide in the specified box. Here we will upload the template file 4RTE.pdb.

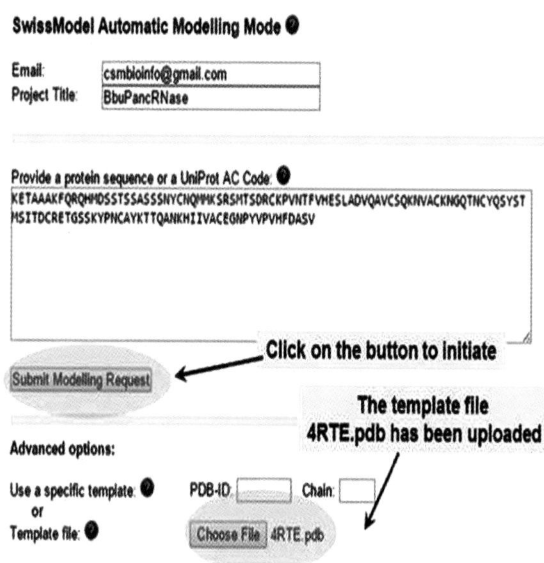

FIGURE 29.3 Window of SWISS-MODEL workspace for providing the input parameters and starting homology modeling.

29.3.7 Job Submit

Click on "Submit Modeling Request". You will need to wait until the work is done. The result will be emailed to the specified email address. Download the predicted structure in *.pdb format.

29.4 OUTPUT

The output page is self-explanatory. The predicted model can be viewed on a Java-enabled system by clicking on the image. The output page also mentions the following outputs:

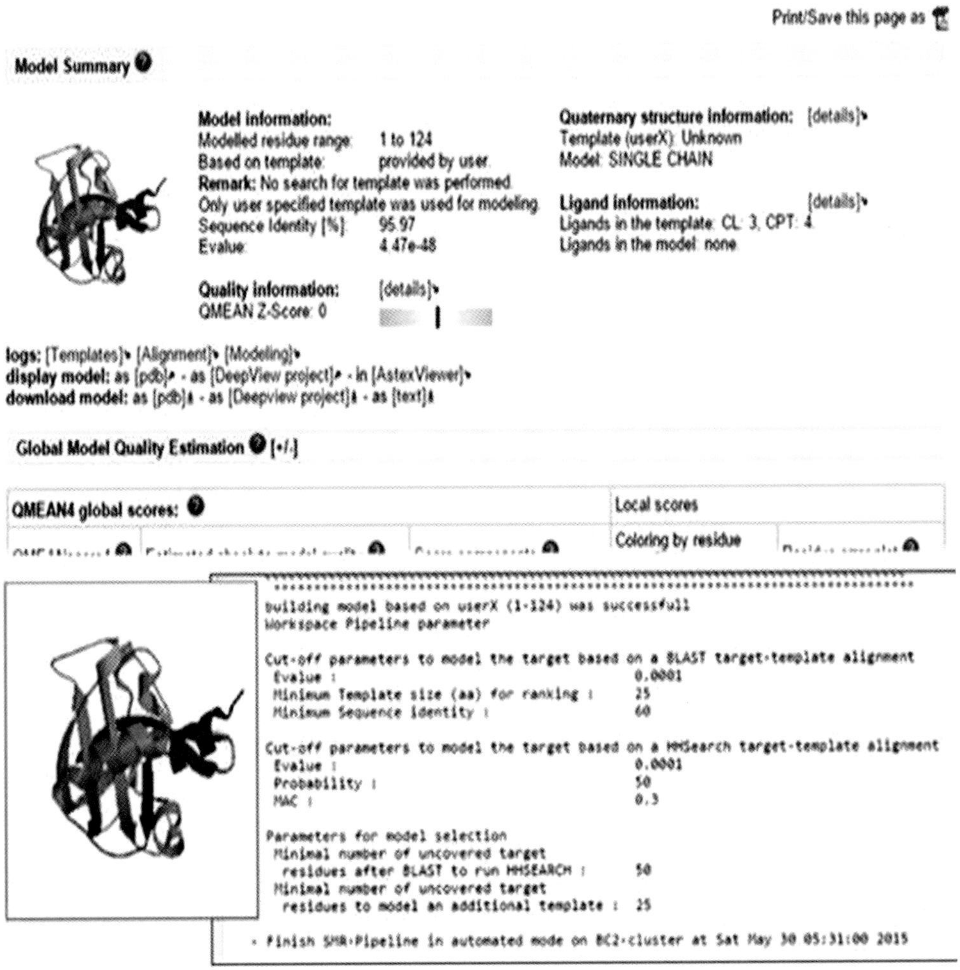

FIGURE 29.4 Important sections of the SWISS-MODEL output. One can download the complete result in PDF format, or can specifically download the sections as required.

29.4.1 Model information

This is the length of the input sequence that has been used for predicting its structure. Sequence identity (95.97%) and E-value (4.47e-48) are also given to indicate the degree of homology between the query and template sequences.

29.4.2 Quality information

QMEAN values (QMEAN4 and QMEAN6) are determined by the program to assess the overall quality of the model and the legitimacy of the bonds between amino acids in the predicted structure. The quality scores are also determined using methods like atomic empirical mean force potential (ANOLEA) and GROMOS. The detail on these terms can be obtained at the SWISS-MODEL help page: http://swissmodel.expasy.org/workspace/index.php?func=special_help&=#A

29.4.3 Alignment

The pairwise alignment between the query and template sequences is displayed, along with the secondary structure information in the last row in each alignment block.

29.4.4 Ligand information

It is useful if the ligand is included in the model during prediction. Otherwise, the program predicts the possible ligand interacting sites on the predicted structure.

29.4.5 Modelling log

This is a relevant section to get information about the steps during homology modeling. If the template model is not supporting enough, the program uses an *ab initio* approach to model the query sequence. This log also indicates which portions have been built *ab initio*. In our example, the "final total energy" of the predicted model is –7759.335 kJ/mol.

29.4.6 Template log

Information about template search and use are given here.

29.5 VISUALIZING THE PREDICTED STRUCTURE

a. The predicted structure can be visualized using a molecular visualization program such as Cn3D (http://www.ncbi.nlm.nih.gov/Structure/CN3D/cn3d.shtml) or RasMol (please refer to Chapter 27).
b. Open the downloaded pdb file (Model_1.pdb, for instance) to visualize and assess the structural features.

29.6 INTERPRETATION OF RESULTS

The results obtained should be judged for the quality of the predicted structure using the following main parameters:

- *Pairwise alignment*: If the degree of homology (indicated by percent identity and E-value) between the query and template sequence is high, the likelihood of accurate prediction of the structure is also high.
- *Global model quality assessment*: Check parameters such as QMEAN4 or QMEAN6. The packing quality of the model, minimum energy represented as favorable energy environment for each of the amino acid (green colored graph), is shown in the graph of ANOLEA. Lower energy indicates stability of the predicted structure.
- *Parameters given under template selection log*: Very useful for judging the cut-off parameters set for BLAST and HHSearch during modeling.

Finally, after obtaining the model, the predicted structure must be validated using suitable tools (see Chapter 32) such as ProCheck, WhatCheck, What If, and so on. The correctness of the dihedral angle of rotation (peptide torsion angles Phi(ϕ) and Psi(ψ)) of the amino acid residues can be checked using Ramachandran's Plot (RaptorX or any suitable tool).

29.7 QUESTIONS

1. What are the differences between the homology modeling based on *ab initio* approaches of tertiary structure prediction methods for tertiary structure prediction of protein?
2. Predict the tertiary structure of the same peptide (bubaline keratin) by homology modeling using a human homolog.
3. Given the following amino acid sequence, can you determine its structure through homology modeling, using a suitable homologous model?
   >Query
   MALKSLVLLSLLVLVLLLVRVQPSLGKETAAAKFERQH
   MDSSTSAASSSNYCNQMMKSRNLTKDRCKPVNTFVHESLADV
   QAVCSQKNVACKNGQTNCYQSYSTMSITDCRETGSSKYPNCA
   YKTTQANKHIIVACEGNPYVPVHFDASV
4. What are the parameters for evaluating the tertiary structure obtained through homology modeling? Evaluate the tertiary structure obtained in the previous question.
5. Enlist the factors that determine which template is best suited for homology modeling.

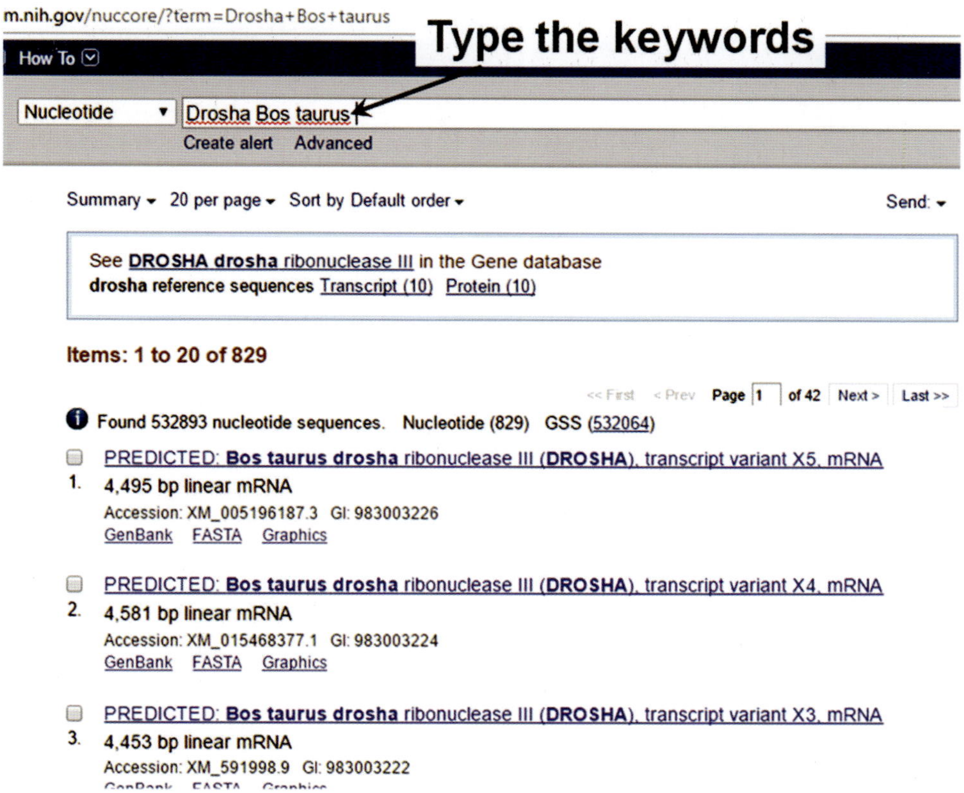

FIGURE 1.1 Main search window of NCBI Nucleotide page and list of hits for nucleotide sequences of taurine *Drosha* (gene/mRNA).

Basic Applied Bioinformatics, First Edition. Chandra Sekhar Mukhopadhyay, Ratan Kumar Choudhary and Mir Asif Iquebal.

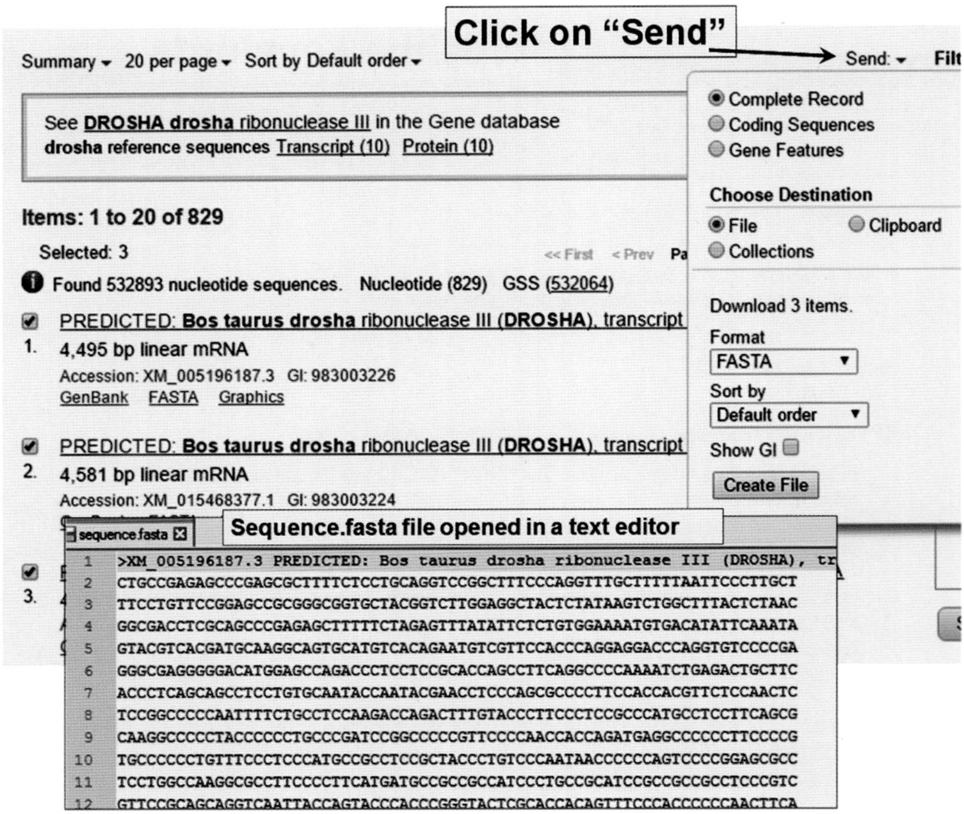

FIGURE 1.2 Click on the "Send to" button to download and save (in a text file) the first three Drosha mRNA sequences in "Summary" format.

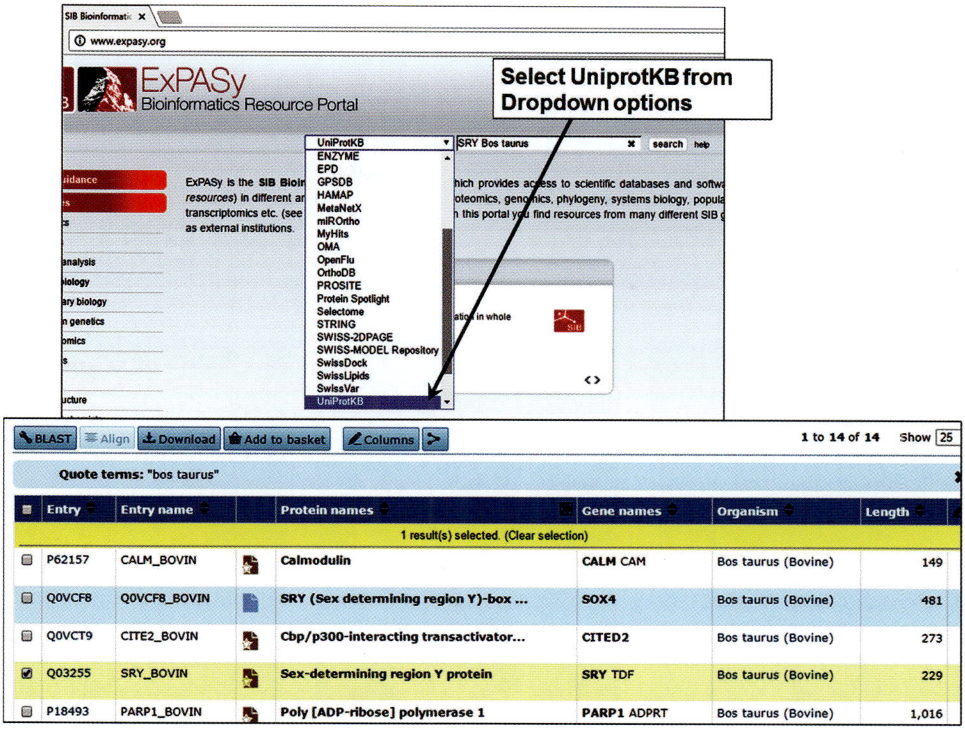

FIGURE 2.2 Click on the specific entry to open it in a separate window.

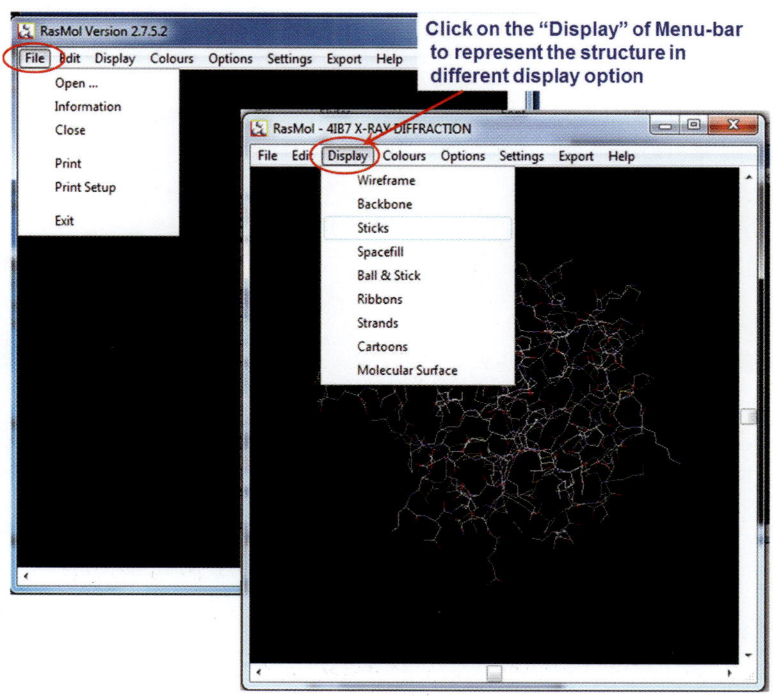

FIGURE 4.1 Graphical user interface (GUI) of RasMol and the drop-down menu to open, modify or alter the display of the peptide.

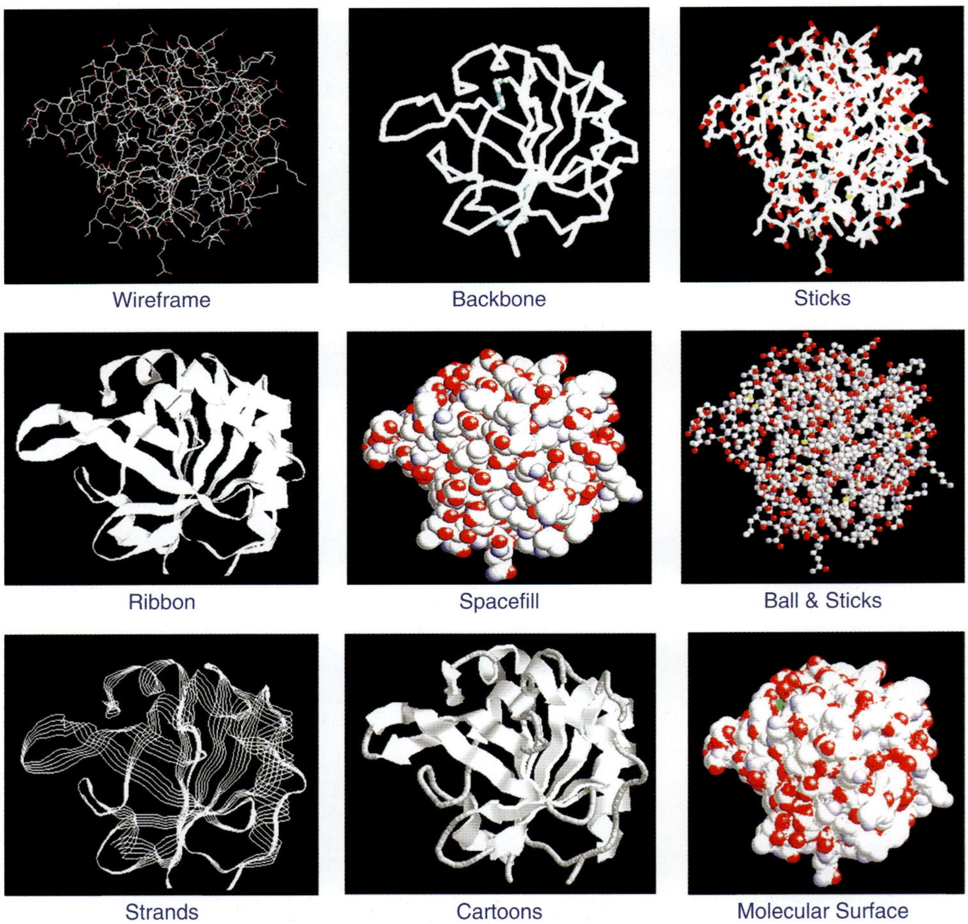

FIGURE 4.2 A single peptide, displayed in 'Wireframe', 'Backbone', 'Sticks', 'Spacefill', 'Ball and Stick', 'Ribbons', 'Strands', 'Cartoons' and 'Molecular surface' patterns.

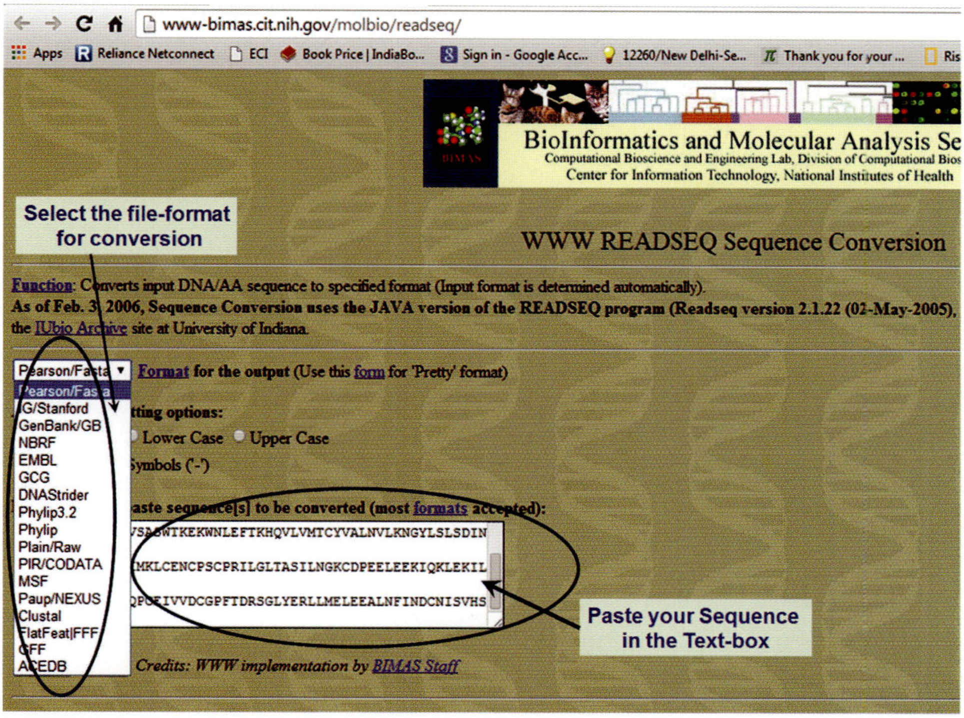

FIGURE 5.1 Homepage of the *ReadSeq* biosequence format conversion tool.

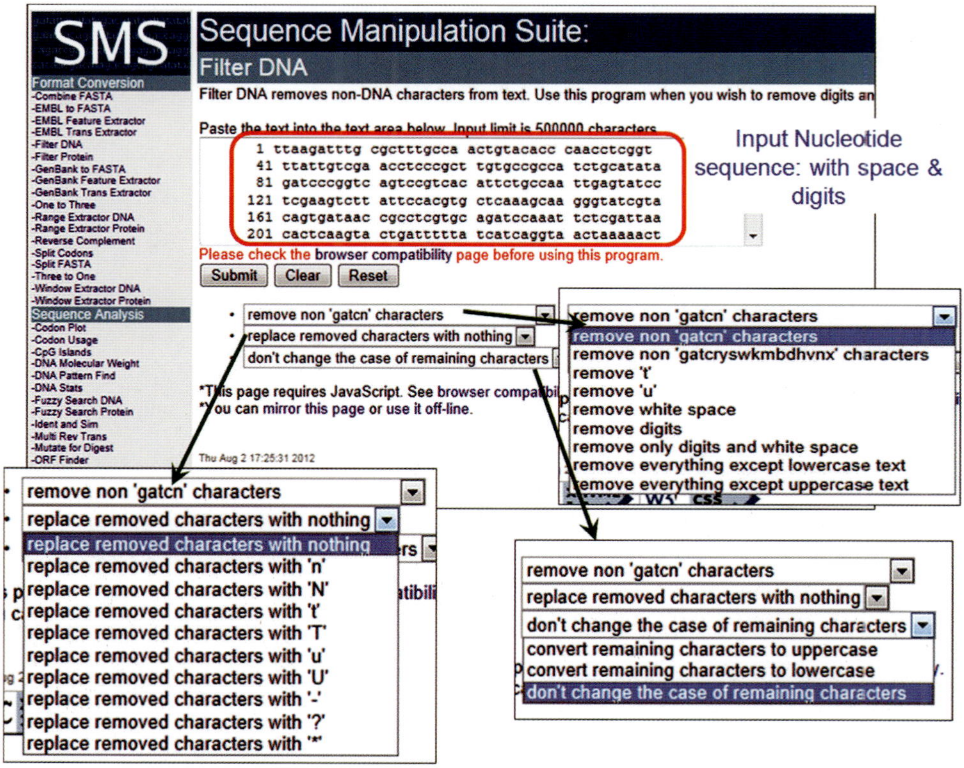

FIGURE 6.3 "Filter DNA" input page, along with various options as control parameters.

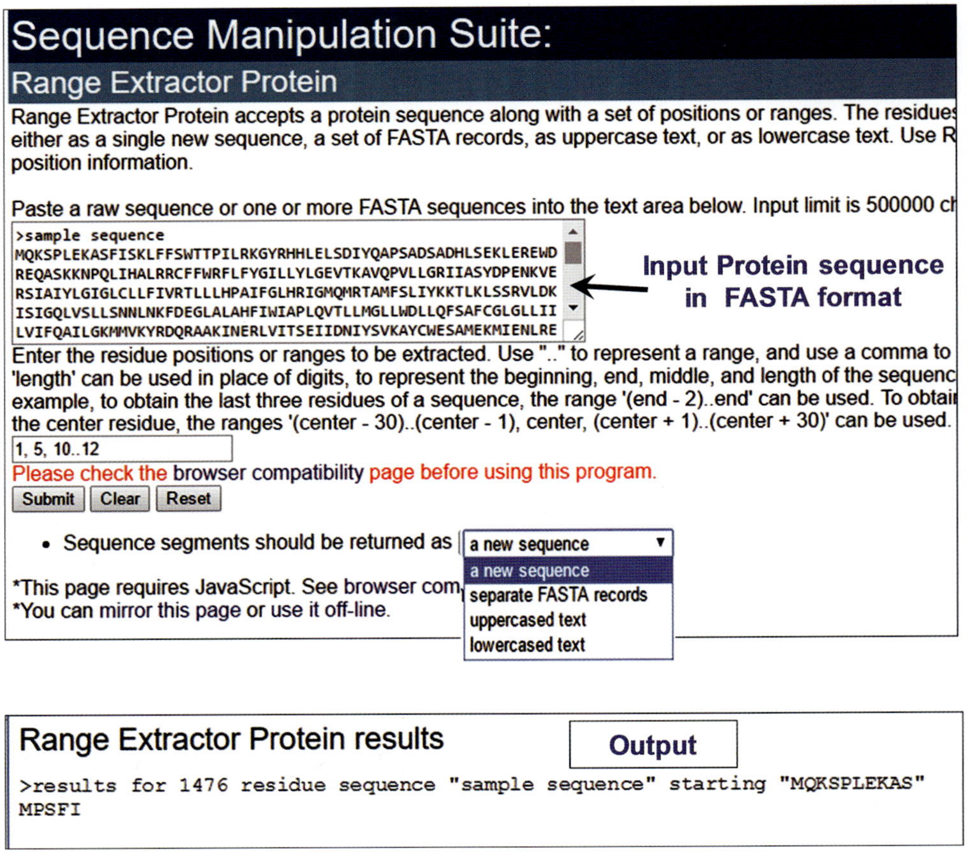

FIGURE 6.4 "Range Extractor Protein" input page and the corresponding output page with extracted sequences.

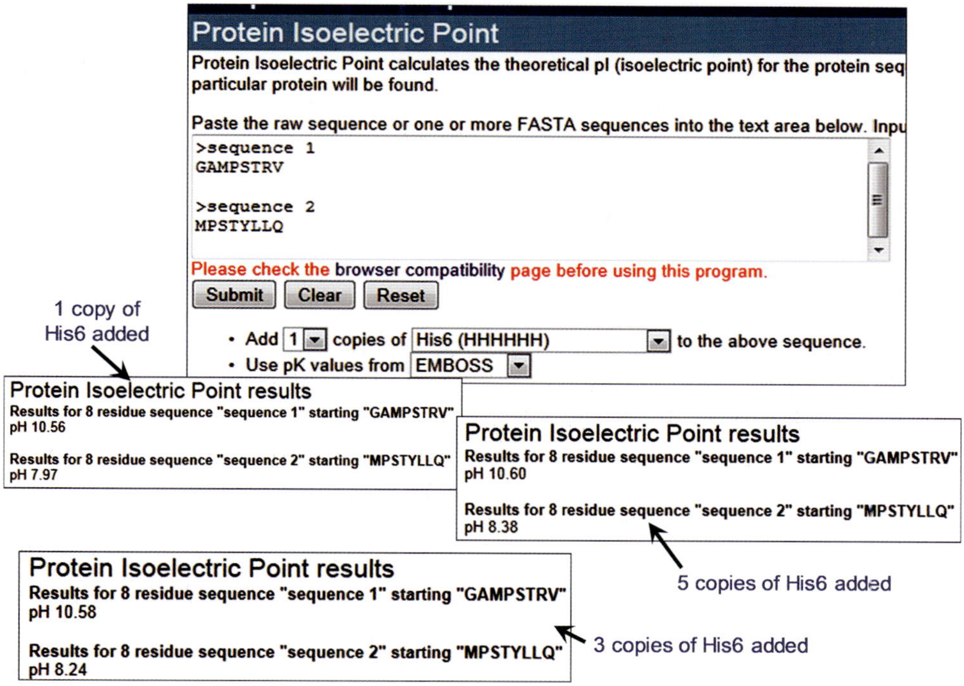

FIGURE 6.6 "Protein Isoelectric Point" input page and the corresponding output page with results, with respect to the parameters.

FIGURE 7.1 A short nucleotide sequence (oligo) can be searched in the input sequence for determining specific RE sites present in the oligos.

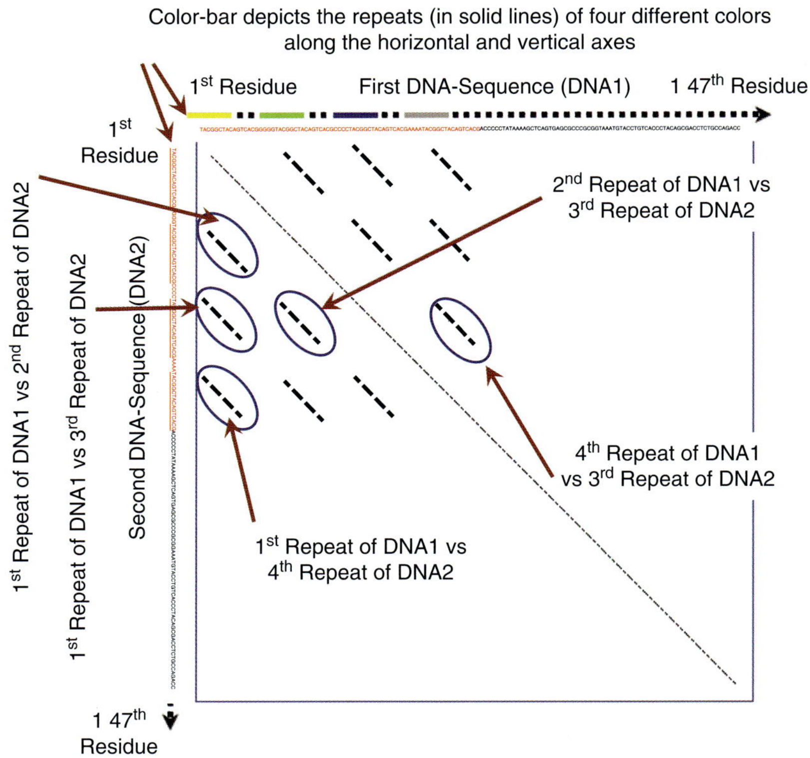

FIGURE 8.2 Interpretation of dot plot based on the same repeat sequence (shown above) which has been placed along both axes. The four different colors (yellow, green, blue and gray) have been shown to indicate the 1st, 2nd, 3rd and 4th repeat of "TACGGCTACAGTCACG".

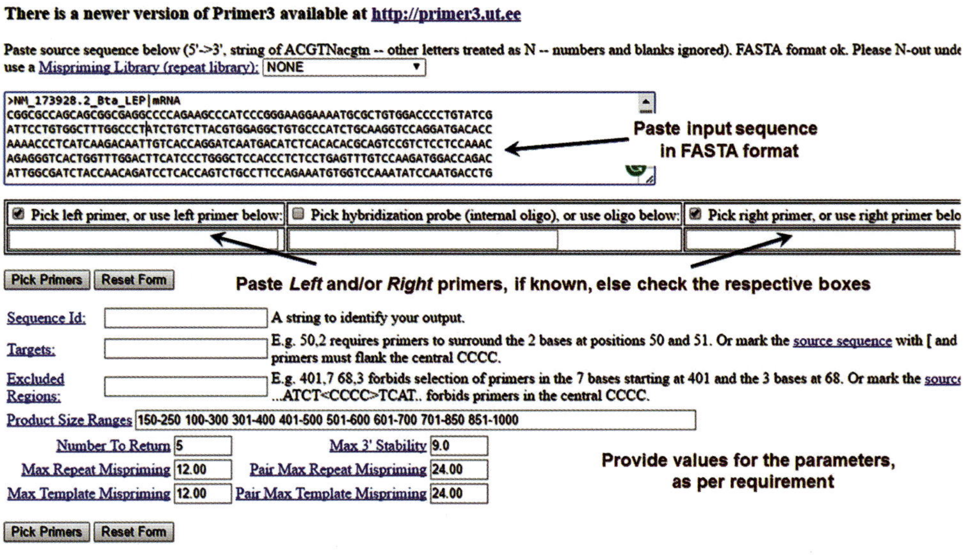

FIGURE 18.1 Setting the parameters of the Primer3 online tool for primer designing.

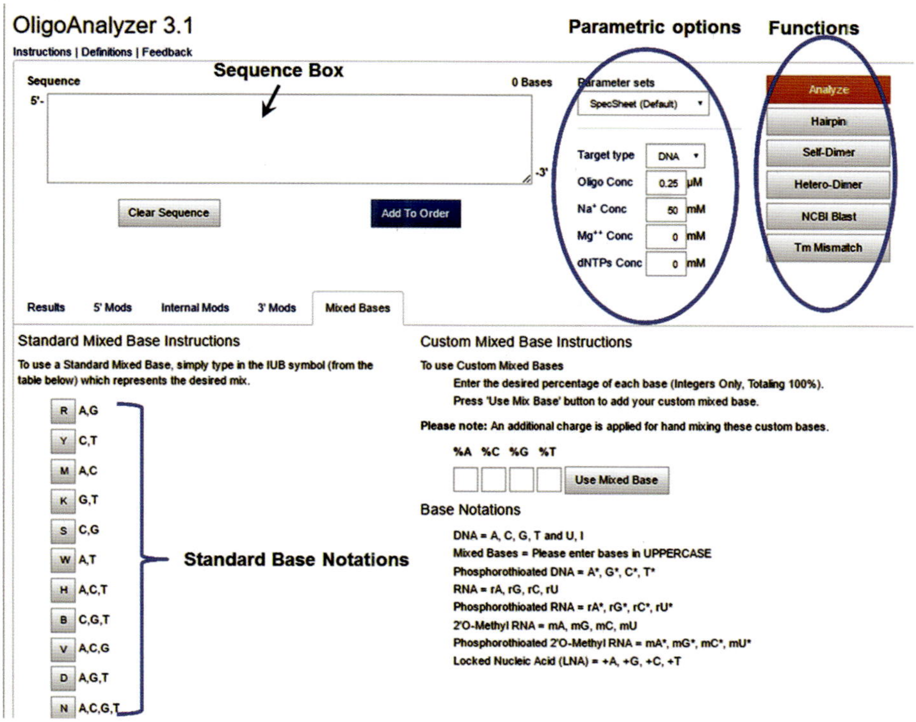

FIGURE 19.1 Homepage of Oligoanalyzer 3.1, indicating different parameters and functions for the output of the function "Analyze".

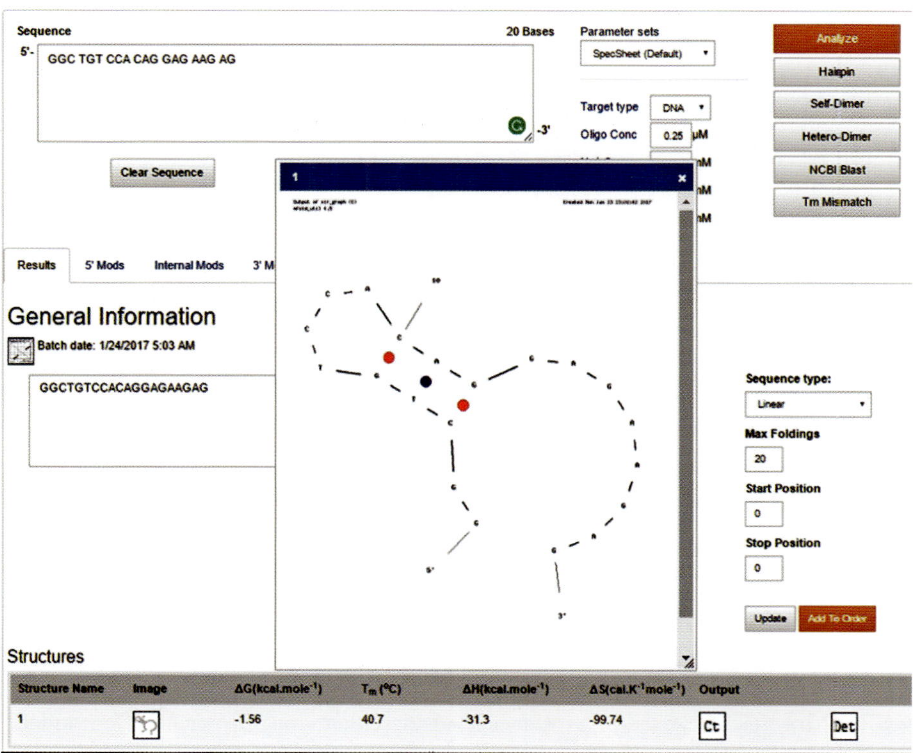

FIGURE 19.2 Output of the function "Hairpin" of the Oligoanalyzer 3.1 tool, displaying the possible hairpins and the related thermodynamic values.

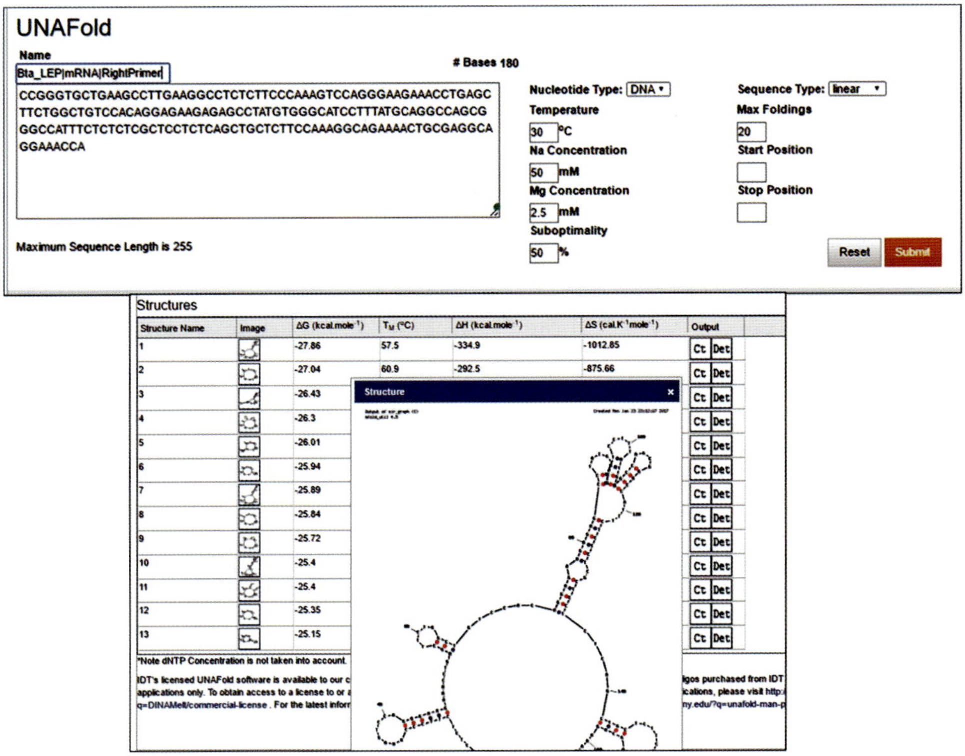

FIGURE 19.3 Prediction of secondary structure in the amplicon using the UNAFold tool of IDT.

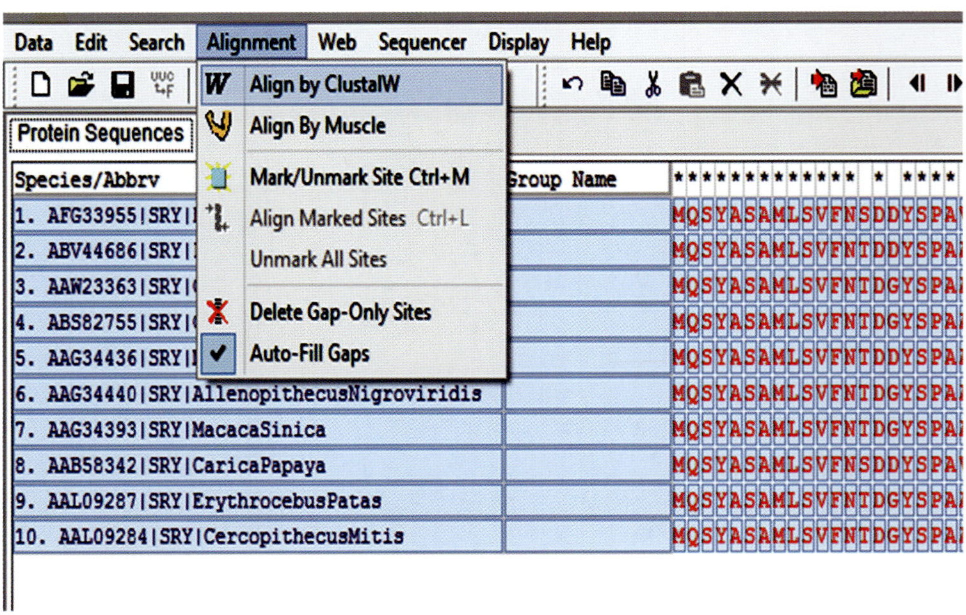

FIGURE 26.2 Aligning the input sequences using either ClustalW or Muscle available in MEGA7 interface.

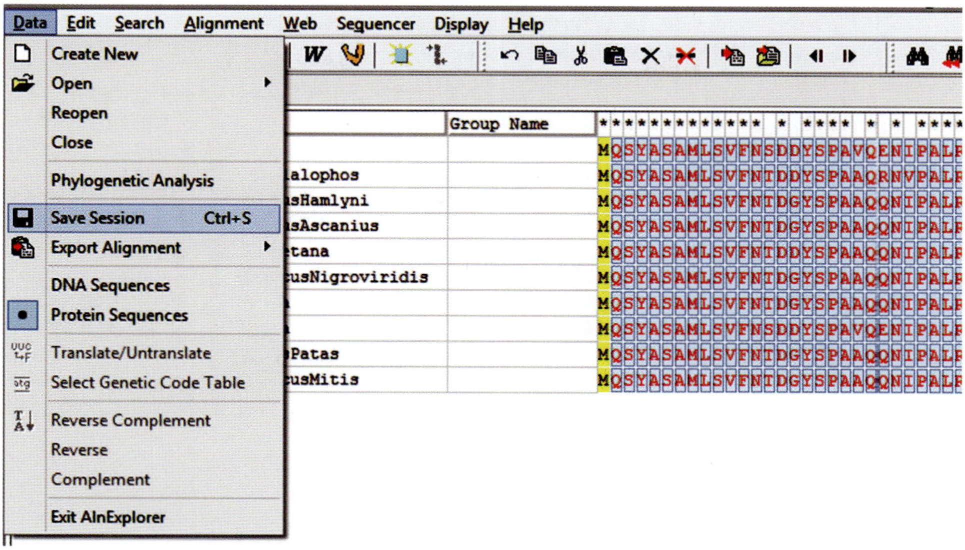

FIGURE 26.3 Exporting the alignment file and saving the alignment session for further use.

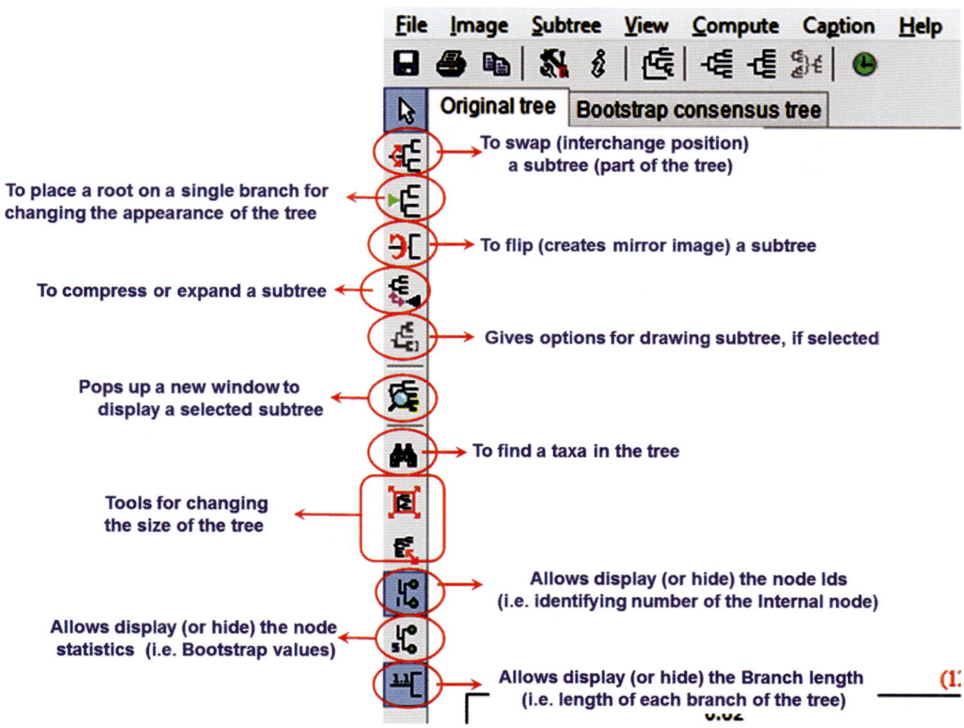

FIGURE 26.7 Controlling the tree display parameters using the left-hand-side buttons.

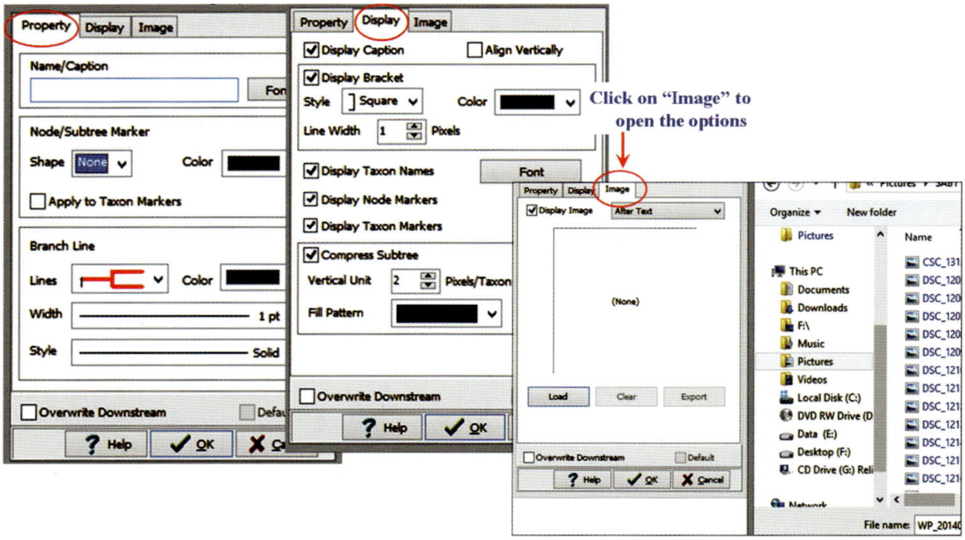

FIGURE 26.8 Insertion of figures for the external nodes (species name).

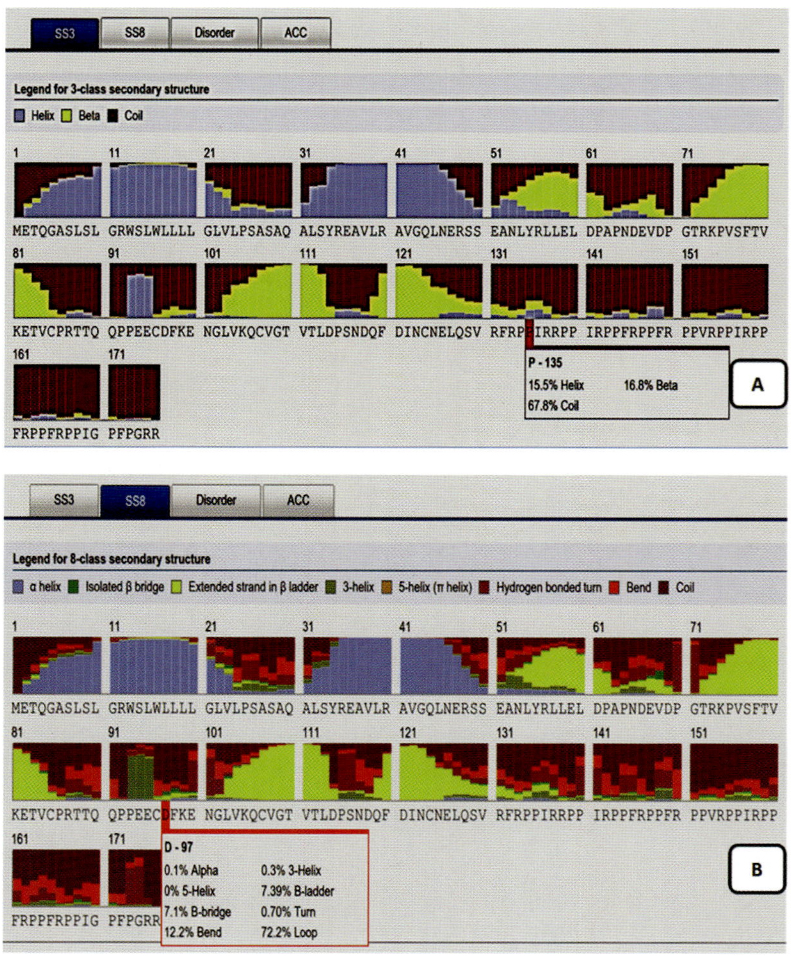

FIGURE 30.4 3 Class SS3 and 8 Class SS8 secondary structural element contribution to the 3D structure.

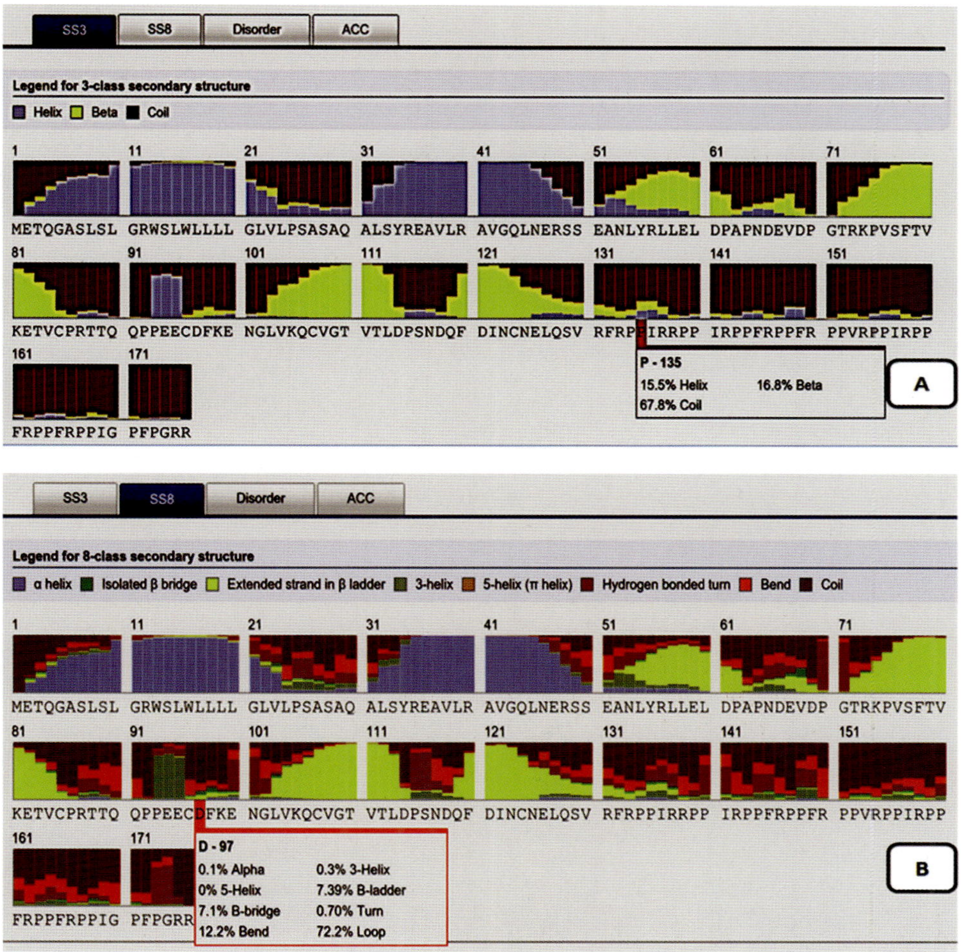

FIGURE 30.5 Conformationally ordered and disordered contribution of the residues in the 2D and 3D structure (C). Contribution of each residue in solvent accessibility (D).

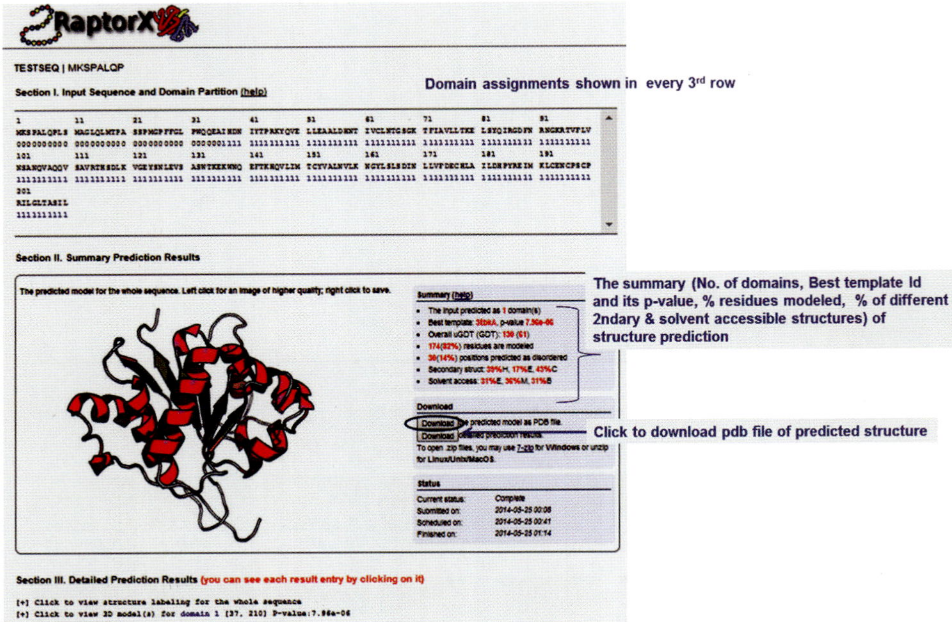

FIGURE 31.2 Results windows of RaptorX, indicating assignment of protein domain and 3D prediction results.

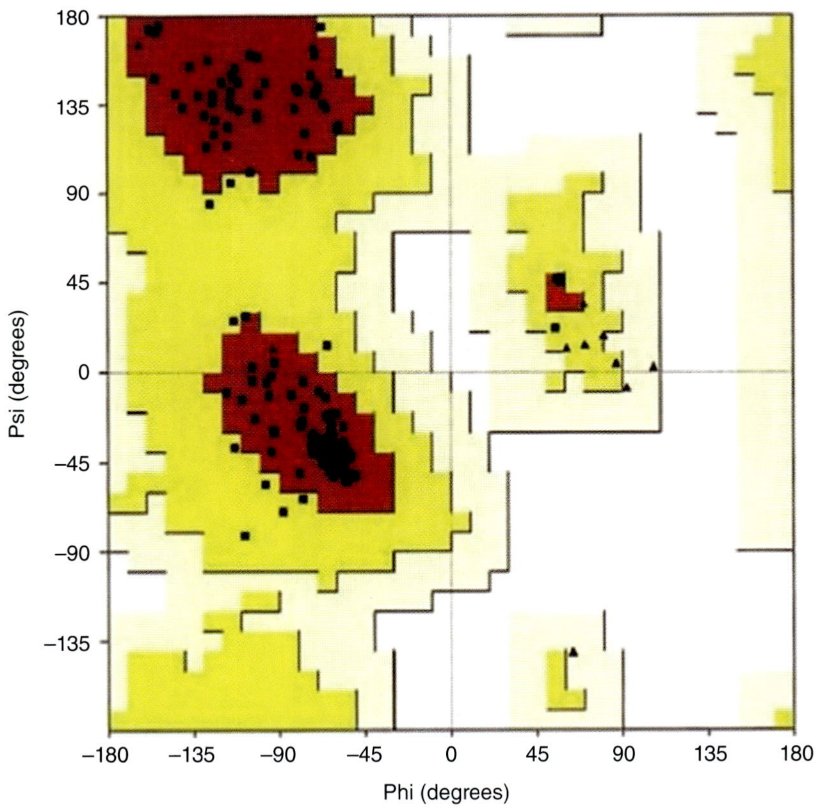

FIGURE 32.4 Ramachandran plot for a typical protein structure. The different regions were taken from the observed phi-psi distribution for 121 870 residues from 463 known X-ray protein structures.

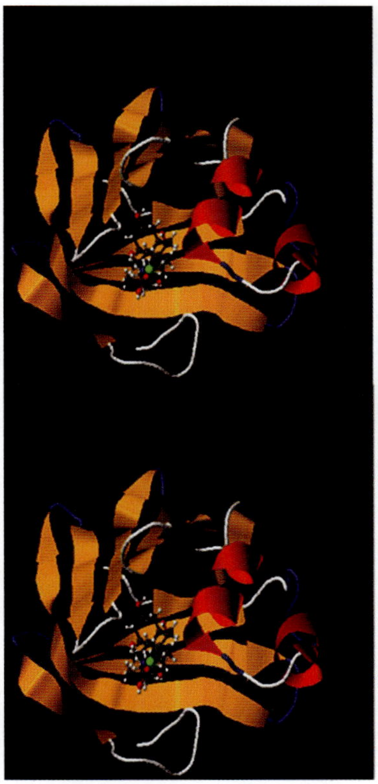

Show	Cluster	Element	FullFitness (kcal/mol)	Estimated ΔG (kcal/mol)
●	0	0	-437.97	-10.77
○	0	1	-437.94	-10.76
○	0	2	-427.32	-9.55
○	0	3	-418.70	-8.82
○	0	4	-417.48	-8.75
○	0	5	-412.26	-9.41
○	0	6	-411.32	-9.01
○	0	7	-406.22	-8.96
○	0	8	-403.94	-8.77
○	1	0	-418.27	-9.67
○	1	1	-418.23	-9.67
○	1	2	-414.20	-9.52
○	1	3	-413.48	-9.59
○	1	4	-413.48	-9.59
○	1	5	-411.57	-9.27
○	1	6	-405.92	-8.96
○	1	7	-403.69	-8.70

(a)

First binding pose: The fitness of this pose of the ligand and the free energy of the complex are –437.97 kcal/mole & –10.77 kcal/mole respectively.

Show	Cluster	Element	FullFitness (kcal/mol)	Estimated ΔG (kcal/mol)
○	0	0	-437.97	-10.77
●	0	1	-437.94	-10.76
○	0	2	-427.32	-9.55
○	0	3	-418.70	-8.82
○	0	4	-417.48	-8.75
○	0	5	-412.26	-9.41
○	0	6	-411.32	-9.01
○	0	7	-406.22	-8.96
○	0	8	-403.94	-8.77
○	1	0	-418.27	-9.67
○	1	1	-418.23	-9.67
○	1	2	-414.20	-9.52
○	1	3	-413.48	-9.59
○	1	4	-413.48	-9.59
○	1	5	-411.57	-9.27
○	1	6	-405.92	-8.96
○	1	7	-403.69	-8.70

(b)

Second binding pose: The fitness of this pose of the ligand and the free energy of the complex are –437.94 kcal/mole & –10.76 kcal/mole respectively.

FIGURE 35.3 Fitness of ligand and free energy of docked complex of the first and second binding poses, shown as "A" and "B".

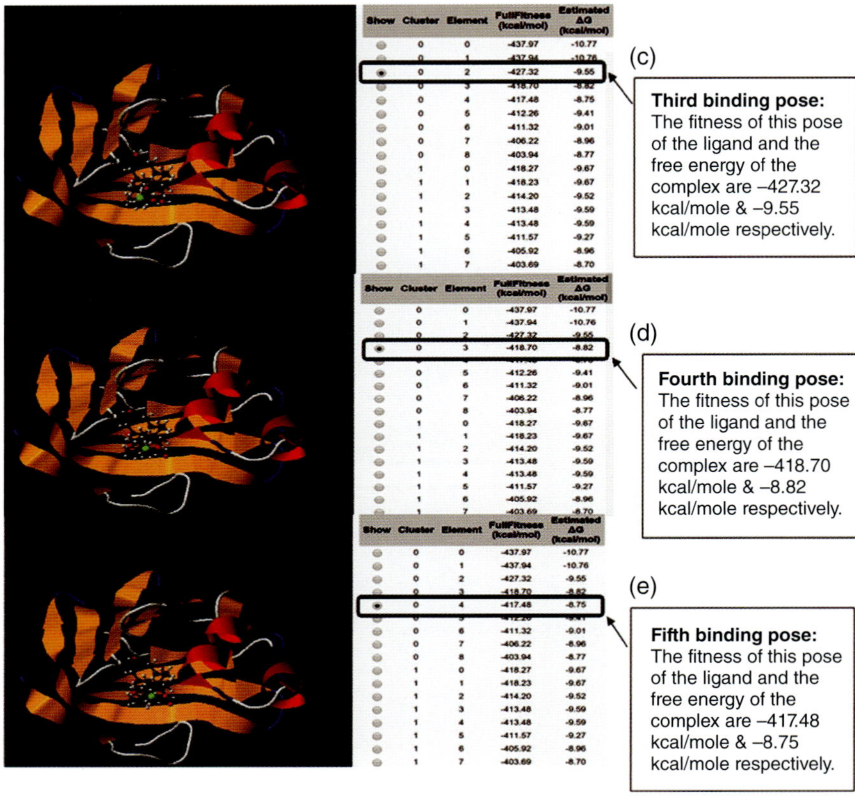

FIGURE 35.4 Fitness of ligand and free energy of docked complex of the third, fourth and fifth binding poses, shown as "C", "D" and "E".

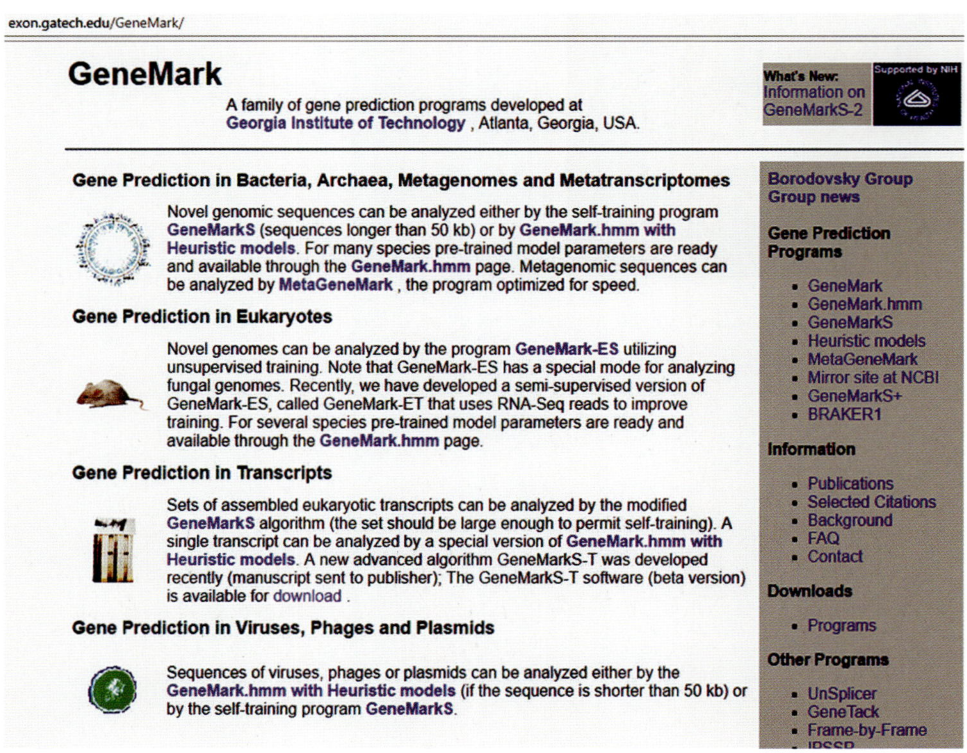

FIGURE 36.1 Homepage of the GeneMark online tool.

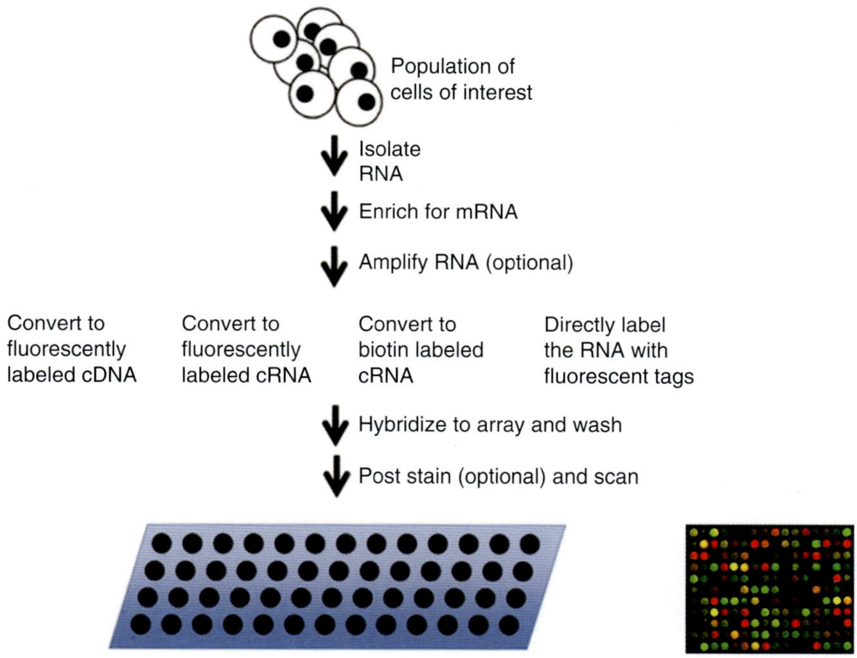

FIGURE 39.2 Application of microarray for gene expression analysis. Fluorescently labeled cDNA or cRNA is hybridized with probes, and the image is scanned through a scanner. Based upon the intensity of the signal, up regulated (red dots) and down regulated (green dots) genes are detected.

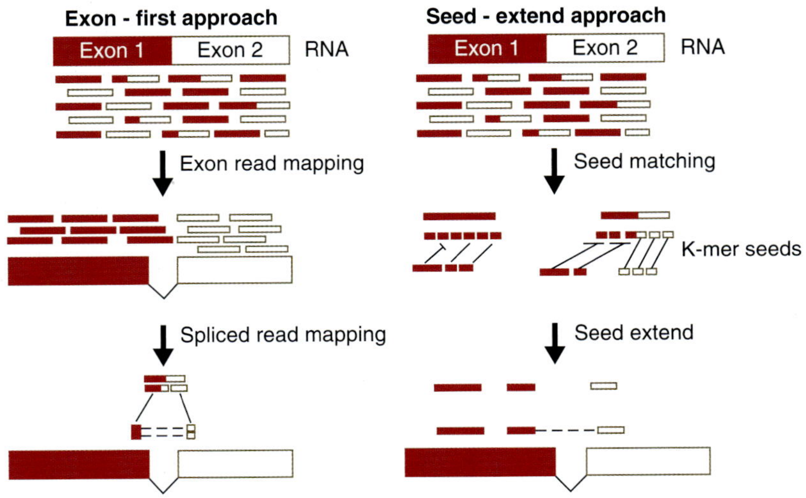

FIGURE 42.1

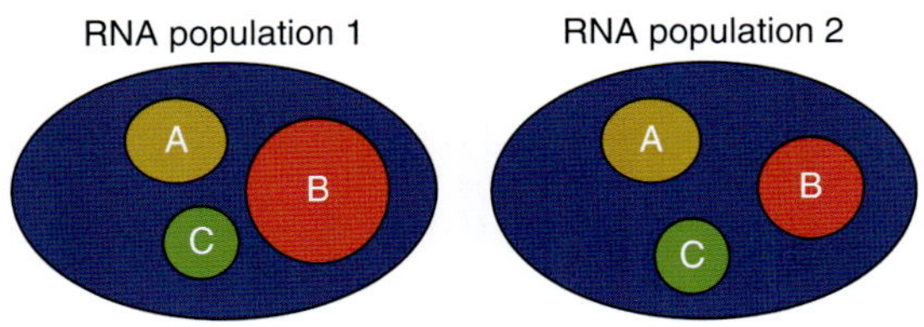

FIGURE 42.2

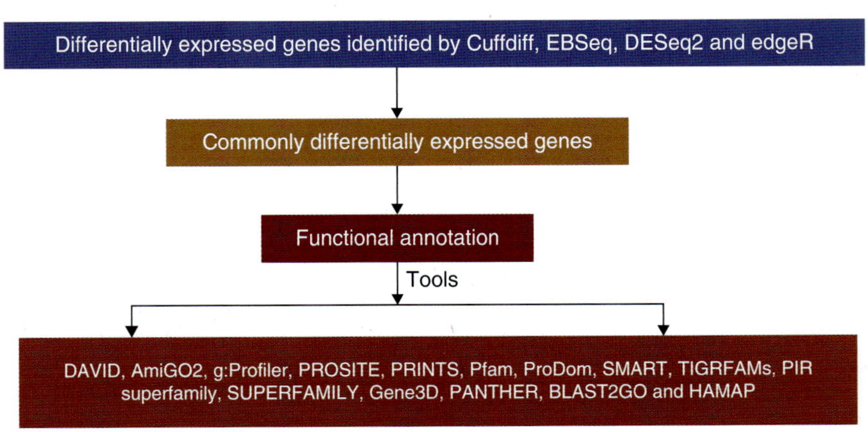

FIGURE 43.1

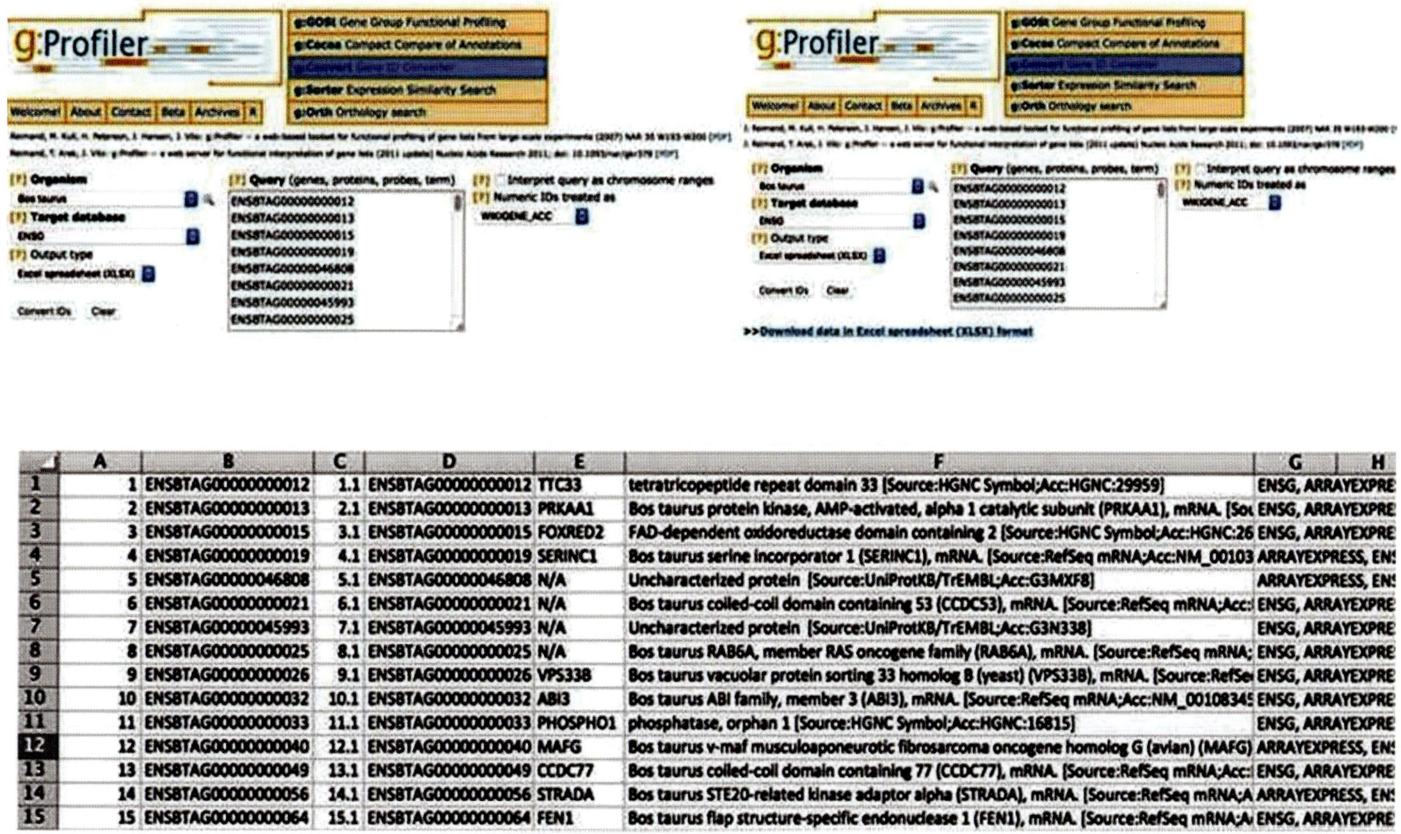

	A	B	C	D	E	F	G	H
1	1	ENSBTAG00000000012	1.1	ENSBTAG00000000012	TTC33	tetratricopeptide repeat domain 33 [Source:HGNC Symbol;Acc:HGNC:29959]	ENSG, ARRAYEXPRE	
2	2	ENSBTAG00000000013	2.1	ENSBTAG00000000013	PRKAA1	Bos taurus protein kinase, AMP-activated, alpha 1 catalytic subunit (PRKAA1), mRNA. [Sou	ENSG, ARRAYEXPRE	
3	3	ENSBTAG00000000015	3.1	ENSBTAG00000000015	FOXRED2	FAD-dependent oxidoreductase domain containing 2 [Source:HGNC Symbol;Acc:HGNC:26	ENSG, ARRAYEXPRE	
4	4	ENSBTAG00000000019	4.1	ENSBTAG00000000019	SERINC1	Bos taurus serine incorporator 1 (SERINC1), mRNA. [Source:RefSeq mRNA;Acc:NM_00103	ARRAYEXPRESS, EN	
5	5	ENSBTAG00000046808	5.1	ENSBTAG00000046808	N/A	Uncharacterized protein [Source:UniProtKB/TrEMBL;Acc:G3MXF8]	ARRAYEXPRESS, EN	
6	6	ENSBTAG00000000021	6.1	ENSBTAG00000000021	N/A	Bos taurus coiled-coil domain containing 53 (CCDC53), mRNA. [Source:RefSeq mRNA;Acc:	ENSG, ARRAYEXPRE	
7	7	ENSBTAG00000045993	7.1	ENSBTAG00000045993	N/A	Uncharacterized protein [Source:UniProtKB/TrEMBL;Acc:G3N338]	ENSG, ARRAYEXPRE	
8	8	ENSBTAG00000000025	8.1	ENSBTAG00000000025	N/A	Bos taurus RAB6A, member RAS oncogene family (RAB6A), mRNA. [Source:RefSeq mRNA;	ENSG, ARRAYEXPRE	
9	9	ENSBTAG00000000026	9.1	ENSBTAG00000000026	VPS33B	Bos taurus vacuolar protein sorting 33 homolog B (yeast) (VPS33B), mRNA. [Source:RefSe	ENSG, ARRAYEXPRE	
10	10	ENSBTAG00000000032	10.1	ENSBTAG00000000032	ABI3	Bos taurus ABI family, member 3 (ABI3), mRNA. [Source:RefSeq mRNA;Acc:NM_00108345	ENSG, ARRAYEXPRE	
11	11	ENSBTAG00000000033	11.1	ENSBTAG00000000033	PHOSPHO1	phosphatase, orphan 1 [Source:HGNC Symbol;Acc:HGNC:16815]	ENSG, ARRAYEXPRE	
12	12	ENSBTAG00000000040	12.1	ENSBTAG00000000040	MAFG	Bos taurus v-maf musculoaponeurotic fibrosarcoma oncogene homolog G (avian) (MAFG)	ARRAYEXPRESS, EN	
13	13	ENSBTAG00000000049	13.1	ENSBTAG00000000049	CCDC77	Bos taurus coiled-coil domain containing 77 (CCDC77), mRNA. [Source:RefSeq mRNA;Acc:	ENSG, ARRAYEXPRE	
14	14	ENSBTAG00000000056	14.1	ENSBTAG00000000056	STRADA	Bos taurus STE20-related kinase adaptor alpha (STRADA), mRNA. [Source:RefSeq mRNA;A	ARRAYEXPRESS, EN	
15	15	ENSBTAG00000000064	15.1	ENSBTAG00000000064	FEN1	Bos taurus flap structure-specific endonuclease 1 (FEN1), mRNA. [Source:RefSeq mRNA;A	ENSG, ARRAYEXPRE	

FIGURE 43.2

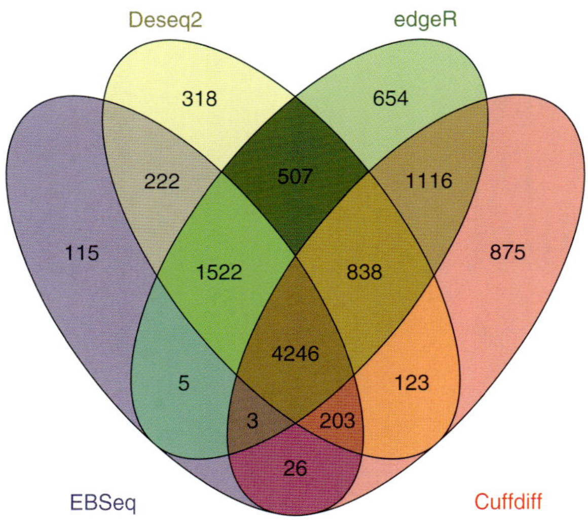

Deseq2
edgeR
318
654
222
507
1116
115
1522
838
875
4246
5
123
3
203
26
EBSeq
Cuffdiff

FIGURE 43.3

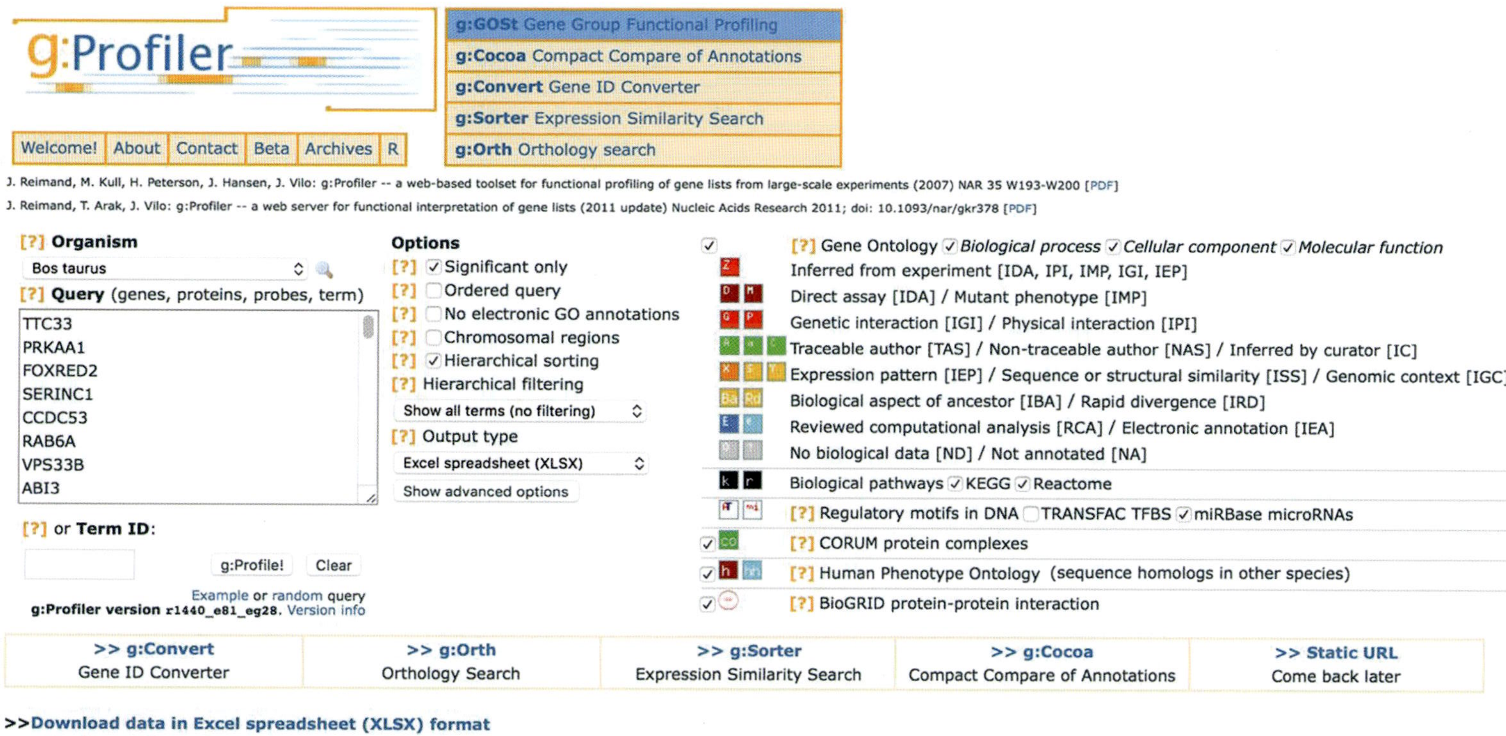

g:Profiler
g:GOSt Gene Group Functional Profiling
g:Cocoa Compact Compare of Annotations
g:Convert Gene ID Converter
g:Sorter Expression Similarity Search
g:Orth Orthology search
Welcome! About Contact Beta Archives R
J. Reimand, M. Kull, H. Peterson, J. Hansen, J. Vilo: g:Profiler -- a web-based toolset for functional profiling of gene lists from large-scale experiments (2007) NAR 35 W193-W200 [PDF]
J. Reimand, T. Arak, J. Vilo: g:Profiler -- a web server for functional interpretation of gene lists (2011 update) Nucleic Acids Research 2011; doi: 10.1093/nar/gkr378 [PDF]
[?] Organism
Bos taurus
[?] Query (genes, proteins, probes, term)
TTC33
PRKAA1
FOXRED2
SERINC1
CCDC53
RAB6A
VPS33B
ABI3
[?] or Term ID:
g:Profile!
Clear
Example or random query
g:Profiler version r1440_e81_eg28. Version info
Options
[?] Significant only
[?] Ordered query
[?] No electronic GO annotations
[?] Chromosomal regions
[?] Hierarchical sorting
[?] Hierarchical filtering
Show all terms (no filtering)
[?] Output type
Excel spreadsheet (XLSX)
Show advanced options
[?] Gene Ontology Biological process Cellular component Molecular function
Inferred from experiment [IDA, IPI, IMP, IGI, IEP]
Direct assay [IDA] / Mutant phenotype [IMP]
Genetic interaction [IGI] / Physical interaction [IPI]
Traceable author [TAS] / Non-traceable author [NAS] / Inferred by curator [IC]
Expression pattern [IEP] / Sequence or structural similarity [ISS] / Genomic context [IGC]
Biological aspect of ancestor [IBA] / Rapid divergence [IRD]
Reviewed computational analysis [RCA] / Electronic annotation [IEA]
No biological data [ND] / Not annotated [NA]
Biological pathways KEGG Reactome
[?] Regulatory motifs in DNA TRANSFAC TFBS miRBase microRNAs
[?] CORUM protein complexes
[?] Human Phenotype Ontology (sequence homologs in other species)
[?] BioGRID protein-protein interaction
>> g:Convert
Gene ID Converter
>> g:Orth
Orthology Search
>> g:Sorter
Expression Similarity Search
>> g:Cocoa
Compact Compare of Annotations
>> Static URL
Come back later
>>Download data in Excel spreadsheet (XLSX) format

FIGURE 43.4

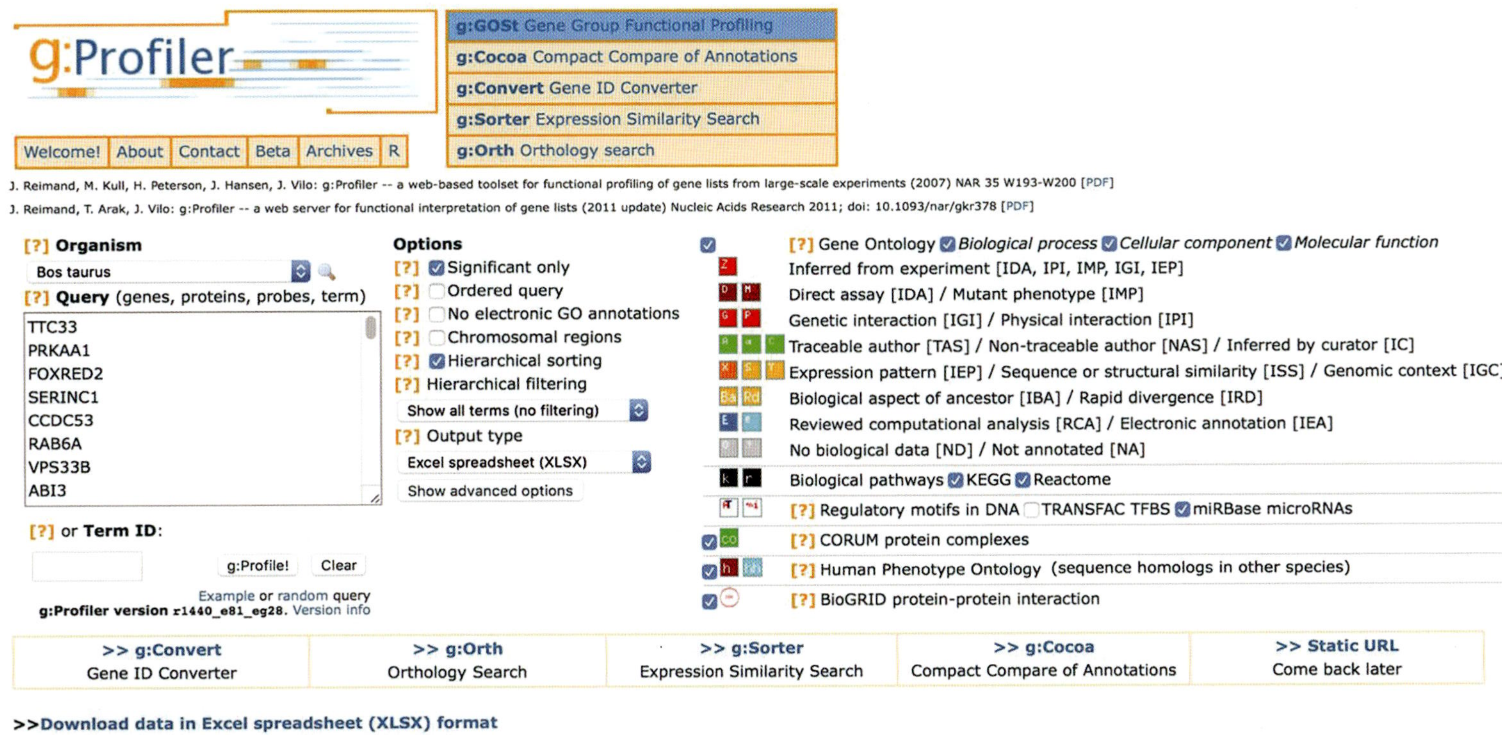
g:Profiler
Welcome! About Contact Beta Archives R
g:GOSt Gene Group Functional Profiling
g:Cocoa Compact Compare of Annotations
g:Convert Gene ID Converter
g:Sorter Expression Similarity Search
g:Orth Orthology search
J. Reimand, M. Kull, H. Peterson, J. Hansen, J. Vilo: g:Profiler -- a web-based toolset for functional profiling of gene lists from large-scale experiments (2007) NAR 35 W193-W200 [PDF]
J. Reimand, T. Arak, J. Vilo: g:Profiler -- a web server for functional interpretation of gene lists (2011 update) Nucleic Acids Research 2011; doi: 10.1093/nar/gkr378 [PDF]
[?] Organism
Bos taurus
[?] Query (genes, proteins, probes, term)
TTC33
PRKAA1
FOXRED2
SERINC1
CCDC53
RAB6A
VPS33B
ABI3
[?] or Term ID:
g:Profile!
Clear
Example or random query
g:Profiler version r1440_e81_eg28. Version info
Options
[?] Significant only
[?] Ordered query
[?] No electronic GO annotations
[?] Chromosomal regions
[?] Hierarchical sorting
[?] Hierarchical filtering
Show all terms (no filtering)
[?] Output type
Excel spreadsheet (XLSX)
Show advanced options
[?] Gene Ontology Biological process Cellular component Molecular function
Inferred from experiment [IDA, IPI, IMP, IGI, IEP]
Direct assay [IDA] / Mutant phenotype [IMP]
Genetic interaction [IGI] / Physical interaction [IPI]
Traceable author [TAS] / Non-traceable author [NAS] / Inferred by curator [IC]
Expression pattern [IEP] / Sequence or structural similarity [ISS] / Genomic context [IGC]
Biological aspect of ancestor [IBA] / Rapid divergence [IRD]
Reviewed computational analysis [RCA] / Electronic annotation [IEA]
No biological data [ND] / Not annotated [NA]
Biological pathways KEGG Reactome
[?] Regulatory motifs in DNA TRANSFAC TFBS miRBase microRNAs
[?] CORUM protein complexes
[?] Human Phenotype Ontology (sequence homologs in other species)
[?] BioGRID protein-protein interaction
>> g:Convert
Gene ID Converter
>> g:Orth
Orthology Search
>> g:Sorter
Expression Similarity Search
>> g:Cocoa
Compact Compare of Annotations
>> Static URL
Come back later
>>Download data in Excel spreadsheet (XLSX) format

FIGURE 43.5

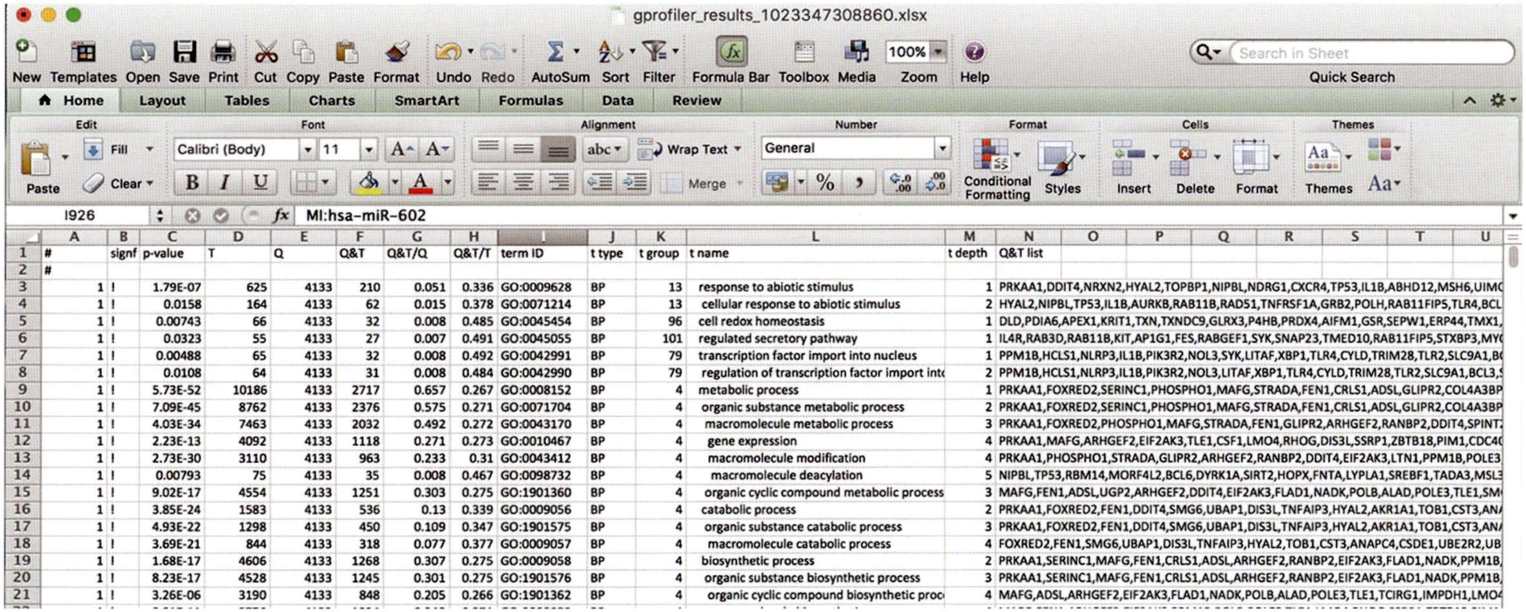

gprofiler_results_1023347308860.xlsx

I926 | MI:hsa-miR-602

#	signf	p-value	T	Q	Q&T	Q&T/Q	Q&T/T	term ID	t type	t group	t name	t depth	Q&T list
#													
1	!	1.79E-07	625	4133	210	0.051	0.336	GO:0009628	BP	13	response to abiotic stimulus	1	PRKAA1,DDIT4,NRXN2,HYAL2,TOPBP1,NIPBL,NDRG1,CXCR4,TP53,IL1B,ABHD12,MSH6,UIM
1	!	0.0158	164	4133	62	0.015	0.378	GO:0071214	BP	13	cellular response to abiotic stimulus	2	HYAL2,NIPBL,TP53,IL1B,AURKB,RAB11B,RAD51,TNFRSF1A,GRB2,POLH,RAB11FIP5,TLR4,BCL
1	!	0.00743	66	4133	32	0.008	0.485	GO:0045454	BP	96	cell redox homeostasis	1	DLD,PDIA6,APEX1,KRIT1,TXN,TXNDC9,GLRX3,P4HB,PRDX4,AIFM1,GSR,SEPW1,ERP44,TMX1,
1	!	0.0323	55	4133	27	0.007	0.491	GO:0045055	BP	101	regulated secretory pathway	1	IL4R,RAB3D,RAB11B,KIT,AP1G1,FES,RABGEF1,SYK,SNAP23,TMED10,RAB11FIP5,STXBP3,MY
1	!	0.00488	65	4133	32	0.008	0.492	GO:0042991	BP	79	transcription factor import into nucleus	1	PPM1B,HCLS1,NLRP3,IL1B,PIK3R2,NOL3,SYK,LITAF,XBP1,TLR4,CYLD,TRIM28,TLR2,SLC9A1,B
1	!	0.0108	64	4133	31	0.008	0.484	GO:0042990	BP	79	regulation of transcription factor import int	2	PPM1B,HCLS1,NLRP3,IL1B,PIK3R2,NOL3,LITAF,XBP1,TLR4,CYLD,TRIM28,TLR2,SLC9A1,BCL3,
1	!	5.73E-52	10186	4133	2717	0.657	0.267	GO:0008152	BP	4	metabolic process	1	PRKAA1,FOXRED2,SERINC1,PHOSPHO1,MAFG,STRADA,FEN1,CRLS1,ADSL,GLIPR2,COL4A3BP
1	!	7.09E-45	8762	4133	2376	0.575	0.271	GO:0071704	BP	4	organic substance metabolic process	2	PRKAA1,FOXRED2,SERINC1,PHOSPHO1,MAFG,STRADA,FEN1,CRLS1,ADSL,GLIPR2,COL4A3BP
1	!	4.03E-34	7463	4133	2032	0.492	0.272	GO:0043170	BP	4	macromolecule metabolic process	3	PRKAA1,FOXRED2,PHOSPHO1,MAFG,STRADA,FEN1,GLIPR2,ARHGEF2,RANBP2,DDIT4,SPINT
1	!	2.23E-13	4092	4133	1118	0.271	0.273	GO:0010467	BP	4	gene expression	4	PRKAA1,MAFG,ARHGEF2,EIF2AK3,TLE1,CSF1,LMO4,RHOG,DIS3L,SSRP1,ZBTB18,PIM1,CDC4
1	!	2.73E-30	3110	4133	963	0.233	0.31	GO:0043412	BP	4	macromolecule modification	4	PRKAA1,PHOSPHO1,STRADA,GLIPR2,ARHGEF2,RANBP2,DDIT4,EIF2AK3,LTN1,PPM1B,POLE3
1	!	0.00793	75	4133	35	0.008	0.467	GO:0098732	BP	4	macromolecule deacylation	5	NIPBL,TP53,RBM14,MORF4L2,BCL6,DYRK1A,SIRT2,HOPX,FNTA,LYPLA1,SREBF1,TADA3,MSL
1	!	9.02E-17	4554	4133	1251	0.303	0.275	GO:1901360	BP	4	organic cyclic compound metabolic process	3	MAFG,FEN1,ADSL,UGP2,ARHGEF2,DDIT4,EIF2AK3,FLAD1,NADK,POLB,ALAD,POLE3,TLE1,SM
1	!	3.85E-24	1583	4133	536	0.13	0.339	GO:0009056	BP	4	catabolic process	2	PRKAA1,FOXRED2,FEN1,DDIT4,SMG6,UBAP1,DIS3L,TNFAIP3,HYAL2,AKR1A1,TOB1,CST3,AN
1	!	4.93E-22	1298	4133	450	0.109	0.347	GO:1901575	BP	4	organic substance catabolic process	3	PRKAA1,FOXRED2,FEN1,DDIT4,SMG6,UBAP1,DIS3L,TNFAIP3,HYAL2,AKR1A1,TOB1,CST3,AN
1	!	3.69E-21	844	4133	318	0.077	0.377	GO:0009057	BP	4	macromolecule catabolic process	4	FOXRED2,FEN1,SMG6,UBAP1,DIS3L,TNFAIP3,HYAL2,TOB1,CST3,ANAPC4,CSDE1,UBE2R2,UB
1	!	1.68E-17	4606	4133	1268	0.307	0.275	GO:0009058	BP	4	biosynthetic process	2	PRKAA1,SERINC1,MAFG,FEN1,CRLS1,ADSL,ARHGEF2,RANBP2,EIF2AK3,FLAD1,NADK,PPM1B,
1	!	8.23E-17	4528	4133	1245	0.301	0.275	GO:1901576	BP	4	organic substance biosynthetic process	3	PRKAA1,SERINC1,MAFG,FEN1,CRLS1,ADSL,ARHGEF2,RANBP2,EIF2AK3,FLAD1,NADK,PPM1B,
1	!	3.26E-06	3190	4133	848	0.205	0.266	GO:1901362	BP	4	organic cyclic compound biosynthetic proc	4	MAFG,ADSL,ARHGEF2,EIF2AK3,FLAD1,NADK,POLB,ALAD,POLE3,TLE1,TCIRG1,IMPDH1,LMO

FIGURE 43.6

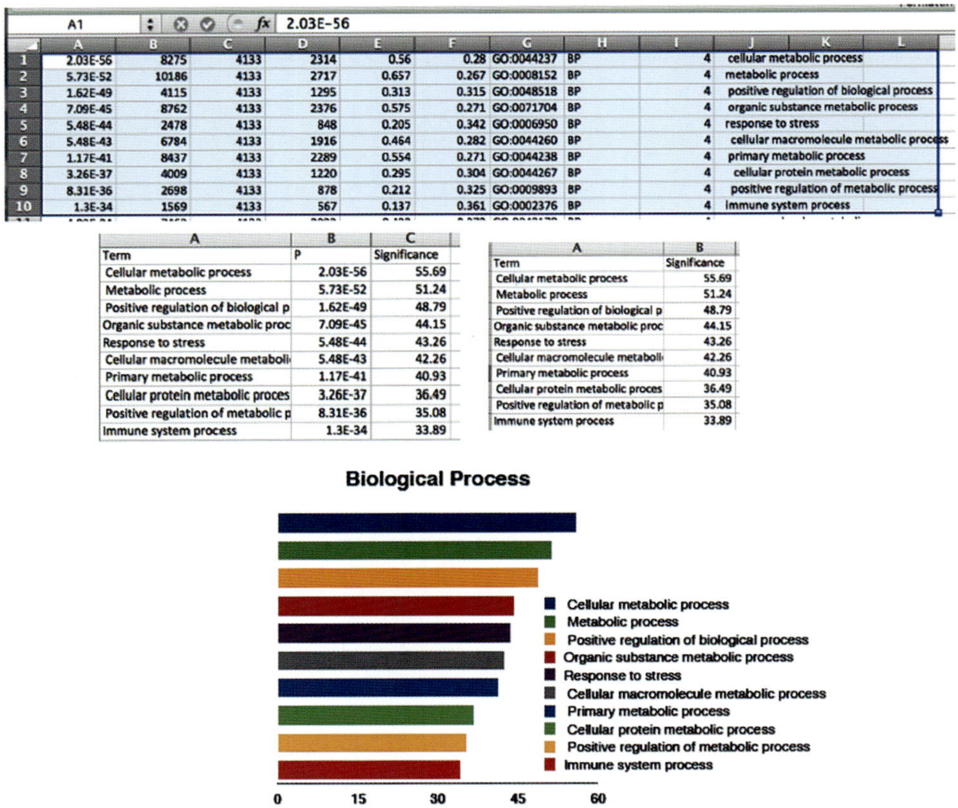

FIGURE 43.7

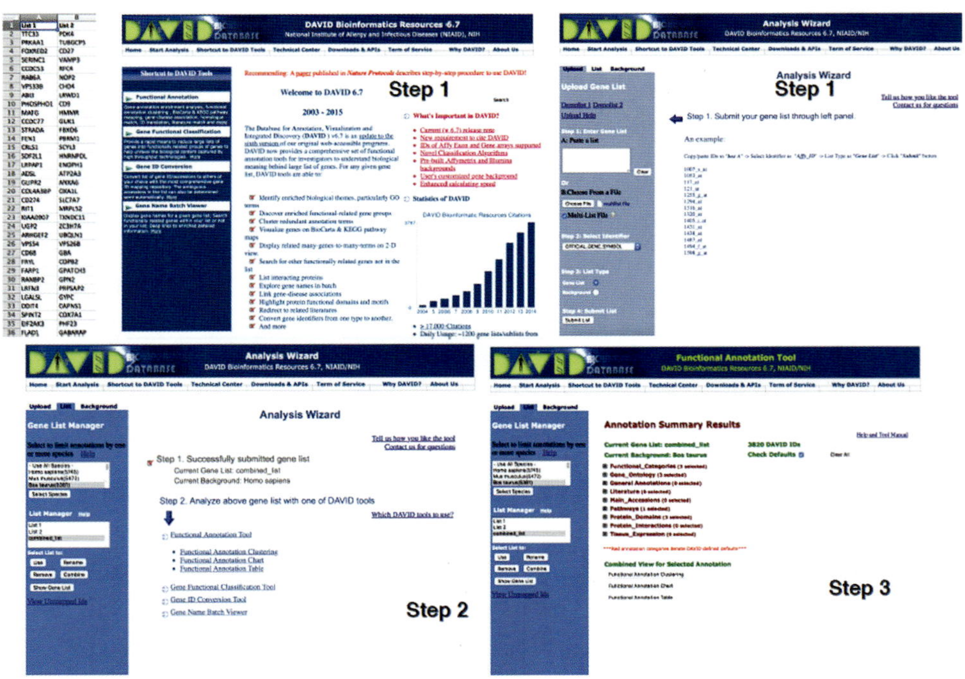

FIGURE 43.8

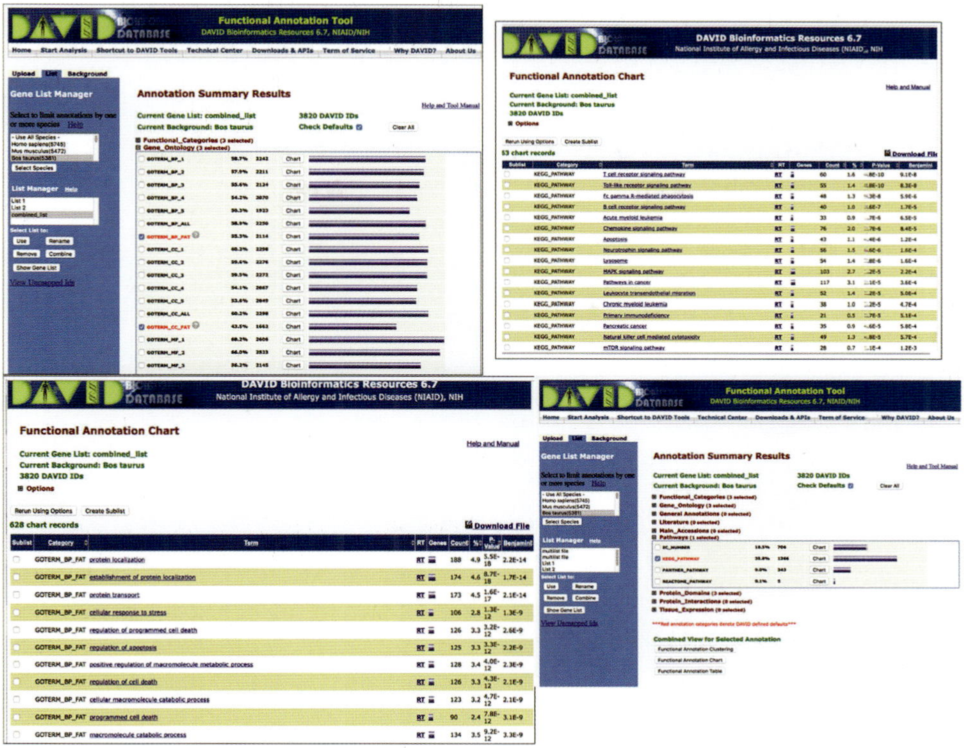

FIGURE 43.9

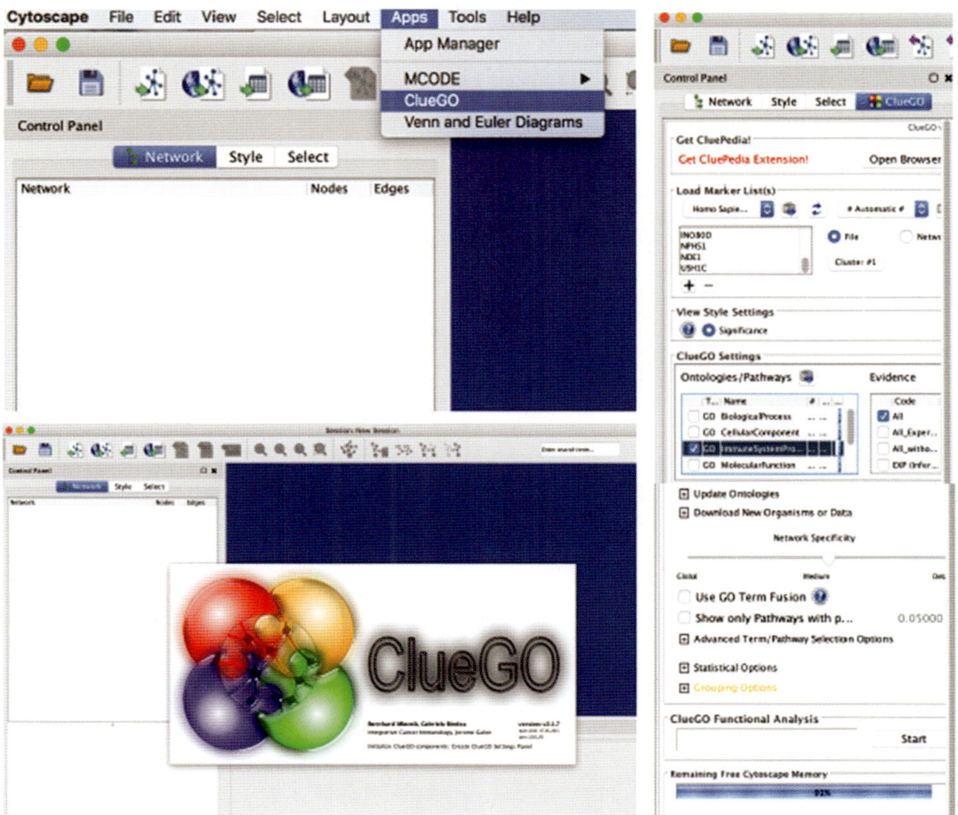

FIGURE 43.10

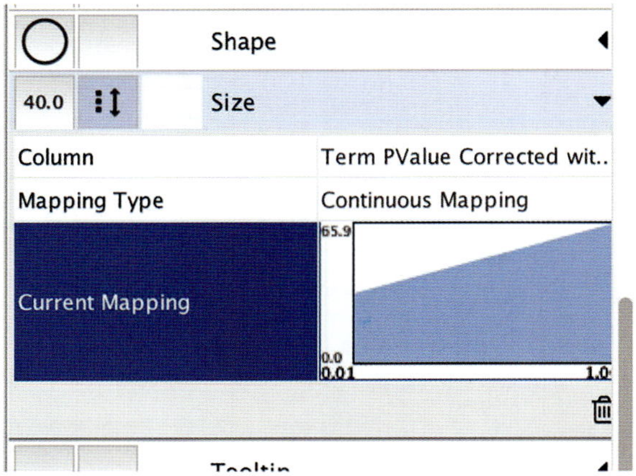

FIGURE 43.11

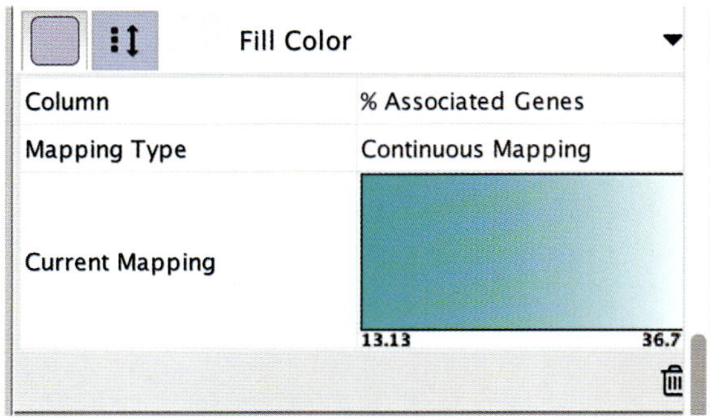

FIGURE 43.12

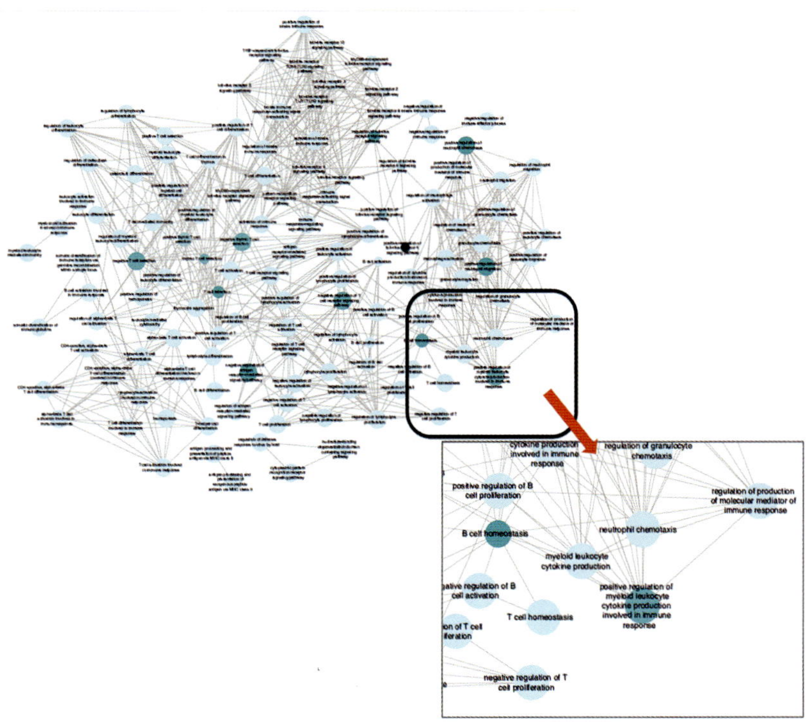

FIGURE 43.13

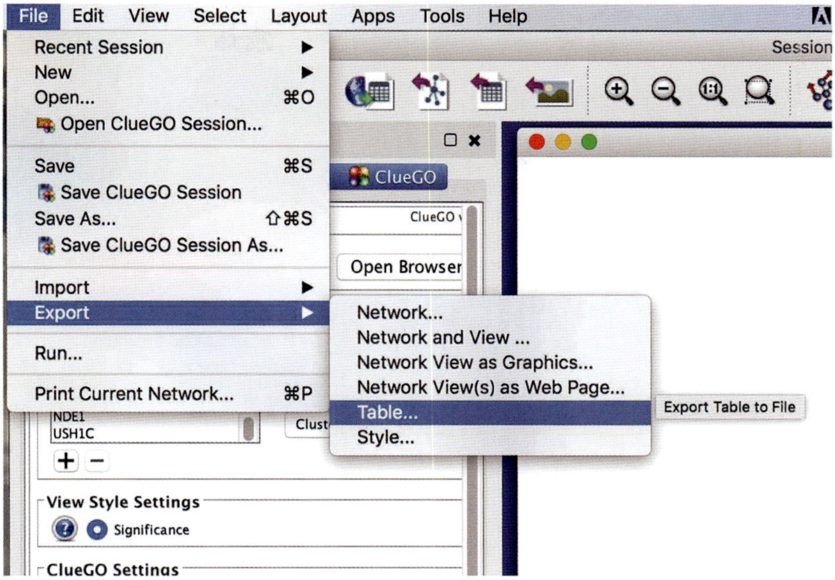

FIGURE 43.14

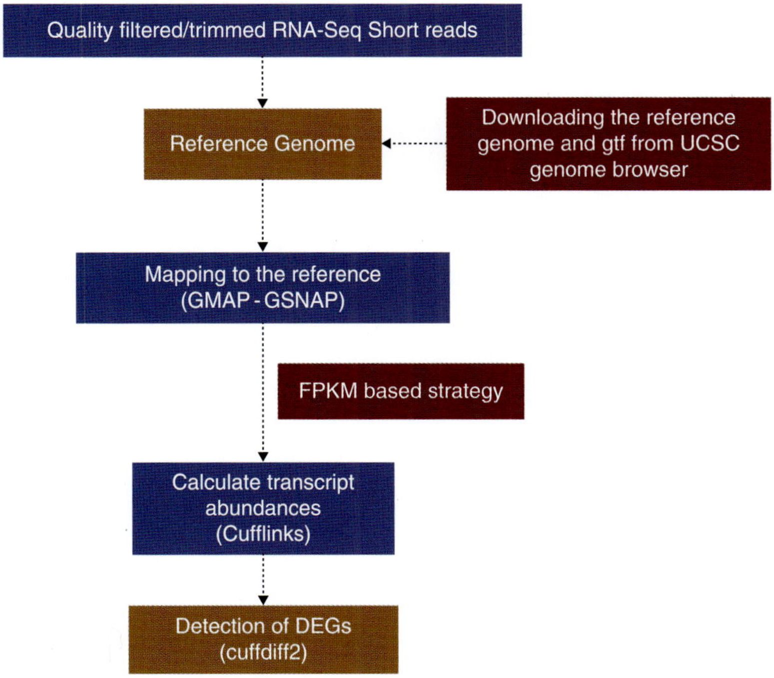

FIGURE 44.4 Workflow for identifying DEGs using Cufflinks.

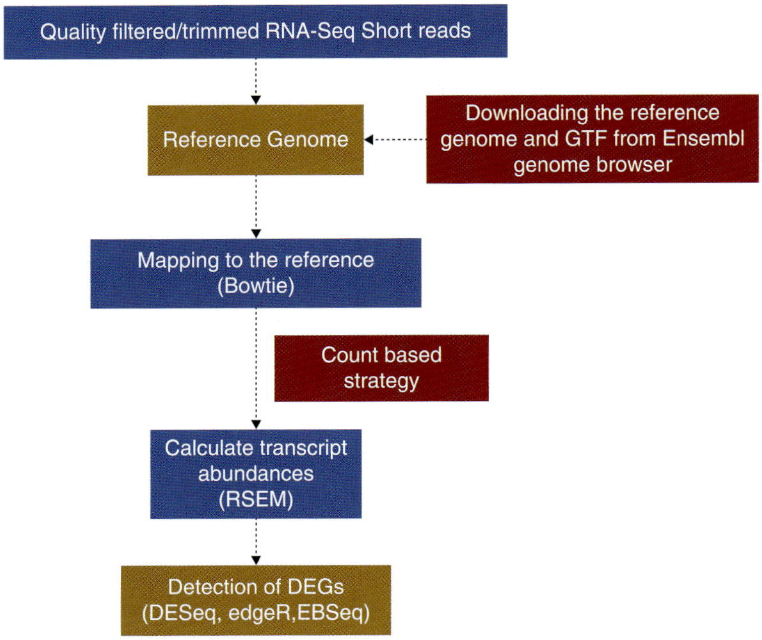

FIGURE 44.15 Workflow for identifying DEGs using RSEM and DE packages.

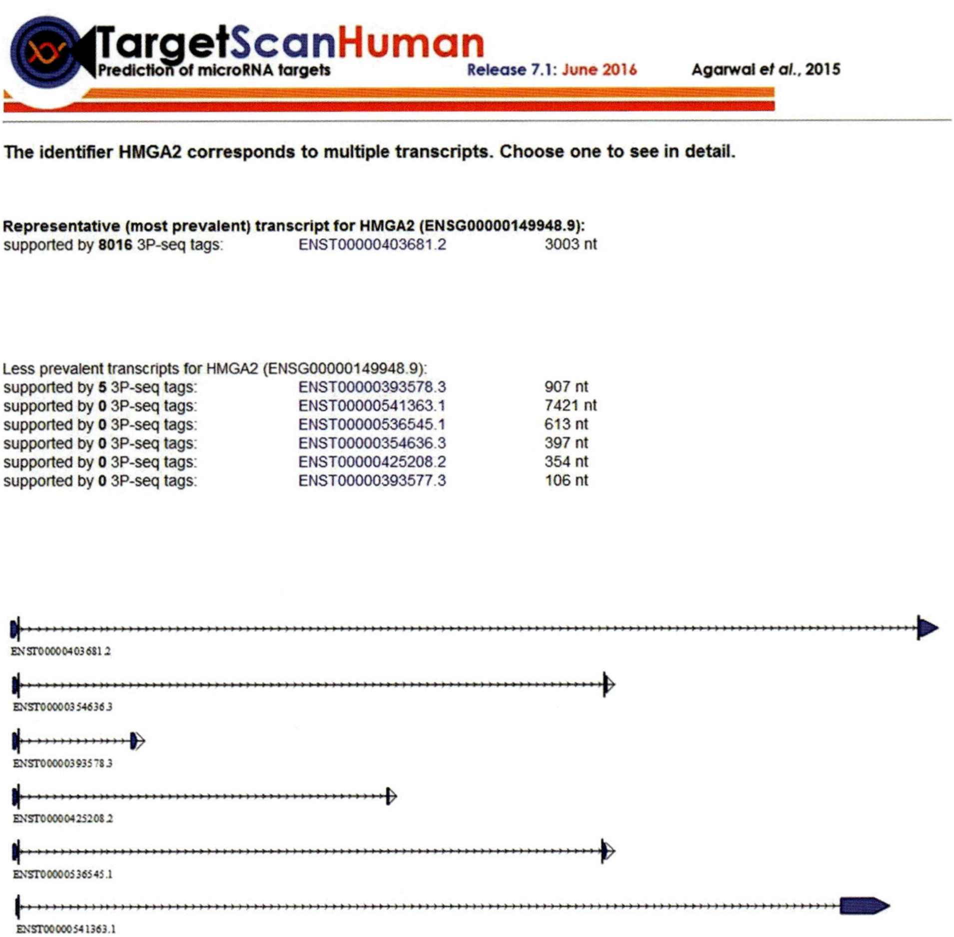

FIGURE 46.2 Output page showing multiple transcripts in the TargetScan tool.

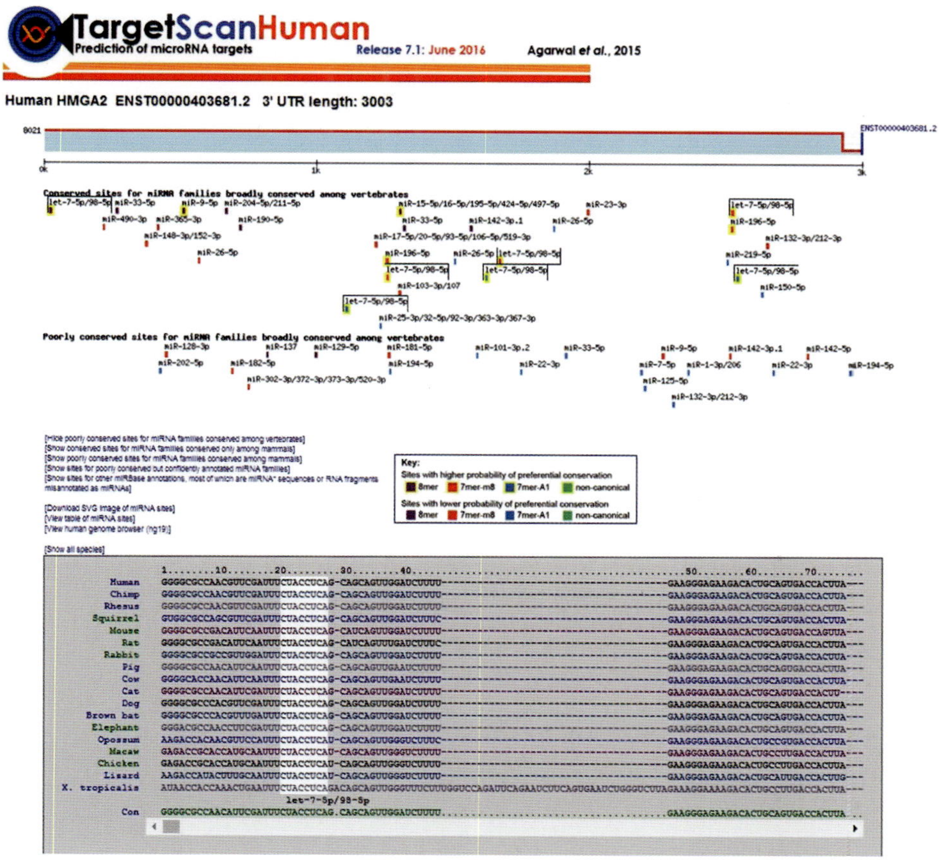

FIGURE 46.3 Output page of the TargetScan tool, showing conserved sites fcr miRNA families.

Conserved

	Predicted consequential pairing of target region (top) and miRNA (bottom)	Site type	Context++ score	Context++ score percentile	Weighted context++ score	Conserved branch length	P_{CT}
Position 21-28 of HMGA2 3' UTR hsa-let-7e-5p	5' ...GCCAACGUUCGAUUUCUACCUCA... ||||||| 3' UUGAUAUGUUGGAGGAUGGAGU	8mer	-0.63	99	-0.63	4.512	0.94
Position 21-28 of HMGA2 3' UTR hsa-let-7b-5p	5' ...GCCAACGUUCGAUUUCUACCUCA... ||||||| 3' UUGGUGUGUUGGAUGAUGGAGU	8mer	-0.63	99	-0.63	4.512	0.94
Position 21-28 of HMGA2 3' UTR hsa-let-7c-5p	5' ...GCCAACGUUCGAUUUCUACCUCA... ||||||| 3' UUGGUAUGUUGGAUGAUGGAGU	8mer	-0.63	99	-0.63	4.512	0.94
Position 21-28 of HMGA2 3' UTR hsa-let-7i-5p	5' ...GCCAACGUUCGAUUUCUACCUCA... ||||||| 3' UUGUCGUGUUUGAUGAUGGAGU	8mer	-0.63	99	-0.63	4.512	0.94
Position 21-28 of HMGA2 3' UTR hsa-let-7f-5p	5' ...GCCAACGUUCGAUUUCUACCUCA... ||||||| 3' UUGAUAUGUUAGAUGAUGGAGU	8mer	-0.63	99	-0.63	4.512	0.94
Position 21-28 of HMGA2 3' UTR hsa-let-7d-5p	5' ...GCCAACGUUCGAUUUCUACCUCA... ||||||| 3' UUGAUACGUUGGAUGAUGGAGA	8mer	-0.64	99	-0.64	4.512	0.94
Position 21-28 of HMGA2 3' UTR hsa-miR-4458	5' ...GCCAACGUUCGAUUUCUACCUCA... ||| ||||||| 3' AAGAAGGUGUGGAUGGAGA	8mer	-0.67	99	-0.67	4.512	0.94
Position 21-28 of HMGA2 3' UTR hsa-miR-4500	5' ...GCCAACGUUCGAUUUCUACCUCA... ||||||| 3' UUCUUUGAUGAUGGAGU	8mer	-0.63	99	-0.63	4.512	0.94
Position 21-28 of HMGA2 3' UTR hsa-let-7g-5p	5' ...GCCAACGUUCGAUUUCUACCUCA... ||||||| 3' UUGACAUGUUUGAUGAUGGAGU	8mer	-0.63	99	-0.63	4.512	0.94
Position 21-28 of HMGA2 3' UTR hsa-let-7a-5p	5' ...GCCAACGUUCGAUUUCUACCUCA... ||||||| 3' UUGAUAUGUUGGAUGAUGGAGU	8mer	-0.63	99	-0.63	4.512	0.94

FIGURE 46.4 Detailed table of all the conserved sites.

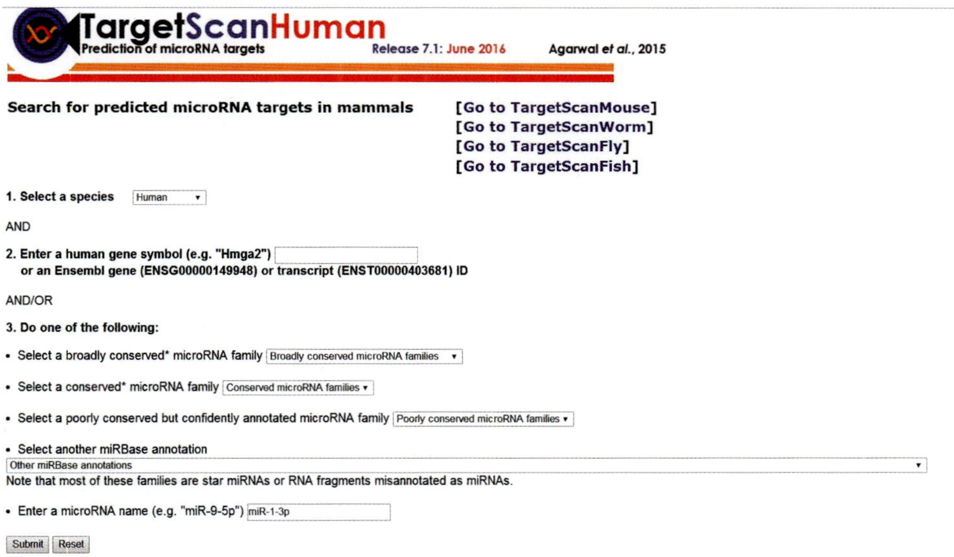

FIGURE 46.5 Input page of the TargetScan tool.

TargetScanHuman
Prediction of microRNA targets
Release 7.1: June 2016
Agarwal *et al.*, 2015

Human | miR-1-3p/206

883 transcripts with conserved sites, containing a total of **976** conserved sites and **336** poorly conserved sites.

Genes with only poorly conserved sites are not shown. [View top predicted targets, irrespective of site conservation]

Table sorted by cumulative weighted context++ score [Sort table by aggregate P_{CT}]

The table shows at most one transcript per gene, selected for being the most prevalent, based on 3P-seq tags (or the one with the longest 3' UTR, in case of a tie). [Download table]

Target gene	Representative transcript	Gene name	Number of 3P-seq tags supporting UTR + 5	Link to sites in UTRs	Conserved sites				Poorly conserved sites				6mer sites	Representative miRNA	Cumulative weighted context++ score	Total context++ score	Aggregate P_{CT}	Previous TargetScan publication(s)
					total	8mer	7mer-m8	7mer-A1	total	8mer	7mer-m8	7mer-A1						
CORO1C	ENST00000261401.3	coronin, actin binding protein, 1C	1970	Sites in UTR	2	2	0	0	0	0	0	0	2	hsa-miR-206	-1.08	-1.09	0.96	2005, 2007, 2009, 2011
SMIM14	ENST00000295958.5	small integral membrane protein 14	601	Sites in UTR	2	2	0	0	3	0	1	2	3	hsa-miR-206	-1.03	-1.38	0.98	2007, 2009, 2011
ARPC3	ENST00000228825.7	actin related protein 2/3 complex, subunit 3, 21kDa	323	Sites in UTR	1	1	0	0	2	1	1	0	0	hsa-miR-1-3p	-0.99	-1.30	< 0.1	
PTPLAD1	ENST00000261875.5	protein tyrosine phosphatase-like A domain containing 1	11697	Sites in UTR	2	1	1	0	1	0	1	0	1	hsa-miR-206	-0.97	-1.11	0.95	2003, 2007, 2009, 2011
ARCN1	ENST00000534182.2	archain 1	1139	Sites in UTR	1	1	0	0	1	0	1	0	0	hsa-miR-1-3p	-0.91	-0.91	0.95	2005, 2007, 2009, 2011
TAGLN2	ENST00000368096.1	transgelin 2	2192	Sites in UTR	2	0	2	0	1	0	0	1	1	hsa-miR-1-3p	-0.85	-0.85	0.89	2005, 2007, 2009, 2011
GJA1	ENST00000282561.3	gap junction protein, alpha 1, 43kDa	1732	Sites in UTR	2	2	0	0	0	0	0	0	1	hsa-miR-206	-0.84	-0.84	0.93	2003, 2005, 2007, 2009, 2011
ERMP1	ENST00000381506.3	endoplasmic reticulum metallopeptidase 1	53	Sites in UTR	2	0	2	0	1	1	0	0	0	hsa-miR-1-3p	-0.81	-0.81	ORF	
SERP1	ENST00000239944.2	stress-associated endoplasmic reticulum protein 1	3242	Sites in UTR	3	0	0	3	0	0	0	0	2	hsa-miR-1-3p	-0.77	-0.80	0.97	2005, 2007, 2009, 2011
TMSB4X	ENST00000451311.2	thymosin beta 4, X-linked	5	Sites in UTR	1	1	0	0	0	0	0	0	0	hsa-miR-206	-0.74	-0.74	0.77	2009, 2011
MMD2	ENST00000406755.1	monocyte to macrophage differentiation-associated 2	5	Sites in UTR	1	1	0	0	1	1	0	0	0	hsa-miR-1-3p	-0.69	-0.69	0.86	2005, 2007, 2009, 2011
BDNF	ENST00000439476.2	brain-derived neurotrophic factor	2696	Sites in UTR	3	1	2	0	0	0	0	0	0	hsa-miR-1-3p	-0.68	-0.95	0.98	2003, 2005, 2007, 2009, 2011
SLC10A7	ENST00000264986.3	solute carrier family 10, member 7	77	Sites in UTR	2	1	1	0	0	0	0	0	2	hsa-miR-1-3p	-0.68	-0.90	0.88	2009, 2011
G6PD	ENST00000393562.2	glucose-6-phosphate dehydrogenase	9	Sites in UTR	3	0	3	0	0	0	0	0	0	hsa-miR-1-3p	-0.68	-0.68	> 0.99	2011
SRI	ENST00000265729.2	sorcin	278	Sites in UTR	1	1	0	0	0	0	0	0	0	hsa-miR-1-3p	-0.65	-0.65	0.89	2011
GLCCI1	ENST00000223145.5	glucocorticoid induced transcript 1	654	Sites in UTR	2	2	0	0	0	0	0	0	2	hsa-miR-206	-0.65	-0.73	0.92	2005, 2007, 2009, 2011

FIGURE 46.6 Detailed information about the target gene symbol in the TargetScan tool.

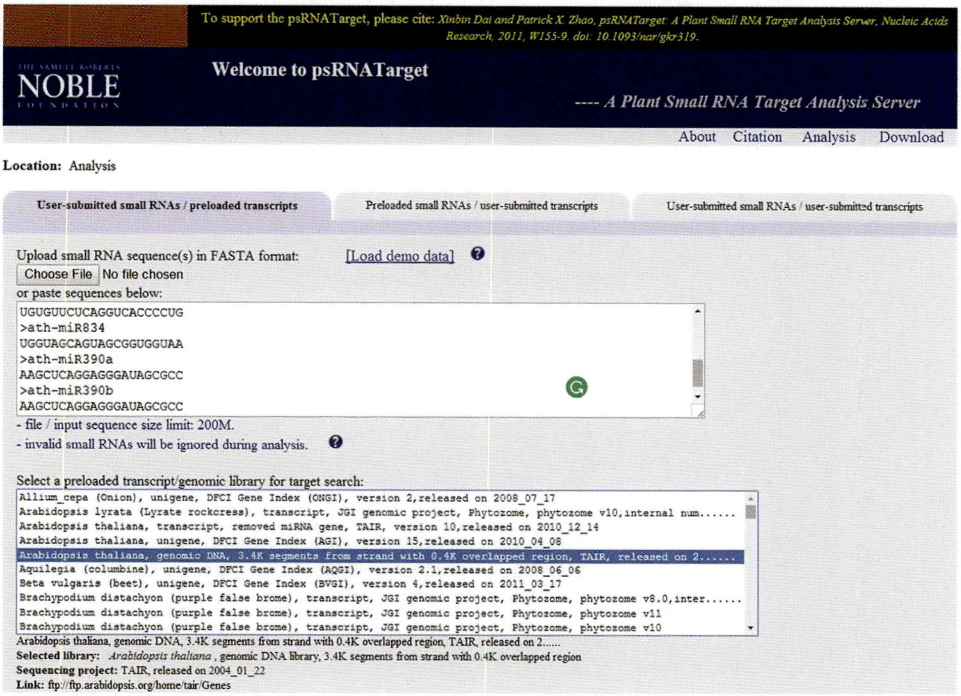

FIGURE 46.7 Input page of the psRNATarget tool.

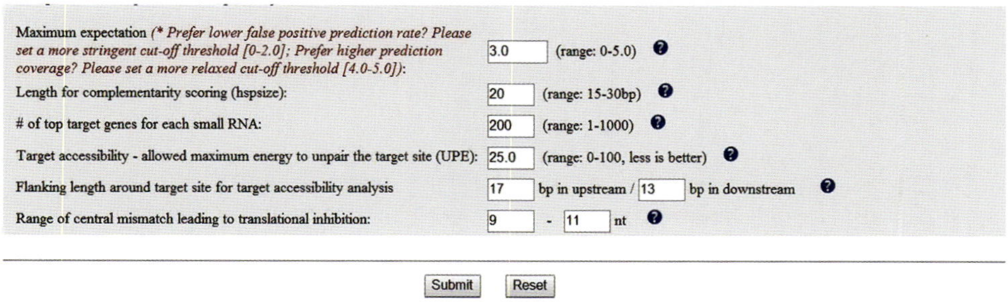

FIGURE 46.8 Input page of the psRNATarget tool for other parameters.

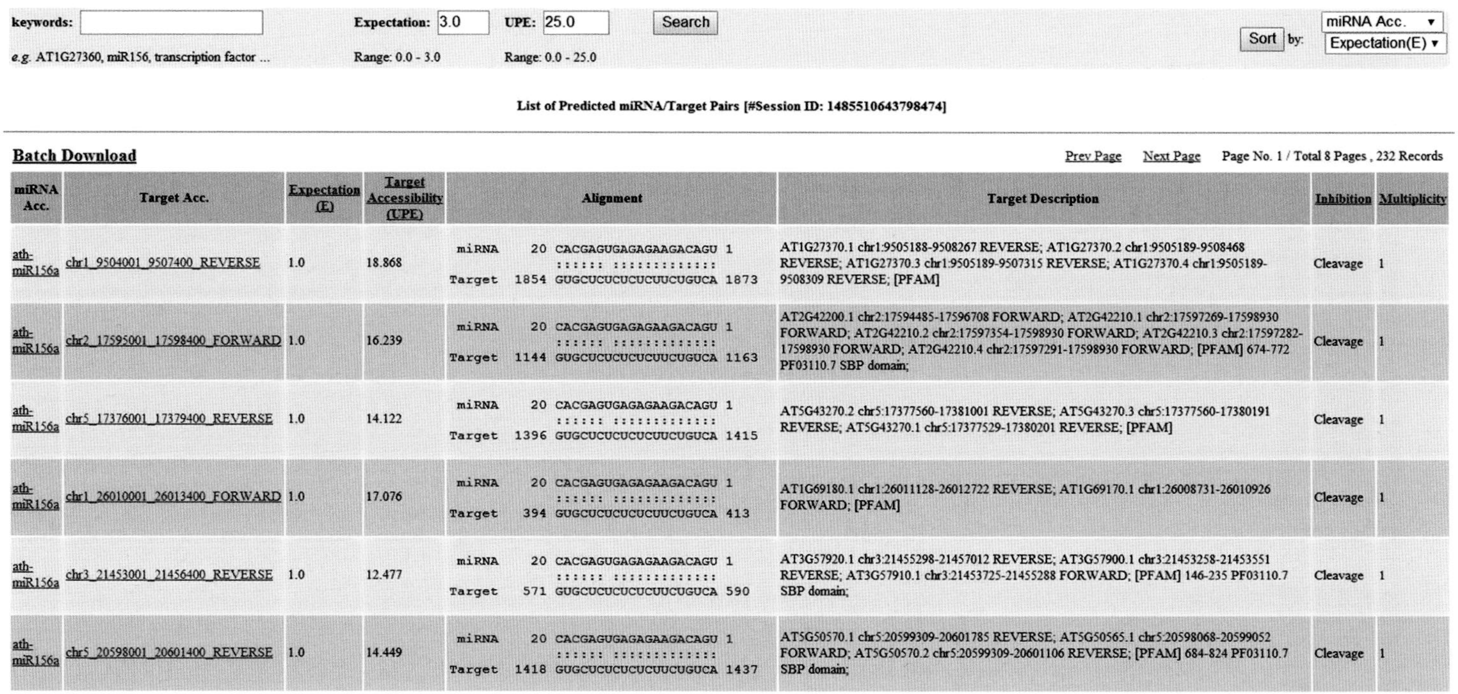

keywords: e.g. AT1G27360, miR156, transcription factor ...

Expectation: 3.0 Range: 0.0 - 3.0

UPE: 25.0 Range: 0.0 - 25.0

Search

Sort by: miRNA Acc. Expectation(E)

List of Predicted miRNA/Target Pairs [#Session ID: 1485510643798474]

Batch Download

Prev Page Next Page Page No. 1 / Total 8 Pages , 232 Records

miRNA Acc.	Target Acc.	Expectation (E)	Target Accessibility (UPE)	Alignment	Target Description	Inhibition	Multiplicity
ath-miR156a	chr1_9504001_9507400_REVERSE	1.0	18.868	miRNA 20 CACGAGUGAGAGAAGACAGU 1 :::::: ::::::::::::: Target 1854 GUGCUCUCUCUCUUCUGUCA 1873	AT1G27370.1 chr1:9505188-9508267 REVERSE; AT1G27370.2 chr1:9505189-9508468 REVERSE; AT1G27370.3 chr1:9505189-9507315 REVERSE; AT1G27370.4 chr1:9505189-9508309 REVERSE; [PFAM]	Cleavage	1
ath-miR156a	chr2_17595001_17598400_FORWARD	1.0	16.239	miRNA 20 CACGAGUGAGAGAAGACAGU 1 :::::: ::::::::::::: Target 1144 GUGCUCUCUCUCUUCUGUCA 1163	AT2G42200.1 chr2:17594485-17596708 FORWARD; AT2G42210.1 chr2:17597269-17598930 FORWARD; AT2G42210.2 chr2:17597354-17598930 FORWARD; AT2G42210.3 chr2:17597282-17598930 FORWARD; AT2G42210.4 chr2:17597291-17598930 FORWARD; [PFAM] 674-772 PF03110.7 SBP domain;	Cleavage	1
ath-miR156a	chr5_17376001_17379400_REVERSE	1.0	14.122	miRNA 20 CACGAGUGAGAGAAGACAGU 1 :::::: ::::::::::::: Target 1396 GUGCUCUCUCUCUUCUGUCA 1415	AT5G43270.2 chr5:17377560-17381001 REVERSE; AT5G43270.3 chr5:17377560-17380191 REVERSE; AT5G43270.1 chr5:17377529-17380201 REVERSE; [PFAM]	Cleavage	1
ath-miR156a	chr1_26010001_26013400_FORWARD	1.0	17.076	miRNA 20 CACGAGUGAGAGAAGACAGU 1 :::::: ::::::::::::: Target 394 GUGCUCUCUCUCUUCUGUCA 413	AT1G69180.1 chr1:26011128-26012722 REVERSE; AT1G69170.1 chr1:26008731-26010926 FORWARD; [PFAM]	Cleavage	1
ath-miR156a	chr3_21453001_21456400_REVERSE	1.0	12.477	miRNA 20 CACGAGUGAGAGAAGACAGU 1 :::::: ::::::::::::: Target 571 GUGCUCUCUCUCUUCUGUCA 590	AT3G57920.1 chr3:21455298-21457012 REVERSE; AT3G57900.1 chr3:21453258-21453551 REVERSE; AT3G57910.1 chr3:21453725-21455288 FORWARD; [PFAM] 146-235 PF03110.7 SBP domain;	Cleavage	1
ath-miR156a	chr5_20598001_20601400_REVERSE	1.0	14.449	miRNA 20 CACGAGUGAGAGAAGACAGU 1 :::::: ::::::::::::: Target 1418 GUGCUCUCUCUCUUCUGUCA 1437	AT5G50570.1 chr5:20599309-20601785 REVERSE; AT5G50565.1 chr5:20598068-20599052 FORWARD; AT5G50570.2 chr5:20599309-20601106 REVERSE; [PFAM] 684-824 PF03110.7 SBP domain;	Cleavage	1

FIGURE 46.9 Result page of the psRNATarget tool.

Protein Structure Prediction Using Threading Method

CHAPTER 30

CS Mukhopadhyay and HK Manku
School of Animal Biotechnology, GADVASU, Ludhiana

30.1 INTRODUCTION

Threading or fold recognition is the method for protein tertiary structure prediction, and is implemented when no significant homologs are searched. The underlying theory assumes the existence of a limited number of distinct protein folds. The threading method thus searches the structural analogs of the query sequence in a library of representative structures through comparative fold recognition. The threading approach relies on fitting the target peptide sequence to one or more entries of a library of protein structures (e.g., protein domains, peptide chains or conserved protein cores). The best fit is evaluated on the basis of an appropriate energy function (determined by a bio-computational algorithm) to determine the best possible templates. In this way, folds are recognized simultaneously, and the whole structure can be predicted.

30.2 OBJECTIVE

To predict the tertiary structure of a protein sequence using the fold recognition method.

30.3 PROCEDURE

3D structure prediction using the threading method is explained by using the RaptorX server:

a. Query submission: Submit a query sequence for protein structure prediction via the URL http://raptorx.uchicago.edu/StructurePrediction/predict/.
b. Click on "Submit a new job".
c. Create a job identification title so that you can access your results page in the future.
d. Enter a valid email ID.
e. Paste your query sequence or upload the sequence in FASTA format, and submit your query.

Basic Applied Bioinformatics, First Edition. Chandra Sekhar Mukhopadhyay, Ratan Kumar Choudhary and Mir Asif Iquebal.

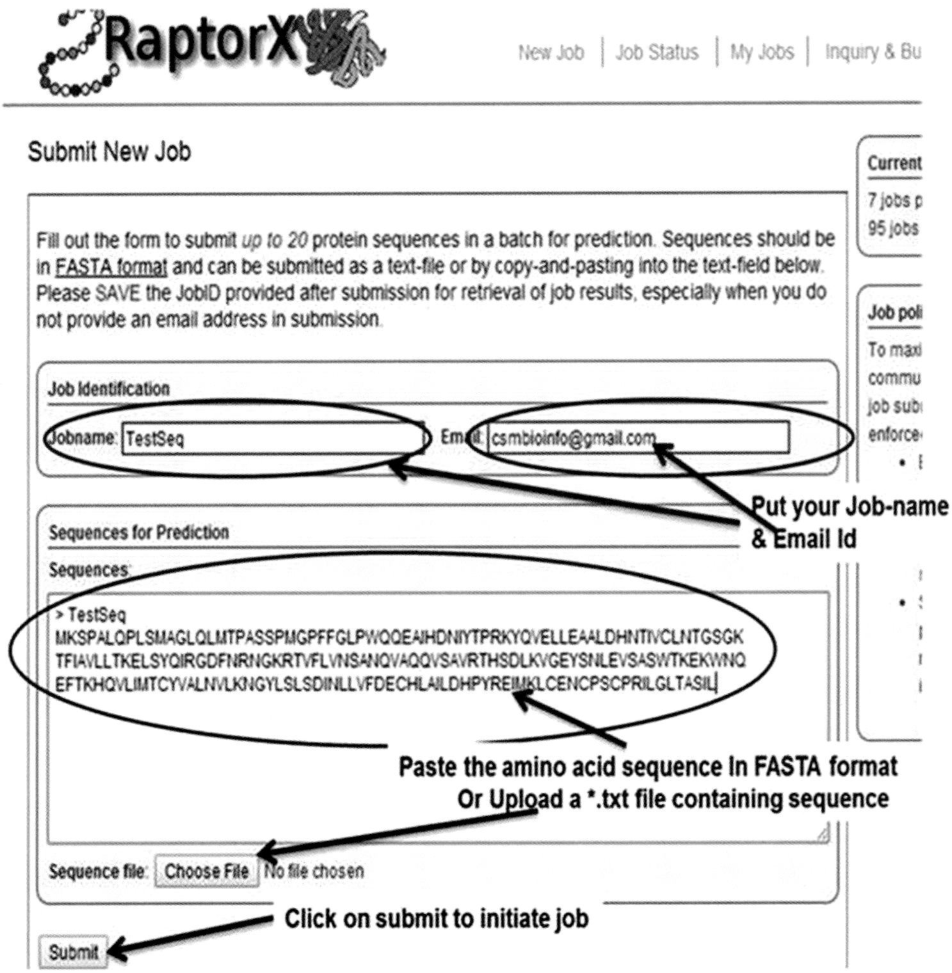

FIGURE 30.1 Home window of RaptorX for job submission.

30.4 RESULTS AND INTERPRETATION

The Results window displays the results as follows:

Section 1 shows the amino acid residues of the query sequence which are modeled and non-modeled. Modeled residues are labeled as 1, whereas non-modeled residues are labeled as 0 (shown in (A) of Figure 30.2). Modeled residues are considered as a functional part of the protein, and non-modeled residues are considered as side loops. Moreover, the result shows the best template selection by GDT (Global Distance Test) score. For a protein with > 100 residues, GDT > 50 is a good indicator (shown in (B) of Figure 30.2). In this example protein sequence, accession no. P79362 in SwissProt has been taken for tertiary structure prediction.

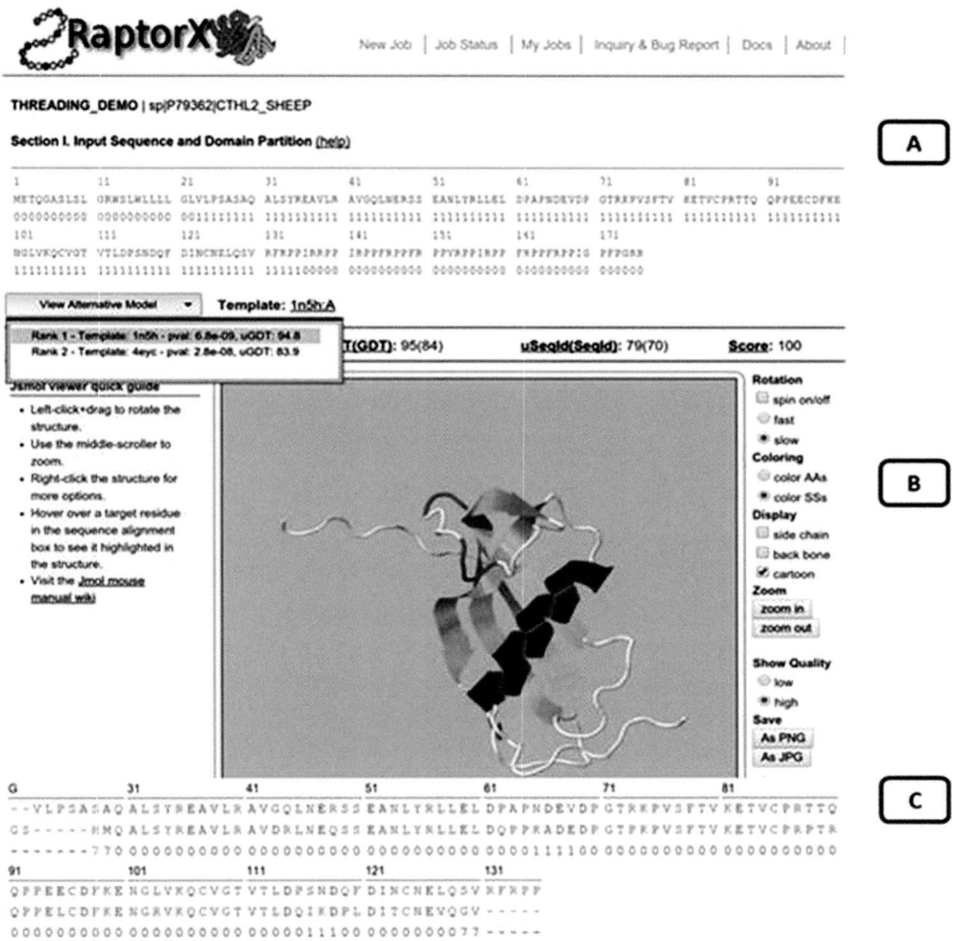

FIGURE 30.2 (A) The modeled and non-modeled residues; (B) 3D cartoon view of selected template; (C) the target-template alignment view.

30.4.1 The target-template alignment view

This alignment (shown in (C) of Figure 30.2) is used for constructing the 3D model. The chemical nature of the residues in the alignment dictates the colors of each position:

i. Blue = Acidic
ii. Magenta = Basic
iii. Red = Hydrophobic
iv. Green = Hydroxyl + Amine.

30.4.2 Tertiary structure prediction

The predicted results indicated that one domain that had the best matching template; 1n5hA had the P-value of $6.78e^{-09}$. The P-value is an indicator of the goodness of the best template representing the predicted secondary structure. Overall, 64% of the residues were modeled.

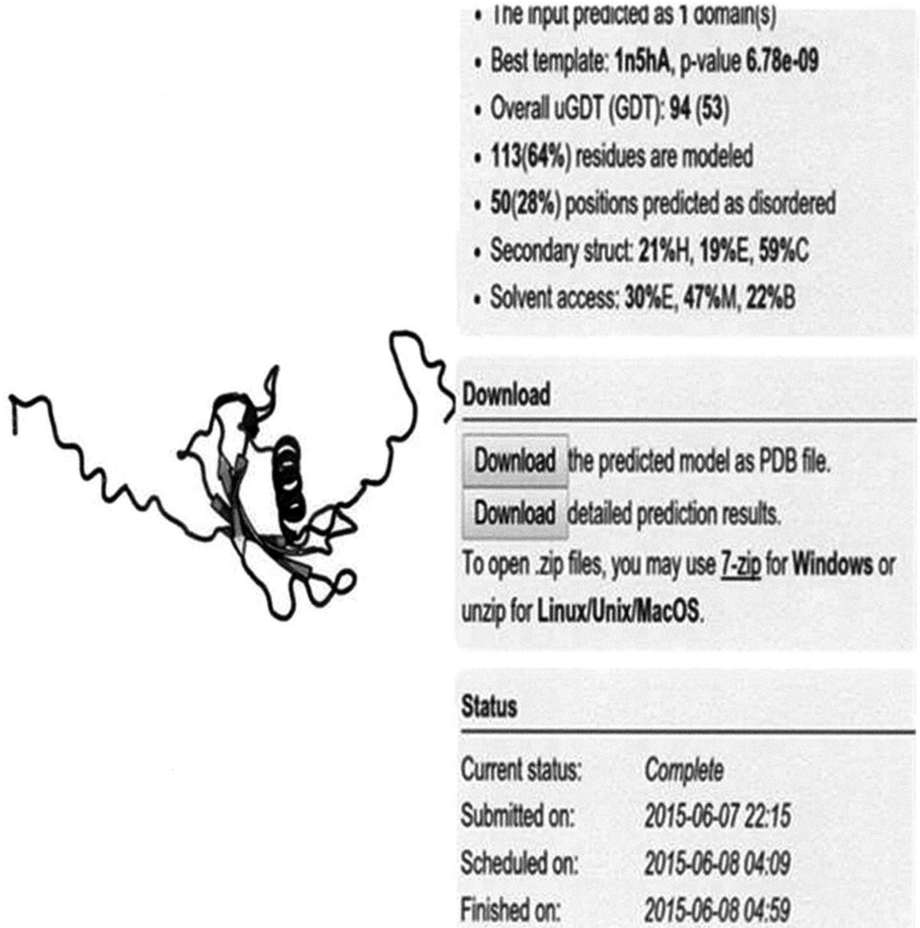

FIGURE 30.3 3D cartoon view of tertiary structure predicted by RaptorX server.

30.4.3 Secondary structure prediction

The user can switch between the pairs of models (three state and eight state models), or use both of them for secondary structure prediction. Hovering over a residue will display the exact distribution of secondary structure classes in a pop-up box appearing next to the residue. The three-state secondary structure types are also abbreviated as "H", "E", and "C", which represent helix, beta sheet, and loop, respectively.

This results window shows the three-class secondary structure result. A single block contains ten residues, which together form a histogram-like shape in a block, where each bar indicates the percentage of helix, coil, and beta strand. For example, P at the 135th position contributes 15.5% helix, 18.8% beta strand, and 67.8% coil (as shown in (A) in Figure 30.4). Similarly, an eight-class secondary structure result has been demonstrated, showing the percentage of alpha helix, isolated beta bridge,

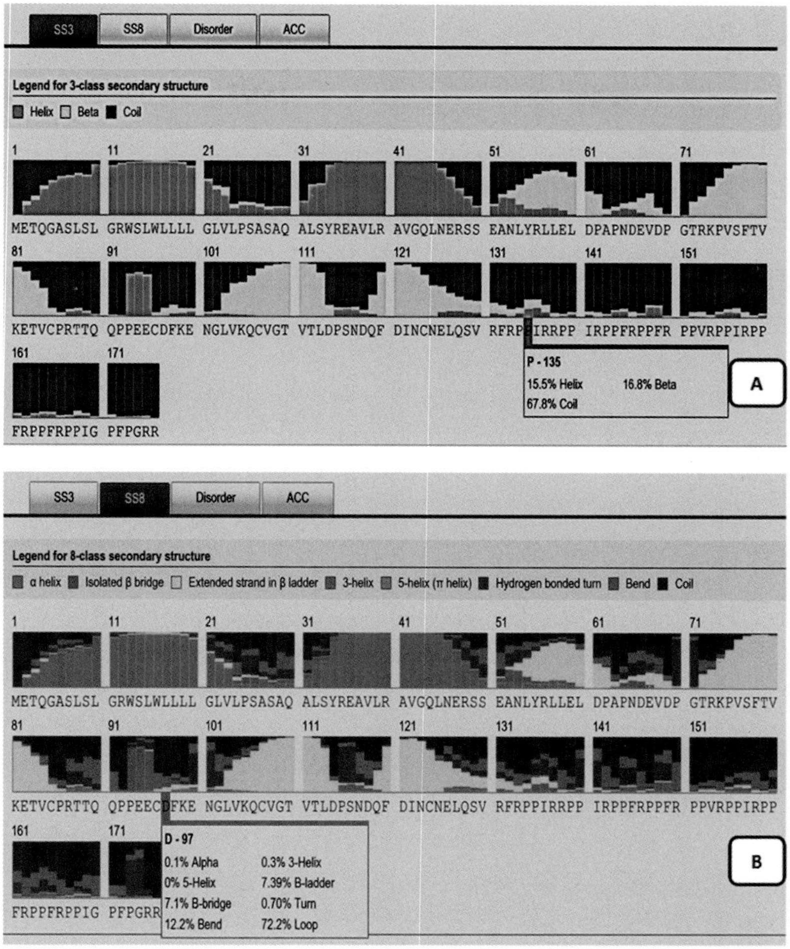

FIGURE 30.4 3 Class SS3 and 8 Class SS8 secondary structural element contribution to the 3D structure. (*See insert for colour representation of the figure.*)

extended strand in the beta ladder, 3-helix, 5-helix, hydrogen bonded turn, bend, and coil (as shown in (B) in Figure 30.4).

In addition to the structural element prediction, RaptorX also predicts the percentages of the residues which contribute to the disordered conformational randomness of the secondary structure. Moreover, it gives the percentages of the residues which are involved in the formation of the solvent-accessible surface. Maximum exposure of the residue means a greater contribution towards the solvent-accessible surface. For example, I at the 122nd position is 81.8% buried, 15.4% in the medium region, and 2.7% exposed. This means that this residue makes the smallest contribution towards the formation of the solvent-accessible surface.

The threading method is beneficial for the identification of the conserved domains. The predicted tertiary structure can be further analyzed for model quality validation using Procheck.

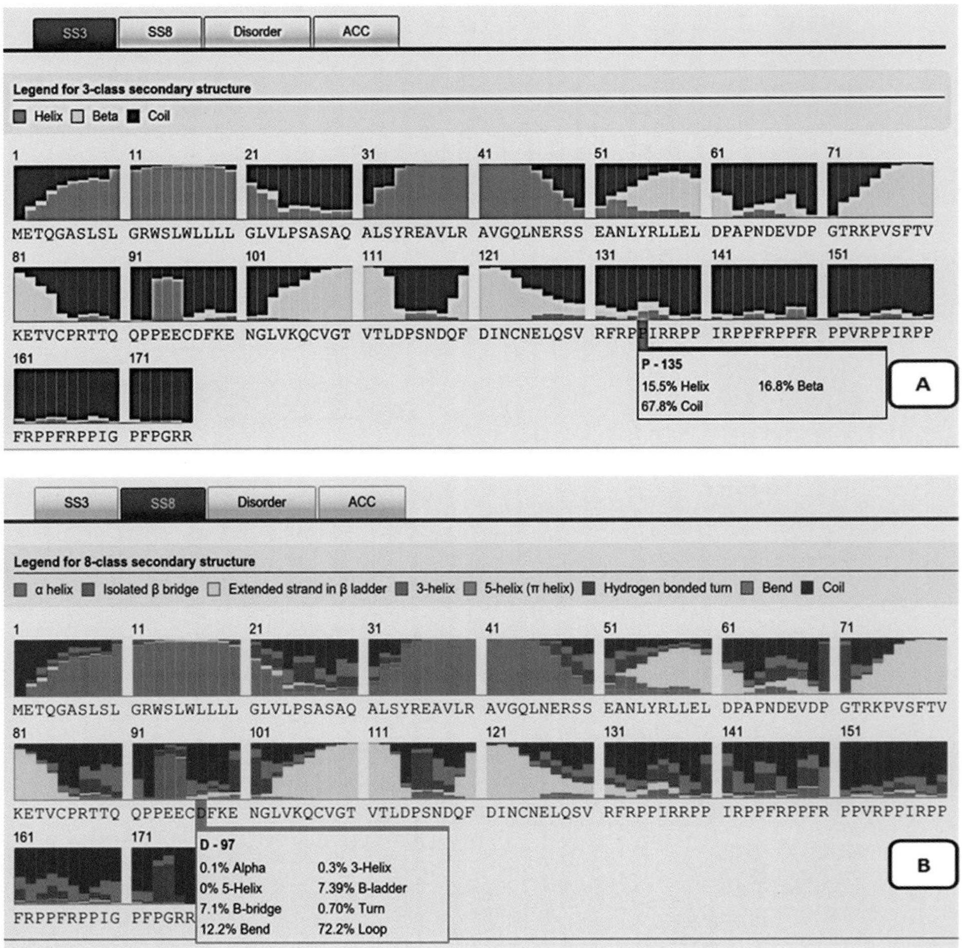

FIGURE 30.5 Conformationally ordered and disordered contribution of the residues in the 2D and 3D structure (C). Contribution of each residue in solvent accessibility (D). (*See insert for colour representation of the figure.*)

30.5 QUESTIONS

1. Why choose a threading method for tertiary structure prediction when the homology modeling method is already available?
2. Which template will be the best among the top five templates?
3. What is solvent accessibility?
4. Predict the tertiary structure prediction for AB973433 protein using the fold recognition method.

Prediction of Tertiary Structure of Protein: *Ab Initio* Approach

CHAPTER 31

CS Mukhopadhyay and RK Choudhary
School of Animal Biotechnology, GADVASU, Ludhiana

31.1 INTRODUCTION

Protein folding is dictated by the physical forces acting on the atoms of the protein. In general, the most accurate way of formulating the protein-folding or structure prediction problem is in terms of an all-atom model subject to the physical forces. Energy functions compatible with the protein representation are considered during the *ab initio* approach. Faster algorithms are developed to search the best-fitting formation, in order to minimize the energy function while predicting the tertiary structure *ab initio*.

31.1.1 Principle

To predict the protein structure based on physical principles (rather than comparative homology using the previously reported structures). The structural conformations that minimize the energy function are evaluated for the structures that the protein is likely to adopt under native conditions.

31.2 OBJECTIVE

To predict the tertiary structure of a peptide using an *ab initio* approach with the online tool RaptorX.

31.3 PROCEDURE (RAPTORX)

a. Suppose we need to predict the structure of the given protein sequence (Bubaline Dicer I; NCBI Protein Accession Number BAP00765.1):
   >BAP00765|Bbu_Dicer1
   M K S P A L Q P L S M A G L Q L M T P A S S P M G P F F G L P W Q Q E A
   I H D N I Y T P R K Y Q V E L L E A A L D H N T I V C L N T G S G K T F
   I A V L L T K E L S Y Q I R G D F N R N G K R T V F L V N S A N Q V A Q

Basic Applied Bioinformatics, First Edition. Chandra Sekhar Mukhopadhyay, Ratan Kumar Choudhary and Mir Asif Iquebal.

Q V S A V R T H S D L K V G E Y S N L E V S A S W T K E K W N L E F T K
H Q V L V M T C Y V A L N V L K N G Y L S L S D I N L L V F D E C H L A
I L D H P Y R E I M K L C E N C P S C P R I L G L T A S I L N G K C D P
E E L E E K I Q K L E K I L K S N A E T A T D L V V L D R Y T S Q P C E
I V V D C G P F T D R S G L Y E R L L M E L E E A L N F I N D C N I S V
H S K E R D S T L I S K Q I L S D C R A V L V V L G P W C A D K V A G M
M V R E L Q K H I K H E Q E E L H R K F L L F T D T F L R K I H A L C E
E H F S P A S L D L K F V T P K V I K L L E I L R K Y K P Y E R Q Q F E
S V E W Y N N R N Q D N Y V S W S D S E D D E E D E E I E E K E K P E T
N F P S P F T N I L C G I I F V E R R Y T A V V L N R L I K E A G K Q D
P E L A Y I S S N F I T G H G I G K N Q P R N K Q M E A E F R K Q E E V
L R K F R A H E T N L L I A T S I V E E G V D I P K C N L V V R F D L P
T E Y R S Y V Q S K G R A R A P I S N Y V M L A D T D K I K S F E E D L
K T Y K A I E K I L R N K C S K S V D T G E A D T E P V V D D D D V F P
P Y V L R P E D G P R V T I N T A I G H V N R Y C A R L P S D P F T H L
A P K C R T R E L P D G T F Y S T L Y L P I N S P L R A S I V G P P M S
C I R L A E R V V A L I C C E K L H K I G E L D D H L M P V G K E T V K
Y E E E L D L H D E E E T S V P G R P G S T K R R Q C Y P K A I P E C L
R E S Y P R P G Q P C Y L Y V I G M V L T T P L P D E L N F R R R K L Y
P P E D T T R C F G I L T A K P I P Q I P H F P V Y T R S G E V T I S I
E L K K S G F T L S L Q M L E L I T R L H Q Y I F S H I L R L E K P A L
E F K P T D A D S A Y C V L P L N V V N D S S T L D I D F K F M E D I E
K S E A R I G I P S T K Y S K E T P F V F K L E D Y Q D A V I I P R Y R
N F D Q P H R F Y V A D V Y T D L T P L S K F P S P E Y E T F A E Y Y K
T K Y N L D L T N L N Q P L L D V D H T S S R L N L L T P R H L N Q K G
K A L P L S S A E K R K A K W E S L Q N K Q I L V P E L C A I H P I P A
S L W R K A V C L P S I L Y R L H C L L T A E E L R A Q T A S D A G V G
V R S L P V D F R Y P N L D F G W K K S I D S K S F I S I A N S S S A E
N E N Y C K H S T I V V P E N A A H Q G A N R T S P L E N H D Q M S V N
C R T L F S E S P G K L Q I E V S T D L T A I N G L S Y N K S L A N G S
Y D L A N R D F C Q G N H L N Y Y K Q E I P V Q P T T S Y P I Q N L Y N
Y E N Q P K P S D E C T L L S N K Y L D G N A D T S T S D G S P V T A A
V P G T T E T G E A P P D R T A S E Q S P S P G Y S S R T L G P N P G L
I L Q A L T L S N A S D G F N L E R L E M L G D S F L K H A I T T Y L F
C T Y P D A H E G R L S Y M R S K K V S N C N L Y R L G K K K G L P S R
M V V S I F D P P V N W L P P G Y V V N Q D K S N T E K W E K D E M T K
D C M L A N G K L D D D F E E E E E E E E D L M W R A P K E D A D D E D
D F L E Y D Q E H I K F I D N M L M G S G A F V K K I S L S P F S A T D
S A Y E W K M P K K S S L G S L P F S S D F E D F D Y S S W D A M C Y L
D P S K A V E E D D F V V G F W N P S E E N C G V D T G K Q S I S Y D L
H T E Q C I A D K S I A D C V E A L L G C Y L T S C G E R A A Q L F L C
S L G L K V L P V I K R T D R E K A M C P T R E N F T S Q Q K N L S G S
R A A A S G A G Y R A S V L K D L E Y G C L K I P P R C M F D H P D A D
R T L R H L I S G F E N F E K K I N Y R F K N K A Y L L Q A F T H A S Y
H Y N T I T D C Y Q R L E F L G E P I M D Y L I T K H L Y E D P R Q H S

P G V L T D L R S A L V N N T I F A S L A V K Y D Y H K Y F K A V S P E
L F H V I D D F V Q F Q L E K N E M Q G M D S E L R R S E E D E E K E E
D I E V P K A M G D I F E S L A G A I Y M D S G M S L E T V W Q V Y Y P
M M R P L I E K F S A N V P R S P V R E L L E M E P E T A K F S P A E R
T Y D G K V R V T V E V V G K G K F K G V G R S Y R I A K S A A A R R A
L R S L K A N Q P Q V P N S

b. Open RaptorX: http://raptorx.uchicago.edu/
c. Click on the "Submit" button in the first server option "RaptorX structure prediction", or directly paste this uniform resource locator (URL) in the space for the URL: http://raptorx.uchicago.edu/StructurePrediction/predict/
d. Type your job identification and email in the respective spaces, as shown in the previous chapter (Figure 30.1).
e. Paste the input sequence (in FASTA format) in the sequence box under "Sequences for Prediction".

31.4 JOB STATUS

Once the job has been started, the status can be seen in real time on the screen as below:

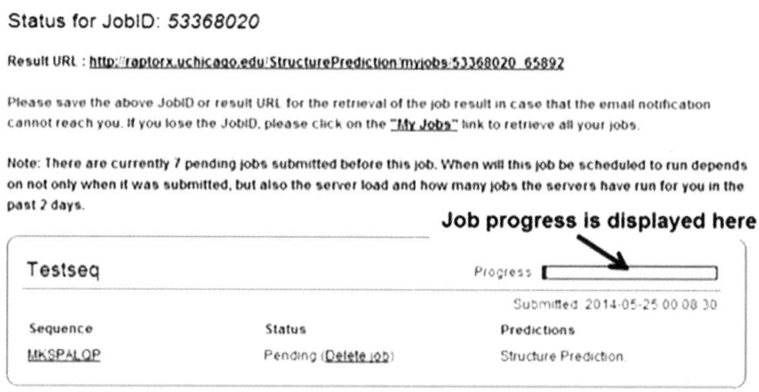

FIGURE 31.1 Job progress box of RaptorX after job submission.

31.5 OUTPUT AND INTERPRETATION OF RESULTS

Open the specified email ID to get the results:

a. Predicted structure in *.png format.
b. Predicted structure in *.pdb format.
c. Ligand binding site detail as *domain_pocket.txt file.

Click on the link provided in the mail to open the detailed result in a new window.

31.5.1 Section I: input sequence and domain partition

The whole protein sequence is partitioned into domains (here six domains), depending on the available template structures in PDB. The domains are indicated by assigning a domain number in every third row of sequence blocks.

i. Up to 20 peptide sequences in FASTA format can be submitted in one run.
ii. No residues other than the standard IUPAC recommended one-word symbols of amino acids (i.e., “A”, “R”, “N”, “D”, “C”, “Q”, “E”, “G”, “H”, “I”, “L”, “K”, “M”, “F”, “P”, “S”, “T”, “W”, “Y”, “V”) can be present in the amino acid sequence(s).
iii. Limitation for number of amino acids: the total number of amino acids should be more than 26 in any one sequence and less than 2000 for all of the peptide sequences.
iv. One user can have a maximum of 500 sequences in the queue at any point in time.
v. If the sequence length is more than 2000 amino acids, or if the sequence belongs to a special project that requires more resources, the user needs to contact the RaptorX personnel.

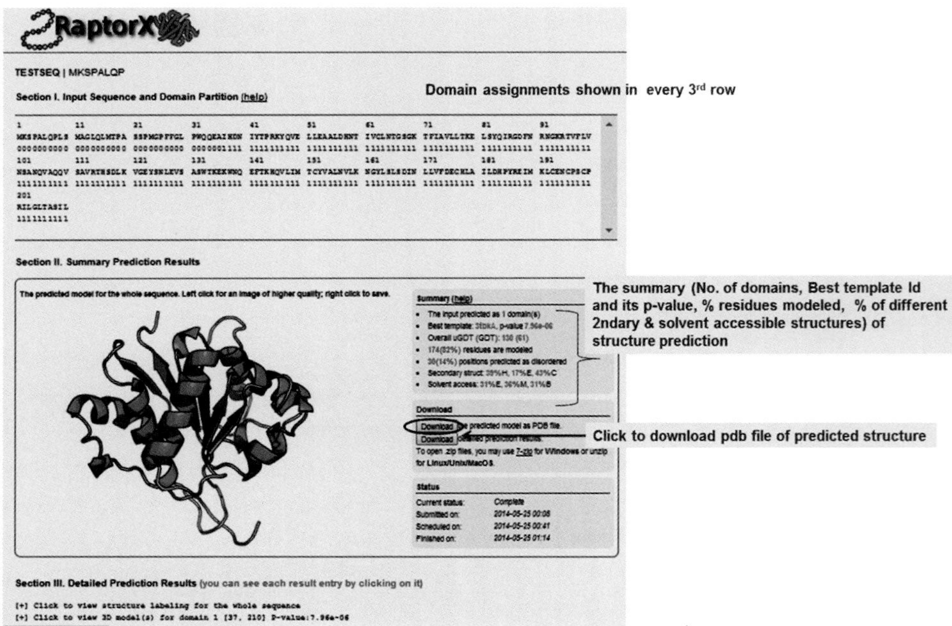

FIGURE 31.2 Results windows of RaptorX, indicating assignment of protein domain and 3D prediction results. (*See insert for colour representation of the figure.*)

31.5.2 Secondary structure prediction

The proportions of secondary structures (e.g., (Helix (H), Beta-sheet (E) and Loop (L)) are given as percentages. In our study, the result shows “Secondary struct: 40%H, 10%E, 49%C”.

31.5.3 Solvent accessibility (ACC)

Three grades of solvent accessibility are designated by Buried (B: cut-off value 10%), Medium (M: range 10–42%) and Exposed (E: cut-off value: 42%). The present example shows: “Solvent access: 32%E, 38%M, 29%B”.

31.5.4 Quality assessment of predicted tertiary structure

Many parameters are checked to determine the quality of the predicted structure.

31.5.5 Probability of obtaining a random structure instead of the best one (P value)

The smaller the value, the better the prediction. If the P-value is more than 10E-3 and 10E-4 for alpha and beta proteins respectively, the predicted sequence is to be discarded.

31.5.6 Alignment score (score)

The higher the score, the better is the prediction.

31.5.7 Number of identical residues (uSeqId and SeqId)

This indicates the number of identical residues in the alignment. Here, the "u" of "uSeqId" stands for un-normalized. The higher the value, the better it is considered to be. In general, for a peptide sequence of 200 residues, the cut-off value of normalized uSeqId is 30%.

31.5.8 Global distance test (GDT)

This estimates the modeling error from a score determined from the residues with modeling error.

31.5.9 Pocket multiplicity

This parameter is considered for binding site prediction in order to assess the quality of predicted pocket. Multiplicity higher than 40% is considered good.

31.6 QUESTIONS

1. Predict the tertiary structure of the caprine beta-defensin (GenBank Acc. No. ABF71365.1) using an *ab initio* approach
2. Predict the tertiary structure of the same peptide (caprine beta-defensin) by homology modeling, using a human homolog.
3. Can you compare the results obtained from the above two methods?
4. What parameters are considered in RaptorX for determining the goodness of the predicted structure?

Validation of Predicted Tertiary Structure of Protein

CHAPTER 32

CS Mukhopadhyay and RK Choudhary
School of Animal Biotechnology, GADVASU, Ludhiana

32.1 INTRODUCTION

The tertiary structure of a protein can be experimentally determined by nuclear magnetic resonance and X-ray crystallography, which are expensive and labor-intensive. Biocomputationally predicted tertiary structures must be validated for the permissibility of dihedral angles.

32.2 OBJECTIVE

To validate the predicted tertiary structure of a protein using two online tools, namely the WHAT IF web interface and MOLprabity.

32.3 PROCEDURE (WHAT IF TOOL FOR VALIDATING THE 3D STRUCTURE PREDICTION RESULTS)

a. Input file: download the .pdb file of the modeled structure: after completion of structure prediction.
b. Open the WHAT IF home page to check the reliability of the predicted structure. URL: http://swift.cmbi.ru.nl/servers/html/index.html.
c. Click on "Build/check/repair model" in the left-hand pane (Figure 32.1).
d. Click on "Protein Model Check" (again, 3rd third option in the main window); this will enable the user to run a set of the protein structure prediction model checking option.
e. Click on "Browse" button to upload the predicted model file (*.pdb format).
f. Click on "Send" button to submit the structure for analysis.

Basic Applied Bioinformatics, First Edition. Chandra Sekhar Mukhopadhyay, Ratan Kumar Choudhary and Mir Asif Iquebal.

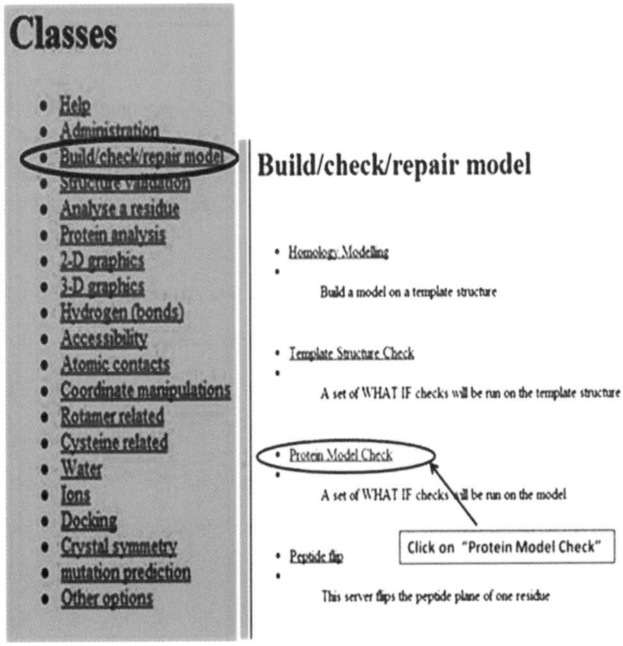

FIGURE 32.1 Homepage of WHAT IF web interface; click on "Build/check/repair model" link (in the left-hand pane) to initiate validation of the predicted tertiary structure of the peptide.

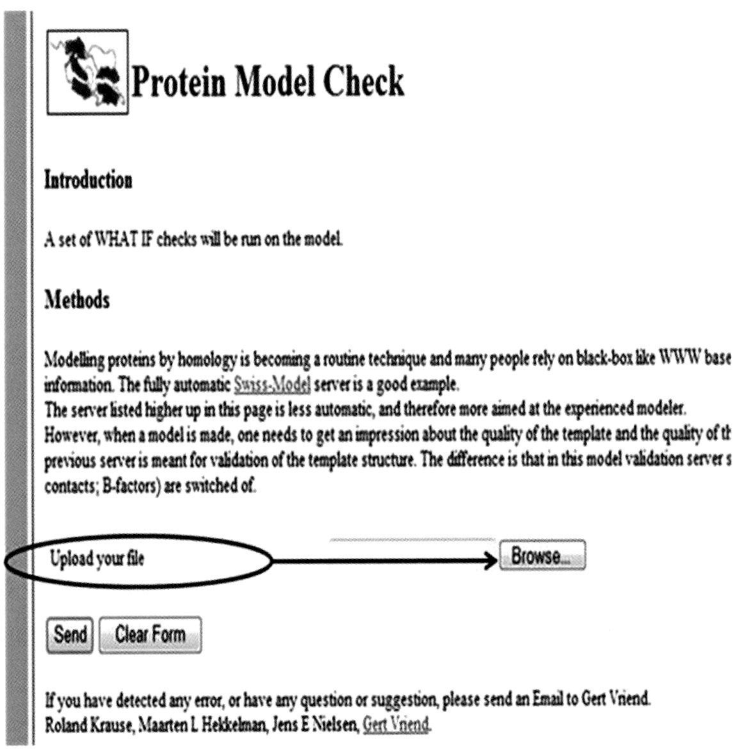

FIGURE 32.2 Click on the "Upload your file" button to browse the input file and then click on "Send" button to upload the file to the server.

32.4 INTERPRETATION OF RESULTS OF WHAT IF

The "WHAT IF" output provides the necessary guidelines and annotations for the terms used. The result is quite user-friendly. The important considerations in this regard are as follows:

a. The program indicates the errors or uncommon findings as "error" and "warning". The user needs to take note of each of the errors and warnings.
b. Explanatory notes are given as and when required to elucidate statistical analyses or plots.
c. The amino acids are mentioned as "residues", along with their positional values within brackets as residue number.
d. Hydrogen atoms are rarely included (but can be, if there is an explicit request to include them) in the error checking and analyses.
e. After studying the whole report, the "aberrant" structural features should be checked, and rectified if possible.

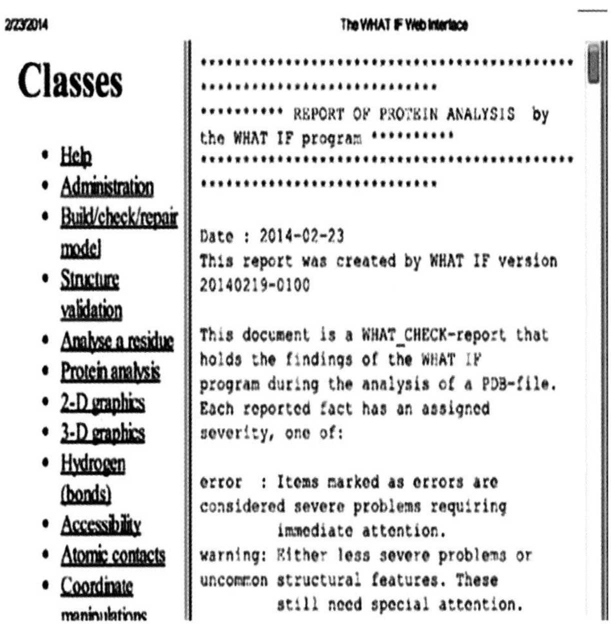

FIGURE 32.3 Output of the WHAT IF analysis of the predicted tertiary structure of the peptide.

32.5 MOLPROBITY TOOL FOR RAMACHANDRAN PLOT

MOLprobity is an online tool for all-atom contacts and structure validation for proteins and nucleic acids.

Ramachandran plot: used to visualize the backbone dihedral angles (ψ against φ) of amino acid residues in protein structure.
Dihedral angles: The dihedral angles in a peptide backbone are:

i. Phi(ϕ): rotation about the N-C(α) bond, involving the C(O)-N-C(α)-C(O) bonds.
ii. Psi(ψ): rotation about the C(α)-C(O) bond, involving the N-C(α)-C(O)-N bonds.
iii. Omega (ω): rotation about the C(O)-N bond, involving the C(α)-C(O)-N-C(α) bonds.

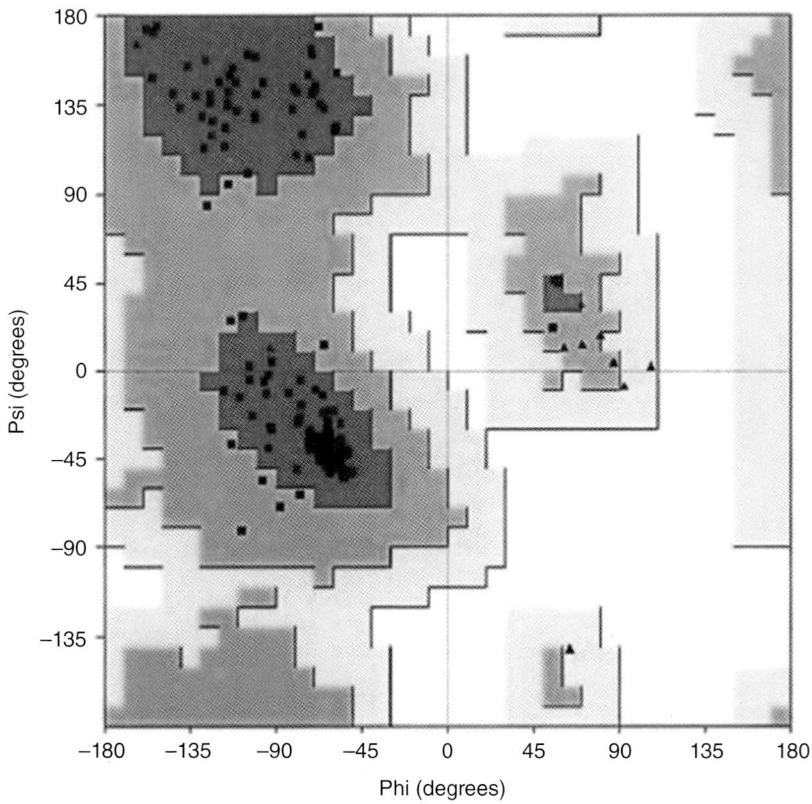

FIGURE 32.4 Ramachandran plot for a typical protein structure. The different regions were taken from the observed phi-psi distribution for 121 870 residues from 463 known X-ray protein structures. (*See insert for colour representation of the figure.*)

32.5.1 Procedure (Ramachandran plot analysis using MOLprobity)

a. Open the MOLProbity homepage. URL: http://molprobity.biochem.duke.edu/.
b. Input file: the.pdb file (downloaded after structure prediction) is uploaded.
c. Click "Analyze geometry without all atom contacts". Alternatively, another good tool for this purpose is RaptorX (http://mordred.bioc.cam.ac.uk/~rapper/rampage.php).

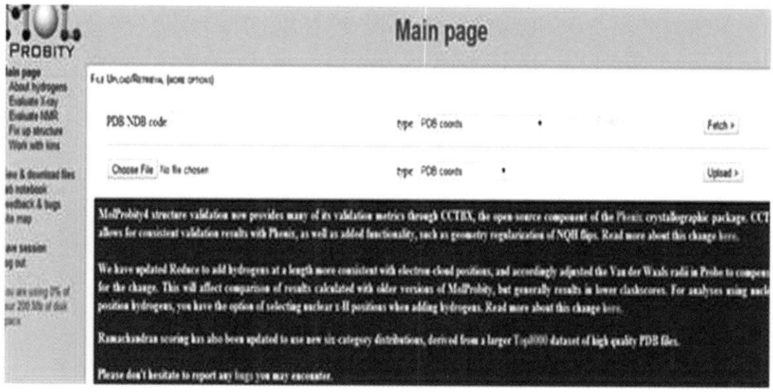

FIGURE 32.5 The homepage of MOLProbity tool for Ramachandran plot analysis.

32.6 INTERPRETATION OF RAMACHANDRAN PLOT ANALYSIS

a. Squares: denote phi/psi angular ratios of each non-glycine residue.
b. Triangles: phi/psi ratios for glycine residues.
c. Red-shaded regions: indicate the most favored regions of the Ramachandran plot; therefore, the amino acid residues falling in these regions are allowed.
d. Yellow shades: stand for the additional allowed regions (or the "generous region").
e. White regions: corresponds to the "non-favorable" regions of the plot.
f. The top left section: for β-strand secondary structure.
g. The middle left section: for α-helix.
h. The small middle right section: corresponds to the left-handed helix.

More than 90% of the non-glycine residues fall within the red regions of the plot.

The **Ramachandran plot** is a 2D graphical depiction of the plot between the peptide torsion angles Phi(ϕ) and Psi(ψ), in degrees. The dihedral angle of rotation about the alpha-carbon to carbonyl-carbon bond (i.e., Psi(ψ)) is plotted against the dihedral angle of rotation about the alpha-carbon to nitrogen bond (i.e., Phi(ϕ)) along the *X* and *Y* axes, respectively. It is used as a powerful tool to detect errors in predicted protein structure. Glycine residues are separately identified, as there is no restriction imposed to glycine to the regions of the plot, like those with a side-chain. The differently colored regions represent the favored, allowed, and "generously allowed" regions to validate predicted structures.

32.7 QUESTIONS

1. Validate the tertiary structures obtained from the analyses in previous chapters for tertiary structure prediction.
2. What are the parameters shown on a Ramachandran plot? Explain how the Ramachandran plot is interpreted.

3. What is the principle of the Ramachandran plot? Why are the phi and psi angles important in validating the protein structure?
4. Why does glycine get special amino acid status in the Ramachandran plot?
5. What will be your comments on the validity of a predicted tertiary protein structure, if 25% of the non-glycine residues fall in the white region?

Molecular Docking and Binding Site Prediction

SECTION VII

Prediction of Transcription Binding Sites

CHAPTER 33

S Jain[1], S Panwar[2] and A Kumar[3]
[1]Department of Applied Sciences & Humanities, Jai Parkash MukandLal Innovative Engineering and Technology Institute, Yamuna Nagar, Haryana, India
[2]Department of Genetics and Plant Breeding, Chaudhary Charan Singh University, Uttar Pradesh, India
[3]Department of Nutrition Biology, Central University of Haryana, Haryana, India

33.1 INTRODUCTION

Transcription factors are crucial for sequence-specific control of transcriptional regulation. Classically, the computational prediction of transcription factor binding sites (TFBS) depends on position weight matrices (PWMs) (Wingender *et al.*, 2001), which give weights to each nucleotide at each position. These models strongly suggest that each nucleotide participates independently in the corresponding DNA–protein interaction and does not account for flexible length motifs.

33.2 OBJECTIVE

To predict the transcription binding site by using the TRANSFAC and MATCH tools

33.3 TRANSFAC

TRANSFAC is a database of **TRANS**cription regulatory **FAC**tors, and is maintained at GBF Braunschweig (Wingender *et al.*, 2000). It combines the data regarding transcription factors, their DNA binding sites, sources of the factors and systematic classification of transcription factors. All the experimental results are accessible mainly through the FACTORS and the SITES table (Frech *et al.*, 1997).

The data regarding binding proteins and the DNA sequences that are recognized by these proteins are maintained by the FACTORS and the SITES table, respectively.

Basic Applied Bioinformatics, First Edition. Chandra Sekhar Mukhopadhyay, Ratan Kumar Choudhary and Mir Asif Iquebal.

Furthermore, many transcription factors can be classified according to the respective DNA binding domains and/or their dimerization domains; therefore, the CLASS table has been introduced to TRANSFAC. Tiny TRP, a browsing tool for TRANSFAC, is the only solution that requires the linked databases in their original format. These links, between TRANSFAC and other databases such as PIR, EMBL, PROSITE, and so on, are crucial for the use of TRANSFAC.

33.3.1 Procedure

Enrique Blanco has discussed the procedure in the "Practical" online tutorial (http://genome.crg.es/courses/Bioinformatics2003_promoters/).

33.3.1.1 Accessing the TRANSFAC database

a. Go to the TRANSFAC database and choose the search in TRANSFAC 6.0 (Figure 33.1). The URL is http://www.biobase-international.com/product/transcription-factor-binding-sites.
b. Select the Factor table (Figure 33.1).
c. Type the factor name TBP (TATA binding protein).
d. Provide a Factor Name (FA) as searching field and then submit.
e. Choose (T00794) to find a description of the factor in humans.
f. On the left, "BS" (for binding sites) and "MX" (for matrices) will be there. Choose one of the sites for assessment.

33.3.1.2 Building a model from a set of actual sites

a. Actual TBP sites are collected from TRANSFAC.
b. Go to the CLUSTALW web server at EBI.
c. Bring up the collection of 23 TBP sites.
d. Switch on the boxes:
   ALIGNMENT = fast
   COLOR ALIGNMENT = yes
   OUTPUT FORMAT = aln wo/numbers
e. Click on "Run".

33.3.1.3 Open the WebLogo server

a. After placing the sequence alignment in the input box, "activate DNA/RNA" in the "Sequence type" box.
b. Submit the query.
c. The resulting representation of TBP sites as shown in Figure 33.2.

33.3.1.4 Obtaining the TRANSFAC position weight matrices

a. Go to the TRANSFAC database and choose the search in TRANSFAC 6.0.
b. Pick the matrix table (Figure 33.3).
c. Put in the factor name TATA.
d. Set Factor Name (NA) as searching field and submit the query. There are two entries: M00252 and M00216.

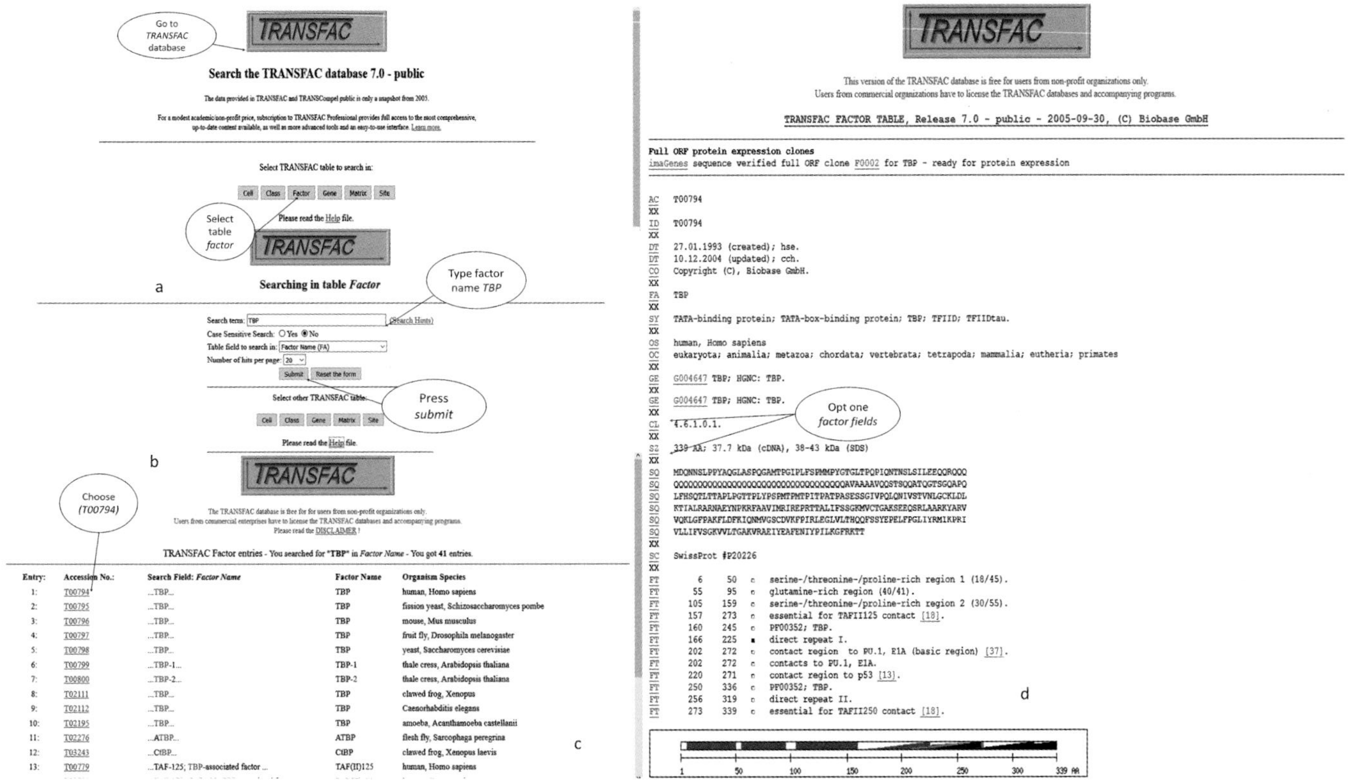

FIGURE 33.1 (a) TRANSFAC database search; (b) FACTOR table search; (c) TRANSFAC Factor entries; (d) output of TRANSFAC Factor table.

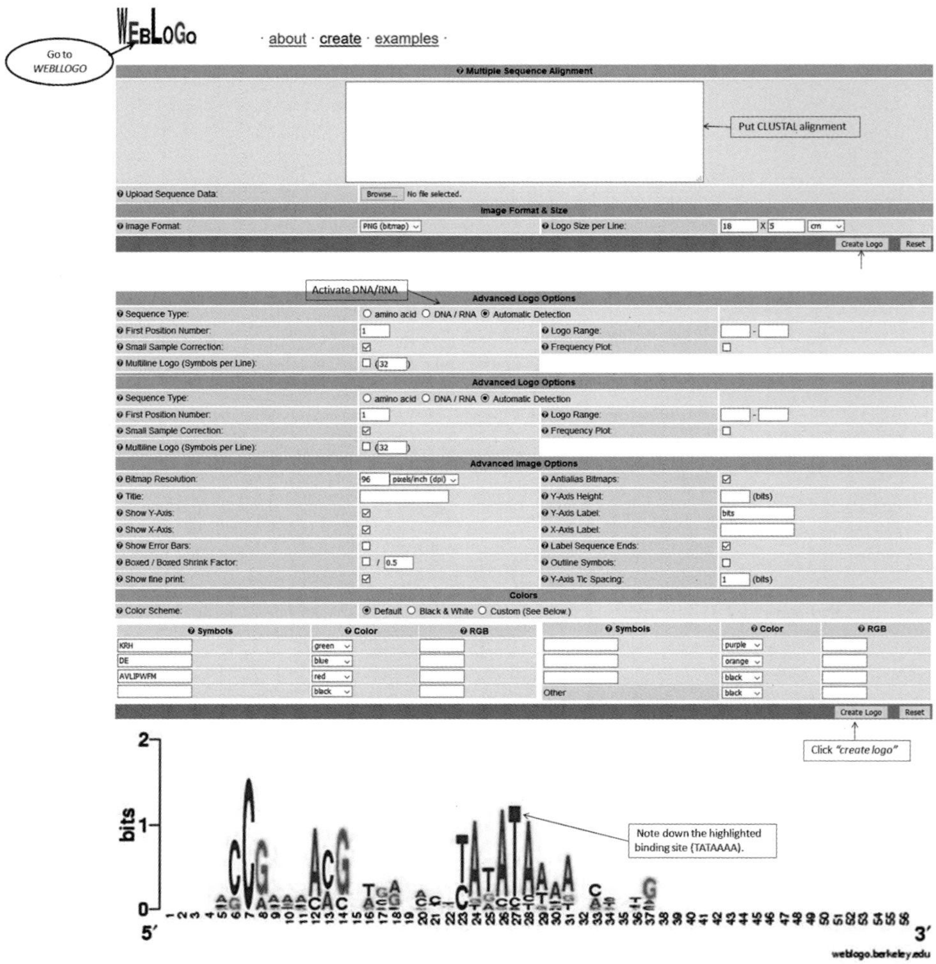

FIGURE 33.2 Creating sequences logos using the web interface.

e. After repeating the procedures, keep the windows containing the matrices (M00252 matrix and those for SP1 and c/EBP).
f. Compare the core of the matrix with the previous sequence logo.
g. Compare both to the TATA box binding site in the ABS entry.

33.3.2 Key features/benefits

a. Quickly access detailed reports of 41 000+ transcription factor binding site, 18 000+ miRNA target sites and 1100+ miRNA reports (Ying *et al.*, 2013), 22 000+ transcription factors, 13 837 000+ ChIP fragments and 273,000+ promoters.
b. Molecular mechanisms that enable transcription factors to orchestrate with gene expression *in vivo*.

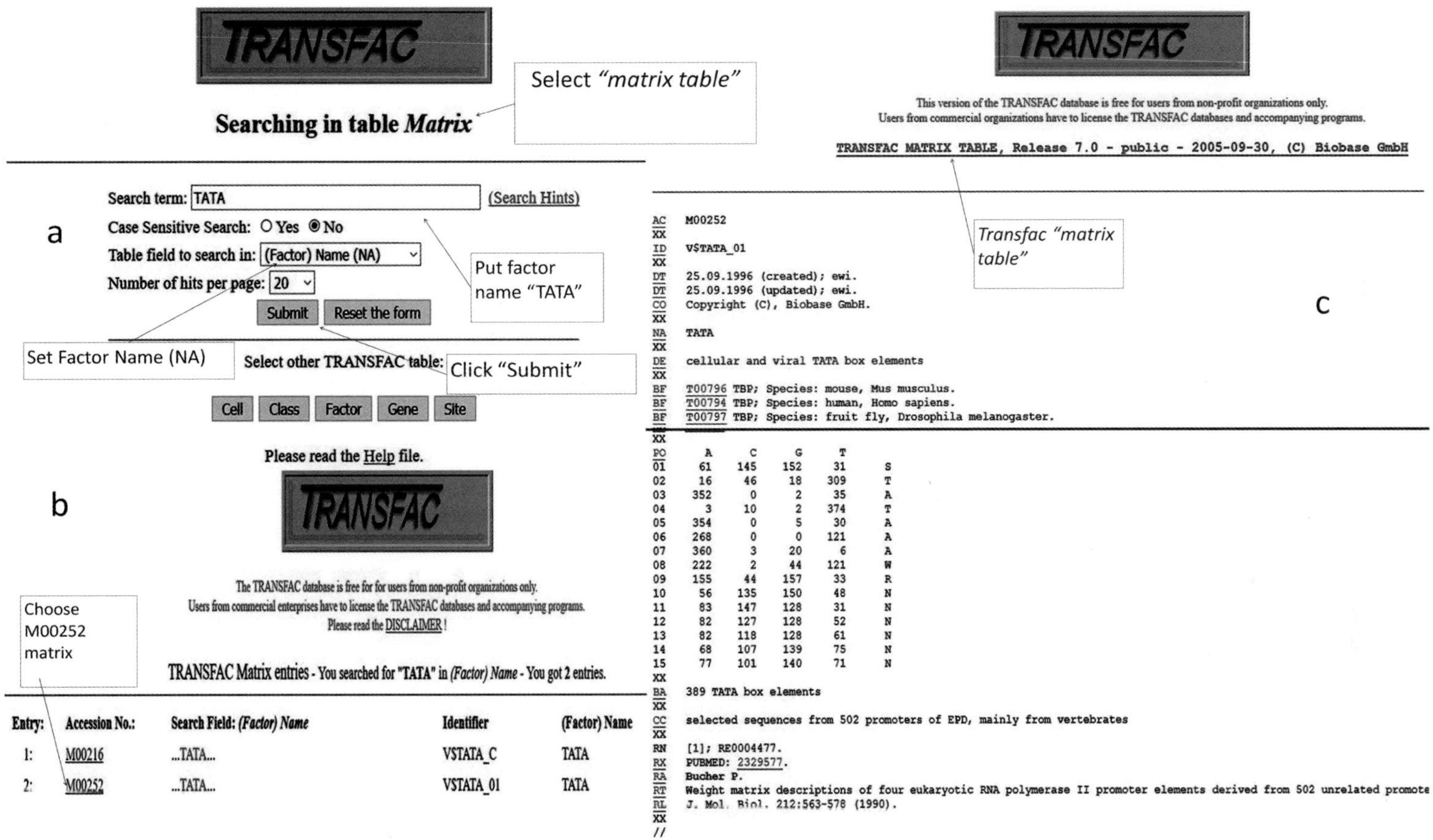

FIGURE 33.3 (a) Searching Transfac matrix table; (b) TRANSFAC Matrix entries; (c) output of TRANSFAC Matrix table.

33.3.3 Access options

An online subscription provides access to the TRANSFAC web interface. However, a download subscription provides access to flat files containing data for factors, matrices, binding sites, genes, ChIP fragments and other supporting information, as well as command line access to the MATCH tool.

33.4 BINDING SITES SEARCHING USING THE MATCH TOOL

The MATCH tool is used for searching binding sites for transcription factors in any sequence, using the mononucleotide weight matrix library from TRANSFAC.

33.4.1 Procedure

33.4.1.1 Enter a name for search

Open the MATCH server to analyze promoter regions with TRANSFAC matrices: http://www.gene-regulation.com/cgi-bin/pub/programs/match/bin/match.cgi.

Enter a name for the search, since MATCH will store the result under that name. It will use the default as the result name.

33.4.1.2 Select a sequence

There are three options for selecting a sequence for a search:

a. Select among the sequences entered for a previous search.
b. Select an example sequence, such as the 5′ flanking region of the tyrosine aminotransferase (TAT) gene of Rat (EMBL: M34257).
c. Enter a name for the new sequence. Store the sequence with that name so that it can be used again for a later search.
d. Then insert the sequence. The formats accepted are FASTA, TRANSFAC, EMBL, GenBank, IG, and RAW.

Select a group of matrices or a profile to run MATCH vertebrates, insects, plants, fungi, bacteria, and nematodes. The term "profile" refers to a set of weight matrices obtained from the TRANSFAC library.

a. *Matrix selection*: select the inserted group of matrices from the library. Several groups can be combined with one search. Specify the set of matrices (based on matrix similarity), either to restrict the search to the use of high-quality matrices only, or to include user-defined matrices. For any group of matrices, the cut-offs for core and matrix similarity can be specified.
b. *Profile selection*: One can either use a predefined profile or create one on the MATCH profiler page. To create a profile with the TRANSFAC search engine, use the following steps:
    i. Select the TRANSFAC query form "MATRIX SEARCH".
    ii. Intended entries boxes should be marked for inclusion in a MATCH search.
    iii. A box with the text Run Match with marked entries will be displayed. Mark this box also.
    iv. Click on "Show marked entries/Start MATCH". A selection of sites will be seen among the user-defined profile.

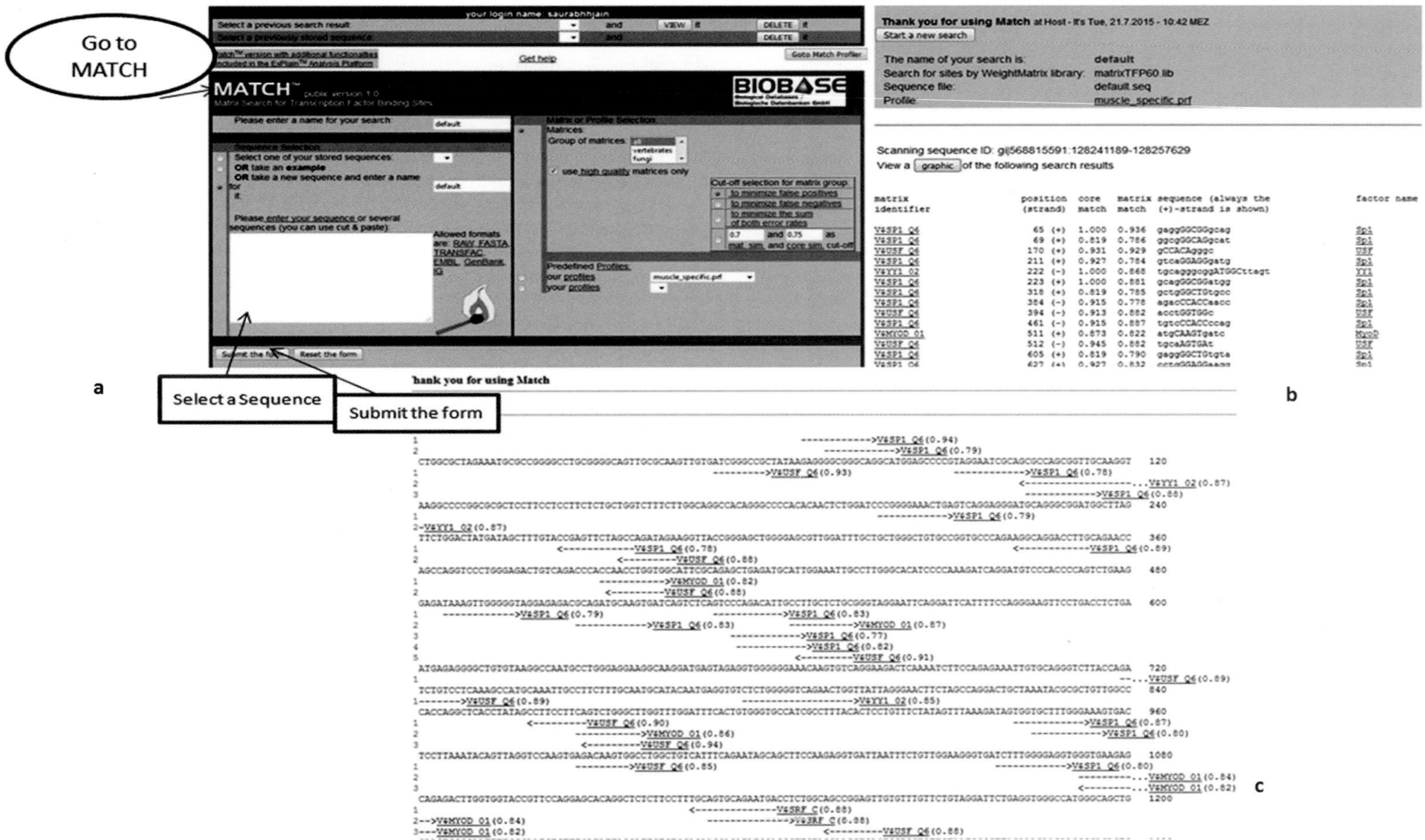

FIGURE 33.4 (a) MATCH user interface; (b) results page of MATCH output; (c) a simple visual representation of locations of the found matches.

c. *Predefined profiles*: various tissue-specific profiles such as immune cell-specific, cell cycle-specific, muscle-specific, and liver-specific are provided by MATCH.

33.4.1.3 Submit the form

The result page tabulates all matches found in the input sequence. The output of the program is limited to 500 000 matches per sequence. The outcomes are represented in Figure 33.4 with the following columns:

a. Respective matrix, linked with each identifier, linked to the TRANSFAC entry.
b. Score for core similarity (core match).
c. Score for matrix similarity (matrix match).
d. Matching sequence.
e. The matrix representing the name of the factor whose binding site is matching.

The last three lines of the result page give the total length of all the searched sequences, along with a total number of sites that have been found, and the frequency of sites per nucleotide.

33.5 QUESTIONS

1. What experimental methods are used to predict the transcription factor binding site? Explain in detail.
   Hint: TRANSFAC and MATCH are used to predict the TFBS. Consult section 33.3 and 33.4 for detail procedure.
2. Have any transcription factors been experimentally demonstrated to regulate the human, mouse or rat RNF43 genes? Justify with the proper experimental set-up.
   Hint: Refer TRANSFAC and follow instructions given in section 33.3.1.
3. How many matrices are available for the human, mouse and rat FOXA family members? Create a profile for these matrices.
   Hint: See IV point of section 33.3.1.
4. How will you predict the transcription factor target site for the promoter region of IFNB1 using bioinformatics tools?
   Hint: Follow the procedure explained in section 33.4.1.
5. Define the experimental set-up of transcription factor target site prediction for the promoter region of Trp63 gene.
   Hint: Consult section 33.4.1.

Prediction of Translation Initiation Sites

CHAPTER 34

S Jain[1], S Panwar[2] and A Kumar[3]
[1]Department of Applied Sciences & Humanities, Jai Parkash MukandLal Innovative Engineering and Technology Institute, Haryana, India
[2]Department of Genetics and Plant Breeding, Chaudhary Charan Singh University, Uttar Pradesh, India
[3]Department of Nutrition Biology, Central University of Haryana, Haryana, India

34.1 INTRODUCTION

The correct recognition of translation initiation sites (TIS) can help us to understand the gene structure and its product. The computational identification of TIS is the main constituent of the gene prediction system and, therefore, has utmost importance in genome annotation. Lots of data mining methods have been employed to identify TIS in transcripts such as mRNA, EST and cDNA sequences. All these methods are based on the scanning model (Kozak, 1989), which states that, in eukaryotes, the first "AUG" (start codon) at the 5′ prime of the mRNA transcript is usually the exact TIS. However, exceptions can occur via the process of leaky scanning, re-initiation and internal initiation of translation, which results in another AUG being the true TIS.

The consensus motif GCCRCCatgG around the TIS was probably the first effort to identify TIS with statistical meaning (Salamov *et al.*, 1998). The general approach for answering the TIS prediction difficulty is to create the numerical data from the cDNA sequences and, subsequently, apply computational methods.

34.2 OBJECTIVE

To predict the translation initiation site by exploiting the NetStart and TIS Miner tools.

34.2.1 The Kozak sequence

The Kozak consensus sequence was originally demarcated as ACCAUGG, based on the effect of single amino acid change around the translation initiation codon (AUG) of the preproinsulin gene. Consequently, it was extended to "GCCGCCACCAUGG",

Basic Applied Bioinformatics, First Edition. Chandra Sekhar Mukhopadhyay, Ratan Kumar Choudhary and Mir Asif Iquebal.

based on the mutation and survey study of 699 vertebrate transcripts. Further, expression of preproinsulin and alpha-globin in the cells showed that a purine (generally "A") in position –3 is essential for efficient initiation of translation, and in its absence, a "G" at position +4 is essential.

34.3 PROCEDURE

34.3.1 Tools used in translation initiation site prediction

Kozak (1987) proposed the first method to identify TIS. The weight matrix is applicable for the modeling of conserved sequence in the vicinity of TIS. Nevertheless, Pedersen and Nielsen (1997) introduced the NetStart system (the first real automated system) and employed the artificial neural network (ANN) to identify TIS in the mRNA transcripts. Salzberg (1997) used a conditional probability (CP) matrix to model TIS. The work was subsequently carried out by Li and Jiang (2004), who developed a new Edit-Kernel approach called TIS hunter.

34.3.1.1 NetStart 1.0

In this method, the artificial neural network predicts which AUG triplet in the mRNA sequence is the start codon. The trained network correctly classifies 88% of *Arabidopsis* and 85% of vertebrate "AUG" triplets in a reading frame. The steps are as follows:

a. Check the link: go to http://www.cbs.dtu.dk/services/NetStart/
b. *Input sequences*: this can done in the following two ways for processing (Figure 34.1):
   i. Paste a nucleotide sequence or a number of sequences in FASTA format into the upper window of the main server page.
   ii. Choose a FASTA file on the hard disk.
   The acceptable input alphabet is "**A**", "**C**", "**G**", "**T**", "**U**" and "**X**" (unknown). All other codes will be converted to **X** before being processed. "T" and "U" are treated as equivalent.
c. *Select organism type*: depending on the origin of input sequences, click on either Vertebrate or *A. Thaliana*. The former is the default setting.
d. *Submit the job*: enter the "Submit" button. The status of the job will be displayed and constantly updated until it terminates, and the server output appears in the browser window.
e. *Output format*: each input sequence will be shown with the predicted translation start site, followed by a table showing the positions and scores of all the positions of ATG in the sequence. Beneath the sequence, the denoted estimated start codon is "i" (initiation). At another position of "ATG," it is "N" (non-start), while all other sequences are denoted by dots (".").The scores are mainly [0.0, 1.0]; however, if the score is higher than 0.5, then it is probably a translation start site. The output format is depicted in Figure 34.2.

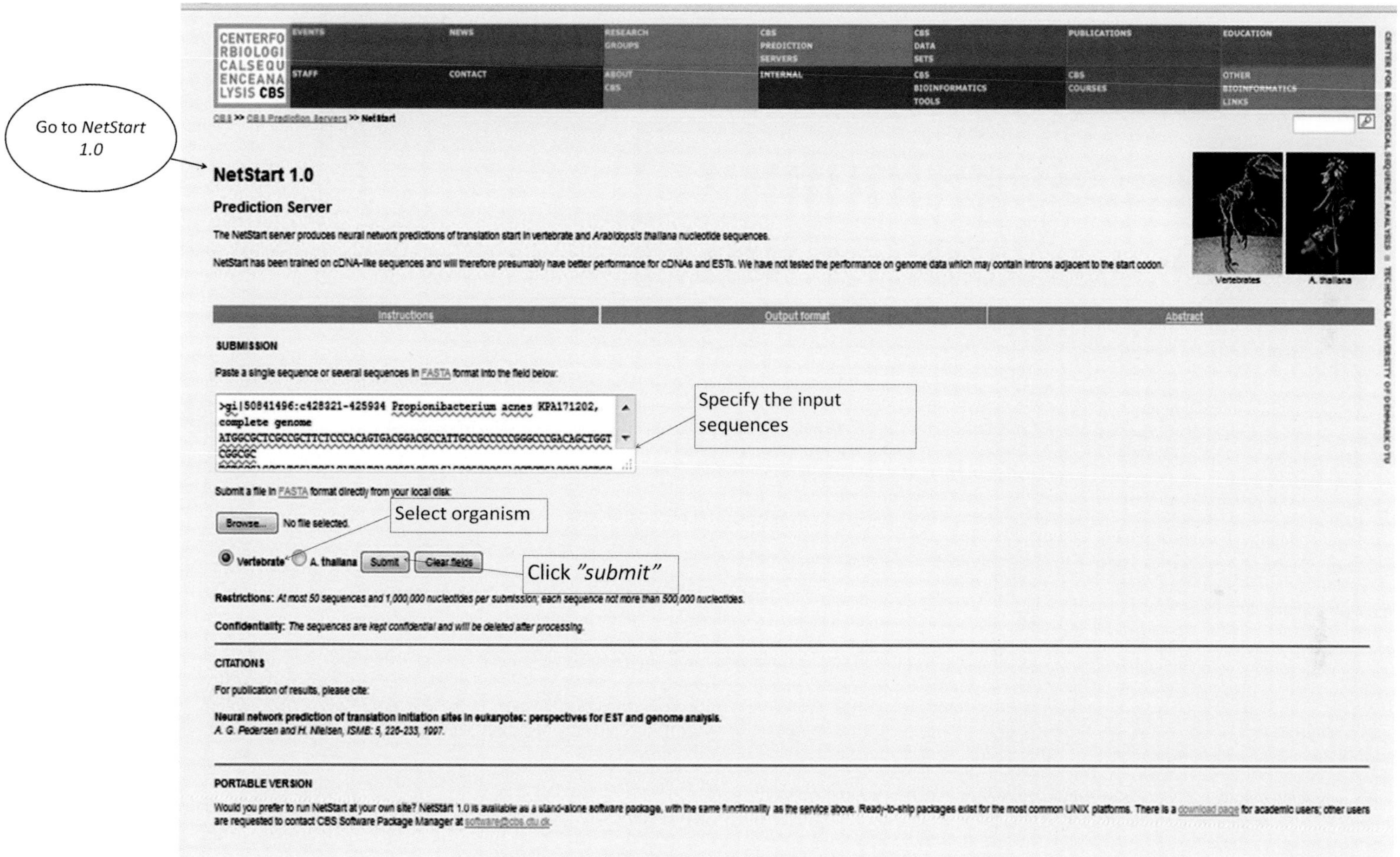

FIGURE 34.1 File format of inserted nucleotide sequence in NetStart 1.0.

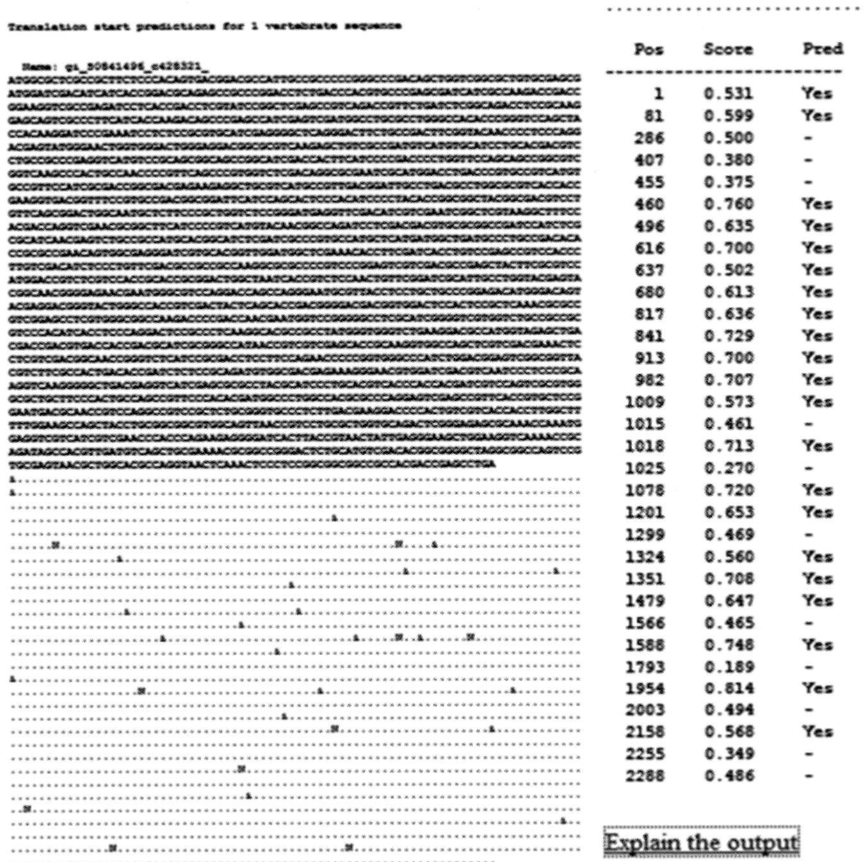

FIGURE 34.2 Output format for translation start predictions for a vertebrate sequence.

34.3.1.2 TIS Miner

This is used for the prediction of translation initiation site(s) in vertebrate DNA/mRNA/cDNA sequences. Training of the TIS Miner was completed on 3312 vertebrate mRNA sequences extracted from GenBank. Pedersen *et al.* (1997) initially analyzed the data and observed 3312 true TIS ATGs and 10063 non-TIS ATGs. The accuracy is 92.45% at 80.19% sensitivity and 96.48% specificity.

a. Go to http://dnafsminer.bic.nus.edu.sg/Tis.html. TIS Miner and Poly (A) Signal Miner are raised from the left panel of the homepage.
b. The nucleotide sequence can be submitted either in raw or in FASTA format. A limit of maximum 50 000 base pairs per sequence per submission is set to avoid a long waiting time for users (Figure 34.3).
c. The number of predictions is defined as the digit of highest-scored candidates of the anticipated functional site. The hexamer poly (A) signal consensus can be opted if anticipating poly (A) signals. The choices are either ATTAAA or any variant of NNTANA-type.

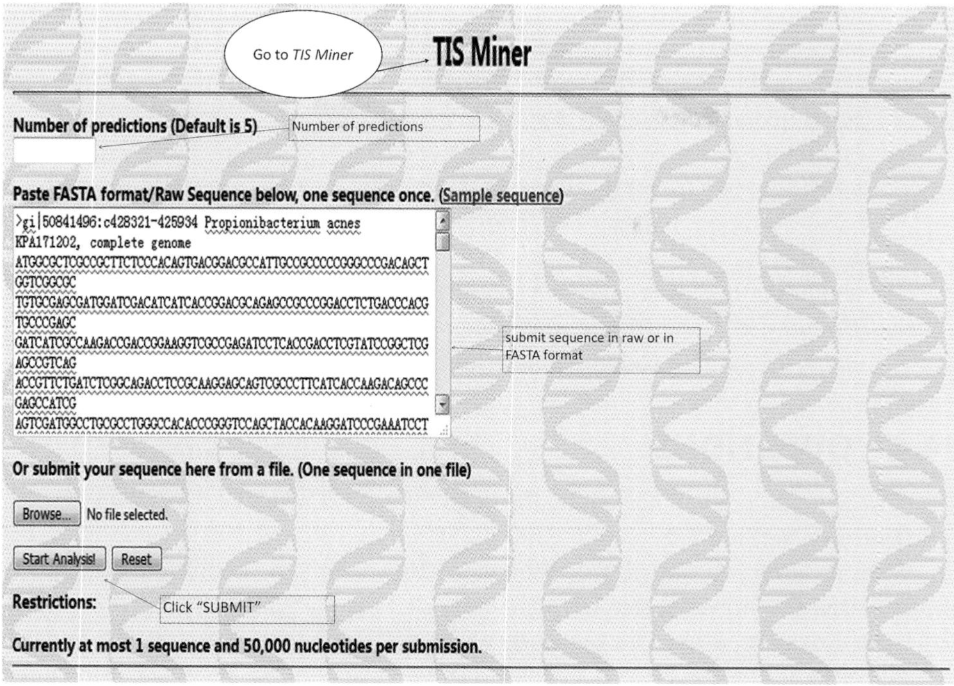

FIGURE 34.3 File format of inserted nucleotide sequence in TIS Miner.

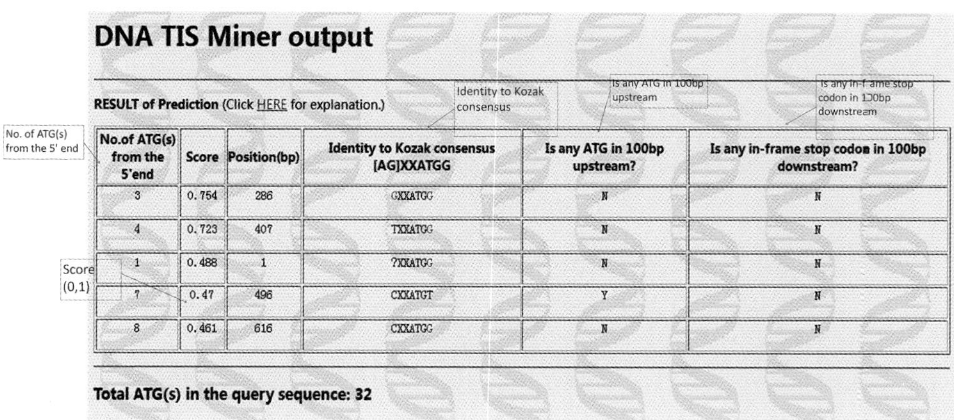

No.of ATG(s) from the 5'end	Score	Position(bp)	Identity to Kozak consensus [AG]XXATGG	Is any ATG in 100bp upstream?	Is any in-frame stop codon in 100bp downstream?
3	0.754	286	GXXATGG	N	N
4	0.723	407	TXXATGG	N	N
1	0.488	1	?XXATGG	N	N
7	0.47	496	CXXATGT	Y	N
8	0.461	616	CXXATGG	N	N

FIGURE 34.4 Output format for TIS Miner

d. Submit the query by pressing "SUBMIT".

e. *Output format*: The output page of TIS miner is summarized below and shown in Figure 34.4.

   i. *No. of ATG(s) from the 5′ prime*. The *i* means that the corresponding candidate is the *i*th candidate ATG from the 5′ end. Normally, a sequence may include several candidates of the functional site.

   ii. *Score*. The anticipated scores range (0, 1) corresponds to the exact TIS and is supported by vector machine (SVM). The higher the score, the greater the likelihood of being an accurate TIS. If the score is higher than 0.6 at a threshold value of 0.6, then it is anticipated to be accurate TIS.

iii. *Position (bp)*. This indicates the position of the corresponding candidate in the submitted nucleic acid sequence.

iv. *Identity to Kozak consensus [AG] XXATGC*: a "G" residue has a tendency to follow a true TIS, while either the "A" or "G" residue is usually found three bases upstream of a true TIS. Thus, the candidate "ATG" fits this consensus.

v. *Is any ATG in 100 bp upstream?* This column shows whether an ATG exists within 100 bp upstream of the candidate.

vi. *The presence of in-frame stop codon 100 bp downstream*: This just presents any in-frame stop codon within 100 bp downstream.

34.4 QUESTIONS

1. How will you predict the start codon in an mRNA sequence of *Arabidopsis*?
   Hint: use NetStart 1.0.
2. What do you mean by Kozak Sequence? How will you predict the translation initiation sites in vertebrate DNA?
   Hint: Kozak sequence is a sequence which occurs on eukaryotic mRNA and has the consensus (gcc)gccRccAUGG; use TIS Miner and follow the usage instructions explained in this chapter.
3. Explain in detail about using the Net Start 1.0 system to classify TIS on a genomic scale.
   Hint: See Section 34.3.1.1 and 34.3.1.2.
4. Elaborate the complete experimental set up of TIS Miner system for TIS prediction in a mRNA sequence extracted from GenBank.
   Hint: See Section 34.3.1.2.
5. Briefly define every column of the output format table of an inserted sequence in the TIS Miner system.
   Hint: Consult 5 point of the section 34.3.1.2.

Molecular Docking

CHAPTER 35

CS Mukhopadhyay and HK Manku
School of Animal Biotechnology, GADVASU, Ludhiana

35.1 INTRODUCTION

Docking is an attempt to determine whether two molecules interact with each other, and to find the best match between these two molecules. Biocomputational "docking", thus, predicts the preferred orientation of one molecule (e.g., protein) bound to another molecule (a ligand) in a lock-and-key manner. The particular orientation necessitates overall minimum free energy (ΔG). Docking is necessary because it is a key to rational and sensible drug design. In the process of docking, all intermolecular forces (i.e., H-bonding, hydrophobicity, dipole–dipole interaction, Van der Waals forces, electrostatic interactions and intra-molecular forces) and the bond features (i.e., bond length, bond angle, dihedral angle) are taken into account. Docking can be rigid, or there are flexible types. The docking studies are categorized broadly into protein–ligand docking; protein–protein docking; and protein–nucleic acid docking.

35.1.1 Software used for docking

Some of the useful software tools used for docking include Sanjeevani; GOLD; ICM; AUTO DOCK; GLIDE; GRAMM-X; FlexX; and SwissDock.

35.2 OBJECTIVE

To find the best binding poses of the receptor–ligand complex, based on energy minima of the system.

Basic Applied Bioinformatics, First Edition. Chandra Sekhar Mukhopadhyay, Ratan Kumar Choudhary and Mir Asif Iquebal.

35.3 PROCEDURE

35.3.1 Target or receptor selection and preparation

The first step of docking is to select a PDB file (protein file for docking with a ligand) and prepare that file for docking. A preparation step is required because PDB structures often contain water molecules, which play no essential role in coordinating to the ligand. The selected receptor should be biologically active and stable.

In this practical example, a file, 1EIO.pdb, has been downloaded from the Protein Data Bank (http://www.rcsb.org/pdb/). We will get a complex protein, "ileal lipid binding protein", that is already docked with the ligand glycocholic acid. The docked. pdb file has been visualized in the UCSC chimera tool (tool for structure visualization) (https://www.cgl.ucsf.edu/chimera/), and ligand and all the water molecules have been removed to prepare the receptor.

35.3.2 Location of binding site

The active site within the receptor needs to be first identified. The receptor may have multiple active sites, but *one site* of interest is to be selected (as the most druggable site).

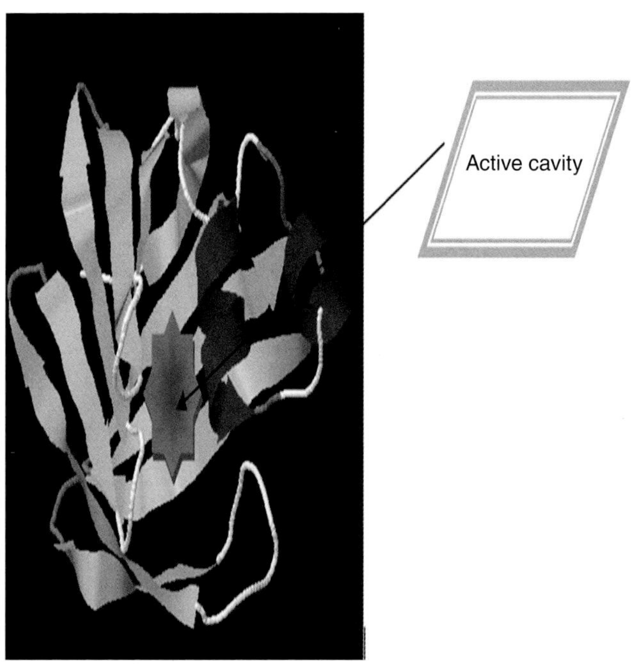

FIGURE 35.1 The identified active site or cavity within the receptor is marked as a star.

35.3.3 Ligand selection and preparation

A reasonable 3D structure is required as a starting point. The ligand can be obtained from databases like ZINC (http://zinc.docking.org/) or PubChem (https://pubchem.ncbi.nlm.nih.gov/), or can be sketched using a tool such as ACS/Chemsketch (downloadable from http://www.acdlabs.com/resources/freeware/chemsketch/) or ChemDraw (commercial software with free trial: http://www.cambridgesoft.com/software/overview.aspx). In this practical example, the ligand has been extracted from the complex and saved as a *.Mol2 file separately from the receptor.

35.3.4 Docking

Open the SwissDock server (www.swissdock.ch/docking) and upload the prepared receptor molecule and the ligand molecule (*.pdb and *.Mol2 files, respectively).

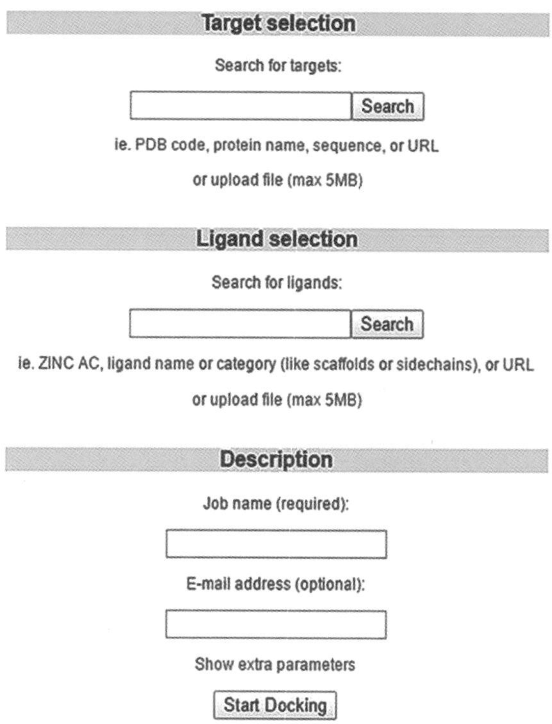

FIGURE 35.2 The "Submit Docking" tab at the top of the homepage of the SwissDock online tool takes you to this page. Upload the target and ligand files by clicking on the appropriate buttons.

After uploading the files, please provide a short description and the email ID where you will receive the results. Click on the "Start Docking" button. This will first identify and analyze the active site; then the ligand will be docked onto the receptor, and the interactions will be checked. The scores generated by the scoring function enable us to identify the best fit ligand.

35.4 RESULT AND INTERPRETATION

The result pages will display the binding modes of the docked complex, along with their most fitting score and free energy score. The best binding mode can be determined through the free energy score; the lower the free energy, the better the binding mode.

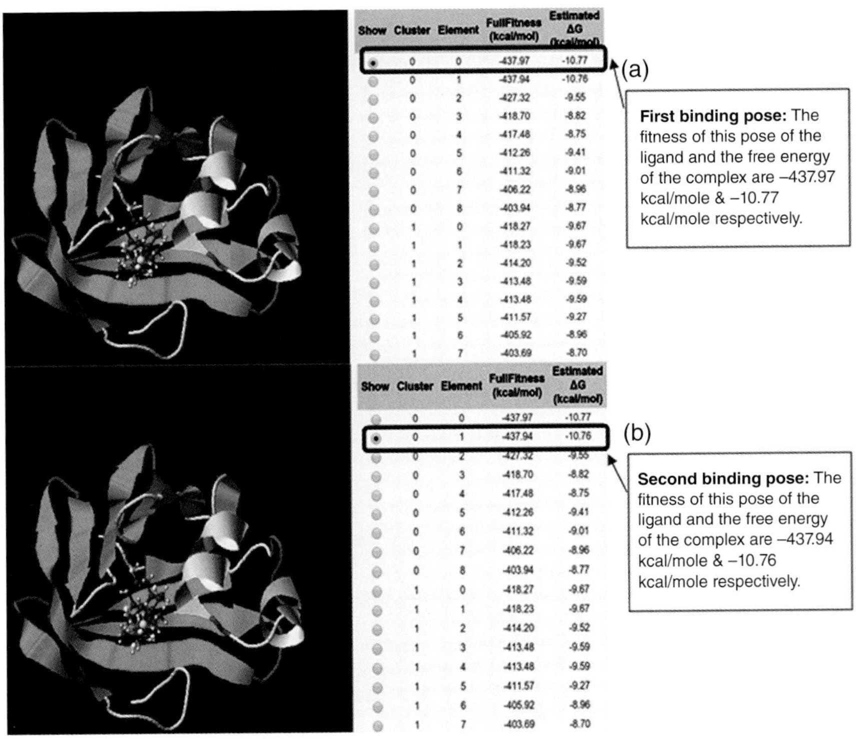

FIGURE 35.3 Fitness of ligand and free energy of docked complex of the first and second binding poses, shown as "A" and "B". (*See insert for colour representation of the figure.*)

The free energy (ΔG) values among all the docked poses can be compared. It is evident that first two docked complexes have very similar energy, compared with the other three docked poses, so these two poses can be considered for further analysis for potential drug discovery.

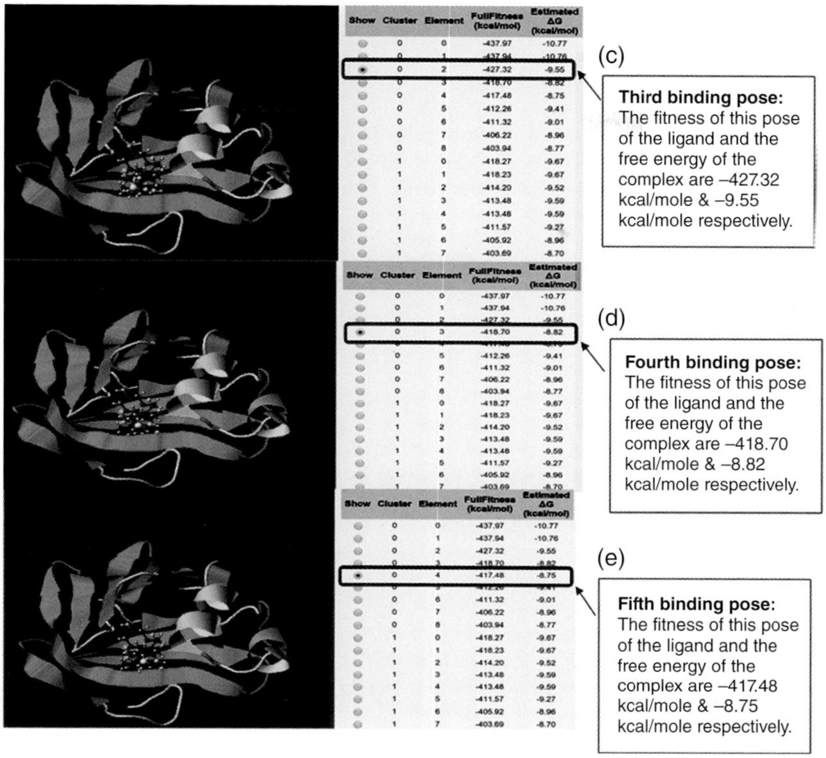

FIGURE 35.4 Fitness of ligand and free energy of docked complex of the third, fourth and fifth binding poses, shown as “C”, “D” and “E”. (*See insert for colour representation of the figure.*)

35.5 QUESTIONS

1. How will you prepare a cleaned receptor.pdb file?
2. How will you prepare the ligand molecule for docking?
3. Which file format for the ligand molecule will accept the SwissDock server?
4. How will you choose the pose if the top ten poses have minute energy differences?
5. Carry out the docking process for your choice of receptor and ligand molecule.
6. Write your interpretation of the docked complex in your own words on the basis of free energy.

Genome Annotation

SECTION VIII

Genome Annotation in Prokaryotes

CHAPTER 36

CS Mukhopadhyay and RK Choudhary
School of Animal Biotechnology, GADVASU, Ludhiana

36.1 INTRODUCTION

The genomic machinery harbors several signals, elements and conserved regions, namely, promoter signals, transcription start and termination signals (start codon "ATG" and termination codons: Ochre: "TAA", Amber: "TAG", Opal: "TGA"), codons, exons, intervening introns (delimited by exon–intron boundaries), etc. Use of suitable software enables the researchers to identify and annotate the various regions of the genome. In this chapter, we will use the GeneMark program to learn gene finding and genome annotation in prokaryotes.

GENEMARK PROGRAM FOR GENE FINDING IN PROKARYOTES

This is a program for *ab initio* gene prediction of prokaryotes and eukaryotes. The suite of programs available in GeneMark includes the following:

a. Gene prediction in bacteria, archaea, metagenomes and meta-transcriptomes:
   - GeneMark-S
   - "GeneMark.hmm with Heuristic models"
   - MetaGeneMark
b. Gene prediction in Eukaryotes:
   - GeneMark-ES
   - GeneMark.hmm
c. Gene prediction in Transcripts:
   - GeneMarkS
   - "GeneMark.hmm with Heuristic models"
d. Gene prediction in Viruses, Phages and Plasmids:
   - "GeneMark.hmm with Heuristic models"
   - GeneMark-S

Basic Applied Bioinformatics, First Edition. Chandra Sekhar Mukhopadhyay, Ratan Kumar Choudhary and Mir Asif Iquebal.

36.2 OBJECTIVE

To annotate the partial genome of a prokaryotic organism, using the GeneMark.hmm (Lukashin and Borodovsky, 1998) online tool.

36.3 PROCEDURE

a. Download the nucleotide sequence from Nucleotide database (NCBI) and save it in Notepad in FASTA format:
   i. The genomic sequence of *Yersinia pestis* (AL590842.1: *Yersinia pestis* CO92 complete genome) has been downloaded from NCBI (http://www.ncbi.nlm.nih.gov/nuccore/AL590842.1).
   ii. Now the sequence is to be saved (in FASTA format) in a .txt file (Notepad), using the "send to" option available on the right-hand side of the top of the page.
b. Open GeneMark.hmm for Prokaryotes (Version 2.8) (Figure 36.1), using the URL: http://exon.gatech.edu/gmhmmp.cgi (optionally, you can go directly to the homepage of GeneMark: http://opal.biology.gatech.edu/GeneMark/).
c. Click "GeneMark hmm" on the left-hand side of the shortcut menu (arrow) and select "prokaryotes".
d. Browse the sequence file by clicking on the "Browse..." button, or paste the sequence into the box provided.

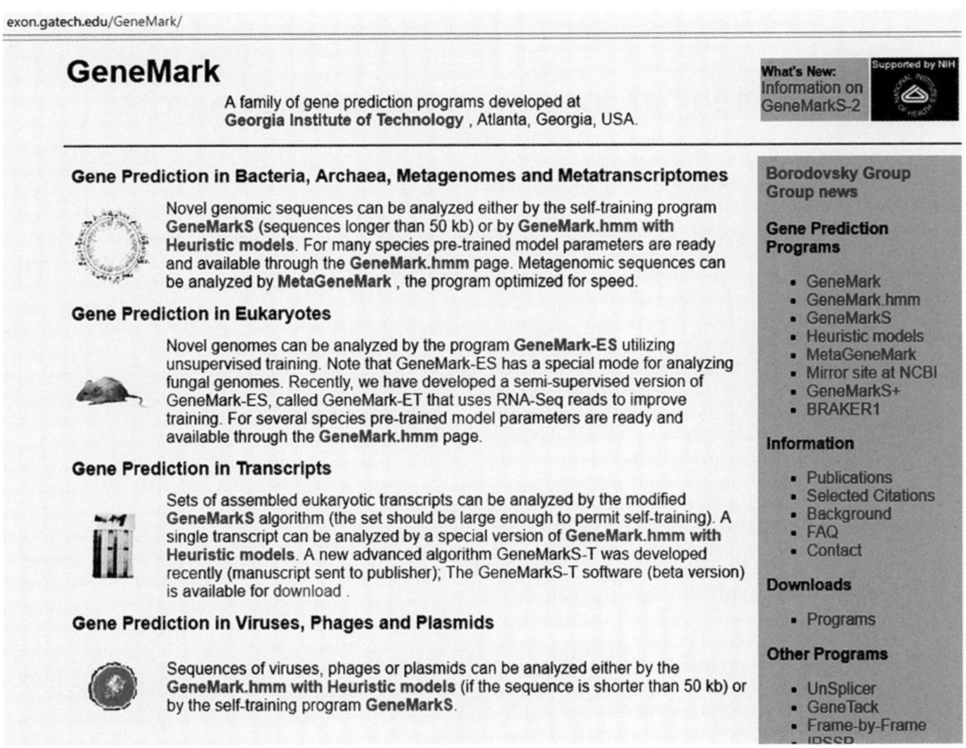

FIGURE 36.1 Homepage of the GeneMark online tool. (*See insert for colour representation of the figure.*)

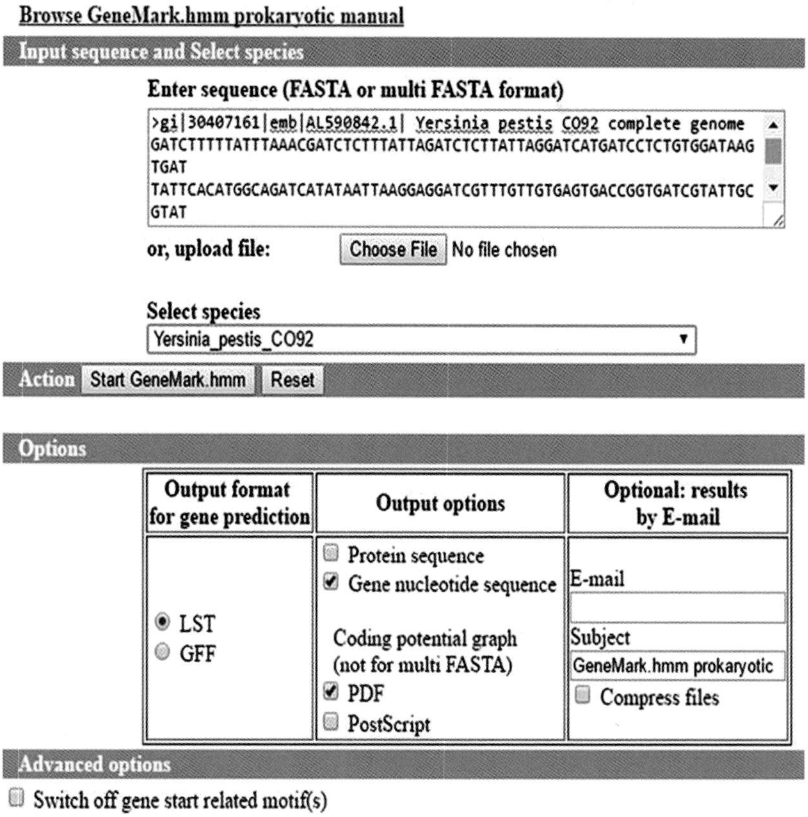

FIGURE 36.2 Specifying the parameters in GeneMark.hmm for prokaryotes for gene finding and annotation.

e. Select the species from the drop-down option. If the exact species is not there, select a similar type of species.
f. Check the boxes located at the bottom, as per the output requirements:
   i. to obtain the *in silico* translations of the predicted genes;
   ii. to get the nucleotide sequences of the predicted genes;
   iii. to produce on-screen PDF graphics;
   iv. to generate PostScript graphics (via email).
g. Click "Start GeneMark.hmm" to start gene searching under the "Action" tab (See Figure 36.2).

36.4 INTERPRETATION OF GENEMARK OUTPUT

a. The program gives the results in tabular format: Gene (as serial number), Strand (positive or negative strand), Left End and Right End (start and end nucleotide number of the gene), Gene length, and Class.
b. This is followed by the predicted sequence of the translated amino acids and the nucleotide sequence of the gene (if the options "Translate predicted genes into proteins" and "Sequences of predicted genes" have been checked) (see Figure 36.3).

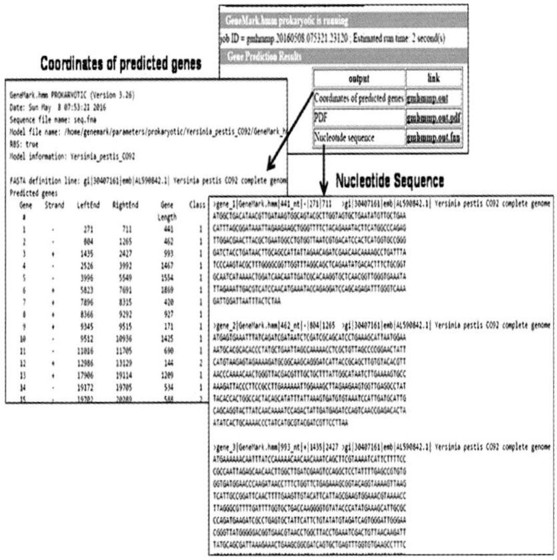

FIGURE 36.3 The output pages of the GeneMark online tool for prokaryotic gene prediction.

36.5 QUESTIONS

1. Download the given sequence form NCBI, predict the possible genes and annotate them:
   a. BX248333.1
   b. NZ_KK354537.1
2. Compare the genome annotation results of RAST and GeneMark using the sequence from NCBI BX248333.1 (accession number).
3. What are the salient points to be considered while annotating a given DNA sequence of a prokaryote?
4. How can we identify the novel genes which are missed by a genome annotation/prediction tool?
5. Please annotate the cloning vector pRB223, using suitable tools.

Genome Annotation in Eukaryotes

CHAPTER 37

CS Mukhopadhyay and RK Choudhary
School of Animal Biotechnology, GADVASU, Ludhiana

37.1 INTRODUCTION

GENSCAN, an HMM algorithm-based online program, is used to identify complete gene structures in genomic DNA, and to predict the location of genes and their exon–intron boundaries in genomic sequences of vertebrates, *Arabidopsis* and maize. GENSCAN was developed by Christopher Burge of the Department of Mathematics, Stanford University (Burge and Karlin, 1997; Burge, 1998).

37.2 OBJECTIVE

To predict the putative gene sequence(s) in a given input nucleotide sequence and annotate the sequence.

37.3 PROCEDURE

a. Download a sequence (fewer than 1 million base pairs) from NCBI Nucleotide, and save in Notepad in FASTA format: here, chromosome 1 (CM000409.1) sequence of duck-billed platypus (*Ornithorhynchus anatinus*) has been downloaded from NCBI (http://www.ncbi.nlm.nih.gov/nuccore/CM000409.1).
b. The original sequence is more than 1 megabase in size, so it needs to be trimmed from any termini to approximately 1 megabase in size (using *Notepad* ++). The user needs to subject the input sequence to repeat-masker to remove low-complexity, repeat regions in the input sequence.
c. Open the GENSCAN web server: http://genes.mit.edu/GENSCAN.html.
d. Set the parameters:
    i. *Organism*: select the appropriate option from “Vertebrate”, “Arabidopsis”, or “Maize”, available in the drop-down options with “Organism”. Here, we will select “Vertebrate”.

Basic Applied Bioinformatics, First Edition. Chandra Sekhar Mukhopadhyay, Ratan Kumar Choudhary and Mir Asif Iquebal.

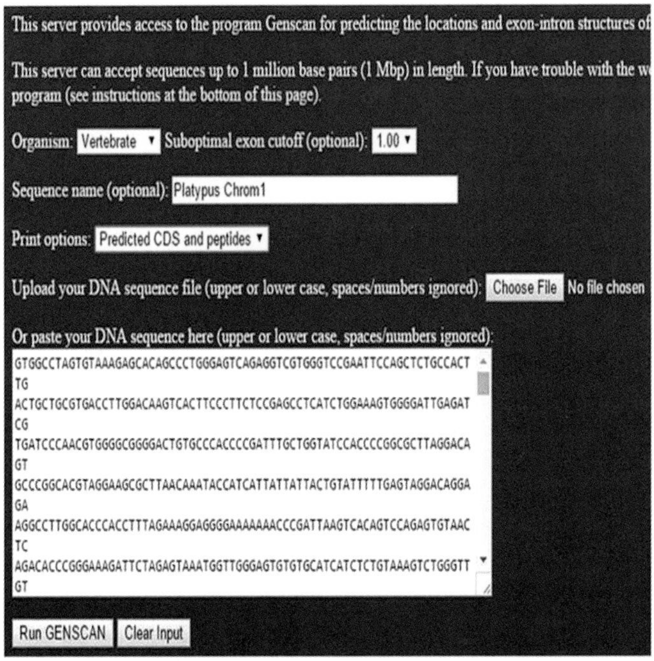

FIGURE 37.1 Homepage of the online GENSCAN software.

ii. *Suboptimal exon cutoff*: values ranging from 0.01 to 1.00. This is the probability value of finding the exon of a gene, and is an optional parameter which, by default, is set to 1.00. It can be reduced; however, the reliability of predicted exons is also reduced. The probability should not be reduced below 0.50.

iii. *Sequence name*: A text box is provided to type the name of the sequence. This is also optional, and is used to name the sequence for ease of identification.

iv. *Print options*: presents two output or result options: "Predicted peptides only" and "Predicted CDS and peptides". The second option will give the predicted amino acid, followed by the encoding nucleotide sequences.

v. *Browse button*: to upload the input nucleotide sequence for gene prediction.

e. Browse to upload the sequence using the "Browse..." button.

f. Click "Run GENSCAN" to start the analysis (Figure 37.1).

37.4 INTERPRETATION OF GENSCAN OUTPUT

- The GENSCAN output appears in a new window on the same web page. The GENSCAN version, date and time of run are shown at the top.
- This is followed by the size of input sequence, G/C percentage, which gives the predicted exons in a tabular form in the next section as:
  Gn.ExTypeS.Begin ...End.Len FrPh I/Ac Do/T CodRgP.... Tscr.. (Figure 37.2)
- It also gives the results for the suboptimal exons with probability 1.
- Finally, the predicted amino acid sequence(s) and the respective coding nucleotide sequence(s) are given.

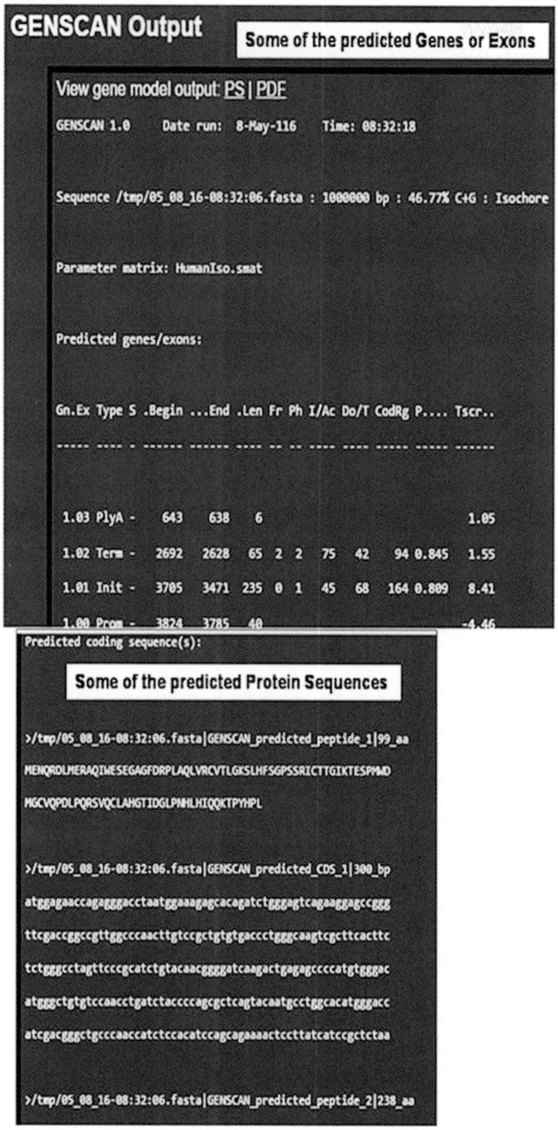

FIGURE 37.2 Output page of the GENSCAN software.

The terms used in the output of GENSCAN are as follows (Source: http://www.biomedcentral.com/content/supplementary/1471-2164-11-156-s2/Additionalfile2/GENSCAN_output/GENSCAN%20output%20EG926217.htm):

a. Gn.Ex → gene number, exon number (for reference purpose).
b. Type: Init → Initial exon (ATG to 5′ splice site).
c. Intr → Internal exon (3′ splice site to 5′ splice site).
d. Term → Terminal exon (3′ splice site to stop codon).
e. Sngl → Single-exon gene (start codon “ATG” to any one of the stop codons).

f. Prom → Promoter (TATA box/transcription initiation site).
g. PlyA → poly-A signal (consensus sequence : AATAAA).
h. S → DNA strand (+ = input strand; − = opposite strand).
i. Begin → beginning of exon or signal (numbered on input strand).
j. End → end point of exon or signal (numbered on input strand).
k. Len → length of exon or signal (bp).
l. Fr → reading frame.
m. Ph → net phase of exon.
n. I/Ac → initiation signal or 3′ splice site score.
o. Do/T → 5′ splice site or termination signal score.
p. CodRg → coding region score.
q. P → probability of exon (sum over all parses containing exon).
r. Tscr → exon score (depends on length, I/Ac, Do/T and CodRg scores).

A detailed explanation regarding GENSCAN output is available at http://genome.crg.es/courses/Bioinformatics2003_genefinding/results/GENSCAN.html.

37.4.1 Some points to remember while using GENSCAN

a. This tool cannot handle data larger than one million bases, so please limit the input sequence size to 1 MB.
b. The user needs to mask the repeat sequences prior to submitting to GENSCAN.
c. It is not to be used for prokaryotic and yeast sequences.
d. It can predict internal exons more accurately than the terminal exons.

37.5 QUESTIONS

1. Discuss the output parameters obtained from GENSCAN.
2. Predict and annotate the genes of the taurine Y chromosome.
3. What are the key elements of the eukaryotic gene that are taken into account while predicting genes? What will be your strategy to predict eukaryotic genes from a given sequence, if no tools are available?
4. Download the sex chromosomes of mouse (*Mus musculus*) and predict the genes in both chromosomes. Which genes are in common in both chromosomes?

SECTION IX

Advanced Biocomputational Analyses

Concepts of Real-Time PCR Data Analysis

CHAPTER 38

RK Choudhary
School of Animal Biotechnology, GADVASU, Ludhiana

38.1 INTRODUCTION

Real-time quantitative polymerase chain reaction (RT-qPCR) is used for various applications in basic research, including analyses of gene expression, detection of single nucleotide polymorphism (SNP), and cancer screening. In contrast to conventional PCR, where amplicon is identified by running gel after completion of PCR, in RT-qPCR, the end production is visualized in "real time" as the product is amplified in the PCR machine. Real-time detection of amplified product is made possible by the addition of fluorescent dyes in a primer/probe/reaction mixture that reports amplified DNA following each cycle. The intensity of the fluorescent signal is proportional to the amount of template cDNA.

38.2 GETTING STARTED WITH RT-qPCR

38.2.1 How it works

To understand how RT-qPCR works, let us consider an example of a sample's amplification plot (Figure 38.1). Features of amplification plots are:

a. The *X*-axis represents the number of PCR cycles, and the *Y*-axis represents fluorescence generated from the PCR reaction.
b. Phases of amplification plot comprising two phases:
   - exponential: the amount of qPCR-product is doubled in each cycle;
   - non-exponential phase: the reaction reaches a plateau, due to exhaustion of one or more PCR-ingredients in the reaction mixture.

Basic Applied Bioinformatics, First Edition. Chandra Sekhar Mukhopadhyay, Ratan Kumar Choudhary and Mir Asif Iquebal.

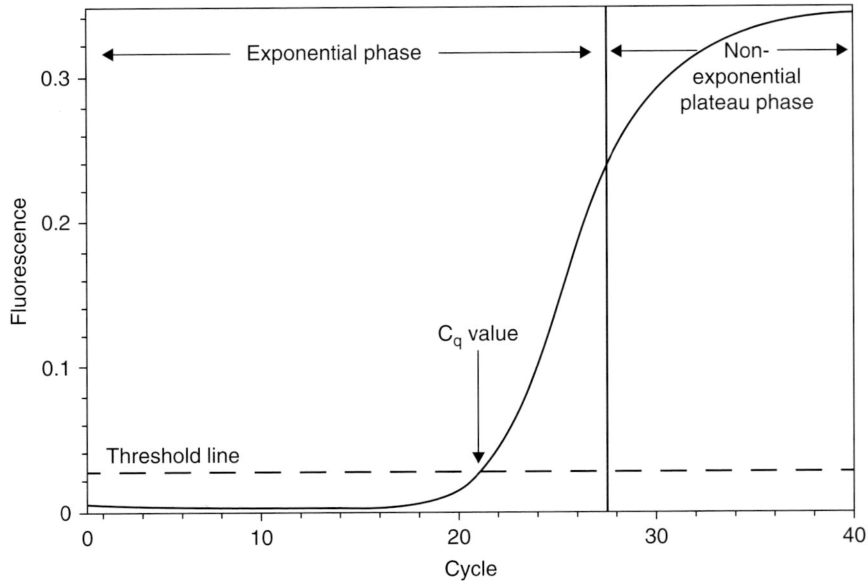

FIGURE 38.1 Amplification plot of RT-qPCR.

c. The threshold cycle (C_t) or quantitation cycle (C_q) value: initially, although the product is amplified exponentially, fluorescence is non-detectable (from 1–18 cycles in this figure) unless a threshold level of PCR products is achieved. The cycle number (here, 21 cycles of PCR) at which the accumulated product is detectable is called the threshold cycle or C_t value (or C_q = 21 cycles in figure 38.1).
d. C_t or C_q value is determined by the amount of template present in the reaction. The higher the amount of template, the earlier the C_t, value and vice versa. This forms the basis of quantification of nucleic acid in RT-qPCR.

38.2.2 Hallmarks of RT-qPCR

For accurate and reproducible quantification, an optimal PCR assay is required. An optimized PCR assay should have:

a. Linear standard curve, determined by coefficient of determination (R^2) = 0.98.
b. High amplification efficiency = 95–105%.
c. The consistency of C_t value across all replicates (usually three replicates).

To access PCR optimization, a ten-fold serial dilution of cDNA is employed. A perfect doubling occurs after each amplification cycle; then the span of two consecutive fluorescence will be 2^n = dilution factor. If the dilution factor is ten-fold then $2^n = 10$ or $n = \log_2(10) = 3.32$. This means that the C_t value should be separated by 3.32 cycles, if amplification is perfect. An amplification efficiency of < 90% or > 105% needs optimization of qPCR.

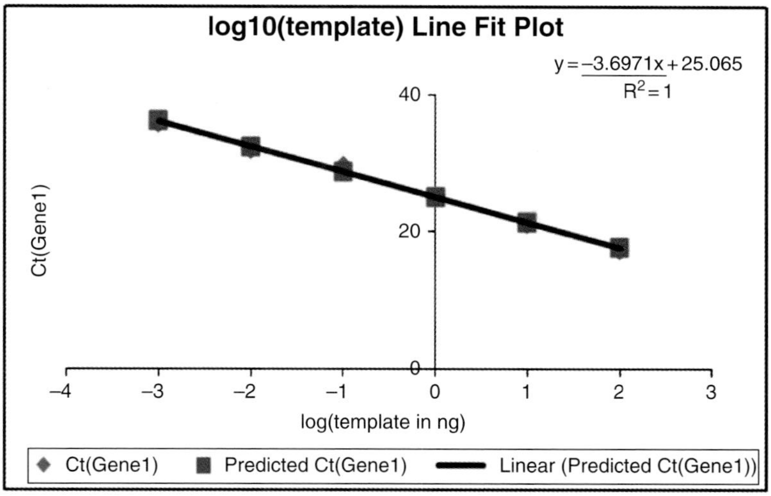

FIGURE 38.2 Construction of standard curve. Construct standard curves for both target and reference genes individually, by plotting C_t values (through the Y-axis) against the $\log_{10}{}^{\text{(template amount or dilution)}}$ along the X-axis.

38.3 PCR FLUORESCENCE CHEMISTRY

Select the appropriate fluorescent dye to monitor the amplification signal. Two types of fluorescent reporters are available:

- DNA-binding dyes, e.g., SYBR green.
- fluorescent dye-labeled oligonucleotide probes/primers (molecular beacons, TaqMan, scorpion, LUX, Eclips and other newer dyes).

Of these, TaqMan hydrolysis probes are the most common dye-labeled chemistries used.

SYBR Green is the most commonly used DNA-binding dye used in real-time qPCR chemistry. SYBR is a fluorophore that binds to the minor groove of double-stranded DNA. It is used in quantitative PCR (qPCR) to determine the amount of amplicon generated following each cycle. When SYBR green dye is free in solution that is unbound to DNA, it emits a small signal. However, when the dye is bound with double-stranded DNA, it emits a fluorescence signal that is 1000 times more intense.

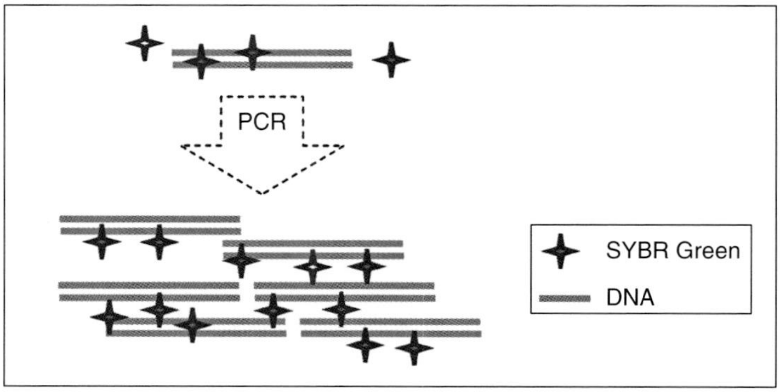

FIGURE 38.3 SYBR Green fluorophore binds with double-stranded DNA (PCR amplicon). The amount of DNA amplified is proportional to fluorescence intensity.

The SYBR Green1 assay includes PCR primers that amplify a target gene or region of the gene, and SYBR green dye to detect amplified product. Students can refer to the previous chapter for information on primer designing. A SYBR Green assay reaction mixture contains the following reagents:

a. PCR master mix with SYBR green;
b. template (your cDNA samples);
c. a pair of primers (gene-specific).

Identify melting temperature for gene-specific primer and construct a standard curve to know your PCR efficiency, before you proceed with your real assay of entire samples.

38.4 RT-QPCR DATA ANALYSIS: GENE EXPRESSION ANALYSIS

Real-time qPCR is a method to determine the amount of nucleic acid in a sample. However, merely knowing the amount of nucleic acid (say, 10 000 mRNA molecule of prolactin) in one sample is not meaningful. Rather, biologists need to know how much more mRNA is present in normal vs. diseased sample, or fold change of mRNA in normal vs. diseased sample. There are two methods for identification of nucleic acid in samples – absolute and relative quantification.

38.4.1 Absolute quantification

This compares the C_t value of test samples with a standard curve. The result of this analysis would be copy number or μg of mRNA per cell, or per μg of total RNA. A known amount of cDNA or DNA or PCR product could be used to make a standard curve. This standard curve generates a regression line on which basis the amount of unknown samples can be determined (Figure 38.4). The absolute quantification method is used when we are interested in determining the intrinsic property of a sample. The quantity of virus particles per ml of blood, for example, is a quantitative description of a sample.

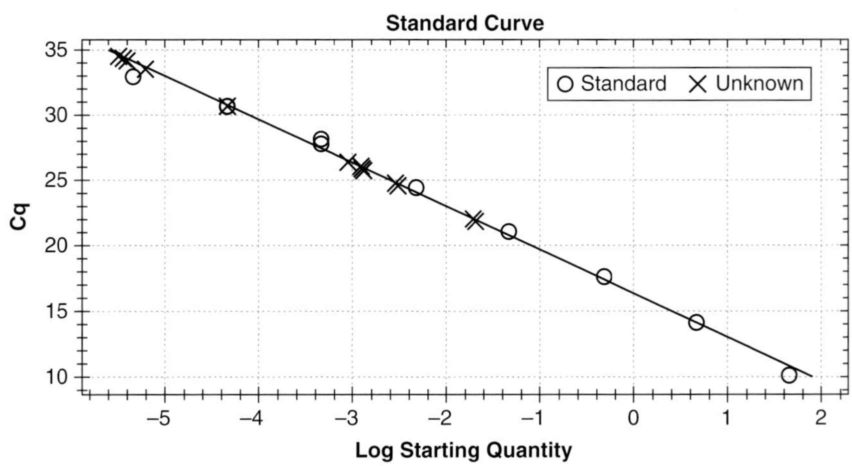

FIGURE 38.4 Absolute quantification of the transcript using the standard curve method. Using a known amount of DNA, a standard curve is made, and unknown samples are plotted on a regression line of known samples.

38.4.2 Relative quantification

In relative quantification, the result is a ratio of Sample A vs. Sample B, expressed in terms of fold change. Suppose a researcher wants to know the relative amount (fold change) of c-Myc gene (an oncogene) in a normal sample vs. a suspected case of cancer sample. There are two approaches to this:

a. Relative quantification against unit mass (say, cell number or μg of RNA).
b. Relative quantification, normalized to reference gene.

Example of relative quantification against unit mass: to determine the relative expression of the *c-Myc* gene in normal vs. cancer tissue, obtain RNA from an equal number of cells or tissue mass. Determine C_t value of the test sample (cancer cells) from a calibrator sample (normal cells). If the efficiency of PCR reaction is 100%, then the amount of product amplified in each cycle is doubled (i.e., E = 2).

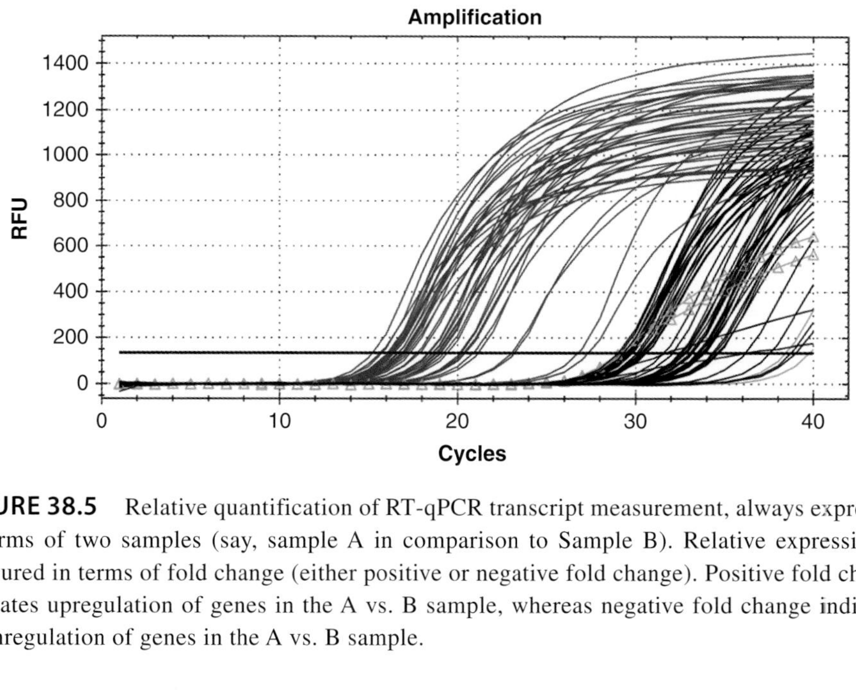

FIGURE 38.5 Relative quantification of RT-qPCR transcript measurement, always expressed in terms of two samples (say, sample A in comparison to Sample B). Relative expression is measured in terms of fold change (either positive or negative fold change). Positive fold change indicates upregulation of genes in the A vs. B sample, whereas negative fold change indicates downregulation of genes in the A vs. B sample.

$$\text{Ratio}_{(\text{test/calibrator})} = E^{\text{Ct (calibrator)} - \text{Ct (test)}}$$
$$\text{Ratio}_{(\text{test/calibrator})} = 2^{\text{Ct (calibrator)} - \text{Ct (test)}}$$
$$\text{Ratio}_{(\text{test/calibrator})} = E^{\Delta \text{Ct}}, \text{ where } \Delta C_t = C_t(\text{calibrator}) - C_t(\text{test})$$

a. Cancer cells = 50 ng of RNA; normal cell = 50 ng of RNA – keep same amount of RNA (remember the concept of unit mass!).
b. C_t of cancer cell (test) = 12 and C_t of normal cell (calibrator) = 15.
c. Then, $\text{Ratio}_{(\text{test/calibrator})} = 2^{(15-12)} = 8$

This means there is eight-fold higher expression of the *c-Myc* gene in the cancer sample.

Relative quantification normalized to reference genes (e.g., beta-actin, GAPDH) circumvents the use of an equal amount of starting material (RNA) from the samples. This method is preferred if the starting material is in limited quantity. However, knowledge of reference genes, in this case, is needed. In contrast to two C_t values that were needed with relative quantification with unit mass, four C_t values are needed.

	Test	Calibrator
Target gene	C_t (target, test)	C_t (target, calibrator)
Reference gene	C_t (Reference, test)	C_t (Reference, calibrator)

This can be analyzed using three different methods:

a. Livak method or $2^{-\Delta\Delta Ct}$ method (Livak and Schmittgen, 2001).
b. ΔC_t method using reference gene (Schmittgen and Livak, 2008).
c. Pfaffl method (Pfaffl, 2001).

This chapter describes only the Livak method (or $2^{-\Delta\Delta Ct}$ method).

$$\begin{aligned}
\Delta C_t\ (\text{test}) &= C_t(\text{target, test}) - C_t(\text{reference, test}) \\
\Delta C_t\ (\text{calibrator}) &= C_t(\text{target, calibrator}) - C_t(\text{reference, calibrator}) \\
\Delta\Delta C_t &= \Delta C_t(\text{test}) - \Delta C_t(\text{calibrator})
\end{aligned}$$

$2^{-\Delta\Delta Ct}$ = normalized expression ratio

See the example below to understand this better:

Sample	C_t (*c-Myc*) = Target	C_t (GAPDH) = Reference
Normal (calibrator)	15	16.5
Cancer (test)	12	15.9

$$\begin{aligned}
\Delta C_t(\text{calibrator}) &= C_t(\text{target, calibrator}) - C_t(\text{reference, calibrator}) \\
&= 15 - 16.5 = -1.5 \\
\Delta C_t(\text{test}) &= C_t(\text{target, test}) - C_t(\text{reference, test}) \\
&= 12 - 15.9 = -3.9 \\
\Delta\Delta C_t &= -3.9 - (-1.5) = -2.4
\end{aligned}$$

Finally, normalized expression ratio $= 2^{-\Delta\Delta Ct} = 2^{-(-2.4)} = 5.3$

Therefore, the tumor cell (test samples) is expressing *c-Myc* gene 5.3 times higher than the normal cell.

An overview of the workflow of real-time qPCR experiment is given in Figure 38.6 (obtained from the Web).

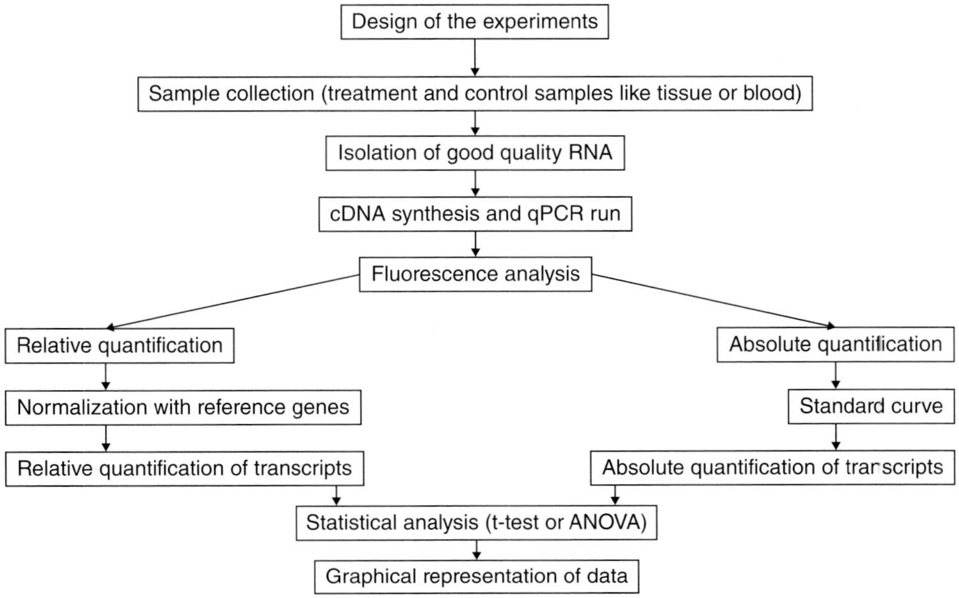

FIGURE 38.6 Analytical flow diagram of the use of real-time PCR.

38.5 QUESTIONS

1. Define the C_t value.
2. What is the difference between PCR and RT-qPCR?
3. Explain the difference between relative and absolute quantification of gene expression in RT-qPCR.
4. Describe how to generate a standard curve in qPCR analyses.

Overview of Microarray Data Analysis

CHAPTER 39

RK Choudhary
School of Animal Biotechnology, GADVASU, Ludhiana

39.1 CONCEPT

Microarray technology is used to study the expression of many genes at a time, which is the simultaneous detection of many genes. Thousands of gene sequences (called probes) are placed on a glass slide (called a gene chip), and a sample containing DNA or RNA or protein (depending upon the types of microarray) called "probes" is placed in contact with gene chip. Hybridization occurs between targets and probes, and produces light that is measured and quantifed.

39.2 GETTING STARTED WITH MICROARRAY

The principle of microarray has been used in different areas of molecular biology (Source: Wikipedia), including: DNA microarrays (cDNA, oligonucleotide microarrays, SNP-chip); MMChips (microRNA); Protein-microarrays; Reverse Phase Protein Microarrays; Tissue-microarrays; Chemical compound microarrays; Antibody microarrays; Carbohydrate arrays (glycoarrays), and so on.

These different types of microarray are spotted arrays on glass, self-assembled arrays and *in situ* synthesized arrays. There are two types of experimental procedures, and they are either one-color or two-color approaches. In the one-color approach, a sample is labeled singly with either Cy3 or Cy5 fluorophore, and hybridized to a single microarray chip. In the two-color microarray, two samples (usually sample and reference) are labeled with two different fluorophores (Cy3 and Cy5), and are hybridized together in a single microarray chip. In a single-color array, the result obtained is an absolute fluorescent signal, unlike that of the two-color array, which produces ratios of fluorescent intensities. A one-color array is simple to perform, but the results of two-color arrays are more robust, due to internal references. However, due to high costs, dye biases, high input RNA, and difficulty in finding an appropriate reference sample with two-color microarray, the single-color array is preferred.

Basic Applied Bioinformatics, First Edition. Chandra Sekhar Mukhopadhyay, Ratan Kumar Choudhary and Mir Asif Iquebal.

39.3 MICROARRAY DATA ANALYSIS: GENE EXPRESSION ANALYSIS

39.3.1 Microarray experimental design

1. *Reference design:* Pooling of RNA samples serves as a common reference RNA(R). The logarithm of the ratio of the label intensities at those spots is used as a measure of relative hybridization. Label the reference, on each array, with the same dye. A limitation of reference design is that half of the hybridizations used for the reference sample may be of no real interest.
2. *Loop design:* These are an alternative to the reference design. Two aliquots for each of the samples need to be arrayed for loop design while performing a cluster analysis.

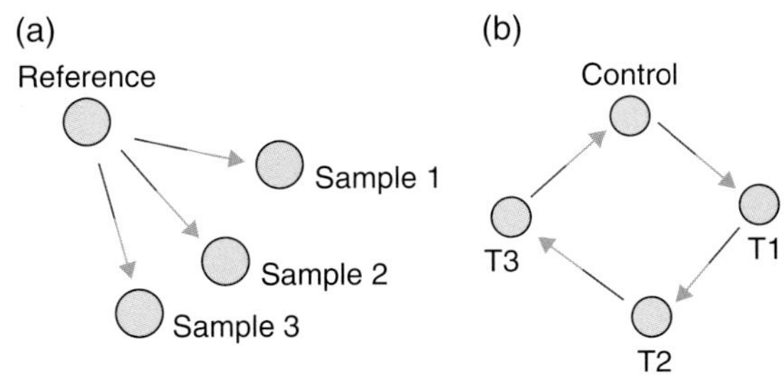

FIGURE 39.1 Reference design (a) and loop design (b) of a two-color microarray. Different colors (red and green here) represent microarray chips. In order to avoid dye bias, the same samples are used twice, with opposing labeling schemes, such as array 1: sample a (labeled with red dye) vs. Sample b (labeled with green dye) and array 2: sample b (labeled with red dye) vs. sample a (labeled with green dye).

39.3.2 Concepts of replicates

1. *Biological replicates:* repeat hybridizations using the same RNA sample. This tells us about variation due to hybridization, imaging, etc.
2. *Technical replicates:* repeat hybridizations using different RNA isolates (other animal/cells from the same group). Technical replicates indicate real variability in the sample.

WHY ARE REPLICATES REQUIRED FOR MICROARRAY EXPERIMENTS?

In replicates, gene expression patterns in two independent samples should be identical (correlation coefficient >0.97) – that is, 3% variation is random.

If the total number of genes is 40 000, then 3% variation $= 40\,000 \times 0.03 = 1200$ genes are false positive.

If experiments are run in duplicate, then $40\,000 \times 0.03 \times 0.03 = 36$ genes out of 40 000 are false positive. Therefore, replication reduces the number of false positives and increases the proportion of true positive results in a microarray experiment.

39.4 STEPS INVOLVED IN MICROARRAY DATA ANALYSIS

a. Image processing
b. Background subtraction
c. Normalization
d. Identify differentially expressed genes
e. Which genes are expressed?
f. Which genes are differentially expressed?
g. Cluster analysis (time series)
h. Integration of differentially expressed genes with functional information: pathways

An overview of microarray is now given.

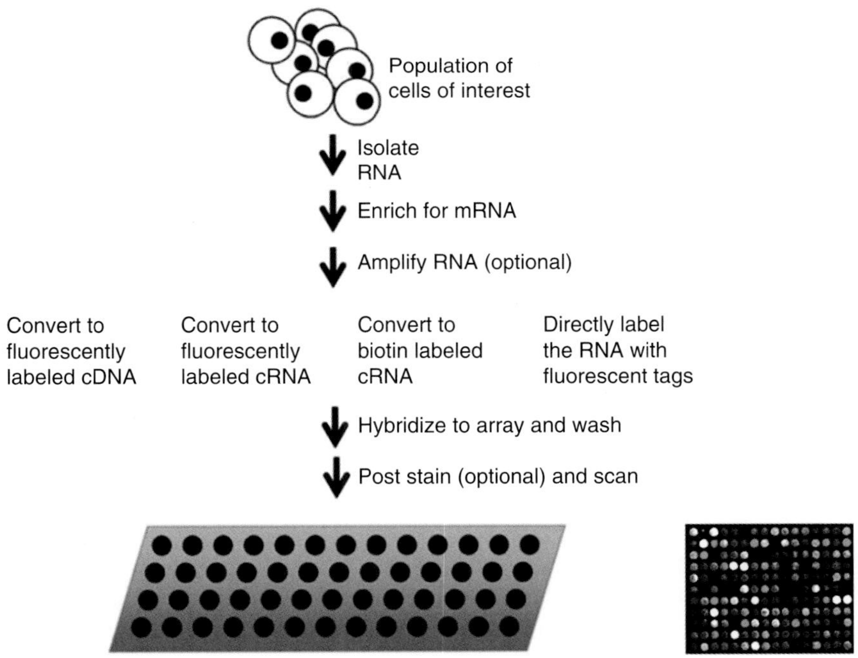

FIGURE 39.2 Application of microarray for gene expression analysis. Fluorescently labeled cDNA or cRNA is hybridized with probes, and the image is scanned through a scanner. Based upon the intensity of the signal, up regulated (red dots) and down regulated (green dots) genes are detected. (*See insert for colour representation of the figure.*)

39.4.1 Image processing

Microarray chips after hybridization are scanned using a microarray scanner, and quality images (.TIFF in 16-bits/pixel) are obtained and then processed using suitable software. The content of the image is characterized by spot shape (morphology), spot intensity, background correction and noise level.

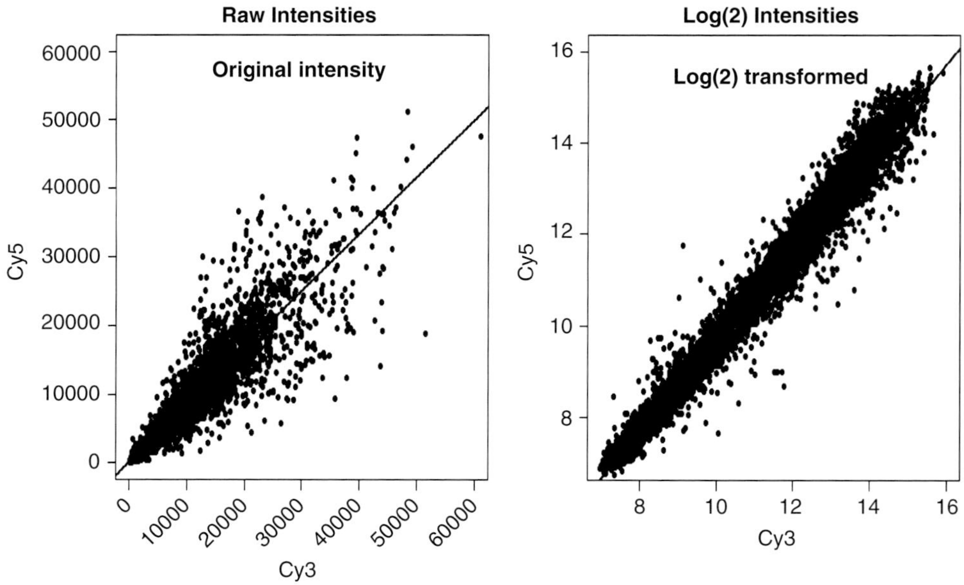

FIGURE 39.3 Data transformation converts the raw signal intensity of each probe-target hybridization into a log scale. Transformation of the data brings values in a normal distribution.

39.4.2 Normalization

- First, microarray expression data are transformed
- Normalization of data does not mean the data that are not normally (Gaussian) distributed, but normalization of microarray data refers to processing of correcting data before comparing gene expression.
- Normalization takes care of efficiencies of incorporation of Cy3 and Cy5 and brings them to comparable levels.
- Normalization also enables comparison of multiple microarray experiments.
- The first step of data normalization is calculation of the background signal:
  - a lower 5% signal can be used as background signal intensity;
  - signal of empty spots on array can be used as background signal.
- Global normalization to raw signal intensity – average ratio of gene expression is 1.
  - Example: Mean green channel intensity of samples is 10 000 units and of red channel is 5000 units; then, the intensity of red channel will be multiplied by two such that mean ratio is 1. If data are log transformed the mean ratio would be 0.
- Another approach to global normalization involves the use of housekeeping genes, the amount of starting quantities of RNA, and difference in the labeling efficiencies. Divide each gene expression value by the mean expression value of all housekeeping genes

39.4.3 Identification of differentially expressed genes

This involves steps in order to remove false positive results. The steps are:

a. Fold change > twofold is most common.
b. P-value: the probability of a result being observed, given that the null hypothesis is true.

c. Type I error (a, "p-value"): false positives.
d. Multiple testing corrections or Bonferroni corrections, or family-wise error rate corrections (Bonferroni correction: set a to desired a/number of tests = 0.05/20 000 = 2.5×10^{-6}).
e. The above correction means that a cumulative critical value of 2.5×10^{-6} is considered, instead of using a $p \leq 0.05$, for each of the genes.
f. However, this is a very small value, and it is hard to get genes to qualify for the test.
g. False Discovery Rate (FDR or q-value): statistically obtained proportion of the false-positives out of the positive results.
h. We assume 10% of our results to be false positive at FDR =10.

39.4.4 Cluster analysis

This is done to see how a group of genes as a cluster varies between the two conditions, thereby dividing the experimental samples into homogeneous groups:

- Supervised clustering:
  - Support Vector Machines (SVM)
  - Artificial Neural Networks (ANN)
- Unsupervised clustering:
  - hierarchical clustering
  - *k*-means clustering
  - self-organizing maps (SOM)
  - principal component analysis (PCA)

39.5 FUNCTIONAL INFORMATION USING GENE NETWORKS AND PATHWAYS

Finally, the investigator needs to understand the underlying system's biology. Differentially expressed genes are identified between the two groups (treatment and control) in order to explore the biological phenomenon. Genes interact with each other and forms gene networks. Gene networks could be described in four hierarchical levels;

a. Part lists – genes, transcription factors, promoters, binding sites.
b. Control logics – interactions between different combinations of regulatory signals.
c. Topology – a graph describing the connections between the parts.
d. Dynamics.

The information that we need to know from genes is about the gene product, its place and time of action and role in physiology. Bioinformatics initiates called "Gene Ontologies", or simply "GO", provide such information. GO teams provide three main domains of a gene, namely:

a. Cellular components – the parts of a cell or its extracellular components.
b. Molecular functions – functions of a gene product.
c. Biological process – sets of molecular events that occur with the particular gene product.

39.6 LIVESTOCK RESEARCH THAT INVOLVED MICROARRAY ANALYSIS (SOME EXAMPLES)

- Understanding the physiology of the mammary glands of bovine (Hu *et al.*, 2009; Moyes *et al.*, 2010) and mammary stem cells (Choudhary *et al.*, 2013).
- Gene expression profile of lactating and non-lactating mammary gland (Suchyta *et al.*, 2003).
- Gene expression profile of Brahman steers (calf) muscle tissue, to understand remodeling of muscle tissue in response to nutritional stress (Byrne *et al.*, 2005).

39.7 APPLICATIONS OF MICROARRAY

- Gene expression analysis.
- Genotyping.
- Transcription factor binding analysis.
- Treatment comparisons.
- Detection of cancer vs normal cells.

39.8 QUESTIONS

1. Explain the importance of replicates in a microarray experiment.
2. What are the various steps involved in microarray data analysis?

Single Nucleotide Polymorphism (SNP) Mining Tools

CHAPTER 40

Mir Asif Iquebal, Sarika and D Kumar
CABiN, ICAR-IASRI, New Delhi, India

40.1 INTRODUCTION

There exist genetic variations among individuals of all organisms, and these genetic variations make individuals look phenotypically different. Single nucleotide polymorphisms (SNPs) are considered the simplest and most abundant type of genetic variations in the genome of organisms. SNPs are the markers of choice in most species for genome-wide association studies (GWAS), phylogenetic analysis, marker-assisted selection and genomic selection (Liu *et al.* 2013). They are the genetic markers of choice due to their high density and stability, and the highly automated techniques which are available for detection of SNPs (Kerstens *et al.*, 2009).

Numerous tools are available online for mining SNPs computationally. SNP mining in NGS data has been well documented using two online open source tools: Stacks (Catchen *et al.*, 2011; Ogden *et al.*, 2013) and GATK (DePristo *et al.*, 2011).

40.2 OBJECTIVE

To learn about SNP mining using Stacks, the Burrows–Wheeler algorithm (BWA) aligner, the Genome analysis toolkit (GATK) and Samtools.

40.3 PROCEDURE

We will learn to install and run the tools STACKS, BWA, GATK, Samtools, and so on, to mine SNPs in given nucleotide sequences.

40.3.1 Stacks

This is a program to study population genetics, and it is designed to work with any restriction-enzyme-based data, such as GBS (Genotyping by Sequencing), CRoPS,

Basic Applied Bioinformatics, First Edition. Chandra Sekhar Mukhopadhyay, Ratan Kumar Choudhary and Mir Asif Iquebal.

FIGURE 40.1 Screenshot of Stacks software: http://creskolab.uoregon.edu/stacks/

RAD-Seq, and ddRAD-Seq. Stacks can identify SNPs within or among populations. It has different modules to generate summary statistics and also compute parameters of population genetics, such as F_{is} and π, within populations, and F_{st} between populations.

The output of Stacks can be exported in VCF (Variant Call Format) and many other standard formats, which we can use in different programs like STRUCTURE and GenePop for downstream analysis. The SNPs predicted by Stacks can also be exported in Phylip format to predict phylogenetic trees by any standard phylogenetic software/tool. Stacks can be used to predict SNPs *de novo*, as well as by a reference-based method. It is a Linux-based program and is available for free download from the Stacks website (http://creskolab.uoregon.edu/stacks/) (Figure 40.1).

40.3.1.1 Stacks installation

1. Untar the compressed file.
   tar xfvz stacks_x.xx.tar.gz
2. Change directory
   cd stacks_x.xx
3. Configure the software.
   ./configure
4. Build the programs from the source code using "make".
   make
5. Become root user.
   make install

> Note: We may also change the install location of the program by using the –prefix command line option in step 3.
>
> ./configure –prefix=/path/to/new/location/
>
> For reference-based SNP mining, install the BWA aligner to align reads to the reference.

40.3.1.2 Install the BWA aligner

1. Download the BWA software (Li and Durbin, 2010).
2. Untar the compressed file
   tar –jxvf bwa_x.xx.tar.bz2
3. Change directory
   cd bwa_x.xx
4. Build the programs from the source code using "make".
   make

Now run ./bwa from the command line to check if BWA is installed correctly. Additionally add the binaries of the softwares to the bashrc.

40.3.1.3 De novo-*based SNP mining using Stacks*

The denovo_map.pl Perl wrapper script (which includes three components – ustacks, cstacks and sstacks) is used for *de novo* SNP mining in Stacks. The following codes have been obtained from http://catchenlab.life.illinois.edu/stacks/comp/denovo_map.php. The codes and its explanations are verbatim as available on the website.

USAGE:

denovo_map.pl {–p path –r path|–s path} –o path [–t] [–m min_cov] [–M mismatches] [–n mismatches] [–T num_threads] [–A type] [–O popmap] [–b batch_id –D desc –a yyyy–mm–dd] [–S –inum] [–e path] [–d] [–h]

p – path to a FASTQ/FASTA file containing parent sequences from a mapping cross.
r – path to a FASTQ/FASTA file containing progeny sequences from a mapping cross.
s – path to a FASTQ/FASTA file containing an individual sample from a population.
o – path to write pipeline output files.
A – if processing a genetic map, specify the cross type, "CP", "F2", "BC1", "DH", or "GEN".
O – if analyzing one or more populations, specify a population map. The population map is passed on to the "populations" program.
T – specify the number of threads to execute.
e – executable path, location of pipeline programs.
d – perform a dry run. Do not actually execute any programs – just print what would be executed.
h – display this help message.

STACK ASSEMBLY OPTIONS:

m – specify a minimum number of identical raw reads required to create a stack.
P – specify a minimum number of identical raw reads required to create a stack in "progeny" individuals.
M – specify the number of mismatches allowed between loci when processing a single individual (default=2).
n – specify the number of mismatches allowed between loci when building the catalog (default=0).
t – remove, or break up, highly repetitive RAD-Tags in the ustacks program.
H – disable calling haplotypes from secondary reads.

DATABASE OPTIONS:

b – batch ID representing this dataset.
B – specify a database to load data into.
D – batch description
a – batch run date, yyyy-mm-dd
S – disable recording SQL data in the database.
i – starting sample_id. This is determined automatically if database interaction is enabled.

```
[NEERAJ@wrk20 ~]$ denovo_map.pl -m 3 -M 3 -n 2 -T 15 -S -b 1 -o ~/Desktop/mango/results_denovo/ -s ~/Desk
top/mango/sequences/Mango_R1.fastq -s ~/Desktop/mango/sequences/Mango_R2.fastq -X "populations:-b 1 -t 10
0 --vcf"
Found 2 sample file(s).
Identifying unique stacks; file   1 of   2 [Mango_R1]
  /usr/local/bin/ustacks -t fastq -f /home/NEERAJ/Desktop/mango/sequences/Mango_R1.fastq -o /home/NEERAJ/
Desktop/mango/results_denovo -i 1 -m 3 -M 3 -p 15 2>&1
Identifying unique stacks; file   2 of   2 [Mango_R2]
  /usr/local/bin/ustacks -t fastq -f /home/NEERAJ/Desktop/mango/sequences/Mango_R2.fastq -o /home/NEERAJ/
Desktop/mango/results_denovo -i 2 -m 3 -M 3 -p 15 2>&1
Generating catalog...
  /usr/local/bin/cstacks -b 1 -o /home/NEERAJ/Desktop/mango/results_denovo -s /home/NEERAJ/Desktop/mango/
results_denovo/Mango_R1 -s /home/NEERAJ/Desktop/mango/results_denovo/Mango_R2  -n 2 -p 15 2>&1
Matching samples to catalog; file   1 of   2 [Mango_R1]
  /usr/local/bin/sstacks -b 1 -c /home/NEERAJ/Desktop/mango/results_denovo/batch_1 -s /home/NEERAJ/Deskto
p/mango/results_denovo/Mango_R1 -o /home/NEERAJ/Desktop/mango/results_denovo -p 15 2>&1
Matching samples to catalog; file   2 of   2 [Mango_R2]
  /usr/local/bin/sstacks -b 1 -c /home/NEERAJ/Desktop/mango/results_denovo/batch_1 -s /home/NEERAJ/Deskto
p/mango/results_denovo/Mango_R2 -o /home/NEERAJ/Desktop/mango/results_denovo -p 15 2>&1
Calculating population-level summary statistics
/usr/local/bin/populations -b 1 -P /home/NEERAJ/Desktop/mango/results_denovo -s -t 15 -b 1 -t 100 --vcf 2
>&1
[NEERAJ@wrk20 ~]$ 
```

FIGURE 40.2 Image of denovo_map.pl script of Stacks to call SNPs *de novo* from RADSeq data.

Please also consult "Stacks", using the following link for downloading Stacks: http://creskolab.uoregon.edu/stacks/

40.3.1.4 Steps for de novo *SNP mining from example RAD-Seq dataset*

Use the denovo_map.pl script to call SNP from Mango example RAD-Seq dataset (Figure 40.2):

denovo_map.pl –m <number> –M <number> –n <number> –T <number> –S –b <number> –o <path/to/output/result/folder> –s <path/to/input/file1> –s <path/to/input/file2> –X "populations:-b <1> –t <number> –vcf"

This script will generate results and place them in the output folder specified by the *–o* option. There are two input files (e.g., Mango_R1.fastq and Mango_R2.fastq) that are specified by *option –s*. The code for analysis is set using a number of options as given below:

- *–m option:* uses three identical raw reads to create a stack;
- *–M option*: indicates permitted mismatches (here, we have three mismatches) allowed between loci when processing a single individual;
- *–n option*: for the number of mismatches permissible between loci when building the catalog (here, two mismatches);

- *–T option:* for executing the threads (here, in this example, with 15 threads);
- *–b option*: batch ID for the run (here, 1);
- *–S option*: whether the records are to be entered in the MySQL database (here, not entered).

Additionally, our code is running a "populations program" (*–X option*) to generate "population genetics statistics" on batch 1 (*–b option*) with 100 threads to run in a parallel section *(–t option)* and generate results in vcf format *(–vcf option)*.

40.3.1.5 Reference-based SNP mining using STACKS

The reference-based SNP mining in Stacks is done using ref_map.pl Perl, which includes three components – pstacks, cstacks and sstacks. This program needs a reference-aligned data file and can take input data that has been aligned using Bowtie or any other aligner (like BWA); output will be in SAM (Sequence alignment/Map) format.

USAGE:

ref_map.pl {–p path –r path|–s path} –o path [–n mismatches] [–m min_cov] [–T num_threads] [–A type] [–O popmap] [–B db –b batch_id –D "desc" –a yyyy–mm–dd] [–S –i id] [–e path] [–d] [–h]

p – path to a SAM/BAM file containing parent sequences.
r – path to a SAM/BAM file containing progeny sequences.
s – path to a SAM/BAM file containing an individual sample from a population.
o – path to write pipeline output files.
n – specify the number of mismatches allowed between loci when building the catalog (default 0).
T – specify the number of threads to execute.
m – specify the minimum depth of coverage to report a stack in pstacks (default = 1).
A – if processing a genetic map, specify the cross type: "CP", "F2", "BC1", "DH", or "GEN".
O – if analyzing one or more populations, specify a population map. The population map is passed on to the *populations* program. See the manual for more information.
e – executable path, location of pipeline programs.
h – display this help message.
d – turn on debug output.

DATABASE OPTIONS:

B – specify a database to load data into.
b – batch ID representing this dataset in the database.
D – batch description
a – batch run date, yyyy-mm-dd
S – disable recording SQL data in the database.
i – starting sample_id; this is determined automatically if database interaction is enabled.

Source: Stacks: http://creskolab.uoregon.edu/stacks/and http://www.vcru.wisc.edu/simonlab/bioinformatics/programs/stacks/ref_map.pl.txt. The codes and related annotations are verbatim, as these are available on the source page.

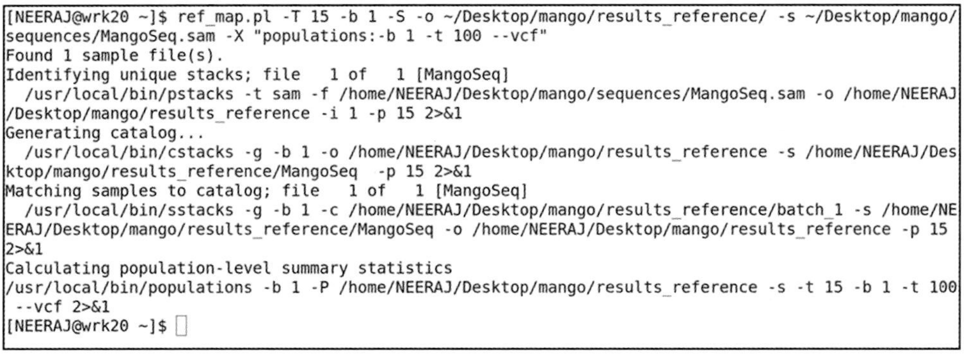

```
[NEERAJ@wrk20 ~]$ ref_map.pl -T 15 -b 1 -S -o ~/Desktop/mango/results_reference/ -s ~/Desktop/mango/
sequences/MangoSeq.sam -X "populations:-b 1 -t 100 --vcf"
Found 1 sample file(s).
Identifying unique stacks; file   1 of   1 [MangoSeq]
  /usr/local/bin/pstacks -t sam -f /home/NEERAJ/Desktop/mango/sequences/MangoSeq.sam -o /home/NEERAJ
/Desktop/mango/results_reference -i 1 -p 15 2>&1
Generating catalog...
  /usr/local/bin/cstacks -g -b 1 -o /home/NEERAJ/Desktop/mango/results_reference -s /home/NEERAJ/Des
ktop/mango/results_reference/MangoSeq  -p 15 2>&1
Matching samples to catalog; file   1 of   1 [MangoSeq]
  /usr/local/bin/sstacks -g -b 1 -c /home/NEERAJ/Desktop/mango/results_reference/batch_1 -s /home/NE
ERAJ/Desktop/mango/results_reference/MangoSeq -o /home/NEERAJ/Desktop/mango/results_reference -p 15
2>&1
Calculating population-level summary statistics
/usr/local/bin/populations -b 1 -P /home/NEERAJ/Desktop/mango/results_reference -s -t 15 -b 1 -t 100
 --vcf 2>&1
[NEERAJ@wrk20 ~]$ 
```

FIGURE 40.3 Image of ref_map.pl script of STACKS to call SNPs reference based from RAD-Seq data.

40.3.1.6 Steps for reference-based SNP mining from example RAD-Seq data

The ref_map.pl perl wrapper script is used to run reference-based SNP mining in Stacks software (Figure 40.3):

a. First index the genome by bwa index:
   bwa index <path/to/reference/file.fa>
b. Align the reads to the reference sequence using bwa mem:
   bwa mem <path/to/reference/file.fa> <path/to/input/sequence/file.fastq>
   <path/to/input/sequence/file2.fastq> > <path/to/output/file.sam>
c. Then call the SNPs using the ref_map.pl:
   ref_map.pl –T <number> –b <number> –S –o <path/to/results/folder> –s <path/to/input/file.sam> –X "populations:–b <number> –t <number> –vcf"

This perl script will generate results in the output folder specified by the –o option from a reference aligned input file – e.g., MangoSeq.sam, specified by the –s option with 15 threads to execute, specified by –T option and batch id 1, for the run specified by –b option. Additionally, it can run a populations program (–X option) to generate population genetics statistics on batch 1 (–b option), with 100 threads to run in parallel section (–t option), and generate results in VCF format (–vcf option).

40.3.2 Genome Analysis Toolkit (GATK)

This is an organized programming framework (DePristo *et al.*, 2011), which is designed to develop effective and durable analysis tools for next-generation DNA sequencing, using the functional programming theme of MapReduce (Dean and Ghemawat, 2008; DePristo *et al.* 2011). The GATK has a variety of tools which primarily focus on SNP discovery and genotyping, and has a strong emphasis on data quality assurance. It has a robust architecture, a powerful processing engine, and high-performance computing features which make it suitable to be used for a project of any size (https://www.broadinstitute.org/gatk/index.php). A reference-based SNP mining program that uses the Linux operating system, it is available for free download from the GATK website (Figure 40.4) (https://www.broadinstitute.org/gatk/index.php).

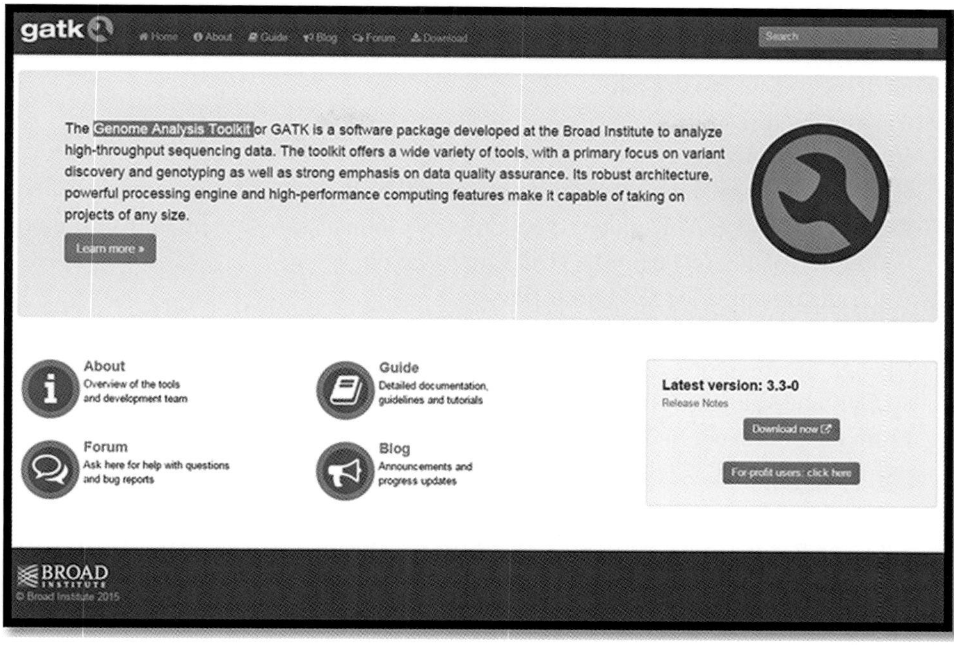

FIGURE 40.4 Screenshot of GATK software website: https://www.broadinstitute.org/gatk/index.php.

In order to use GATK, it is first necessary to install the BWA, Samtools, and Picard tools.

40.3.2.1 BWA installation

a. Untar the downloaded BWA.tar file (Li and Durbin, 2010):
   tar -jxvf bwa-x.xx.tar.bz2
b. Change the directory:
   cd bwa-x.xx
c. Build the program from the source code, using the "make" command:
   make
d. Execute the command, ./bwa from the command line (to check if it is properly installed).
e. Additionally, add the bwa binary to the path to make it available on the command line.

40.3.2.2 Samtools installation

a. Untar the downloaded Samtools.tar file (Li *et al.*, 2009a):
   tar –jxvf samtools-x.xx.tar.bz2
b. Change the directory:
   cd samtool-x.xx
c. Build the program from the source code, using the "make" command
   make
d. From the command line, run ./samtools (to check if it is properly installed).
e. Additionally, add the samtools binary to the path to make it available on the command line.

40.3.2.3 Picard tools and GATK installation

a. Untar the downloaded Picard tools zip file:
   unzip picard-tools-x.xx.zip
b. Change the directory:
   cd picard-tools-x.xx
c. From the command line, run java –jar picard.jar –h (to check if it is properly installed).
   Now download GATK (latest version) from https://software.broadinstitute.org/gatk/download/, and begin the GATK installation:
d. Untar the downloaded GATK tar file:
   tar -jxvf GenomeAnaysisTK-x.xx.tar.bz2
e. Change the directory:
   cd GenomeAnalysisTK-x.xx
f. From the command line, run java –jar GenomeAnalysisTK.jar –h (to check if it is properly installed).

40.3.3 Steps for SNP mining from chickpea data using GATK pipeline

To call SNPs using the GATK toolkit, we have to pre-process the input bam files. Steps for pre-processing the bam files are described below:

a. Sort the bam file using the SortSam tool in Picard-tools:
   java –jar <path/to/SortSam.jar> INPUT = <path/to/input/file.bam> OUTPUT = <path/to/output/file.bam > SORT_ORDER = <type > VALIDATION_STRINGENCY = <SILENT,LENIENT or STRICT>
b. Mark the duplicates in the bam file using "MarkDuplicates":
   java –jar < path/to/AddOrReplaceReadGroups.jar > INPUT = <path/to/Markduplicated/file.bam > OUTPUT = <path/to/output/file.bam > RGID = group-name RGLB = lib name RGPL = platformname RGPU = unit number RGSM = sample name VALIDATION_STRINGENCY = <SILENT,LENIENT or STRICT>
c. Add read group information using "AddOrReplaceReadGroups":
   java –jar < path/to/MarkDuplicates.jar > INPUT = <path/to/sortedfile.bam > OUTPUT = <path/to/output/file.bam > METRICS_FILE = <path/to/metrics.txt>
d. Build the bam file index using "BuildBamIndex":
   java –jar < path/to/BuildBamIndex.jar > INPUT = <path/to/input/file.bam VALIDATION_STRINGENCY = <SILENT,LENIENT or STRICT>

With these processed Bam files, we now can proceed further to call SNPs by GATK using the "UnifiedGenotyper" program, as described below:

java –jar<path/to/GenomeAnalysisTK.jar –T UnifiedGenotyper –R<path/to/reference/sequence.fa –I<path/to/input/file.bam –I<path/to/input/file2.bam –o<path/to/output/file.vcf VALIDATION_STRINGENCY=<SILENT,LENIENT or STRICT>

The script used in the GATK pipeline calls SNPs from two input bam files (specified by the –I option), using the reference sequence specified by the –R option, and outputs results to the output folder specified by the –o option in VCF format, using the "UnifiedGenotyper" tool. The user can select the level of stringency required, from Silent, Lenient, and Strict options.

The GATK script, running on an example dataset, is shown in Figure 40.5.

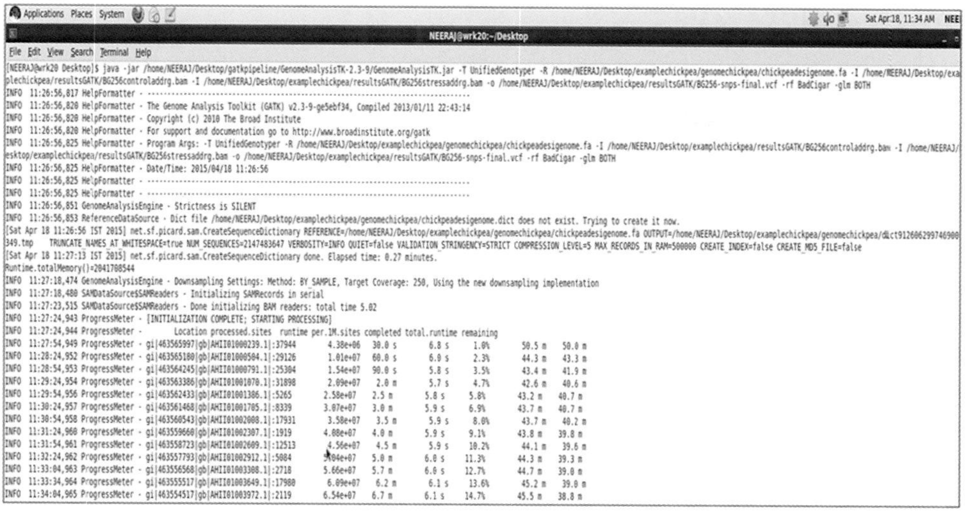

FIGURE 40.5 Image of GATK command used to mine SNPs from an example dataset.

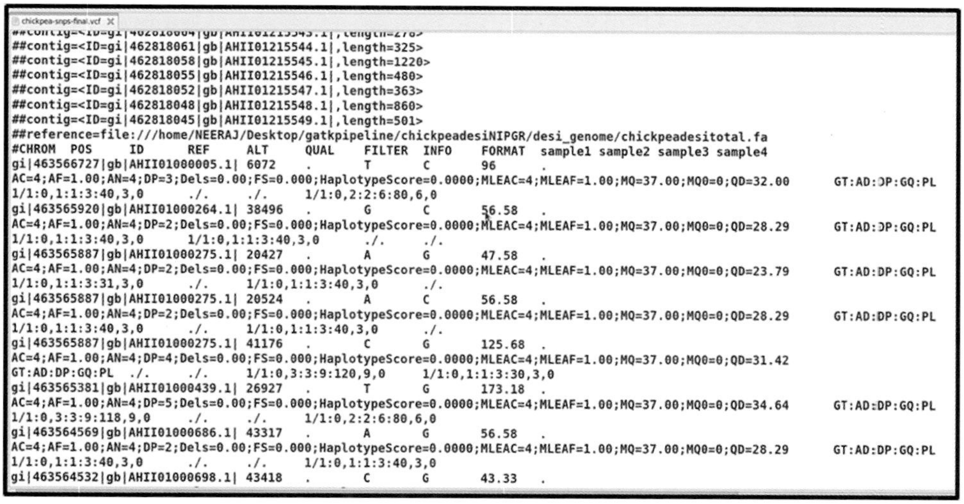

FIGURE 40.6 Result of GATK SNPs mining from an example dataset.

40.4 INTERPRETATION OF RESULTS

The Variant Call Format (VCF) has its first six columns representing observed variation, and is explained as follows (see Figure 40.6):

a. CHROM and POS: This gives the contig with position on which the variant occurs.
b. ID: Shows the dbSNP RefSeq (rs) identifier of the SNP.
c. REF and ALT: the reference base and alternative base, which vary in the samples or in the population.
d. QUAL: The quality value – namely, Phred-scaled probability that a REF/ALT polymorphism exists at this site given sequencing data.
e. FILTER: The VCFs produced carry both the passing and failing filter records.

40.5 QUESTIONS

1. What are the various tools/applications for SNP mining?
2. Trace the sequence NC_003062.2 from the public domain database, and search for the SNPs in this sequence by the *de novo* approach.
3. How many SNP calls did you get?
4. How many SNPs passed for the given sequence?

In Silico Mining of Simple Sequence Repeats (SSR) Markers

CHAPTER 41

Mir Asif Iquebal, Sarika and D Kumar
CABiN, ICAR-IASRI, New Delhi, India

41.1 INTRODUCTION

Microsatellites are simple sequence repeats (SSRs), where repeat units are di-, tri- tetra- or penta-nucleotides. A common repeat motif in birds is $(AC)_n$, where the two nucleotides *A* and *C* are repeated *n* number of times (*n* ranges from 8 to 50). They tend to occur in non-coding regions of the DNA, but a few human genetic disorders are caused by microsatellite falling in coding regions. There are many tools for mining microsatellite markers.

41.2 OBJECTIVE

To learn how to mine simple sequence repeats (SSR) markers in a given DNA sequence.

A number of tools for mining microsatellite markers from genome are available in the public domain. Examples include *Repeatmasker* (www.repeatmasker.org/; Smit *et al.*, 1996), *Sputnik* (http://espressosoftware.com/pages/sputnik.jsp; Abajian, 1994) *Tandem Repeats Finder* (TRF) (http://tandem.bu.edu/trf/trf.html; Benson, 1999), *MISA* (http://pgrc.ipkgatersleben.de/misa/; Theil *et al.*, 2003), *SSRIT* (Temnykh *et al*, 2001), and others.

41.3 MISA (MICROSATELLITE IDENTIFICATION TOOL)

This can be found at: http://pgrc.ipk-gatersleben.de/misa/misa.html (Figure 41.1). Requirements for MISA installation are:

a. Windows XP operating system (minimum 512 MB RAM, Pentium IV processor), or Linux-based system.
b. Perl is to be installed.

Basic Applied Bioinformatics, First Edition. Chandra Sekhar Mukhopadhyay, Ratan Kumar Choudhary and Mir Asif Iquebal.

MISA - MIcroSAtellite identification tool

Download MISA

Download misa.ini - example of a specification file

Syntax: misa.pl **<FASTAfile>**

Declarations:

- <FASTAfile> is the filename containing DNA sequences in FASTA format.
- An additional file containing the search parameters is required named "misa.ini".
    - In a single line beginning with "def" a sequence of number pairs is expected, whereas the first number defines the unit sizes and the second number the lower threshold of repeats for that specific unit.
    - In a single line beginning with "int" a single number is expected defining the maximal number of bases between two adjacent microsatellites to be recognised as being a compound microsatellite.

Output:

Results of the microsatellite search are stored in two files:

1. In "<FASTfile>.misa" the localization and type of identified microsatellite(s) are stored in a tablewise manner.
2. The file "<FASTfile>.statistics" summarizes different statistics as the frequency of a specific microsatellite type according to the unit size or individual motifs.

FIGURE 41.1 MISA homepage.

```
#!/usr/bin/perl -w
# Author: Thomas Thiel
# Program name: misa.pl

###_____________________________________________________________________
###
###Program name: misa.pl
###Author:       Thomas Thiel
###Release date: 14/12/01 (version 1.0)
###
###_____________________________________________________________________
###
## _____________________________________________________________________
##
## DESCRIPTION: Tool for the identification and localization of
##              (I)  perfect microsatellites as well as
##              (II) compound microsatellites (two individual microsatellites,
##                   disrupted by a certain number of bases)
##
## SYNTAX:   misa.pl <FASTA file>
##
##    <FASTAfile>    Single file in FASTA format containing the sequence(s).
##
##    In order to specify the search criteria, an additional file containing
##    the microsatellite search parameters is required named "misa.ini", which
##    has the following structure:
##      (a) Following a text string beginning with 'def', pairs of numbers are
##          expected, whereas the first number defines the unit size and the
##          second number the lower threshold of repeats for that specific unit.
##      (b) Following a text string beginning with 'int' a single number defines
##          the maximal number of bases between two adjacent microsatellites in
##          order to specify the compound microsatellite type.
##    Example:
##      definition(unit_size,min_repeats):          1-10 2-6 3-5 4-5 5-5 6-5
##      interruptions(max_difference_for_2_SSRs):   100
##
## EXAMPLE: misa.pl seqs.fasta
##
##
```

FIGURE 41.2 Download misa.pl.

41.3.1 MISA installation

Copy the file (Figure 41.2) and save it in a text document as misa.pl.

Copy the misa.ini file (Figure 41.3) and save it in a text document as misa.ini. After installation of misa.pl and misa.ini, microsatellites can be identified using the ./misa.pl FASTAfile.

41.3.2 Objective

To mine SSRs from a given sequence:

>sequence
AATTCGGCACCAGTAAATTTTCCCAAAGGTTTCAAAAATGAAA
ATTTTGATTTTCCTAATAATGTTTCTTGCTATGTTGCTAGTAA
CAAGTGGGAATAATAATCTAGTAGAGACAACATGCAAGAACAC
ACCAAATTATAATTTGTGTGTGAAAACTTTGTCTTTAGACAAA
AGAAGTGAAAAAGCAGGAGATATTACAACATTAGCATTAATTA
TGGTTGATGCTATTAAATCTAAAGCTAATCAAGCTGCTAATAC
TATTTCAAAACTTAGGCATTCTAATCCTCCTCAAGCTTGGAAA
GATCCTTTGAAGAATTGTGCCTTTTCGTATAAGGTAATTTTAC
CAGCAAGTATGCCAGAAGCATTAGAAGCATTAACAAAAGGTGA
TCCAAAATTTGCAGAAGATGGAATGGTTGGTTCTTCTGGTGAT
GCACAAGAATGTGAAGAATATTTTAAAGCTACAACTATTAAAT
ATTCACCACTTTCTAAATTAAATATAGATGTTCATGAACTTTC
TGATGTTGGTAGAGCCATTGTAAGAAATTTATTGTAATATGTC
ATGTCATAATGTTACATATCGAAAAGTTTTTATAGTTTAGTTT
GATAGACTGTCTGAATTATTATTTTATTCTTGCTAGTAAAAAT
TCGATTCGTCACATTATGATCATCTGTGGTTCATTTTTCTTTT
TTCTACCTCAAATGTTATGTGTGTATCCCCTCTTAATTATTAT
AAGAAAAATATATCATAAATATTTGTACAAGTGTAATACTCTT
ATCCAATATATATGTTKGYCCCCTTCTAAAAAAAAAAAAAAAA
AAAAAAAAAAAA

41.3.3 Procedure

a. Copy and paste the above sequence in notepad and save it as "testfile.fasta".
b. Start the command prompt and change the directory (i.e., specify the path) where your misa.pl and misa.ini have been saved.
c. The input file is to be placed in the same directory where misa.pl and misa.ini are placed – or, give the correct path to your input file.
d. Type the command: <perl misa.pl testfile.fasta> (Figure 41.4).

```
definition(unit_size,min_repeats):                   1-10 2-6 3-5 4-5 5-5 6-5
interruptions(max_difference_between_2_SSRs):        100
```

FIGURE 41.3 Download misa.ini.

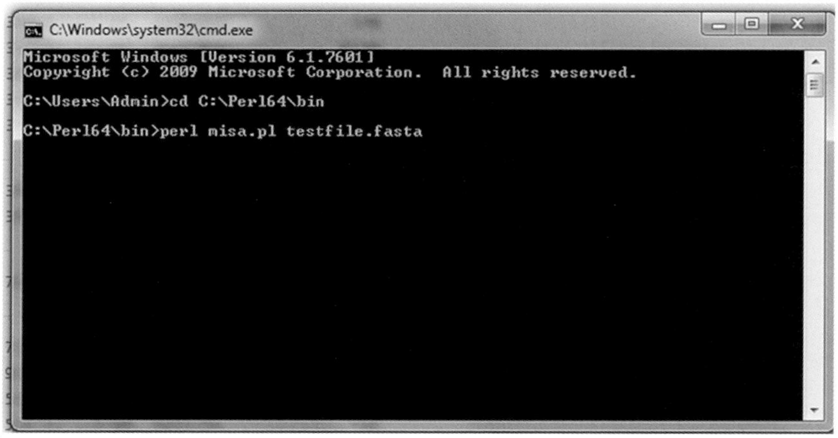

FIGURE 41.4 The command prompt where code is written.

ID	SSR nr.	SSR type	SSR	size	start	end
sequence	1	p1	(A)27	27	802	828

FIGURE 41.5 The output, as seen in testfile.misa.

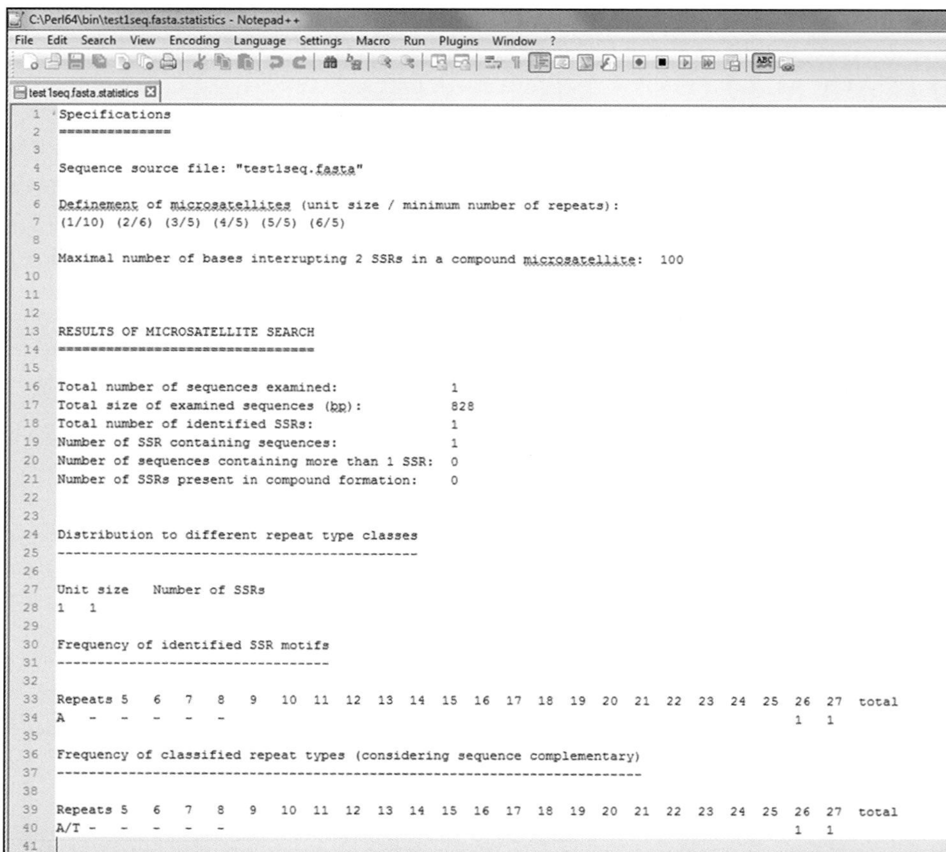

FIGURE 41.6 The output, as seen in testfile.statistics.

41.4 RESULT

a. Two files are generated, namely testfile.misa (Figure 41.5) and testfile.statistics (Figure 41.6).
b. For the given sequence, we get 1 SSR, which is "mono" type repeated 27 times. The start and end positions are 802 and 828, respectively.
c. Primers can be designed for the marker obtained using the Primer3 tool.

41.5 QUESTIONS

1. From NCBI, trace ESTs for "Sesame". How many hits do you get? Copy the top 20 hits to perform SSR mining from the selected hits.
2. How many SSRs do you get from the selected FASTA sequences?
3. What is the number of SSR-containing sequences?
4. How many sequences contain more than 1 SSR?
5. What are the repeat types of the mined SSRs? Discuss in detail.

Basics of RNA-Seq Data Analysis

CHAPTER 42

GVPPSR Kumar, AP Sahoo and A Kumar
Animal Biotechnology Division, IVRI, UP, India

42.1 INTRODUCTION

The complete set of transcripts (transcribed RNA products) in a cell is called the transcriptome (in terms of both type and quantity). Transcriptome analysis helps in understanding the pattern of gene expression to address basic biological questions, and unravels biological pathways and molecular mechanisms that regulate cell fate, development and disease progression. The transcriptome is analyzed through RNA-sequencing (RNA-seq) or microarray experiments. RNA-sequencing involves sequencing of the entire transcriptome, using next-generation sequencing (NGS) platforms.

The data generated through various platforms go into secondary analysis, which mainly involves alignment (if it is reference-based) or assembly (if it is *de novo*). In this book, RNA-seq data generated are analyzed through distinct pipelines to identify differentially expressed genes under different experimental conditions. The basics of RNA-sequencing data analysis will be discussed in this chapter.

42.2 AIM OF AN RNA-SEQ EXPERIMENT

a. To quantify RNA abundance.
b. Annotating the transcription start site, 5′ and 3′ ends and splicing patterns of genes.
c. To quantify the changing expression levels of each transcript during development, and under different conditions.
d. To identify variants on the transcripts.

With proper depth/coverage, NGS addresses the limitations of a microarray experiment, such as: high background levels owing to cross-hybridization; limited dynamic range of detection owing to both background and saturation of signals; and reliance upon existing knowledge about genome sequence.

Basic Applied Bioinformatics, First Edition. Chandra Sekhar Mukhopadhyay, Ratan Kumar Choudhary and Mir Asif Iquebal.

WHAT IS COVERAGE – AND HOW TO CALCULATE COVERAGE?

Coverage is the number of times a base is sequenced in the data generated. For example, if the genome is 3×10^9 bp, and the transcriptome (protein coding) is 2% of the genome, the transcriptome estimates to 6×10^7 bp. If each base is to be covered 50 times, then the total bases to be sequenced should be $50 \times 6 \times 10^7$ bp $\approx$ 3 Gb – that is, for a coverage of 50×, the *RNA-Seq* (mRNA sequencing) data generated should be 3 Gb.

42.2.1 Sequence alignment

The secondary analysis primarily involves mapping of the reads to the reference genome (reference-based assembly) or a reference transcriptome (*de novo* assembly). Mapping is aligning the read sequence to portions of the genome. This mapping/alignment is nothing but a pairwise alignment. The most accurate and sensitive method of pairwise alignment is dynamic programming, but this is time-consuming and cannot be used for aligning the NGS reads to the genome. This warrants fast sequence alignment strategies.

42.3 FAST SEQUENCE ALIGNMENT STRATEGIES

The faster alignment strategies that have evolved either use hash table-based indexing (seed extend paradigm with space allowance) or suffix/prefix tree-based indexing (suffix array or Burrows–Wheeler (BW) transformation and FM-index).

There are several aligners in vogue. Some of these are summarized below:

42.3.1 Short read aligners (Table 42.1)

TABLE 42.1 Example and purpose of short read aligners.

Aligner	Purpose	Strategy
Bowtie	Fast but gaps not allowed	BW transformation and FM index
BWA	small gaps (indels)	BW transformation and FM index
GSNAP	Large gaps (introns)	A double lookup scheme
Bowtie 2	Takes care of gaps	BW transformation and FM index

42.3.2 Long read aligners (Table 42.2)

TABLE 42.2 Example and purpose of long read aligners.

Aligner	Purpose	Strategy
BLAST	Many reference genome	Heuristic method
BLAT	Large gaps (introns)	BW transformation and FM Index
BWA	Small gaps (indels)	BW transformation and FM Index
Exonerate	Ease of use	Bounded sparse dynamic programming
GMAP	Large gaps (introns)	A double lookup scheme
MUMmer	Align two genome	Suffix tree

42.3.3 Challenges in RNA-seq alignment

There are three major challenges in aligning the RNA-seq reads:

a. the reads are short (35–125 *n*);
b. error rates are considerable;
c. Some reads span exon–exon junctions (Garber *et al.*, 2011).

Among the reads generated from the RNA-seq experiment, some cover only the exon regions and some span across the intron junctions covering two exons – termed *junction reads* (Wang *et al.*, 2009). The alignment of the reads can be done to a reference genome or a reference transcriptome. If the alignment is done to a reference transcriptome, short read aligners that do not allow gaps can be used, and, if the reads are aligned to reference genome, short read aligners that allow gaps have to be used. The provision to allow gaps is mainly to align junction reads to the reference genome, which contains introns.

There are two major alignment strategies for RNA-seq alignment – the exon-first approach and the seed-extend approach (Garber *et al.*, 2011).

a. *Exon-first approach*: tools that use this approach are *TopHat*, *MapSplice* and *Splice Map*.
   This is a two-step procedure which initially involves mapping reads continuously, using unspliced read aligners (i.e., aligners that do not allow gaps), followed by mapping the unmapped reads by splitting them into shorter segments and independently mapping these segments to the reference. *TopHat*, followed by *Cufflinks*, is one of the most common pipelines followed in analyzing the RNA-Seq data (Garber *et al.*, 2011).
b. *Seed-extend approach*: tools that use this approach are *GMAP*, *GSNAP*, and *QPALMA*.
   This involves breaking the reads into short seeds, and then combining the candidate regions by the Smith–Waterman algorithm (Smith and Waterman, 1981).

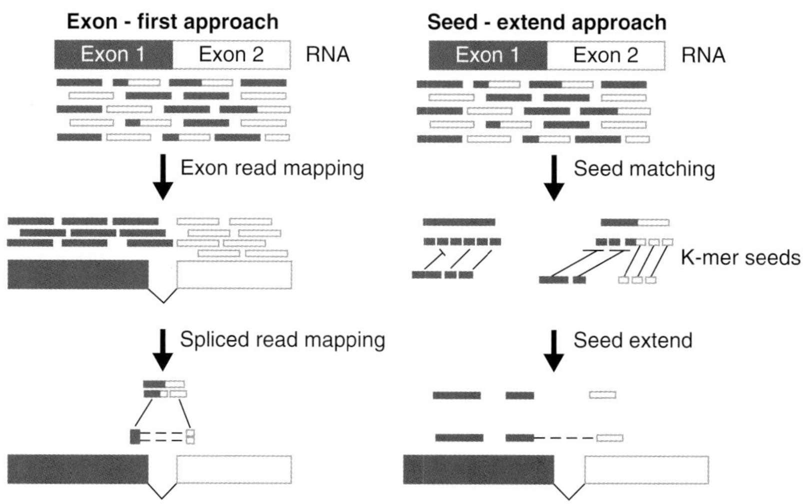

FIGURE 42.1 *(See insert for colour representation of the figure.)*

All the aligners mentioned above generate SAM (sequence alignment mapping) files from RNA-Seq data. *Cufflinks* or *RSEM* use these SAM files (depending on the experimental setup) in DE packages (DESeq2, EBSeq, edgeR) to generate differentially expressed genes. These packages normalize the read counts by using different metrics, such as RPKM/FPKM, TPM, or TMM.

42.3.4 Why is normalization needed?

The RNA-Seq reads are normalized for sequencing for the depth and length of the gene. Sequencing with greater depth will have more reads mapped to each gene, and longer genes will have more reads mapped to them. FPKM and TPM are the metrics used for normalizing for the length of the gene and depth of sequencing. There is a new metric termed TMM, which normalizes for differences in RNA composition between samples. Some parts of the following section are adapted from statquest.org:

42.3.4.1 R/FPKM (Mortazavi et al., 2008)

Reads/fragment per kilobase of exon per million mappable reads. *FPKM* is used for paired-end, and RPKM for single-end reads. "Per million reads" means the counts of fragments are normalized against the library sizes, which allows comparison of the same gene across samples. This value is further normalized per kilobase of exon, by dividing by the total length of all exons in the gene. This allows comparison of the expression of genes of different lengths. *Cufflinks* gives FPKM values, whereas *RSEM* gives both the FPKM and TPM values.

FPKM normalization involves two steps. In step 1, reads are normalized for library sizes, while step 2 involves normalization for the length of the gene. Here we consider four genes, with a variable number of reads for three samples.

a. **Step 1**. Divide the reads of each gene with the total reads of the sample. Total reads for sample 1, 2 and 3 are 7 M, 9 M and 21.2 M, respectively (here millions of reads equated to a scale of tens of reads). By dividing the reads for each gene in each sample with the corresponding total reads, we get the information shown in Table 42.3. (i.e., for sample 1 gene 1, 20/7 = 2.86 and likewise).
b. **Step 2**. Divide the values obtained after step 1 with the gene lengths. Hereafter, step 2 reads are scaled for both length and library size to get the RPKM values. (i.e., for sample 1 gene 1, 2.86/2 = 1.43 and likewise).

42.3.4.2 TPM (Transcripts per million) (Li et al., 2010; Wagner et al., 2012)

This metric corrects for transcript length distribution in an RNA pool and provides better across-sample comparability. Here, the read counts are initially normalized for length, and the total taken for the library sizes after dividing for length is used to normalize for library sizes. RSEM gives TPM.

Calculation of TPM is as follows:

a. **Step 1**. Divide the reads of each gene by the length of each gene:

TABLE 42.3 Information about total reads of samples 1, 2, and 3, and values obtained by dividing the reads for each gene in each sample with the corresponding total reads.

Genes	Sample1	Sample 2	Sample 3
1 (2 kb)	20	24	60
2 (4 kb)	40	50	120
3 (1 kb)	10	16	30
4 (10 kb)	0	0	2
Total	70 (7 M)	90 (9 M)	212 (21.2 M)
Genes	Sample 1 (RPM)	Sample 2 (RPM)	Sample 3 (RPM)
1 (2 kb)	2.86	2.67	2.83
2 (4 kb)	5.71	5.56	5.66
3 (1 kb)	1.43	1.78	1.42
4 (10 kb)	0	0	0.09

TABLE 42.4 Calculation of RPKM by dividing the reads obtained after step 1 for each gene with gene length.

Genes	Sample 1 (RPKM)	Sample 2 (RPKM)	Sample 3 (RPKM)
1 (2 kb)	1.43	1.33	1.42
2 (4 kb)	1.43	1.39	1.42
3 (1 kb)	1.43	1.78	1.42
4 (10 kb)	0	0	0.009
Total normalized reads	4.29	4.5	4.5

RPKM VS TPM

RPKM – the total normalized reads vary from sample to sample (i.e., 4.29, 4.5 and 4.5 for sample 1, 2 and 3, respectively). This makes comparison across genes and across samples difficult to interpret.

TPM – the total normalized reads are the same for all the samples (i.e.,10 for all the samples). This makes comparison across genes and across samples meaningful and easier to interpret.

TABLE 42.5 Total reads of samples 1, 2, 3, and 4.

Genes	Sample 1	Sample 2	Sample 3
1 (2 kb)	20	24	60
2 (4 kb)	40	50	120
3 (1 kb)	10	16	30
4 (10 kb)	0	0	2

TABLE 42.6 Total reads per kb (RPK) of gene for sample 1, 2, and 3 (millions of reads equated to a scale of tens of reads).

Genes	Sample 1 (RPK)	Sample 2 (RPK)	Sample 3 (RPK)
1 (2 kb)	10	12	30
2 (4 kb)	10	12.5	30
3 (1 kb)	10	16	30
4 (10 kb)	0	0	0.2
Total	30 (3 M)	40.5 (4.05 M)	90.2 (9.02 M)

Total reads per kb of gene for samples 1, 2 and 3 are 3 M, 4.05 M, and 9.02 M, respectively (millions of reads equated to a scale of tens of reads).

b. **Step 2**. Divide the values obtained after step 1 with total reads per kb of gene:

TABLE 42.7 Calculation of TPM by dividing the total reads obtained in step 1 sample, with total reads per kb of gene.

Genes	Sample 1 (TPM)	Sample 2 (TPM)	Sample 3 (TPM)
1 (2 kb)	3.33	2.96	3.326
2 (4 kb)	3.33	3.09	3.326
3 (1 kb)	3.33	3.95	3.326
4 (10 kb)	0	0	0.02
Total normalized reads	10	10	10

TABLE 42.8 Comparison of reads of RPKM.

Genes	Sample 1 (RPKM)	Sample 2 (RPKM)	Sample 3 (RPKM)
1 (2 kb)	1.43	1.33	1.42
2 (4 kb)	1.43	1.39	1.42
3 (1 kb)	1.43	1.78	1.42
4 (10 kb)	0	0	0.009
Total normalized reads	4.29	4.5	4.5

TABLE 42.9 Comparison of reads of TPM.

Genes	Sample 1 (TPM)	Sample 2 (TPM)	Sample 3 (TPM)
1 (2 kb)	3.33	2.96	3.326
2 (4 kb)	3.33	3.09	3.326
3 (1 kb)	3.33	3.95	3.326
4 (10 kb)	0	0	0.02
Total normalized reads	10	10	10

42.3.4.3 TMM – Trimmed mean of M value (Robinson and Oshlack, 2010)
The differences in RNA composition between samples are corrected by this metric – that is, in a sample having certain genes that are very highly expressed, there is less scope for the less expressed genes to be sequenced, and RPKM or TPM normalization will yield biased expression values.

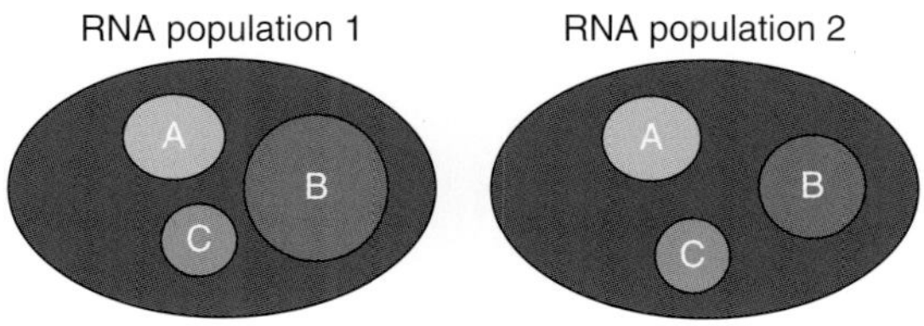

FIGURE 42.2 *(See insert for colour representation of the figure.)*

With the same sequencing depth for populations 1 and 2, A and C will have a lower RPKM in RNA population 1, though the expression of these genes is the same in populations 1 and 2 (Robinson and Oshlack, 2010). TMM is used by the edgeR DE package to normalize counts data.

42.4 QUESTIONS

1. What will be the coverage if 6 Gb data are generated for a transcriptome of 6×10^7 bp and a genome of 3×10^9 bp?
2. The total exon size of a gene is 3000 nt. Calculate the expression levels for this gene in RPKM, in an RNA-seq experiment that contained 50 million mappable reads, with 600 reads falling into exon regions of this gene.
3. Which approach, *Exon-first* or *Seed-extend*, is more appropriate for mapping reads from polymorphic species?

Functional Annotation of Common Differentially Expressed Genes

CHAPTER 43

GVPPSR Kumar, AP Sahoo and A Kumar
Animal Biotechnology Division, IVRI, UP, India

43.1 INTRODUCTION

Cuffdiff predicts Differentially Expressed Genes (DEGs) and gives the gene symbols in the output. However, EBSeq, DESeq2, and edgeR give the output of DEGs in *Ensembl IDs*. These Ensemble IDs are initially converted into gene symbols using g:Convert in g:Profiler. After conversion, it is always better to identify commonly differentially expressed genes across all the packages and further proceed with the analysis. The commonly predicted genes are identified by using the Venny package.

A total of 4246 commonly differentially expressed genes have been identified by all the packages in our analysis.

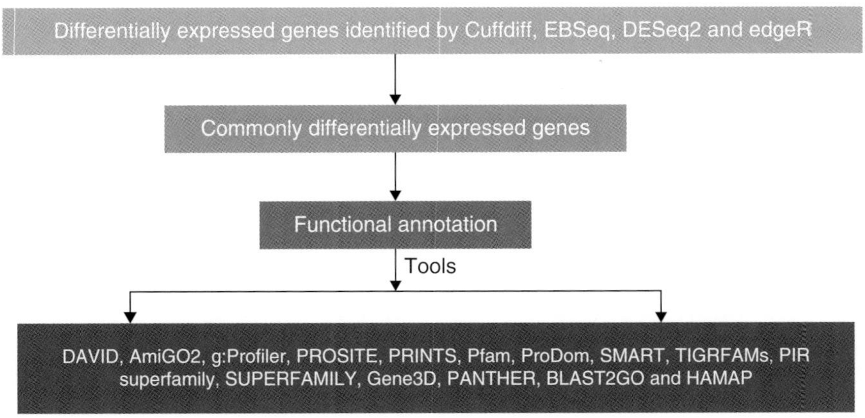

FIGURE 43.1 (*See insert for colour representation of the figure.*)

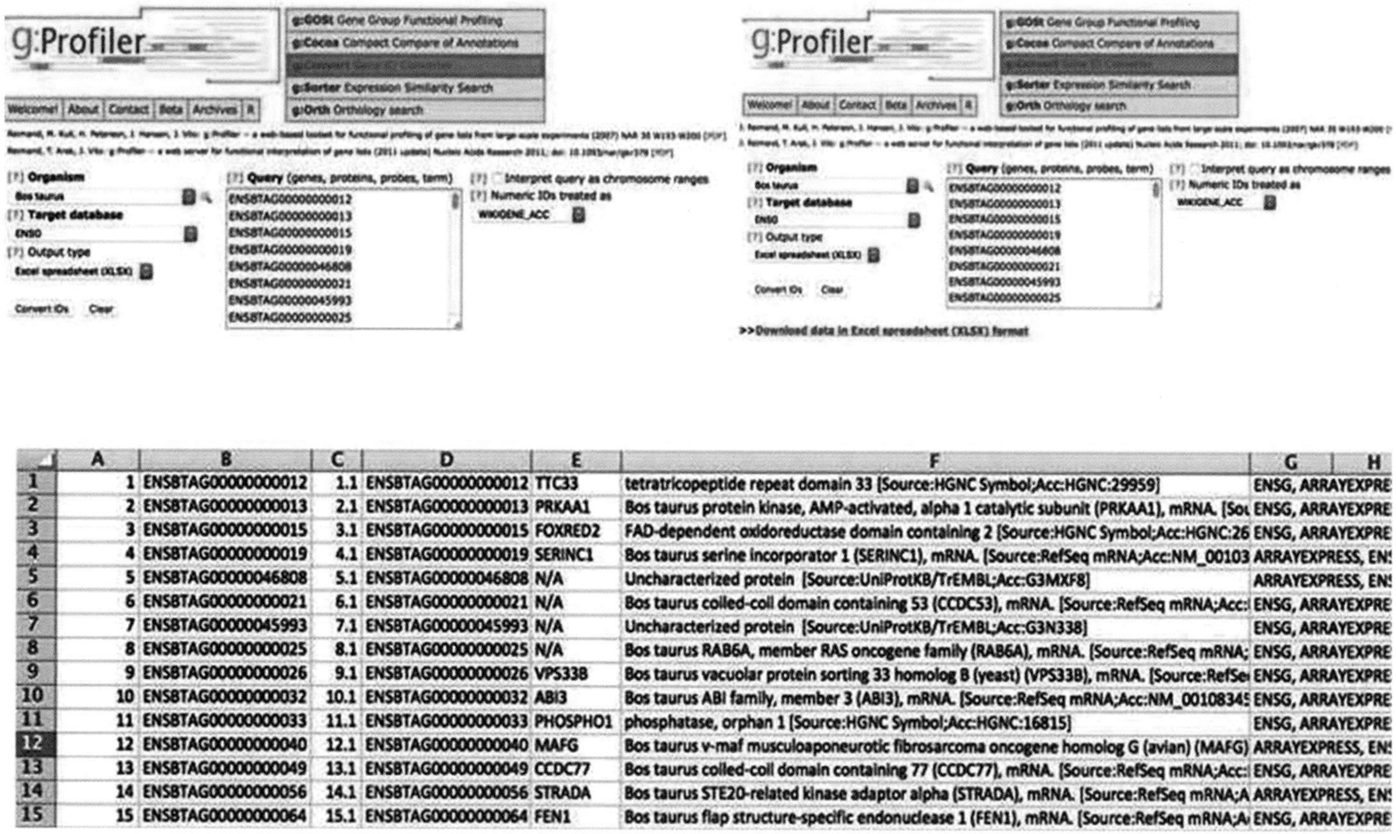

	A	B	C	D	E	F	G	H
1	1	ENSBTAG00000000012	1.1	ENSBTAG00000000012	TTC33	tetratricopeptide repeat domain 33 [Source:HGNC Symbol;Acc:HGNC:29959]	ENSG, ARRAYEXPRE	
2	2	ENSBTAG00000000013	2.1	ENSBTAG00000000013	PRKAA1	Bos taurus protein kinase, AMP-activated, alpha 1 catalytic subunit (PRKAA1), mRNA. [Sou	ENSG, ARRAYEXPRE	
3	3	ENSBTAG00000000015	3.1	ENSBTAG00000000015	FOXRED2	FAD-dependent oxidoreductase domain containing 2 [Source:HGNC Symbol;Acc:HGNC:26	ENSG, ARRAYEXPRE	
4	4	ENSBTAG00000000019	4.1	ENSBTAG00000000019	SERINC1	Bos taurus serine incorporator 1 (SERINC1), mRNA. [Source:RefSeq mRNA;Acc:NM_00103	ARRAYEXPRESS, ENS	
5	5	ENSBTAG00000046808	5.1	ENSBTAG00000046808	N/A	Uncharacterized protein [Source:UniProtKB/TrEMBL;Acc:G3MXF8]	ARRAYEXPRESS, ENS	
6	6	ENSBTAG00000000021	6.1	ENSBTAG00000000021	N/A	Bos taurus coiled-coil domain containing 53 (CCDC53), mRNA. [Source:RefSeq mRNA;Acc:	ENSG, ARRAYEXPRE	
7	7	ENSBTAG00000045993	7.1	ENSBTAG00000045993	N/A	Uncharacterized protein [Source:UniProtKB/TrEMBL;Acc:G3N338]	ENSG, ARRAYEXPRE	
8	8	ENSBTAG00000000025	8.1	ENSBTAG00000000025	N/A	Bos taurus RAB6A, member RAS oncogene family (RAB6A), mRNA. [Source:RefSeq mRNA;	ENSG, ARRAYEXPRE	
9	9	ENSBTAG00000000026	9.1	ENSBTAG00000000026	VPS33B	Bos taurus vacuolar protein sorting 33 homolog B (yeast) (VPS33B), mRNA. [Source:RefSe	ENSG, ARRAYEXPRE	
10	10	ENSBTAG00000000032	10.1	ENSBTAG00000000032	ABI3	Bos taurus ABI family, member 3 (ABI3), mRNA. [Source:RefSeq mRNA;Acc:NM_00108345	ENSG, ARRAYEXPRE	
11	11	ENSBTAG00000000033	11.1	ENSBTAG00000000033	PHOSPHO1	phosphatase, orphan 1 [Source:HGNC Symbol;Acc:HGNC:16815]	ENSG, ARRAYEXPRE	
12	12	ENSBTAG00000000040	12.1	ENSBTAG00000000040	MAFG	Bos taurus v-maf musculoaponeurotic fibrosarcoma oncogene homolog G (avian) (MAFG)	ARRAYEXPRESS, ENS	
13	13	ENSBTAG00000000049	13.1	ENSBTAG00000000049	CCDC77	Bos taurus coiled-coil domain containing 77 (CCDC77), mRNA. [Source:RefSeq mRNA;Acc:	ENSG, ARRAYEXPRE	
14	14	ENSBTAG00000000056	14.1	ENSBTAG00000000056	STRADA	Bos taurus STE20-related kinase adaptor alpha (STRADA), mRNA. [Source:RefSeq mRNA;A	ARRAYEXPRESS, ENS	
15	15	ENSBTAG00000000064	15.1	ENSBTAG00000000064	FEN1	Bos taurus flap structure-specific endonuclease 1 (FEN1), mRNA. [Source:RefSeq mRNA;A	ENSG, ARRAYEXPRE	

FIGURE 43.2 *(See insert for colour representation of the figure.)*

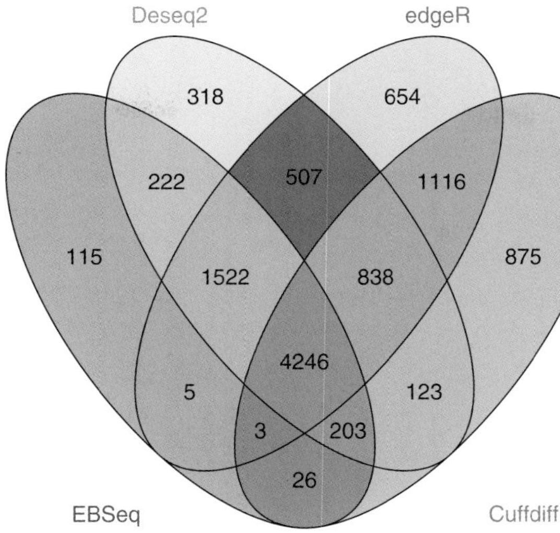

FIGURE 43.3 *(See insert for colour representation of the figure.)*

43.2 FUNCTIONAL ANNOTATION

Functional annotation is used to determine the gene ontology terms enriched in common differentially expressed genes. Gene ontology (GO) (Ashburner *et al.*, 2000) is an *in silico* approach to amalgamate the methods of presenting the genes and gene product attributes over divergent species. Gene products are categorized into three categories (biological processes, cellular components and molecular functions) in a species-independent manner in the process of assigning the annotations. There are several databases for performing the functional annotation: DAVID; AmiGO2; g:Profiler; PROSITE; PRINTS; Pfam; ProDom; SMART; TIGRFAMs; SUPERFAMILY; PIR superfamily; Gene3D; PANTHER; BLAST2GO; and HAMAP. Here we will be discussing g:Profiler, DAVID, and clueGO.

43.2.1 Functional annotation using g:Profiler (Reimand et al., 2011)

The gene lists resulting from analysis of high-throughput genomic data can be manipulated and characterized by g:Profiler. This is a simple, user-friendly web interface to derive and visualize GO functional pathways from enrichments of the transcription factor binding site up to individual gene levels (Reimand *et al.*, 2007).

43.2.1.1 Step 1

Open http://biit.cs.ut.ee/gprofiler/, paste the gene list, and select *Bos taurus* as the organism (species of interest) and the output type as *Excel* spreadsheet (Figures 43.4 and 43.5)

43.2.1.2 Step 2

Download the *Excel* file to check for the annotations enriched in the differentially expressed genes (see Figure 43.6):

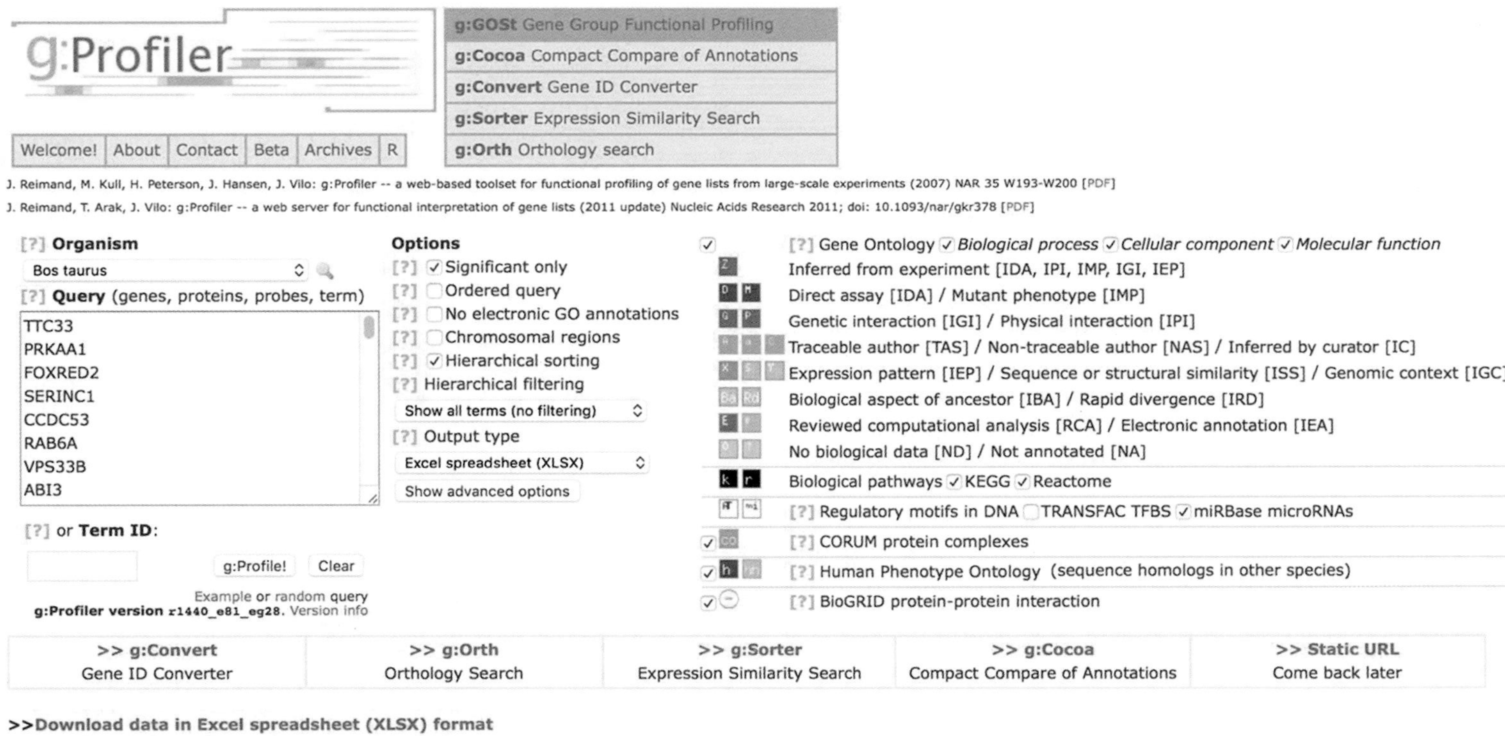

FIGURE 43.4 *(See insert for colour representation of the figure.)*

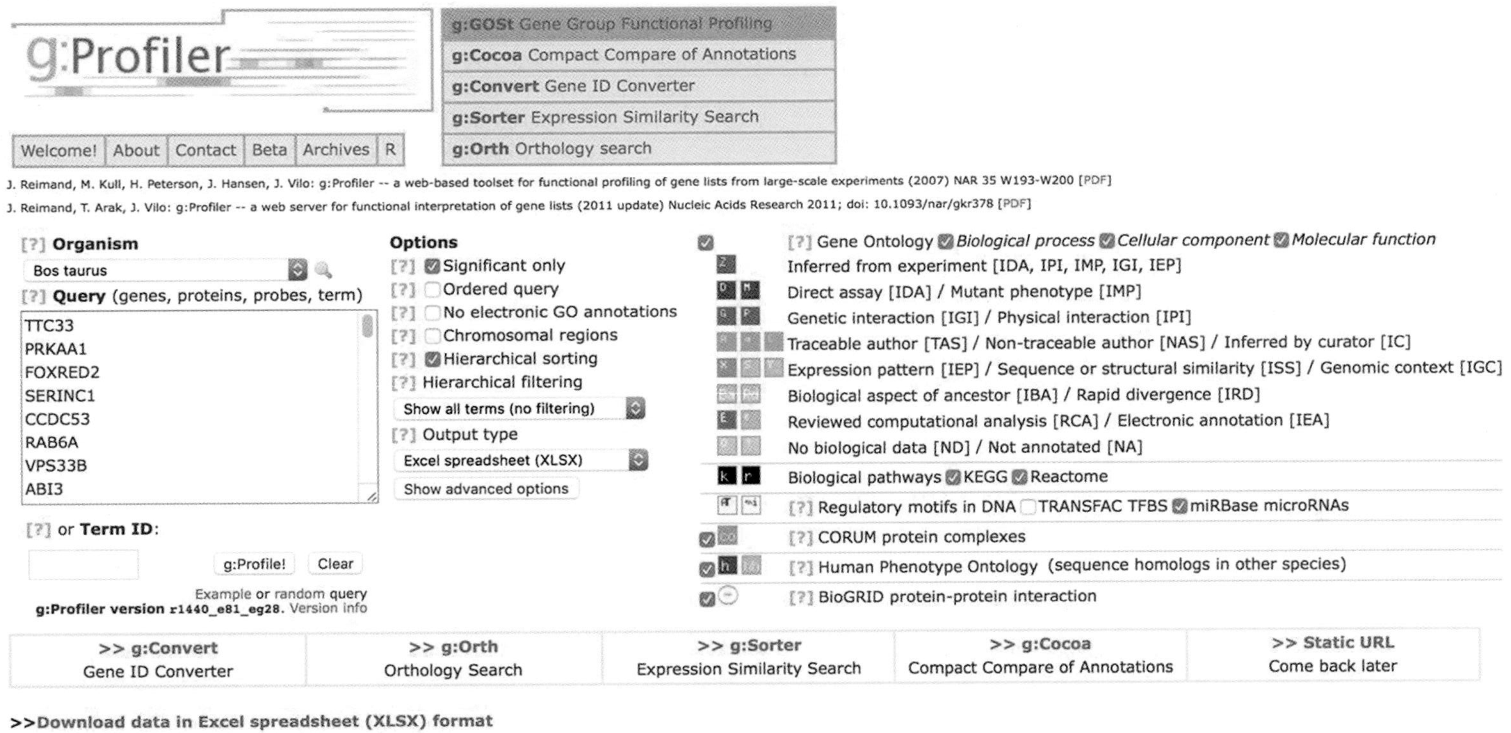

FIGURE 43.5 *(See insert for colour representation of the figure.)*

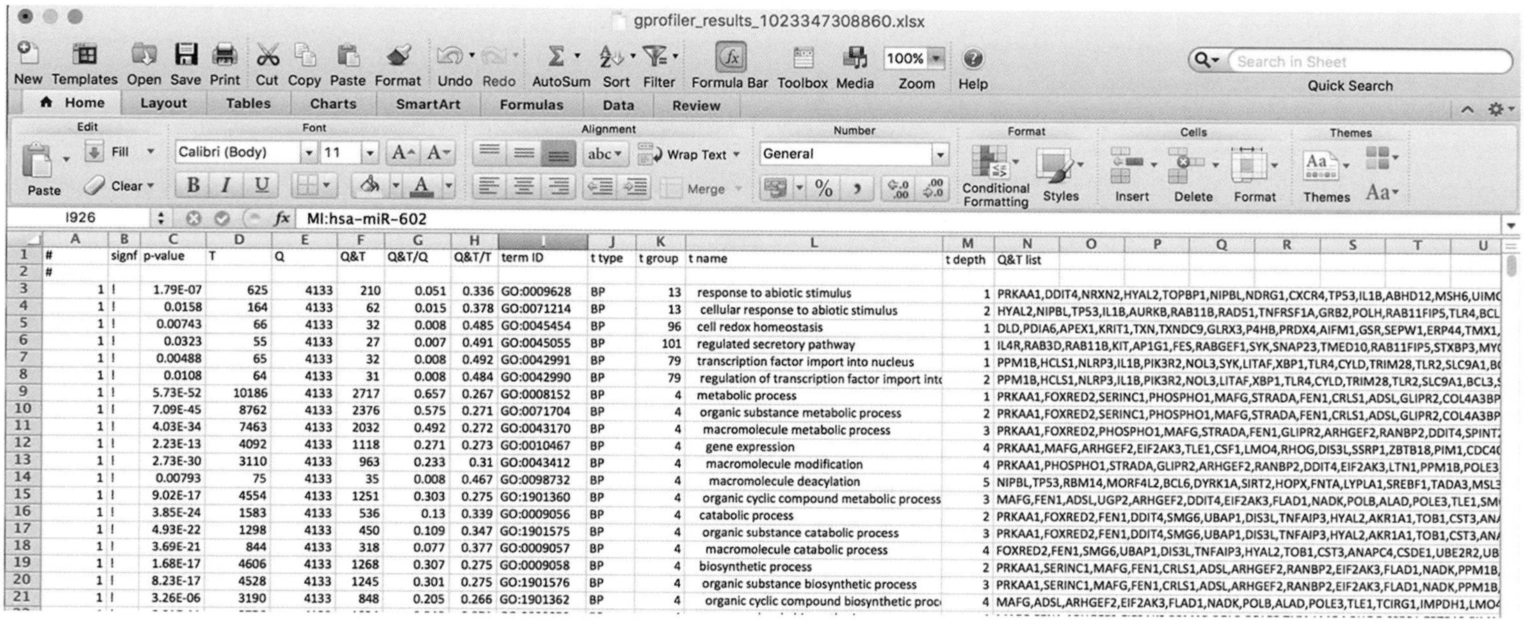

FIGURE 43.6 *(See insert for colour representation of the figure.)*

The output shows the significance of terms and the genes associated with the query (Q) in the term (T). The first term – response to abiotic stimulus (Biological process – BP (t type)) has a term ID of GO:0009628, with a p-value of 1.79E-07. The term has 625 genes associated with it, of which only 210 are enriched in the gene list, out of a total of 4133 genes considered.

43.2.1.3 Step 3: Representing the functional terms graphically

The most common way of representing the functional terms is by choosing the top ten terms (by sorting on the basis of p-value) in each category (Biological processes – BP; Molecular process – MP and cellular component – CC), and representing the term on the *y*-axis and the significance ($-\log_{10}P$) on the *x*-axis, as shown below for the biological processes. The same can be done for all the categories (Figure 43.7).

43.2.1.4 Step 4

Interpretation of the data is completely the researcher's purview.

A1 fx 2.03E-56

	A	B	C	D	E	F	G	H	I	J
1	2.03E-56	8275	4133	2314	0.56	0.28	GO:0044237	BP	4	cellular metabolic process
2	5.73E-52	10186	4133	2717	0.657	0.267	GO:0008152	BP	4	metabolic process
3	1.62E-49	4115	4133	1295	0.313	0.315	GO:0048518	BP	4	positive regulation of biological process
4	7.09E-45	8762	4133	2376	0.575	0.271	GO:0071704	BP	4	organic substance metabolic process
5	5.48E-44	2478	4133	848	0.205	0.342	GO:0006950	BP	4	response to stress
6	5.48E-43	6784	4133	1916	0.464	0.282	GO:0044260	BP	4	cellular macromolecule metabolic process
7	1.17E-41	8437	4133	2289	0.554	0.271	GO:0044238	BP	4	primary metabolic process
8	3.26E-37	4009	4133	1220	0.295	0.304	GO:0044267	BP	4	cellular protein metabolic process
9	8.31E-36	2698	4133	878	0.212	0.325	GO:0009893	BP	4	positive regulation of metabolic process
10	1.3E-34	1569	4133	567	0.137	0.361	GO:0002376	BP	4	immune system process

A	B	C
Term	P	Significance
Cellular metabolic process	2.03E-56	55.69
Metabolic process	5.73E-52	51.24
Positive regulation of biological p	1.62E-49	48.79
Organic substance metabolic proc	7.09E-45	44.15
Response to stress	5.48E-44	43.26
Cellular macromolecule metaboli	5.48E-43	42.26
Primary metabolic process	1.17E-41	40.93
Cellular protein metabolic proces	3.26E-37	36.49
Positive regulation of metabolic p	8.31E-36	35.08
Immune system process	1.3E-34	33.89

A	B
Term	Significance
Cellular metabolic process	55.69
Metabolic process	51.24
Positive regulation of biological p	48.79
Organic substance metabolic proc	44.15
Response to stress	43.26
Cellular macromolecule metaboli	42.26
Primary metabolic process	40.93
Cellular protein metabolic proces	36.49
Positive regulation of metabolic p	35.08
Immune system process	33.89

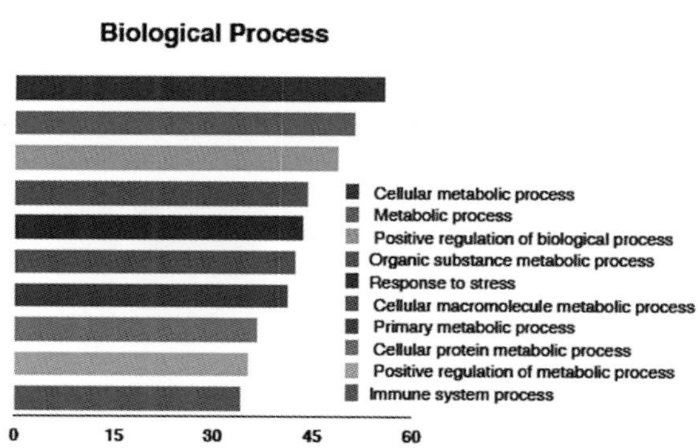

FIGURE 43.7 *(See insert for colour representation of the figure.)*

43.2.2 Functional annotation using DAVID (Database for Annotation, Visualization, and Integrated Discovery) (Huang da et al., 2009)

DAVID is an integrated biological knowledge base and analytic tool, aimed at systematically extracting biological meaning from large gene/protein lists:

43.2.2.1 Step 1

Open https://david.ncifcrf.gov and upload a multi-list file if you have > 3000 genes to be annotated. The multi-list file should be a list1 and list2, separated by a tab (shown in the figure below). Upload this list into DAVID, select the official gene symbol from the drop-down menu (as an identifier), check the radio button against the gene list and submit (see Figure 43.8).

43.2.2.2 Step 2

Select an appropriate background (here it is *Bos taurus*) against which you wish to test your gene list. Create a combined list by clicking "combine" after selecting both the lists, and select the combined list to get the functional annotations (see Figure 43.8).

43.2.2.3 Step 3

Click on the functional annotation tool in the window to get the annotation summary results (see Figure 43.8).

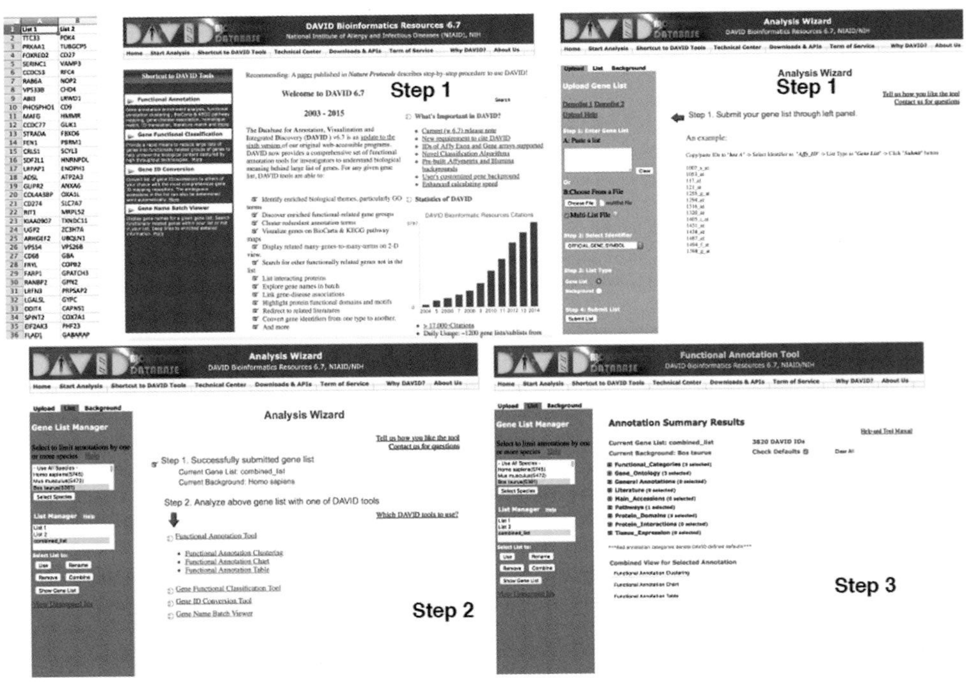

FIGURE 43.8 *(See insert for colour representation of the figure.)*

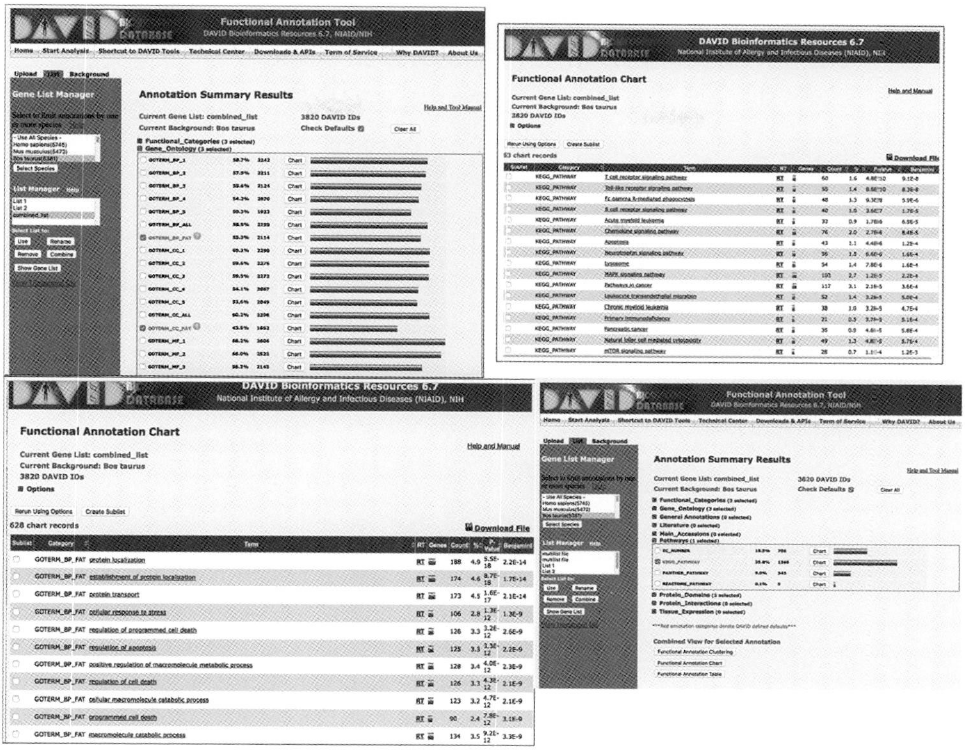

FIGURE 43.9 *(See insert for colour representation of the figure.)*

43.2.2.4 Step 4

The + button can be clicked in the window to get the results. To get the gene ontology terms, click the + button by the side of the gene ontology and then proceed to any particular category – BP, MP or CC. Clicking on the chart option opens up a window with all the specific terms. Here we click on BP to get all the gene ontology terms enriched for biological processes in our differentially expressed genes. The details containing the genes associated with each gene can be downloaded and opened in *Excel* for further use. The same can be done to visualize pathways enriched in the DEGs (see Figure 43.9).

43.2.3 Functional annotation using ClueGO (Bindea et al., 2009)

ClueGO is a Cytoscape plug-in that helps in functional annotation and interpretation of large lists of genes. It integrates KEGG/BioCarta pathways with GO terms to create a functionally organized GO/pathway term network.

43.2.3.1 Step 1

Open the ClueGO app in Cytoscape, paste genes in the window (see Figure 43.10).

43.2.3.2 Step 2

Select a gene ontology or pathway and start. Here we selected immune system processes, as shown in Figure 43.10.

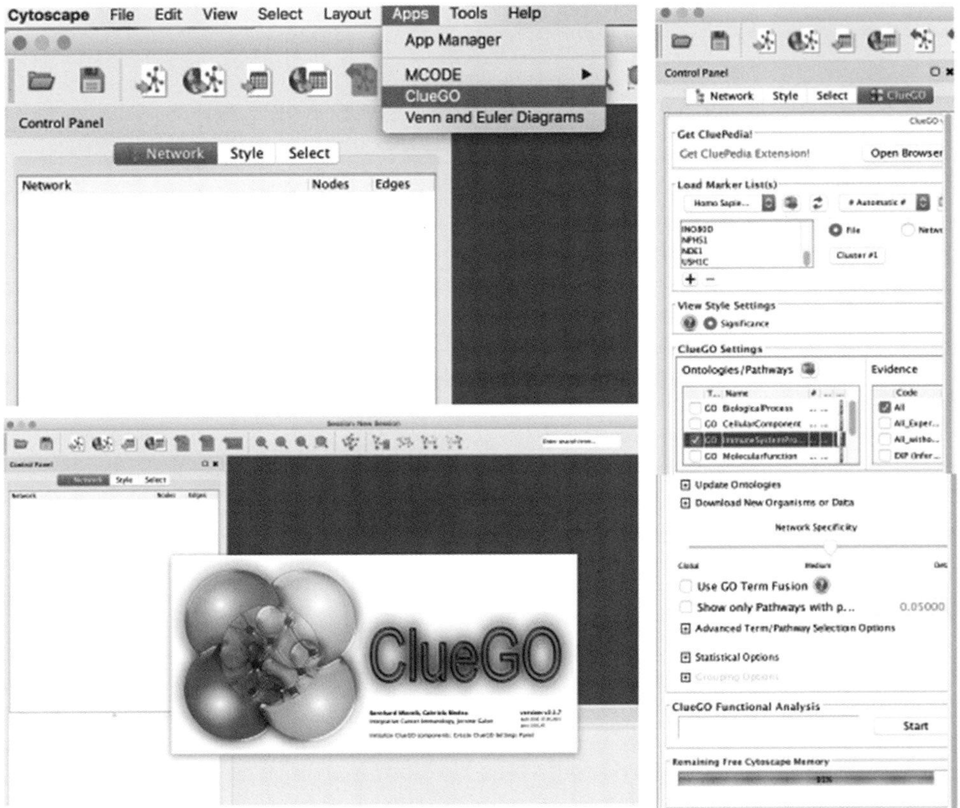

FIGURE 43.10 *(See insert for colour representation of the figure.)*

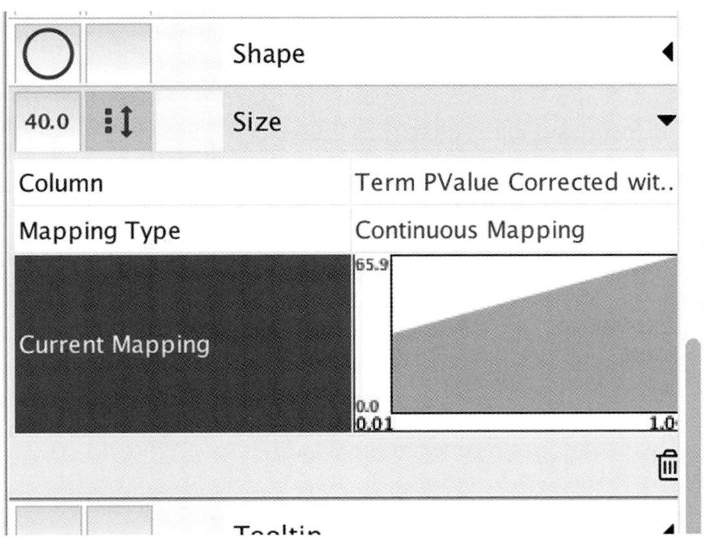

FIGURE 43.11 *(See insert for colour representation of the figure.)*

43.2.3.3 Step 3

Represent a network with GO term as node label, percentage associated genes as node color, and P-Value Corrected with Bonferroni step-down as node size (these parameters are selected as per the requirements of the researcher) (see Figures 43.11, 43.12 and 43.13).

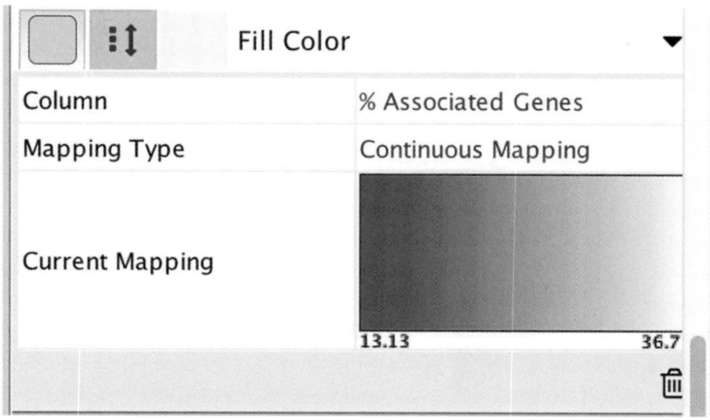

FIGURE 43.12 *(See insert for colour representation of the figure.)*

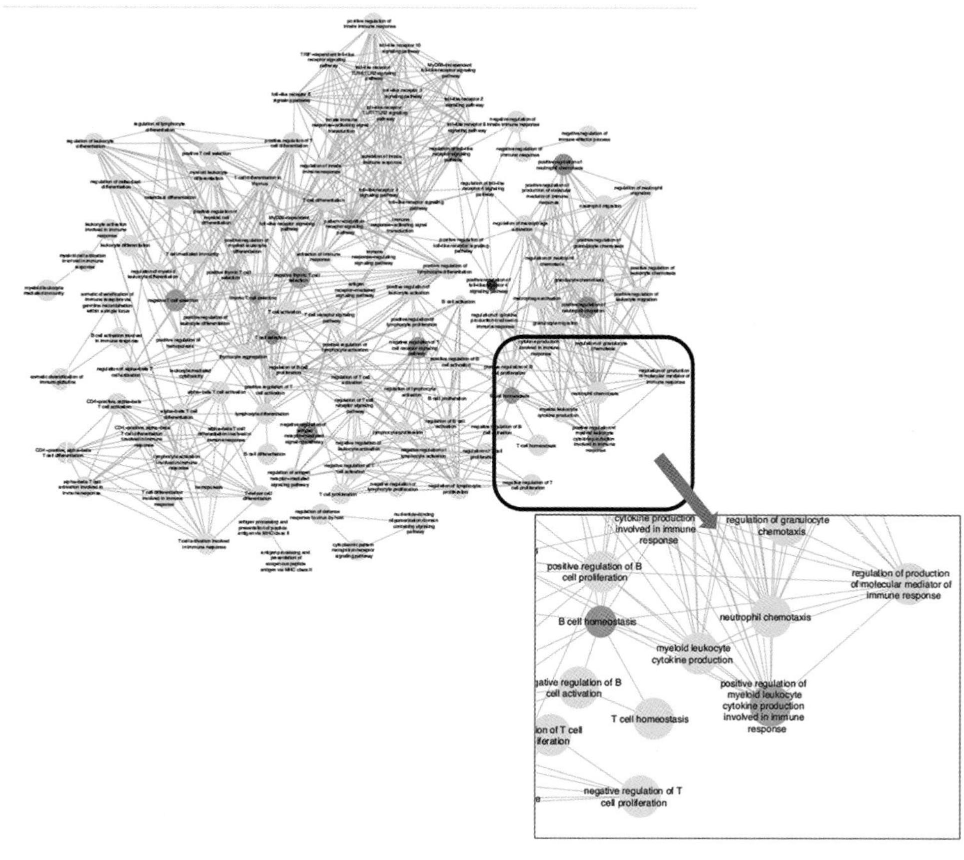

FIGURE 43.13 *(See insert for colour representation of the figure.)*

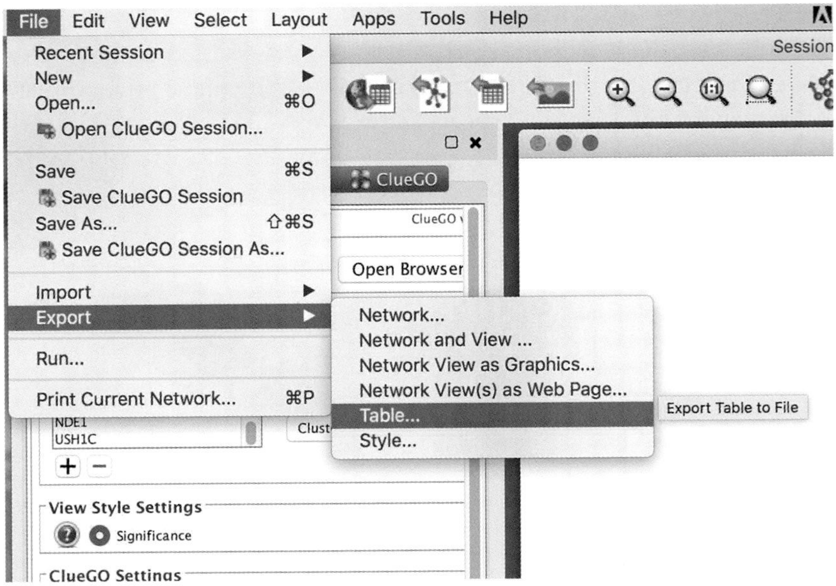

FIGURE 43.14 *(See insert for colour representation of the figure.)*

43.2.3.4 Step 4

The attributes of the network can be exported in a table format (see Figure 43.14).

43.3 QUESTIONS

1. What is gene ontology?
2. Why is there a need for functional annotation of genes obtained in the gene lists from RNA – Seq data analysis?
3. Name five tools for functional annotation of gene lists obtained from RNA – Seq data analysis.

Identification of Differentially Expressed Genes (DEGs)

CHAPTER 44

GVPPSR Kumar, A Kumar and AP Sahoo
Animal Biotechnology Division, IVRI, UP, India

This chapter is discussed in three sections:

- Section I – Quality filtering of data using PRINSEQ
- Section II – Identification of Differentially expressed genes – I (Using Cufflinks)
- Section III – Identification of Differentially expressed genes – II (Using RSEM-DE packages – EBSeq, DESeq2, and edgeR)

44.1 SECTION I. QUALITY FILTERING OF DATA USING PRINSEQ

44.1.1 Introduction

The data generated from most of the platforms are in FASTQ format (i.e., base call data). The data files for this chapter are designated as control.fastq and infected.fastq. Both the fastq files have paired end reads. These data need to be initially checked and quality trimmed for further use. The most commonly used program for quality filtering/trimming is prinseq-lite.pl. There are several options in Prinseq-lite for data trimming and/or filtering. First, trimming is done, followed by execution of the filtering commands. Trimming is commonly done to remove the adapter sequences present in the raw data generated. It is also used to remove the poly A tail at the end of the read.

44.1.2 Quality check analyses using PRINSEQ

From a data set, summary statistics, filtered, reformatted and trimmed quality data can be generated using PRINSEQ. This can be used for all types of sequence data. PRINSEQ can be accessed through a web interface or can be used, standalone.

The command for quality filtering is given below (Figure 44.1):

perl prinseq-lite.pl -fastq control.fastq -out_format 5 -min_len 50 -min_qual_mean 25

For any further help please type:- perl prinseq-lite.pl -h on the command line.

Basic Applied Bioinformatics, First Edition. Chandra Sekhar Mukhopadhyay, Ratan Kumar Choudhary and Mir Asif Iquebal.

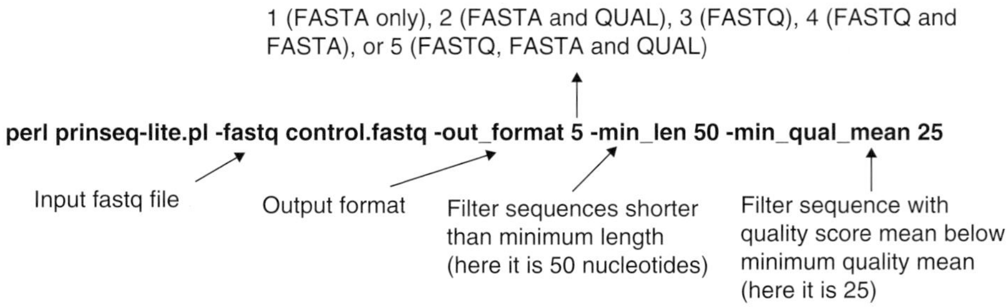

FIGURE 44.1 The basic command for running PRINSEQ-lite.

```
RAVIs-MacBook-Projay-2:prinseqlite ravikumar$ perl prinseq-lite.pl -fastq
control_R1.fastq -out_format 5 -min_len 50 -min_qual_mean 25
Input and filter stats:
        Input sequences: 17,214,799
        Input bases: 1,738,694,699
        Input mean length: 101.00
        Good sequences: 16,748,896 (97.29%)
        Good bases: 1,691,638,496
        Good mean length: 101.00
        Bad sequences: 465,903 (2.71%)
        Bad bases: 47,056,203
        Bad mean length: 101.00
        Sequences filtered by specified parameters:
        min_qual_mean: 465903
```

FIGURE 44.2 Summary statistics after running prinseq-lite.pl.

Prinseqlite is to be run on both the data files as given below:

For the control sample:

perl prinseq-lite.pl –fastq control_R1.fastq –out_format 5 –min_len 50 –min_qual_mean 25

For the infected sample:

perl prinseq-lite.pl –fastq infected_R1.fastq –out_format 5 –min_len 50 –min_qual_mean 25

These quality-filtered data are further analyzed through different pipelines. Here, we initially discuss Cufflinks, and then the RSEM–DE package. The summary statistics, good and bad files generated from control_R1.fastq, are given below.

44.1.3 Summary statistics

A summary of total input sequences, the number of good sequences as per the details provided in the command, the number of bad sequences, and so on, is obtained for each dataset (Figure 44.2).

44.1.4 Good and bad files generated after running Prinseq

With the output format 5, six files (three "good" and three "bad") are generated in the folder from which the command is run (Figure 44.3). The next steps of the analysis pipeline will use the "Good" fastq file.

control_R1_prinseq_bad_9oC5.qual	Today, 8:39 AM	167.5 MB	Document
control_R1_prinseq_bad_sGZq.fasta	Today, 8:39 AM	74.3 MB	FASTA File
control_R1_prinseq_bad_v_I5.fastq	Today, 8:39 AM	147.7 MB	FASTQ...quence
control_R1_prinseq_good_11SG.fastq	Today, 8:39 AM	5.31 GB	FASTQ...quence
control_R1_prinseq_good_d31h.qual	Today, 8:39 AM	6.02 GB	Document
control_R1_prinseq_good_IVqY.fasta	Today, 8:39 AM	2.67 GB	FASTA File
control_R1.fastq	08-Sep-2015, 10:30 AM	4.52 GB	FASTQ...quence

FIGURE 44.3 Six files generated after running prinseq-lite.pl.

44.2 SECTION II. IDENTIFICATION OF DIFFERENTIALLY EXPRESSED GENES – I (USING CUFFLINKS)

44.2.1 Introduction to Cufflinks

"Cufflinks" stands for the suite of software tools as well as the program, which assembles and estimates abundances of transcripts and evaluates differential expression in samples. It accepts mapped reads and assembles them into a parsimonious set of transcripts. "Cuffdiff" then estimates FPKM or RPKM by normalizing for both the library size and gene length (Trapnell *et al.*, 2012).

The prinseq-lite output (the good files) of all the data files can be analyzed either by mapping to the reference genome or by *de novo* assembling the transcriptome. Here, we will be illustrating data analysis using a reference-based approach, by mapping the reads using GMAP-GSNAP.

Before proceeding with GMAP-GSNAP, we need to initially download the GTF and the FASTA file sequence of the reference genome. Note that the reference FASTA

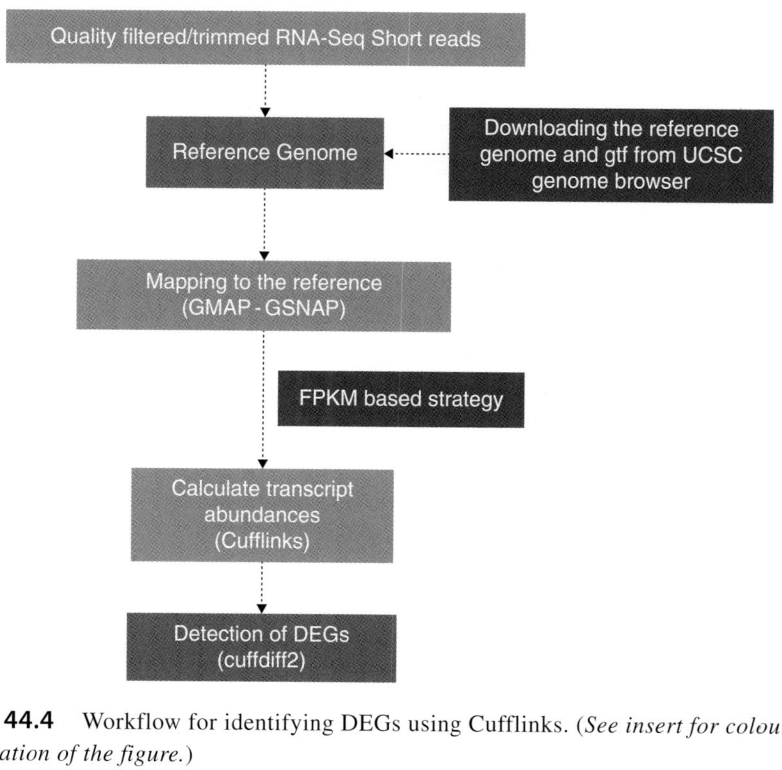

FIGURE 44.4 Workflow for identifying DEGs using Cufflinks. (*See insert for colour representation of the figure.*)

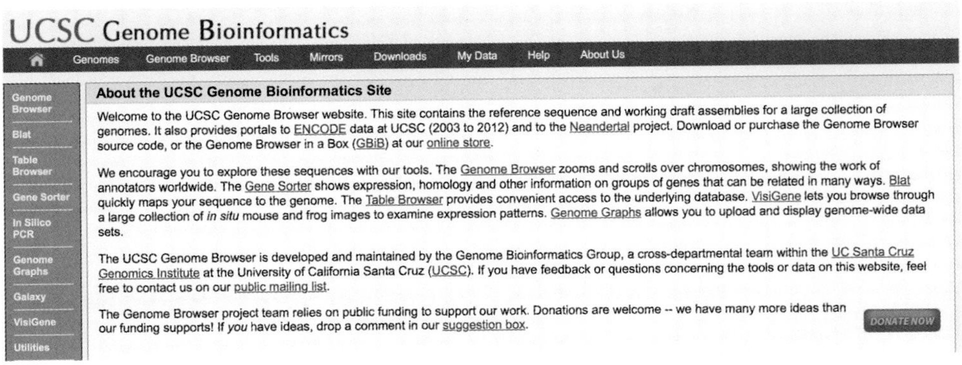

FIGURE 44.5 UCSC genome browser.

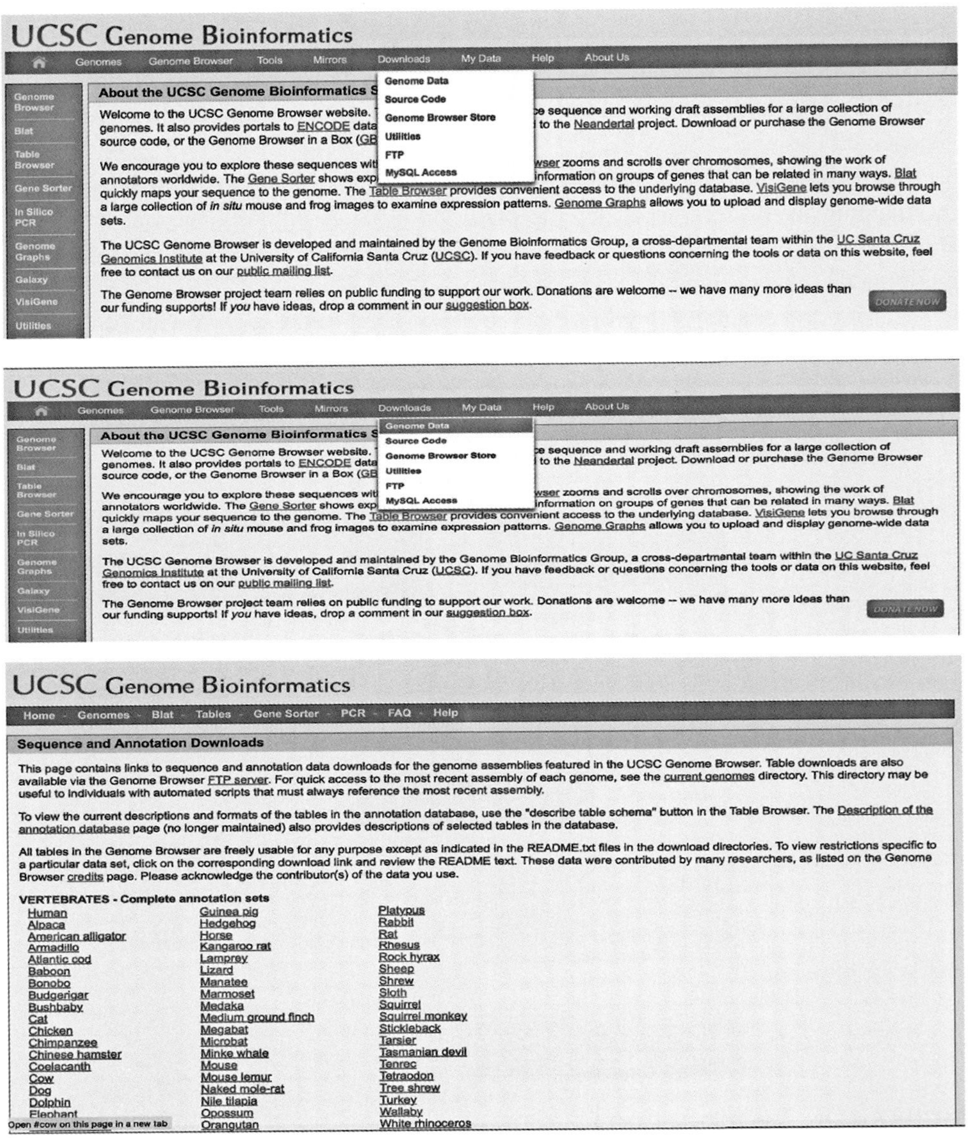

FIGURE 44.6 Click on downloads, genomics data and then select “cow”.

file and the GTF file should be downloaded from the same genome browser. The most commonly used genome browsers are NCBI, UCSC, and Ensembl. Here, we download the GTF and FASTA files from the UCSC genome browser.

44.2.2 Downloading the FASTA file from the UCSC genome browser

(https://genome.ucsc.edu)

Go to the UCSC genome browser, and click on downloads (Figure 44.5) and then on genomics data, to select the species of your interest (Figure 44.6). Here we select the cow to open the cow genome files.

When you click on the bosTau8.fa.gz, you will be able to download a file of 866.1 MB which, on gunzipping, would give a file of 2.72 GB (Figure 44.7).

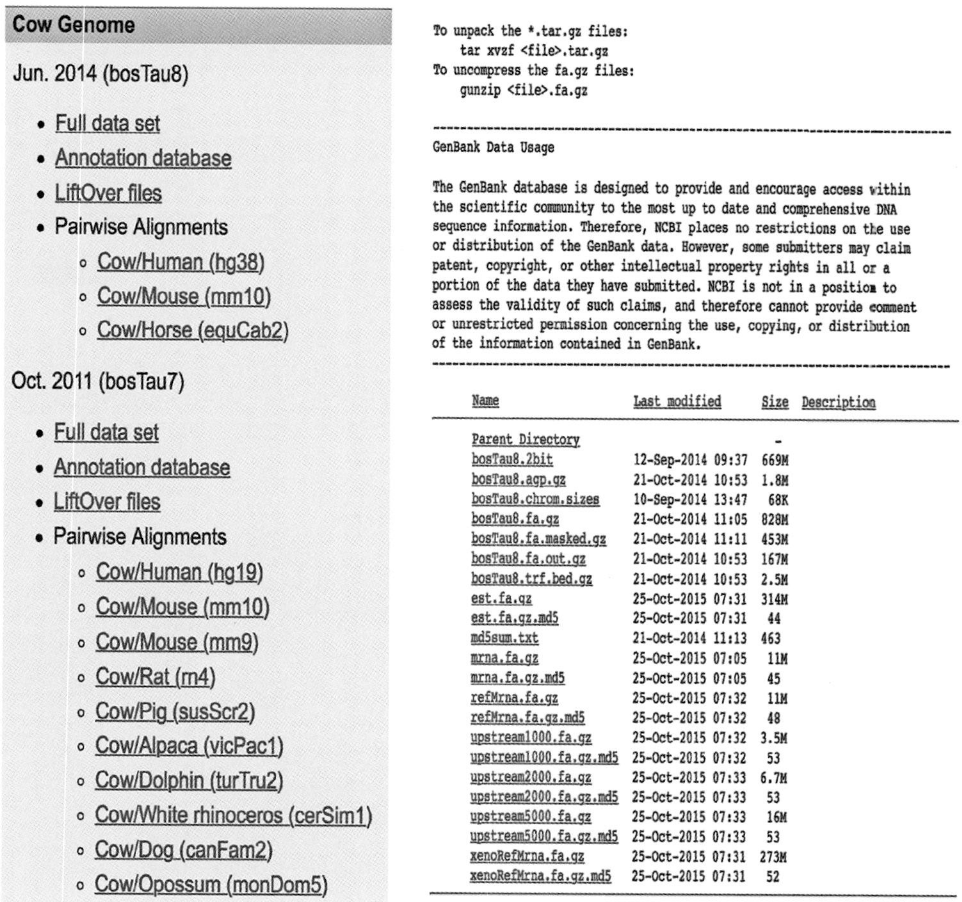
Cow Genome

Jun. 2014 (bosTau8)

- Full data set
- Annotation database
- LiftOver files
- Pairwise Alignments
  - Cow/Human (hg38)
  - Cow/Mouse (mm10)
  - Cow/Horse (equCab2)

Oct. 2011 (bosTau7)

- Full data set
- Annotation database
- LiftOver files
- Pairwise Alignments
  - Cow/Human (hg19)
  - Cow/Mouse (mm10)
  - Cow/Mouse (mm9)
  - Cow/Rat (rn4)
  - Cow/Pig (susScr2)
  - Cow/Alpaca (vicPac1)
  - Cow/Dolphin (turTru2)
  - Cow/White rhinoceros (cerSim1)
  - Cow/Dog (canFam2)
  - Cow/Opossum (monDom5)

```
To unpack the *.tar.gz files:
    tar xvzf <file>.tar.gz
To uncompress the fa.gz files:
    gunzip <file>.fa.gz

--------------------------------------------------------------------------
GenBank Data Usage

The GenBank database is designed to provide and encourage access within
the scientific community to the most up to date and comprehensive DNA
sequence information. Therefore, NCBI places no restrictions on the use
or distribution of the GenBank data. However, some submitters may claim
patent, copyright, or other intellectual property rights in all or a
portion of the data they have submitted. NCBI is not in a position to
assess the validity of such claims, and therefore cannot provide comment
or unrestricted permission concerning the use, copying, or distribution
of the information contained in GenBank.
--------------------------------------------------------------------------
```

Name	Last modified	Size	Description
Parent Directory		-	
bosTau8.2bit	12-Sep-2014 09:37	669M	
bosTau8.agp.gz	21-Oct-2014 10:53	1.8M	
bosTau8.chrom.sizes	10-Sep-2014 13:47	68K	
bosTau8.fa.gz	21-Oct-2014 11:05	828M	
bosTau8.fa.masked.gz	21-Oct-2014 11:11	453M	
bosTau8.fa.out.gz	21-Oct-2014 10:53	167M	
bosTau8.trf.bed.gz	21-Oct-2014 10:53	2.5M	
est.fa.gz	25-Oct-2015 07:31	314M	
est.fa.gz.md5	25-Oct-2015 07:31	44	
md5sum.txt	21-Oct-2014 11:13	463	
mrna.fa.gz	25-Oct-2015 07:05	11M	
mrna.fa.gz.md5	25-Oct-2015 07:05	45	
refMrna.fa.gz	25-Oct-2015 07:32	11M	
refMrna.fa.gz.md5	25-Oct-2015 07:32	48	
upstream1000.fa.gz	25-Oct-2015 07:32	3.5M	
upstream1000.fa.gz.md5	25-Oct-2015 07:32	53	
upstream2000.fa.gz	25-Oct-2015 07:33	6.7M	
upstream2000.fa.gz.md5	25-Oct-2015 07:33	53	
upstream5000.fa.gz	25-Oct-2015 07:33	16M	
upstream5000.fa.gz.md5	25-Oct-2015 07:33	53	
xenoRefMrna.fa.gz	25-Oct-2015 07:31	273M	
xenoRefMrna.fa.gz.md5	25-Oct-2015 07:31	52	

FIGURE 44.6 *(Continued)*

FIGURE 44.7 Zip file and FASTA file of the cow genome.

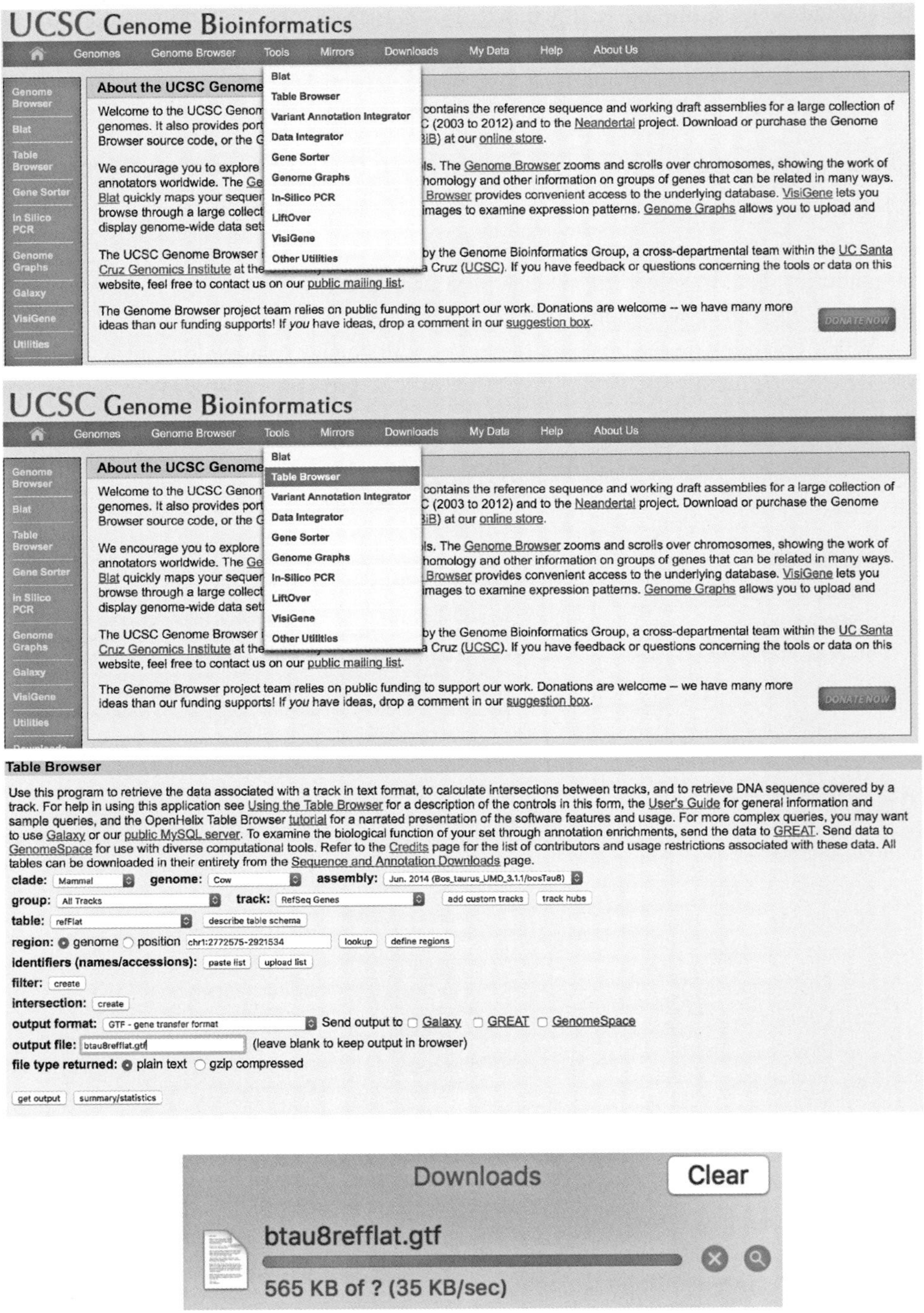

FIGURE 44.8 Downloading the GTF file.

44.2.3 Downloading the GTF file

The GTF file can be downloaded from UCSC by clicking on the table browser, and then selecting the options shown below (Figure 44.8).

44.2.4 Genome mapping and alignment using GMAP-GSNAP

GMAP-GSNAP is a standalone program for mapping and aligning reads to a genome. This program does a fast batch processing of large sequence sets by aligning sequences with minimal startup time and memory requirements. The program generates accurate gene structures without using probabilistic splice site models (Wu and Watanabe, 2005); even plenty of polymorphisms and sequence errors are present in the data. The genome sequence that is downloaded is initially indexed, and this index is further used for mapping the filtered reads to generate the Sequence Alignment/Map (SAM) file(s).

44.2.5 Identifying the differentially expressed genes

Identifying the differentially expressed genes, starting from indexing the genome (Step 1, Figures 44.9 and 44.10), mapping the reads to the indexed genome (Step 2) to generate SAM files, and converting the SAM files to BAM files using Samtools (Step 3), to differential expression using cufflinks suite (Steps 3, 4, 5 and 6, Figures 44.11–44.13), is explained in six steps below:

Step 1: Command for indexing the genome: gmap_build -d btau8 bosTau8.fa.
Here, the FASTA reference genome (bosTau8.fa) is indexed as btau8.
The index files created are as below in the folder btau8.

```
IVRIs-Mac-Pro:gmap apple$ gmap_build -d btau8 bosTau8.fasta
-k flag not specified, so building with default 15-mers
Destination directory not defined with -D flag, so writing to /usr/local/share
Sorting chromosomes in chrom order.  To turn off or sort other ways, use the -s flag.
Creating files in directory /usr/local/share/btau8
Running /usr/local/bin/fa_coords    -o /usr/local/share/btau8.coords -f /usr/local/share/btau8.source
Opening file bosTau8.fasta
  Contig chr1: concatenated at chromosome end: chr1:1..158337067 (length = 158337067 nt)
  Contig chr10: concatenated at chromosome end: chr10:1..104305016 (length = 104305016 nt)
  Contig chr11: concatenated at chromosome end: chr11:1..107310763 (length = 107310763 nt)
  Contig chr12: concatenated at chromosome end: chr12:1..91163125 (length = 91163125 nt)
  Contig chr13: concatenated at chromosome end: chr13:1..84240350 (length = 84240350 nt)
  Contig chr14: concatenated at chromosome end: chr14:1..84648390 (length = 84648390 nt)
  Contig chr15: concatenated at chromosome end: chr15:1..85296676 (length = 85296676 nt)
  Contig chr16: concatenated at chromosome end: chr16:1..81724687 (length = 81724687 nt)
  Contig chr17: concatenated at chromosome end: chr17:1..75158596 (length = 75158596 nt)
  Contig chr18: concatenated at chromosome end: chr18:1..66004023 (length = 66004023 nt)
```

FIGURE 44.9 Indexing the genome using GMAP.

```
IVRIs-Mac-Pro:btau8 apple$ ls
btau8.chromosome          btau8.genomebits128          btau8.ref153positions       btau8.salcpchilddc
btau8.chromosome.iit      btau8.genomecomp             btau8.sachildexc            btau8.salcpexc
btau8.chrsubset           btau8.maps                   btau8.sachildguide1024      btau8.salcpguide1024
btau8.contig              btau8.ref153offsets64meta    btau8.saindex64meta         btau8.sarray
btau8.contig.iit          btau8.ref153offsets64strm    btau8.saindex64strm         btau8.version
```

FIGURE 44.10 Indexing files generated after indexing.

genes.fpkm_tracking	10-Sep-2013 3:40 AM	851 KB	Document
isoforms.fpkm_tracking	10-Sep-2013 3:40 AM	923 KB	Document
skipped.gtf	10-Sep-2013 3:34 AM	Zero bytes	Document
transcriptscontrol.gtf	10-Sep-2013 3:40 AM	29.3 MB	Document

FIGURE 44.11 Files generated after running cufflinks on control BAM file.

genes.fpkm_tracking	10-Sep-2013 3:23 AM	865 KB
isoforms.fpkm_tracking	10-Sep-2013 3:23 AM	943 KB
skipped.gtf	10-Sep-2013 3:16 AM	Zero bytes
transcriptsinfected.gtf	10-Sep-2013 3:23 AM	29.4 MB

FIGURE 44.12 Files generated after running cufflinks on infected BAM file.

assemblies.txt

```
transcriptscontrol_R1.gtf
transcriptscontrol_R2.gtf
transcriptsinfected_R1.gtf
transcriptsinfected_R2.gtf
```

FIGURE 44.13 The assemblies.txt file.

Step 2: Mapping the reads to the genome.

The good fastq files from the prinseq-lite.pl output for control and infected samples are renamed Control_R1.fastq and infected_R1.fastq, respectively.

Note: R1 and R2 paired end reads of the same sample are treated as replicates for further analysis.

Command for mapping:

gsnap –d <genome> –t <nthreads> <fastq_file> > <output_file.sam>

Example:

For the control sample:

gsnap –d btau8 –t 4 control_R1.fastq> control_R1.sam

For the infected sample:

gsnap –d btau8 –t 4 infected_R1.fastq> infected_R1.sam

The end product of the GMAP-GSNAP aligner is a SAM file, which needs to be converted into a BAM file for further analysis in cufflinks. Repeat the same for the other replicates. A total of four SAM files are generated separately for two replicates of each sample.

Step 3: Converting SAM to BAM using Samtools.

Samtools is useful for manipulating alignments in the SAM and BAM formats. It imports from and exports to the SAM format, and does sorting, merging and indexing (Li *et al.*, 2009a, 2009b).

Command for SAM to BAM conversion: ./samtools view –bsh aln.sam >aln.bam

–b: Output in the BAM format. –s: Input in the SAM format. –h: Include header in the output

Example:

For the control sample:

./samtools view –bsh control_R1.sam >control_R1.bam

For the infected sample:

./samtools view –bsh infected_R1.sam >infected_R1.bam

Step 4: Sorting BAM using samtools

Command for sorting: ./samtools sort aln.bam aln.sorted

Example:

For the control sample:

./samtools sort control_R1.bam control_R1_sorted

For the infected sample:

./samtools sort infected_R1.bam infected_R1_sorted

The BAM files generated can be analyzed in two ways:

1. The BAM files can be used to generate a merged assembly of transcripts via cufflinks and cuffmerge. This merged assembly (i.e. merged.gtf) is used in Cuffdiff to generate differentially expressed genes.
2. Cuffdiff can be used directly to generate differentially expressed genes using the BAM files generated.

Step 5 (Option 1): Differential expression using cufflinks, cuffmerge, and cuffdiff.

Command for running Cufflinks on a BAM file (Figures 44.11 and 44.12):

For the control sample:

cufflinks -G btau8refflat.gtf -g btau8refflat.gtf -b bosTau8.fa -u -L CN control_R1_sorted.bam

For the infected sample:

cufflinks -G btau8refflat.gtf -g btau8refflat.gtf -b bosTau8.fa -u -L CN infected_R1_sorted.bam

The transcript.gtf files (Figures 44.10 and 44.11) for each replicate are renamed as per the sample and replicate, and are further used in cuffmerge to generate a merged assembly. This merged assembly is then used in Cuffdiff to generate differentially expressed genes.

Command for running Cuffmerge:

cuffmerge -g btau8refflat.gtf -s bosTau8.fa -p 8 assemblies.txt

assemblies.txt is the file with the list of all the GTFs (transcripts.gtf) for all the replicates of all the samples. The file assemblies.txt is a text file, which looks like the file below (Figure 44.13).

The Cuffmerge command generates a merged.gtf in the merged_asm folder. This file is used in the next Cuffdiff command.

Command for running cuffdiff:

CuffDiff computes differentially expressed genes. The design of experiment should consider at least two contrasting groups of experimental subjects (e.g., healthy vs. diseased) for identifying the differentially expressed genes. CuffDiff should always be run on replicates (i.e., infected vs. control).

cuffdiff merged.gtf control_R1_sorted.bam control_R2_sorted.bam infected_R1_sorted.bam infected_R2_sorted.bam

This command generates many files, out of which, gene_exp.diff is the file to look for the differentially expressed genes.

Step 5 (Option 2): Differential expression using CuffDiff directly from the sorted bam file.

Command:

Cuffdiff –p –N transcripts.gtf

–p: num-threads <int>. –N

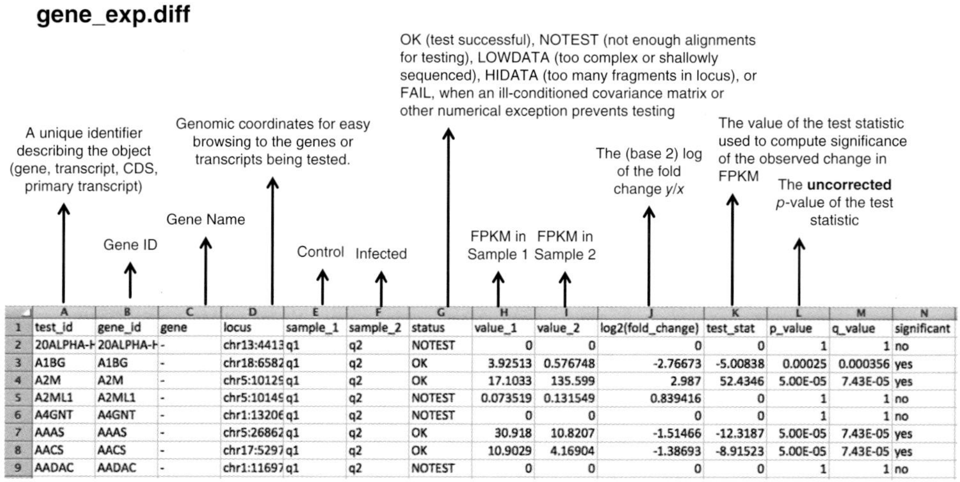

	A	B	C	D	E	F	G	H	I	J	K	L	M	N
1	test_id	gene_id	gene	locus	sample_1	sample_2	status	value_1	value_2	log2(fold_change)	test_stat	p_value	q_value	significant
2	20ALPHA-H	20ALPHA-H	-	chr13:4413	q1	q2	NOTEST	0	0	0	0	1	1	no
3	A1BG	A1BG	-	chr18:6582	q1	q2	OK	3.92513	0.576748	-2.76673	-5.00838	0.00025	0.000356	yes
4	A2M	A2M	-	chr5:10129	q1	q2	OK	17.1033	135.599	2.987	52.4346	5.00E-05	7.43E-05	yes
5	A2ML1	A2ML1	-	chr5:10149	q1	q2	NOTEST	0.073519	0.131549	0.839416	0	1	1	no
6	A4GNT	A4GNT	-	chr1:13206	q1	q2	NOTEST	0	0	0	0	1	1	no
7	AAAS	AAAS	-	chr5:26862	q1	q2	OK	30.918	10.8207	-1.51466	-12.3187	5.00E-05	7.43E-05	yes
8	AACS	AACS	-	chr17:5297	q1	q2	OK	10.9029	4.16904	-1.38693	-8.91523	5.00E-05	7.43E-05	yes
9	AADAC	AADAC	-	chr1:11697	q1	q2	NOTEST	0	0	0	0	1	1	no

FIGURE 44.14 gene_exp.diff file giving the fold change of the genes, along with significance.

44.2.6 Running Cuffdiff for our BAM files

cuffdiff –p 3 –N bostau8refflat.gtf

control_R1_sorted.bam,control_R2_sorted.baminfected_R1_sorted.bam,infected_R2_sorted.bam –o cuffdiff_out

The gene_exp.diff is the file in which to look for the differentially expressed genes. The file contains the fields as marked below (Figure 44.14).

Calculation of Log2fold change for A1BG gene (row 3 in Figure 44.14 above):

Log2fold change = Log2(FPKM infected/FPKM of control)

= Log2(0.576748/3.92513) = –2.76673

44.3 SECTION III. IDENTIFICATION OF DIFFERENTIALLY EXPRESSED GENES – II (USING RSEM-DE PACKAGES EBSEQ, DESEQ2 AND EDGER)

44.3.1 Introduction

RSEM is a cutting-edge RNASeq analysis package that is an end-to-end solution for differential expression, and simplifies the whole process (Li and Dewey, 2011). It also introduces a new more robust unit of RNASeq measurement called TPM. Calculating expression counts using RSEM should be initially taken up. These counts for all the samples and their replicates are further used in differential expression (DE) packages for identifying differential expressed genes (DEGs).

Calculating expression counts using RSEM is explained in nine steps below:

Step 1: Downloading RSEM and installing.

By using the wget command, RSEM can be downloaded using the link below. After unzipping the folder, run "make" to install RSEM.

wget http://deweylab.biostat.wisc.edu/rsem/src/rsem-1.2.19.tar.gz

tar –xvzf rsem-1.2.19.tar.gz

cd rsem-1.2.19/make

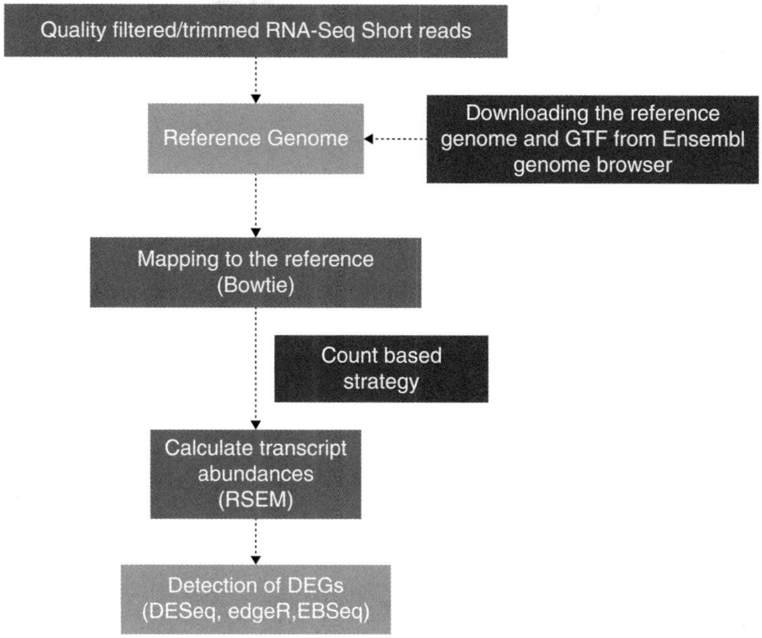

FIGURE 44.15 Workflow for identifying DEGs using RSEM and DE packages. (*See insert for colour representation of the figure.*)

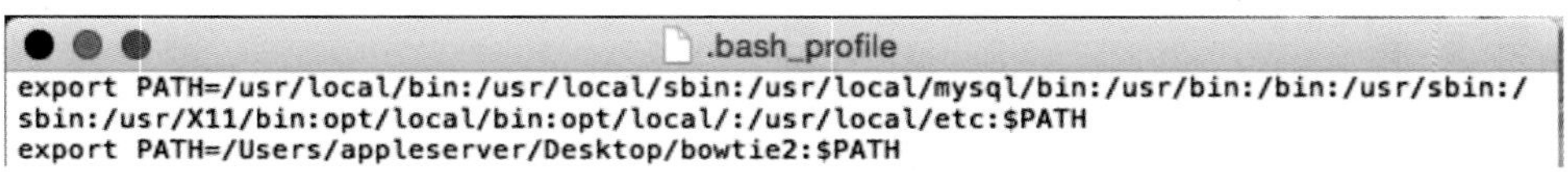

FIGURE 44.16 .bash_profile with the path added.

Step 2: Prerequisites required for running RSEM.

Perl, R, and Bowtie need to be installed. Perl and R are normally present on most computers. Bowtie 2 needs to be added to your path (explained in steps 3 and 4 below).

Step 3: Downloading Bowtie and installing

Download Bowtie from http://sourceforge.net/projects/bowtie-bio/files/bowtie/1.1.1/

Step 4: Copy bowtie in your path or add bowtie path in bash profile.

Copying bowtie in your path:

sudo cp –R/Users/appleserver/Desktop/bowtie2/usr/local/bin

Add bowtie path in bash profile (preferred). Open the.bash_profile (Figure 44.16), add the path below to the file and run the source from the ~/.bash_profile:

export PATH="/Users/ravikumar/Desktop/bowtie2:$PATH"

run source ~/.bash_profile

echo $PATH – to check whether the path has been added

To check whether the path has been added to the.bash_profile, type - echo $PATH (Figure 44.17).

Indicates that the path has been added

```
APPLEs-Mac-Pro:~ appleserver$ source ~/.bash_profile
APPLEs-Mac-Pro:~ appleserver$ echo $PATH
/opt/local/bin:/opt/local/sbin://usr/local/Cellar/circos-0.67-6/bin:/Users/apple
server/Desktop/bowtie2:/opt/local/bin:/opt/local/sbin://usr/local/Cellar/circos-
0.67-6/bin:/Users/appleserver/Desktop/bowtie2:/usr/local/bin:/usr/bin:/bin:/usr/
sbin:/sbin:/bin:/usr/bin:/bin:/usr/bin
```

FIGURE 44.17 Echo $PATH indicating that the path is added.

```
RAVIs-MacBook-Projay-2:~ ravikumar$ wget -m  ftp://ftp.ensembl.org/pub/release-81/fasta/bos_taurus/dna/  &or f in $(find . -name "*.gz")
[1] 13047
--2015-11-03 09:21:45--  ftp://ftp.ensembl.org/pub/release-81/fasta/bos_taurus/dna/
           => 'ftp.ensembl.org/pub/release-81/fasta/bos_taurus/dna/.listing'
Resolving ftp.ensembl.org... 193.62.203.85
Connecting to ftp.ensembl.org|193.62.203.85|:21... connected.
Logging in as anonymous ... Logged in!
==> SYST ... done.    ==> PWD ... done.
==> TYPE I ... done.  ==> CWD (1) /pub/release-81/fasta/bos_taurus/dna ... done.
==> PASV ... done.    ==> LIST ... done.

    [ <=>                                    ] 10,107      --.-K/s   in 0.003s

2015-11-03 09:21:52 (3.37 MB/s) - 'ftp.ensembl.org/pub/release-81/fasta/bos_taurus/dna/.listing' saved [10107]

--2015-11-03 09:21:52--  ftp://ftp.ensembl.org/pub/release-81/fasta/bos_taurus/dna/Bos_taurus.UMD3.1.dna.chromosome.10.fa.gz
           => 'ftp.ensembl.org/pub/release-81/fasta/bos_taurus/dna/Bos_taurus.UMD3.1.dna.chromosome.10.fa.gz'
==> CWD not required.
==> PASV ... done.    ==> RETR Bos_taurus.UMD3.1.dna.chromosome.10.fa.gz ... done.
Length: 31603333 (30M)

 3% [>                                       ] 986,736     74.5KB/s  eta 6m 26s -bash: or: command not found
```

FIGURE 44.18 wget command downloading the genome from the ensemble genome browser.

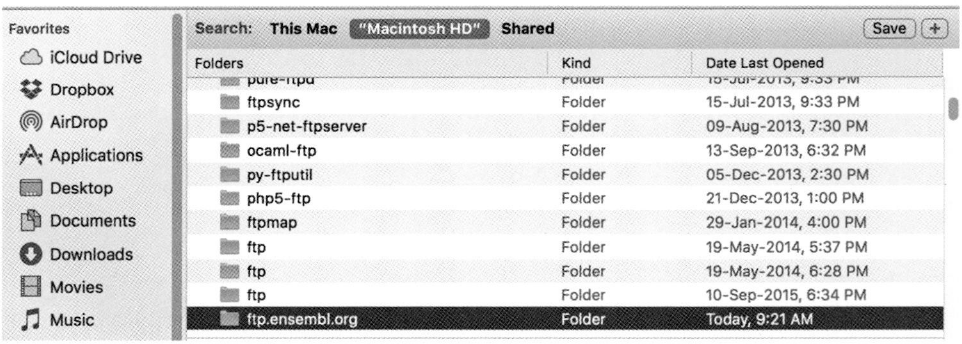

FIGURE 44.19 Folder ftp.ensembl.org created after the download.

Step 5: Downloading the reference, gunzipping and concatenating

Download Bos taurus genome from Ensembl genome browser. An easier alternative is to use the wget command for a direct download on HPC (Figure 44.18):

wget –m ftp://ftp.ensembl.org/pub/release-81/fasta/bos_taurus/dna/&or f in $(find.–name "*.gz")

The folder that is created after the download is ftp.ensembl.org(Figure 44.19). This folder contains FASTA files of all chromosomes (Figure 44.20). These FASTA files are further concatenated into a single file (combined.fa), having all chromosomes.

A direct download of each chromosome from the ftp site can also be done as given below (Figure 44.21). However, this is time-consuming. The first option, downloading using the wget command, is faster.

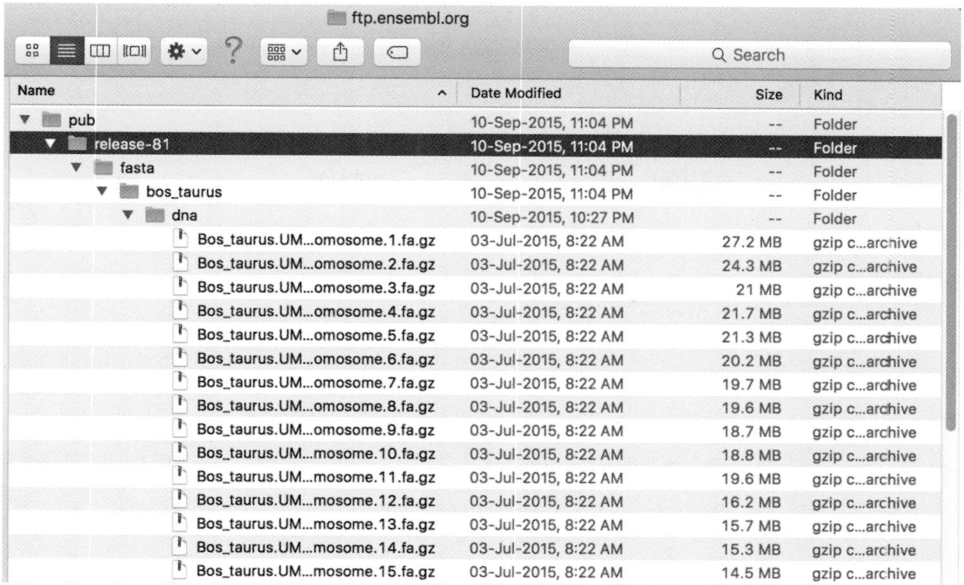

FIGURE 44.20 The chromosome gunzip files in the folder ftp.ensembl.org.

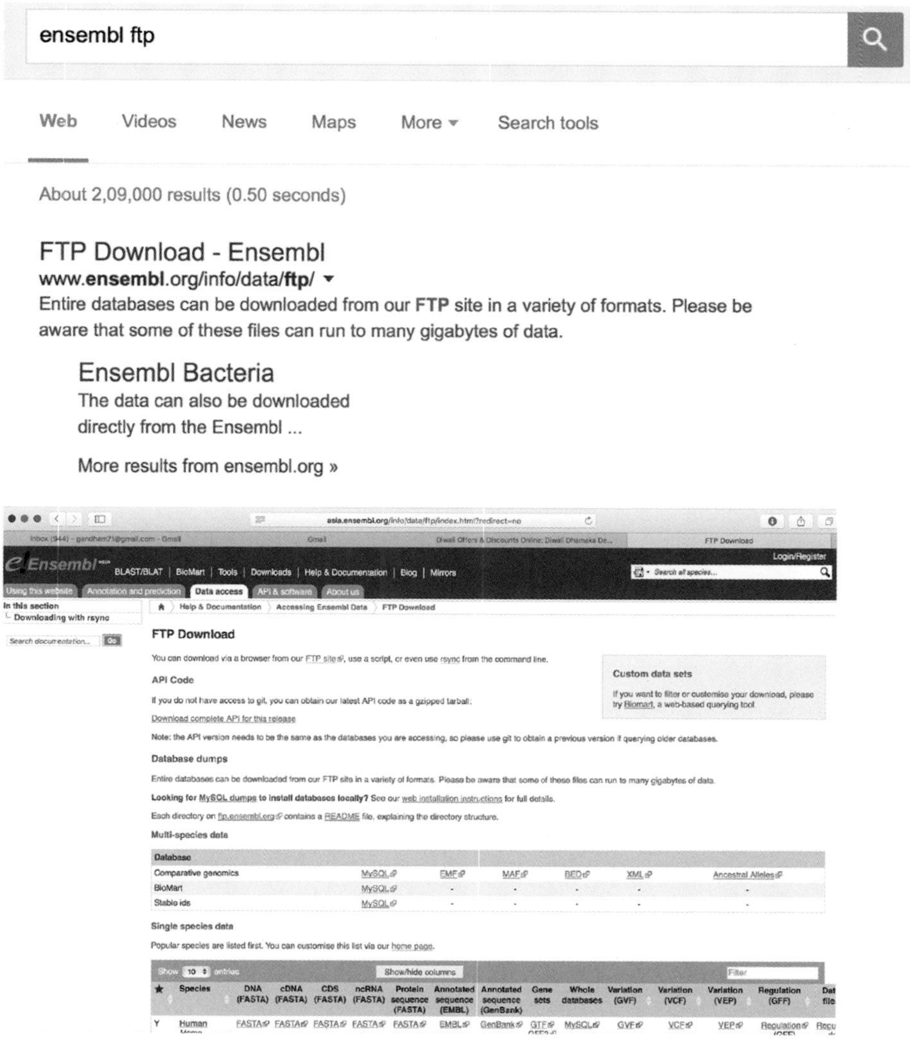

FIGURE 44.21 Direct download from the.ftp site.

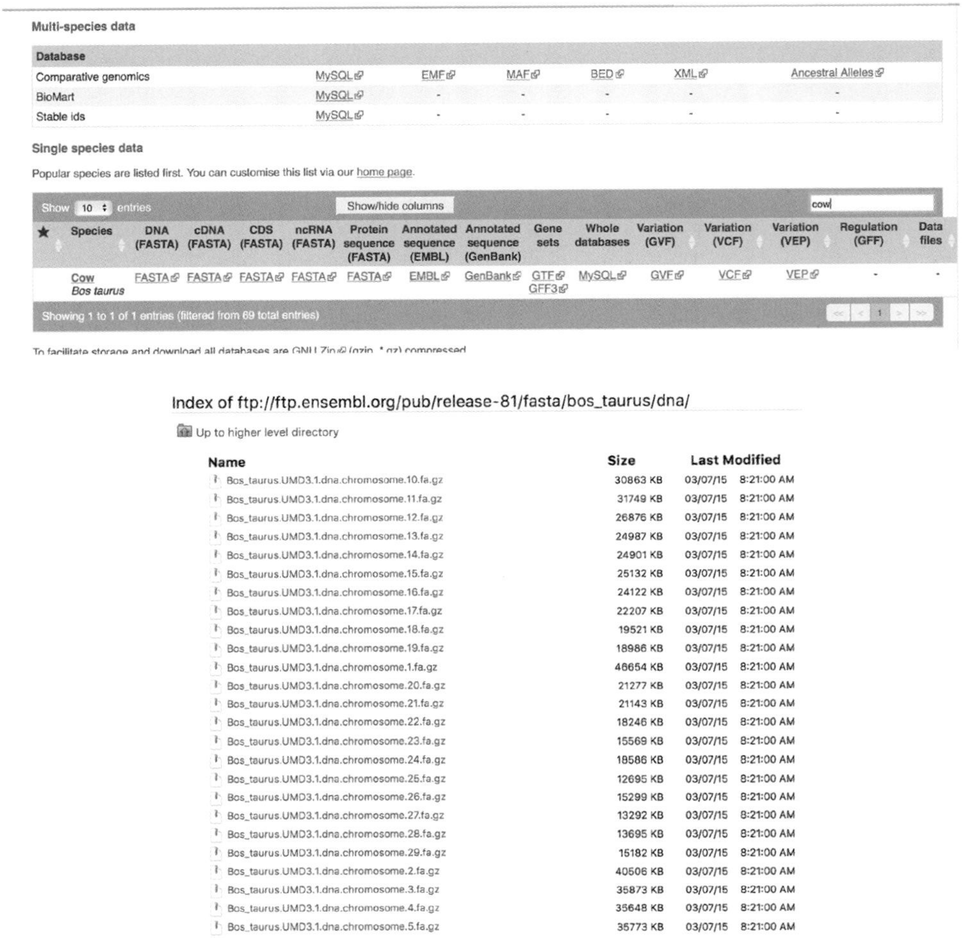

FIGURE 44.21 *(Continued)*

The files downloaded are gunzipped using:

gunzip Bos_taurus.UMD3.1.dna.chromosome.*.fa.gz

Concatenating/combining all the fasta files into a combined fasta file (reference):

cat Bos_taurus.UMD3.1.dna.chromosome.*.fa > combined.fa

Step 6: Download annotation file in gtf format.

Command for downloading the gtf: wget –m ftp://ftp.ensembl.org/pub/release-81/gtf/bos_taurus

The gtf file downloaded needs to be modified for RSEM to extract only the exon annotations. This is done by using an "awk" command to create a filtered.gtf file.

awk command to extract the exon annotations from gtf:

awk '$3 == "exon"' Bos_taurus.UMD3.1.8.1.gtf> filtered.gtf

Step 7: Prepare reference using RSEM

To prepare the reference sequence, run the "rsem-prepare-reference" program.

Command for preparing the reference is simply indexing the reference sequence. This creates 12 files as index files (Figure 44.22) with the name of BT and extension bt2.

BT.1.bt2	11-Sep-2015 2:32 pm	21.6 MB
BT.2.bt2	11-Sep-2015 2:32 pm	12.4 MB
BT.3.bt2	11-Sep-2015 2:32 pm	242 KB
BT.4.bt2	11-Sep-2015 2:32 pm	12.4 MB
BT.chrlist	11-Sep-2015 2:32 pm	371 bytes
BT.grp	11-Sep-2015 2:32 pm	138 KB
BT.idx.fa	11-Sep-2015 2:32 pm	50.1 MB
BT.rev.1.bt2	11-Sep-2015 2:33 pm	21.6 MB
BT.rev.2.bt2	11-Sep-2015 2:33 pm	12.4 MB
BT.seq	11-Sep-2015 2:32 pm	53.5 MB
BT.ti	11-Sep-2015 2:32 pm	14.3 MB
BT.transcripts.fa	11-Sep-2015 2:32 pm	50.1 MB

FIGURE 44.22 Index files created after indexing using bowtie 2.0.

ControlR1.genes.results	11-Sep-2015 4:16 pm	1.8 MB
ControlR1.isoforms.results	11-Sep-2015 4:16 pm	1.9 MB
ControlR1.stat	11-Sep-2015 4:12 pm	--
ControlR1.transcript.bam	11-Sep-2015 4:16 pm	1.96 GB
ControlR1.transcript.sorted.bam	11-Sep-2015 4:22 pm	1.76 GB
ControlR1.transcript.sorted.bam.bai	11-Sep-2015 4:23 pm	1.5 MB

FIGURE 44.23 Six files generated after running the calculate expression command.

	A	B	C	D	E	F	G
1	gene_id	transcript_id(s)	length	effective_length	expected_count	TPM	FPKM
2	ENSBTAG00000000005	ENSBTAT00000000005	2310	2310	884	53.83	40.12
3	ENSBTAG00000000008	ENSBTAT00000000008	1511	1511	5	0.47	0.35
4	ENSBTAG00000000009	ENSBTAT00000000009	1543	1543	0	0	0
5	ENSBTAG00000000010	ENSBTAT00000000010	1348	1348	658	70.05	52.2
6	ENSBTAG00000000011	ENSBTAT00000036775	1579	1579	0	0	0
7	ENSBTAG00000000012	ENSBTAT00000000013	1227	1227	121	14.22	10.6
8	ENSBTAG00000000013	ENSBTAT00000000016,	4432.63	4432.63	291	9.11	6.79
9	ENSBTAG00000000014	ENSBTAT00000000014	1044	1044	526	73.33	54.65
10	ENSBTAG00000000015	ENSBTAT00000000015	2089	2089	39	2.63	1.96

FIGURE 44.24 Expected counts, TPM and FPKM of each of the ensemblIDs.

Step 8: Calculating expression values in counts, TPM and FPKM:

To calculate expression values, the "rsem-calculate-expression" program is used.

The command for running rsem-calculate-expression should be run for each of the replicates (_R1 and _R2) of both the samples. This will generate six files, as shown in Figure 44.23, of which genes.results is the most important file among the six for identifying the differentially expressed genes.

For the control sample:

./rsem-calculate-expression --bowtie2 control_R1.fastq BT ControlR1

There will be six files generated as shown above, and genes.results is the most important file among the six for identifying the differentially expressed genes.

For the infected sample:

./rsem-calculate-expression --bowtie2 infected_R1.fastq BT infectedR1

The output ControlR1.genes.results gives the expected counts, TPM and FPKM for each of the ensemblIDs (Figure 44.24).

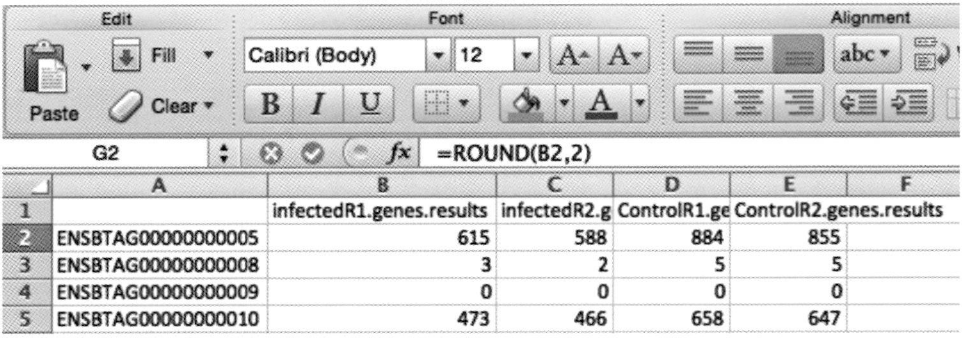

	A	B	C	D	E	F
1		infectedR1.genes.results	infectedR2.g	ControlR1.ge	ControlR2.genes.results	
2	ENSBTAG00000000005	615	588	884	855	
3	ENSBTAG00000000008	3	2	5	5	
4	ENSBTAG00000000009	0	0	0	0	
5	ENSBTAG00000000010	473	466	658	647	

FIGURE 44.25 Combining the counts of all the files and rounding them to the nearest integer.

Step 9: Combining RSEM genes.results of all the files. The expected counts of all the ensemblIDs for all four files (two replicates each of control and infected) are combined (Figure 44.25).

Command for combining the RSEM genes.results of all the files:

```
./rsem-generate-data-matrix *.genes.results > genes.results
```

After rounding these expected counts values to the nearest integer (Figure 44.25), they can be used in programs such as EBSeq, DESeq, or edgeR to identify differentially expressed genes.

44.4 USE OF DE PACKAGES FOR IDENTIFYING THE DIFFERENTIALLY EXPRESSED GENES

(using EBSeq, DESeq2 and edgeR)

44.4.1 Differentially expression using EBSeq (Leng *et al.*, 2013)

EBSeq is an R package for identifying differentially expressed genes (DEGs) across biological conditions. EBSeq uses RSEM counts as input to identify differentially expressed genes. RSEM counts as input to identify differentially expressed genes. Identifying the DEGs using EBSeq is explained in six steps below.

Step 1: Installing EBSeq.

To install, type the following commands in R:

```
source("https://bioconductor.org/biocLite.R")
biocLite("EBSeq")
```

Step 2: Command for loading the package EBSeq (Figure 44.26).

```
>library(EBSeq)
```

Step 3: Command for getting the working directory.

```
>getwd()
```

```
> library(EBSeq)
Loading required package: blockmodeling
Loading required package: gplots

Attaching package: 'gplots'

The following object is masked from 'package:stats':

    lowess

Warning message:
package 'gplots' was built under R version 3.1.3
Warning: unable to access index for repository http://ftp.iitm.ac.in/cran/bin/macosx/mavericks/contrib/3.1
Warning: unable to access index for repository http://ftp.iitm.ac.in/cran/bin/macosx/mavericks/contrib/3.1
Warning: unable to access index for repository http://ftp.iitm.ac.in/cran/bin/macosx/mavericks/contrib/3.1
Warning: unable to access index for repository http://ftp.iitm.ac.in/cran/bin/macosx/mavericks/contrib/3.1
> getwd()
[1] "/Users/appleserver"
Warning message:
In open.connection(con, "r") : too many redirects, aborting ...
```

FIGURE 44.26 Loading the EBSeq package in R.

	A	B	C	D	E	F
1		infectedR1.genes.results	infectedR2.genes.results	ControlR1.genes.results	ControlR2.genes.results	
2	ENSBTAG00000000005	615	588	884	855	
3	ENSBTAG00000000008	3	2	5	5	
4	ENSBTAG00000000009	0	0	0	0	
5	ENSBTAG00000000010	473	466	658	647	
6	ENSBTAG00000000011	1	1	0	0	
7	ENSBTAG00000000012	286	275	121	116	
8	ENSBTAG00000000013	832	821	291	292	
9	ENSBTAG00000000014	362	370	526	523	
10	ENSBTAG00000000015	103	105	39	34	
11	ENSBTAG00000000016	17	16	51	53	
12	ENSBTAG00000000019	739	714	623	644	

FIGURE 44.27 Input file for EBSeq.

Step 4: Command for setting the working library (Figure 44.27).

> setwd()

Set the working directory to RSEM.

Step 5: Input requirement for Gene level DE analysis:

The input file formats supported by EBSeq are.csv,.xls, or.xlsx,.txt (tab delimited). In the input file, rows should be the genes, and columns should be the samples. An example of the data set in .txt format (genesresults.txt) is given in Figure 44.27.

Step 6: Commands to Run EBSeq (the details of each of the commands are given in explaining the commands (https://www.bioconductor.org/packages/3.3/bioc/vignettes/EBSeq/inst/doc/EBSeq_Vignette.pdf)):

```
> x=data.matrix(read.table("genesresults.txt"))
> dim(x)
[1] 24596 4
> str(x)
 num [1:24596, 1:4] 615 3 0 473 1 286 832 362 103 17 ...
 - attr(*, "dimnames")=List of 2
  ..$ : chr [1:24596] "ENSBTAG00000000005" "ENSBTAG00000000008"
"ENSBTAG00000000009" "ENSBTAG00000000010" ...
  ..$ : chr [1:4] "infectedR1.genes.results" "infectedR2.genes.results"
"ControlR1.genes.results" "ControlR2.genes.results"
> Sizes=MedianNorm(x)
```

```
> EBOut=EBTest(Data=x,
+ Conditions=as.factor(rep(c("C1","C2"),each=2)),sizeFactors=Sizes,
+ maxround=5)
Removing transcripts with 75th quantile < = 10
12071 transcripts will be tested
iteration 1 done
time 0.12
iteration 2 done
time 0.13
iteration 3 done
time 0.08
iteration 4 done
> PP=GetPPMat(EBOut)
> str(PP)
 num [1:12071, 1:2] 1 1 0 0 1 ...
 - attr(*, "dimnames")=List of 2
  ..$ : chr [1:12071] "ENSBTAG00000000005" "ENSBTAG00000000010"
"ENSBTAG00000000012" "ENSBTAG00000000013" ...
  ..$ : chr [1:2] "PPEE" "PPDE"
> DEfound=rownames(PP)[which(PP[,"PPDE"]>=.95)]
> str(DEfound)
 chr [1:6528] "ENSBTAG00000000012" "ENSBTAG00000000013"
"ENSBTAG00000000015" "ENSBTAG00000000019" "ENSBTAG00000000021"
"ENSBTAG00000000025" "ENSBTAG00000000026" "ENSBTAG00000000032" ...
> write.table(DEfound,"DE.txt",sep = "\t",quote = F,col.names=F)
> GeneFC=PostFC(EBOut)
> write.table(GeneFC,"FC.txt",sep = "\t",quote = F,col.names=F)
```

Running of EBSeq in R

The output file – FC.txt

The other output file – DE.txt

Explaining the commands: (https://www.bioconductor.org/packages/3.3/bioc/vignettes/EBSeq/inst/doc/EBSeq_Vignette.pdf)

44.4.1.1 Calling of the input file into EBSeq

The object data should be a G × S matrix containing the expression values for each gene and each sample;

where: "G": number of genes

"S": number of samples.

These values should exhibit raw counts, without normalizing over the samples. The dim(X) command gives us the dimensions of the matrix; str(x) command gives the structure of the data; num(x) gives the details of the values of the samples; attr(x) gives the details of row names and column names.

```
> x=data.matrix(read.table("genesresults.txt"))
> dim(x)
[1] 24596 4
> str(x)
 num [1:24596, 1:4] 615 3 0 473 1 286 832 362 103 17 ...
 - attr(*, "dimnames")=List of 2
  ..$ : chr [1:24596] "ENSBTAG00000000005" "ENSBTAG00000000008"
"ENSBTAG00000000009" "ENSBTAG00000000010" ...
  ..$ : chr [1:4] "infectedR1.genes.results" "infectedR2.genes.results"
"ControlR1.genes.results" "ControlR2.genes.results"
```

In our analysis, object "x" is a simulated data matrix containing 24 596 rows of genes and four columns of samples. The genes are named "ENSBTAG0000000000 5", "ENSBTAG00000000008"… (Figure 44.28).

44.4.1.2 Obtaining the library size factor

EBSeq requires the library size factors for each of the samples. This is achieved by the function MedianNorm, which uses the median normalization approach.

> Sizes=MedianNorm(x)

44.4.1.3 Identifying DE genes – running EBSeq to get gene expression estimates

The function EBTest is used to detect DE genes. We define the conditions and size factors.

Explaining the conditions to EBseq:

The object conditions should be a vector of length S that indicates to which condition each sample belongs. For example, if there are two conditions and sample-pair in each, then S = 4 and conditions may be given by as.factor(c("C1","C1","C2","C2")). This means that we have simulated the first two samples to be in condition 1 and the other two in condition 2, and thus defined conditions as:

Conditions=as.factor(rep(c("C1","C2"),each=2))

Normalization using sizeFactors:

Similarly, sizeFactors in the EBTest command is used to define the library size factor of each sample. It could be obtained by summing up the total number of reads per sample. We can opt for median normalization, scaling normalization, upper-quantile normalization or some other such approach. Here, we are doing a median normalization and running the EM algorithm by setting the number of iterations to five via maxround=5, which can be seen in the output of step 6 above (Figure 44.28).

```
> x=data.matrix(read.table("genesresults.txt"))
> dim(x)
[1] 24596     4
> str(x)
 num [1:24596, 1:4] 615 3 0 473 1 286 832 362 103 17 ...
 - attr(*, "dimnames")=List of 2
  ..$ : chr [1:24596] "ENSBTAG00000000005" "ENSBTAG00000000008" "ENSBTAG00000000009" "ENSBTAG00000000010" ...
  ..$ : chr [1:4] "infectedR1.genes.results" "infectedR2.genes.results" "ControlR1.genes.results" "ControlR2.genes.results
> Sizes=MedianNorm(x)
> EBOut=EBTest(Data=x,
+ Conditions=as.factor(rep(c("C1","C2"),each=2)),sizeFactors=Sizes,
+ maxround=5)
Removing transcripts with 75 th quantile < = 10
12071 transcripts will be tested
iteration 1 done

time 0.12

iteration 2 done

time 0.13

iteration 3 done

time 0.08

iteration 4 done

time 0.09

iteration 5 done

time 0.15
```

FIGURE 44.28 Running iterations of EBSeq.

```
> EBOut=EBTest(Data=x,
+ Conditions=as.factor(rep(c("C1","C2"),each=2)),sizeFactors=Sizes,
+ maxround=5)
Calculating the probabilities for the DE genes:
The list of DE genes and the posterior probabilities of being DE are
obtained as follows (Figure 44.29):
> PP=GetPPMat(EBOut)
> str(PP)
 num [1:12071, 1:2] 1 1 0 0 1 &hellip;
 - attr(*, "dimnames")=List of 2
  ..$ : chr [1:12071] "ENSBTAG00000000005" "ENSBTAG00000000010" "ENSBTAG00
000000012" "ENSBTAG00000000013" &hellip;
  ..$ : chr [1:2] "PPEE" "PPDE"
```

PPEE gives the posterior probability of equally expressed and PPDE gives the posterior probability of differentially expressed. This indicates that 12071 genes are differentially expressed (Figure 44.29).

Differentially expressed genes at the 5% level of significance:

To get the DE genes with a probability at the level of significance 5%, we run the DEfound command. DEfound is a list of genes identified with PPDE >=0.95 or FDR <0.05. EBSeq found 6528 genes significantly (P <0.05) differentially expressed (Figure 44.29).

```
> DEfound=rownames(PP)[which(PP[,"PPDE"]>=.95)]
> str(DEfound)
chr [1:6528] "ENSBTAG00000000012" "ENSBTAG00000000013" "ENSBTAG0000000001 5"
"ENSBTAG00000000019" "ENSBTAG00000000021" "ENSBTAG00000000025"
"ENSBTAG00000000026" "ENSBTAG00000000032" …
```

Calculating the fold change:

"PostFC" calculates the posterior fold change for each transcript across conditions (Figure 44.29).

> GeneFC=PostFC(EBOut)

Writing the files:

write.table is used to write the fold changes and the differentially expressed genes into a file. Here the fold changes are saved as fc.txt and DEGs as DE.txt files (Figures 44.30 and 44.31).

44.4.2 Differentially expression using DESeq2 (Love *et al.*, 2014)

This is a differential expression analysis based on the negative binomial distribution. DESeq2 uses RSEM counts as input to identify differentially expressed genes.

```
> PP=GetPPMat(EBOut)
> str(PP)
 num [1:12071, 1:2] 1 1 0 0 1 ...
 - attr(*, "dimnames")=List of 2
  ..$ : chr [1:12071] "ENSBTAG00000000005" "ENSBTAG00000000010" "ENSBTAG00000000012" "ENSBTAG00000000013" ...
  ..$ : chr [1:2] "PPEE" "PPDE"
> DEfound=rownames(PP)[which(PP[,"PPDE"]>=.95)]
> str(DEfound)
 chr [1:6528] "ENSBTAG00000000012" "ENSBTAG00000000013" "ENSBTAG00000000015" "ENSBTAG00000000019" "ENSBTAG00000000021" "ENSBTAG00000000025" "ENSBTAG00000000026
> write.table(DEfound,"DE.txt",sep = "\t",quote = F,col.names=F)
> GeneFC=PostFC(EBOut)
> write.table(GeneFC,"FC.txt",sep = "\t",quote = F,col.names=F)
```

FIGURE 44.29 Identifying DEGs in EBSeq.

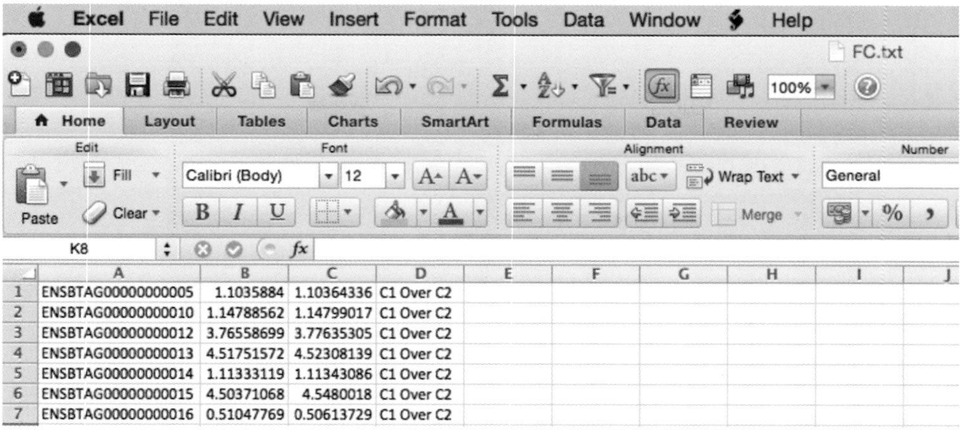

FIGURE 44.30 Fold change of all the ensemblIDs.

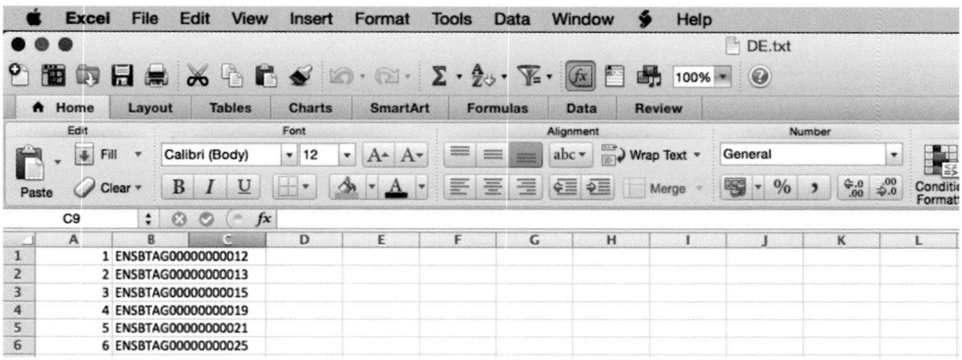

FIGURE 44.31 Significant DE genes.

44.4.2.1 Installing DESeq2

To install, type the following commands in R:

```
>source("https://bioconductor.org/biocLite.R")
>biocLite("DESeq2")
```

44.4.2.2 Command to load the library (Figure 44.32)

>library(DESeq2)

Running DESeq2 in R:

Note: commands for getting the working directory and setting the working directory are the same as for step 3 and step 4 of EBSeq.

Step 5: Input requirement for Gene level DE analysis:

The input file formats supported by DESeq are .csv, .xls, or .xlsx, .txt (tab delimited). In the input file, rows should be the genes and the columns should be the samples.

Example of the data set in .txt format (roundedfn.txt) that is used here (Figure 44.33):

```
> library(DESeq2)
Loading required package: S4Vectors
Loading required package: stats4
Loading required package: BiocGenerics
Loading required package: parallel

Attaching package: 'BiocGenerics'

The following objects are masked from 'package:parallel':

    clusterApply, clusterApplyLB, clusterCall, clusterEvalQ, clusterExport, clusterMap, parApply, parCapply, parLapply,
    parLapplyLB, parRapply, parSapply, parSapplyLB

The following object is masked from 'package:stats':

    xtabs

The following objects are masked from 'package:base':

    anyDuplicated, append, as.data.frame, as.vector, cbind, colnames, do.call, duplicated, eval, evalq, Filter, Find, get,
    intersect, is.unsorted, lapply, Map, mapply, match, mget, order, paste, pmax, pmax.int, pmin, pmin.int, Position, rank,
    rbind, Reduce, rep.int, rownames, sapply, setdiff, sort, table, tapply, union, unique, unlist, unsplit

Loading required package: IRanges

Attaching package: 'IRanges'

The following object is masked from 'package:gplots':

    space

Loading required package: GenomicRanges
Loading required package: GenomeInfoDb
Loading required package: Rcpp
```

FIGURE 44.32 Loading DESeq2 package.

	A	B	C	D	E
1		C_R1	C_R2	T_R1	T_R2
2	ENSBTAG00000000005	884	855	615	588
3	ENSBTAG00000000008	5	5	3	2
4	ENSBTAG00000000009	0	0	0	0
5	ENSBTAG00000000010	658	647	473	466
6	ENSBTAG00000000011	0	0	1	1
7	ENSBTAG00000000012	121	116	286	275

FIGURE 44.33 Example input data set.

Step 6: Commands to Run DESeq2:

```
> counts <- read.table(file = "roundedfn.txt", header = TRUE,row.names=1)
> class(counts)
[1] "data.frame"
> countdata=data.matrix(counts)
> class(countdata)
[1] "matrix"
> Design = data.frame(
+ row.names = colnames(counts),
+ condition = c("Control", "Control", "infected", "infected"),
+ libType = c("single-end", "single-end", "single-end", "single-end"))
> Design
condition libType
C_R1 Control single-end
C_R2 Control single-end
T_R1 infected single-end
T_R2 infected single-end
> dds <- DESeqDataSetFromMatrix(countData = countdata,
+ colData = Design,
+ design = ~ condition)
> dds
class: DESeqDataSet
dim: 24596 4
exptData(0):
assays(1): counts
```

```
rownames(24596): ENSBTAG00000000005 ENSBTAG00000000008 … ENSBTAG00000048316
ENSBTAG00000048317
rowData metadata column names(0):
colnames(4): C_R1 C_R2 T_R1 T_R2
colData names(2): condition libType
> dds <- DESeq(dds)
estimating size factors
estimating dispersions
gene-wise dispersion estimates
mean-dispersion relationship
final dispersion estimates
fitting model and testing
> res <- results(dds)
> resOrdered <- res[order(res$padj),]
> head(resOrdered)
log2 fold change (MAP): condition infected vs Control
Wald test p-value: condition infected vs Control
DataFrame with 6 rows and 6 columns
```

	baseMean	log2FoldChange	lfcSE	stat	pvalue	padj
	<numeric>	<numeric>	<numeric>	<numeric>	<numeric>	<numeric>
ENSBTAG00000000078	2853.089	2.675077	0.06099109	43.86013	0	0
ENSBTAG00000000097	2368.628	2.540755	0.06608276	38.44808	0	0
ENSBTAG00000000133	1774.342	3.147047	0.08124748	38.73408	0	0
ENSBTAG00000000176	2780.405	3.186742	0.06534625	48.76702	0	0
ENSBTAG00000000320	1886.409	3.280391	0.08107852	40.45943	0	0
ENSBTAG00000000347	5612.499	2.256616	0.04303017	52.44265	0	0

FIGURE 44.34 Fold change and significance of ensemblIDs.

> write.table(resOrdered,"DEDEseq2.txt",sep = "\t",quote = F,col.names=F)
The output file is DEDESeq2.txt.

44.4.2.3 Differentially expressed genes at 5% level of significance

To get the DE genes with probability at the 5% level of significance, we select genes with a p value of < 0.05. DESeq found 8249 genes with a p-value or padj < 0.05 (Figure 44.37).

Explaining the commands: https://www.bioconductor.org/packages/3.3/bioc/vignettes/DESeq2/inst/doc/DESeq2.pdf

1. Calling the input file into DESeq2 and defining the dataset:
   The data are first read with read.table to initially create a counts object, which is then read as a data matrix into Countdata. Countdata should be a matrix of read counts, where columns correspond to different samples. Design is an object where we explain to the software the details of the input file, by giving what the columns are and what they represent – namely, control/infected. We also explain the library type of the data that are being called into DESeq2.

```
> counts <- read.table(file = "roundedfn.txt", header = TRUE,row.names=1
> countdata=data.matrix(counts)
> Design = data.frame(
+ row.names = colnames(counts),
+ condition = c("Control", "Control", "infected", "infected"),
+ libType = c("single-end", "single-end", "single-end", "single-end"))
```

```
> counts <- read.table( file = "roundedfn.txt" , header = TRUE,row.names=1)
> class(counts)
[1] "data.frame"
> countdata=data.matrix(counts)
> class(countdata)
[1] "matrix"
> Design = data.frame(
+ row.names = colnames( counts ),
+ condition = c( "Control", "Control", "infected", "infected"),
+ libType = c( "single-end", "single-end", "single-end", "single-end"))
> Design
     condition    libType
C_R1   Control single-end
C_R2   Control single-end
T_R1  infected single-end
T_R2  infected single-end
> dds <- DESeqDataSetFromMatrix(countData = countData,
+ colData = Design,
+ design = ~ condition)
Error in as.matrix(countData) :
  error in evaluating the argument 'x' in selecting a method for function 'as.matrix': Error: object 'countData' not found
> dds <- DESeqDataSetFromMatrix(countData = countdata,
+ colData = Design,
+ design = ~ condition)
> dds
class: DESeqDataSet
dim: 24596 4
exptData(0):
assays(1): counts
rownames(24596): ENSBTAG00000000005 ENSBTAG00000000008 ... ENSBTAG00000048316 ENSBTAG00000048317
rowData metadata column names(0):
colnames(4): C_R1 C_R2 T_R1 T_R2
colData names(2): condition libType
> dds <- DESeq(dds)
estimating size factors
estimating dispersions
gene-wise dispersion estimates
mean-dispersion relationship
final dispersion estimates
fitting model and testing
> res <- results(dds)
> resOrdered <- res[order(res$padj),]
> head(resOrdered)
log2 fold change (MAP): condition infected vs Control
Wald test p-value: condition infected vs Control
DataFrame with 6 rows and 6 columns
                    baseMean log2FoldChange       lfcSE      stat    pvalue      padj
                   <numeric>      <numeric>   <numeric> <numeric> <numeric> <numeric>
ENSBTAG00000000078  2853.089       2.675077 0.06099109  43.86013         0         0
ENSBTAG00000000097  2368.628       2.540755 0.06608276  38.44808         0         0
ENSBTAG00000000133  1774.342       3.147047 0.08124748  38.73408         0         0
ENSBTAG00000000176  2780.405       3.186742 0.06534625  48.76702         0         0
ENSBTAG00000000320  1886.409       3.280391 0.08107852  40.45943         0         0
ENSBTAG00000000347  5612.499       2.256616 0.04303017  52.44265         0         0
> write.table(resOrdered,"DEDEseq2.txt",sep = "\t",quote = F,col.names=F)
```

FIGURE 44.35 Running the DESeq2 package.

	A	B	C	D	E	F	G
1		baseMean	log2FoldChange	lfcSE	stat	pvalue	padj
2	ENSBTAG00000000078	2853.08896	2.675077216	0.06099109	43.8601335	0	0
3	ENSBTAG00000000097	2368.62812	2.540755252	0.06608276	38.44808	0	0
4	ENSBTAG00000000133	1774.34213	3.147046712	0.08124748	38.7340827	0	0
5	ENSBTAG00000000176	2780.40451	3.18674219	0.06534625	48.7670213	0	0

FIGURE 44.36 Fold change and significance of ensemble IDs in the file DEDESeq2.txt.

8244	ENSBTAG00000016784	17.450854	1.4542318	0.67407321	2.15737961	0.0309761	0.04969176
8245	ENSBTAG00000039879	10.9028622	-2.5680861	1.19086761	-2.1564833	0.03104595	0.04979778
8246	ENSBTAG00000002056	85.9534152	-0.6737683	0.31252977	-2.1558531	0.03109514	0.04986458
8247	ENSBTAG00000038477	158.620696	0.4962349	0.23017558	2.15589722	0.03109169	0.04986458
8248	ENSBTAG00000045947	58.8124966	-0.8295954	0.38481976	-2.1558025	0.03109909	0.04986487
8249	ENSBTAG00000015578	55.5163141	0.82574788	0.38313613	2.15523364	0.03114356	0.04993012

FIGURE 44.37 Significant DEGs in DEDEseq2.txt.

With the countData and sample information in Design, we can construct a DESeqDataSet, which is the actual dataset used to identify differentially expressed genes:

```
> dds <- DESeqDataSetFromMatrix(countData = countdata,
+ colData = Design,
+ design = ~ condition)
>dds
class: DESeqDataSet
dim: 24596 4
exptData(0):
assays(1): counts
rownames(24596): ENSBTAG00000000005 ENSBTAG00000000008 … ENSBTAG00000048316
ENSBTAG00000048317
rowData metadata column names(0):
colnames(4): C_R1 C_R2 T_R1 T_R2
colData names(2): condition libType
```

2. Identifying DE genes – Running DESeq2 to get differential gene expression: The DESeq function takes care of normalization, and identifies differentially expressed genes. This will print out a message for the various steps it performs. The estimation of size factors – controlling for differences in the sequencing depth of the samples, the estimation of dispersion values for each gene, and fitting a generalized linear model. The result tables are generated using the function results(), which extracts a results table with log2 fold changes, p values and adjusted p values. We can order our results table by the smallest adjust P value by running the resOrdered function.

```
> dds <- DESeq(dds)
estimating size factors
estimating dispersions
gene-wise dispersion estimates
mean-dispersion relationship
final dispersion estimates
fitting model and testing
> res <- results(dds)
> resOrdered <- res[order(res$padj),]
> head(resOrdered)
log2 fold change (MAP): condition infected vs Control
Wald test p-value: condition infected vs Control
DataFrame with 6 rows and 6 columns
```

	baseMean	log2FoldChange	lfcSE	stat	pvalue	padj
	<numeric>	<numeric>	<numeric>	<numeric>	<numeric>	<numeric>
ENSBTAG00000000078	2853.089	2.675077	0.06099109	43.86013	0	0

FIGURE 44.38 reOrdered command ouput and the various column IDs generated.

Note:

- basemean: the average of the normalized count values, dividing by size factors, taken over all samples in the DESeqDataSet.
- log2FoldChange: the effect size estimate. This tells us how much the gene's expression would be changed in infected samples in comparison with control samples.
- lfcSE: the standard error estimate for the log2 fold change estimate.
- p-value: the probability of a fold change as strong as the observed one, or even stronger.

44.4.3 Differential gene expression using edgeR (Robinson *et al.*, 2010)

Step 1: Installing edgeR.

To install, type the following commands in R:
source("https://bioconductor.org/biocLite.R")
biocLite("edgeR")

Step 2: Command to load the library.

>library(edgeR)

```
> library (edgeR)
Loading required package: limma

Attaching package: 'limma'

The following object is masked from 'package:DESeq2':

    plotMA

The following object is masked from 'package:BiocGenerics':

    plotMA

Warning messages:
1: package 'edgeR' was built under R version 3.1.2
2: package 'limma' was built under R version 3.1.3
```

FIGURE 44.39 Loading the edgeR package in R.

Step 3: Input requirement for Gene level DE analysis.

The input file formats supported by edgeR are .csv,.xls, or.xlsx,.txt (tab delimited). In the input file, the rows should be the genes, and the columns should be the samples.

Example of the data set in.txt format (roundedfnedge.txt) that is used here:

	A	B	C	D	E
1	Gene	C_R1	C_R2	T_R1	T_R2
2	ENSBTAG00000000005	884	855	615	588
3	ENSBTAG00000000008	5	5	3	2
4	ENSBTAG00000000009	0	0	0	0
5	ENSBTAG00000000010	658	647	473	466
6	ENSBTAG00000000011	0	0	1	1

FIGURE 44.40 Input file for edgeR.

(Note: commands for getting the working directory and setting the working directory are the same as step 3 and step 4 of EBSeq)

Step 5: Commands to run edgeR:

```
> raw.data <- read.table(file = "roundedfnedge.txt", header = TRUE)
> counts <- raw.data[, -c(1,ncol(raw.data))]
> head(counts)
C_R1 C_R2 T_R1 T_R2
1 884 855 615 588
2 5 5 3 2
3 0 0 0 0
4 658 647 473 466
5 0 0 1 1
6 121 116 286 275
> rownames(counts) <- raw.data[, 1]
> head(counts)
C_R1 C_R2 T_R1 T_R2
ENSBTAG00000000005 884 855 615 588
ENSBTAG00000000008 5 5 3 2
ENSBTAG00000000009 0 0 0 0
ENSBTAG00000000010 658 647 473 466
ENSBTAG00000000011 0 0 1 1
ENSBTAG00000000012 121 116 286 275
> colnames(counts) <- paste(c(rep("C_R",2), rep("T_R",2)), c(1:2,1:2),
sep="")
> dim(counts)
[1] 24596 4
> colSums(counts)
C_R1 C_R2 T_R1 T_R2
9348648 9150009 10517019 10334348
> colSums(counts)/1e06
C_R1 C_R2 T_R1 T_R2
9.348648 9.150009 10.517019 10.334348
> table(rowSums(counts))[1:30]
> group <- c(rep("C", 2), rep("T", 2))
> cds <- DGEList(counts, group = group)
> names(cds)
[1] "counts" "samples"
> cds <- cds[rowSums(1e+06 * cds$counts/expandAsMatrix(cds$samples$lib.
size, dim(cds)) > 1) >= 3,]
> cds <- calcNormFactors(cds)
> cds$samples
group lib.size norm.factors
group lib.size norm.factors
C_R1 C 9348648 1.2798014
C_R2 C 9150009 1.2807311
T_R1 T 10517019 0.7805419
T_R2 T 10334348 0.7816336
> cds <- estimateCommonDisp(cds)
> names(cds)
[1] "counts" "samples" "common.dispersion" "pseudo.counts" "pseudo.lib.
size" "AveLogCPM"
> cds <- estimateTagwiseDisp(cds)
> names(cds)
[1] "counts" "samples" "common.dispersion" "pseudo.counts" "pseudo.lib.
size" "AveLogCPM"
[7] "prior.n" "tagwise.dispersion"
```

```
> summary(cds$tagwise.dispersion)
Min. 1st Qu. Median Mean 3rd Qu. Max.
1.571e-06 1.571e-06 1.571e-06 1.972e-05 1.571e-06 6.434e-03
> de.tgw <- exactTest(cds,dispersion = "common", pair = c("C", "T")) or
> de.tgw <- exactTest(cds,dispersion = "tagwise", pair = c("C", "T"))
> de.tgw
An object of class "DGEExact"
$table
logFC logCPM PValue
ENSBTAG00000000005 0.008546278 6.205706 8.892866e-01
ENSBTAG00000000010 0.065301786 5.816468 3.023431e-01
ENSBTAG00000000012 1.782004256 4.336904 6.581677e-63
ENSBTAG00000000013 2.043132781 5.820494 1.076701e-219
ENSBTAG00000000014 0.021064053 5.484187 7.743713e-01
10813 more rows …
$comparison
[1] "C" "T"
$genes
NULL
> options(digits = 3)
> topTags(de.tgw, n = 20, sort.by = "p.value")
Comparison of groups: T-C
logFC logCPM PValue FDR
ENSBTAG00000009012 8.89 11.55 0 0
ENSBTAG00000033748 7.81 8.25 0 0
ENSBTAG00000007883 7.81 7.41 0 0
ENSBTAG00000008951 7.55 9.53 0 0
ENSBTAG00000014762 7.52 8.86 0 0
ENSBTAG00000037608 7.39 5.91 0 0
ENSBTAG00000009206 7.26 6.92 0 0
ENSBTAG00000007881 7.25 11.49 0 0
ENSBTAG00000014707 7.02 11.29 0 0
> Z <- topTags(de.tgw, sort.by = "p.value")
> options(digits = 5)
> Z <- topTags(de.tgw, sort.by = "p.value")
>write.table(Z,"DEEdgeR.txt",sep = "\t",quote = F,col.names=F)
> Z <- topTags(de.tgw, n = 10000, sort.by = "p.value")
>write.table(Z,"DEEdgeR.txt",sep = "\t",quote = F,col.names=F)
```

Running of edgeR in R

The output file is DEEdgeR.txt.

Differentially expressed genes at the 5% level of significance:

To get the DE genes at the 5% level of significance, we select genes with p value < 0.05. edgeR found 9113 genes significantly ($P < 0.05$) differentially expressed.

Explaining the commands:

https://www.bioconductor.org/packages/3.3/bioc/vignettes/edgeR/inst/doc/edgeRUsersGuide.pdf

1. Calling the input file into edgeR and defining the dataset.

   The dataset accepted in edgeR should contain counts with the row names as the gene ids and the column names as the sample ids. The given commands take care of reading the data and defining the data set. The counts object is created by reading the input file roundedfnedge.txt, and the row names and column names are defined using rownames(counts) and colnames (counts) function. The group command will group the columns into control and treated.

```
> counts <- raw.data[ , -c(1,ncol(raw.data)) ]
> head(counts)
  C_R1 C_R2 T_R1 T_R2
1  884  855  615  588
2    5    5    3    2
3    0    0    0    0
4  658  647  473  466
5    0    0    1    1
6  121  116  286  275
> rownames( counts ) <- raw.data[ , 1 ]
> head(counts)
                   C_R1 C_R2 T_R1 T_R2
ENSBTAG00000000005  884  855  615  588
ENSBTAG00000000008    5    5    3    2
ENSBTAG00000000009    0    0    0    0
ENSBTAG00000000010  658  647  473  466
ENSBTAG00000000011    0    0    1    1
ENSBTAG00000000012  121  116  286  275
> colnames( counts ) <- paste(c(rep("C_R",2),rep("T_R",2)),c(1:2,1:2),sep="")
> dim(counts)
[1] 24596     4
> colSums( counts )
    C_R1     C_R2     T_R1     T_R2
 9348648  9150009 10517019 10334348
> colSums( counts ) / 1e06
     C_R1      C_R2      T_R1      T_R2
 9.348648  9.150009 10.517019 10.334348
> table( rowSums( counts ) )[ 1:30 ]

   0    1    2    3    4    5    6    7    8    9   10   11   12   13   14   15
9125  490  576  246  278  139  172  137  131  108  101   93   85   67   73   65
  16   17   18   19   20   21   22   23   24   25   26   27   28   29
  52   65   50   53   56   35   41   42   38   37   49   49   30   32
> group <- c(rep("C", 2) , rep("T", 2))
> cds <- DGEList( counts , group = group )
> names( cds )
[1] "counts"  "samples"
> cds <- cds[rowSums(1e+06 * cds$counts/expandAsMatrix(cds$samples$lib.size, dim(cds)) >
1) >= 3, ]
> cds <- calcNormFactors( cds )
> cds$samples
     group lib.size norm.factors
C_R1     C  9348648    1.2798014
C_R2     C  9150009    1.2807311
T_R1     T 10517019    0.7805419
T_R2     T 10334348    0.7816336
> cds <- estimateCommonDisp( cds )
> names( cds )
[1] "counts"           "samples"           "common.dispersion" "pseudo.counts"
[5] "pseudo.lib.size"  "AveLogCPM"
> cds <- estimateTagwiseDisp( cds )
> names( cds )
[1] "counts"            "samples"             "common.dispersion"
[4] "pseudo.counts"     "pseudo.lib.size"     "AveLogCPM"
[7] "prior.n"           "tagwise.dispersion"
> summary( cds$tagwise.dispersion )
     Min.   1st Qu.    Median      Mean   3rd Qu.      Max.
1.571e-06 1.571e-06 1.571e-06 1.972e-05 1.571e-06 6.434e-03
> de.tgw <- exactTest( cds , pair = c( "C" , "T" ) )
> de.poi <- exactTest( cds , dispersion = 1e-06  , pair = c( "C" , "T" ) )
>  de.tgw
An object of class "DGEExact"
$table
```

FIGURE 44.41 Running edgeR in R

```
> raw.data <- read.table(file = "roundedfnedge.txt", header = TRUE)
> counts <- raw.data[, -c(1,ncol(raw.data))]
> head(counts)
> rownames(counts) <- raw.data[, 1]
> head(counts)
>colnames(counts) <- paste(c(rep("C_R",2),rep("T_R",2)),c(1:2,1:2),sep="")
```

```
                         logFC    logCPM         PValue
ENSBTAG00000000005 0.008546278 6.205706  8.892866e-01
ENSBTAG00000000010 0.065301786 5.816468  3.023431e-01
ENSBTAG00000000012 1.782004256 4.336904  6.581677e-63
ENSBTAG00000000013 2.043132781 5.820494 1.076701e-219
ENSBTAG00000000014 0.021064053 5.484187  7.743713e-01
10813 more rows ...

$comparison
[1] "C" "T"

$genes
NULL
> cds <- calcNormFactors( cds )
> cds$samples$lib.size * cds$samples$norm.factors
[1] 11964413 11718701  8208974  8077674
> cds$samples$lib.size * cds$samples$norm.factors
[1] 11964413 11718701  8208974  8077674
> cds$samples$lib.size * cds$samples$norm.factors
[1] 11964413 11718701  8208974  8077674
> options( digits = 3 )
>  topTags( de.tgw , n = 20 , sort.by = "p.value" )
Comparison of groups:  T-C
                   logFC logCPM PValue FDR
ENSBTAG00000009012  8.89  11.55      0   0
ENSBTAG00000033748  7.81   8.25      0   0
ENSBTAG00000007883  7.81   7.41      0   0
ENSBTAG00000008951  7.55   9.53      0   0
ENSBTAG00000014762  7.52   8.86      0   0
ENSBTAG00000037608  7.39   5.91      0   0
ENSBTAG00000009206  7.26   6.92      0   0
ENSBTAG00000007881  7.25  11.49      0   0
ENSBTAG00000014707  7.02  11.29      0   0
ENSBTAG00000031214  7.01   6.82      0   0
ENSBTAG00000040586  6.97   7.94      0   0
ENSBTAG00000017251  6.93   6.68      0   0
ENSBTAG00000035224  6.90   7.42      0   0
ENSBTAG00000039037  6.82   7.35      0   0
ENSBTAG00000039861  6.81  11.38      0   0
ENSBTAG00000037554  6.76   7.65      0   0
ENSBTAG00000013991  6.75   5.75      0   0
ENSBTAG00000008471  6.74  11.86      0   0
ENSBTAG00000009768  6.73  11.74      0   0
ENSBTAG00000007901  6.68   8.46      0   0
> Z <- topTags( de.tgw , sort.by = "p.value" )
> options( digits = 5 )
> Z <- topTags( de.tgw , sort.by = "p.value" )
> write.table(Z,"DEEdgeR.txt",sep = "\t",quote = F,col.names=F)
> Z <- topTags( de.tgw , n = 10000 , sort.by = "p.value" )
> write.table(Z,"DEEdgeR.txt",sep = "\t",quote = F,col.names=F)
```

FIGURE 44.41 *(Continued)*

```
> dim(counts)
> colSums(counts)
> colSums(counts)/1e06
> table(rowSums(counts))[1:30]
```

2. Identifying DE genes – running to get differential gene expression.
   The function DGEList()coverts the count matrix into an edgeR object. In addition to the count matrix, we define a group variable that tells edgeR about the sample groups, which is supplied to DGEList. The elements that the object contains can be seen by using the names() function. Normalization factors can also be estimated.

```
> group <- c(rep("C", 2), rep("T", 2))
> cds <- DGEList(counts, group = group)
> names(cds)
```

The low count reads need to be filtered out to detect differential expression. In edgeR, only those genes that have at least one read per million in at least three samples are kept for further analysis. After filtering, normalization factors, which correct for the different compositions of the samples, are calculated. The product of the actual library sizes, and these factors, give effective library sizes.

```
> cds <- cds[rowSums(1e+06 * cds$counts/expandAsMatrix(cds$samples$lib.
size, dim(cds)) > 1) >= 3,]
> cds <- calcNormFactors(cds)
> cds$samples
```

The following commands are used to estimate common dispersion and tagwise (i.e., genewise) dispersion:

```
> cds <- estimateCommonDisp(cds)
> names(cds)
> cds <- estimateTagwiseDisp(cds)
> names(cds)
> summary(cds$tagwise.dispersion)
```

The pair-wise test for differential expression between two groups is performed by the function exactTest(). One of the lists of elements generated in the output of exactTest() is a table of results. However, the table from exactTest() does not contain p-values adjusted for multiple testing. These can be obtained by using the function topTags(). This takes the account from ExacTest() and adjusts the raw p-values (by False Discovery rate (FDR) correction) to return the top differentially expressed genes. But for a column of adjusted p-values sorted in increasing order, the output is similar to that of exactTest(). The sort.by argument sorts the table by p-value or fold-change. The topTags() function generates the top differentially expressed genes. If the n parameter is set to the total number of genes, the entire topTags() results table can be saved. Write.table writes the output to a txt file (Figure 44.42)

```
> de.tgw <- exactTest(cds, pair = c("C", "T"))
> de.tgw
> options(digits = 3)
```

	A	B	C	D	E
1		logFC	logCPM	Pvalue	FDR
2	ENSBTAG00000009012	8.89123504	11.5455284	0	0
3	ENSBTAG00000033748	7.8135702	8.25003701	0	0
4	ENSBTAG00000007883	7.80792535	7.4053768	0	0
5	ENSBTAG00000008951	7.55495149	9.53008799	0	0
6	ENSBTAG00000014762	7.51886554	8.85743405	0	0
7	ENSBTAG00000037608	7.39231146	5.9087098	0	0
8	ENSBTAG00000009206	7.26001683	6.91744327	0	0

FIGURE 44.42 Fold change and significance of ensemblIDs in DEEdgeR.txt.

```
> topTags(de.tgw, n = 20, sort.by = "p.value")
> Z <- topTags(de.tgw, sort.by = "p.value")
> options(digits = 5)
> Z <- topTags(de.tgw, n = 10000, sort.by = "p.value")
> write.table(Z,"DEEdgeR.txt",sep = "\t", quote = F, col.names=F)
```

44.5 QUESTIONS

1. Why is there a need to filter and trim RNA-Seq data?
2. The gtf file downloaded from Ensembl cannot be used directly in RSEM. Why?
3. If cufflinks : RPKM; then RSEM : ______ ; and edgeR : _________.

Estimating MicroRNA Expression Using the *miRDeep2* Tool

CHAPTER 45

GVPPSR Kumar, A Kumar and AP Sahoo
Animal Biotechnology Division, IVRI, UP, India

45.1 INTRODUCTION

For detection of miRNA from NGS data, several software tools have been developed to support the data analysis. These include: miRTRAP; DSAP; miRExpress; mirTools; miRDeep; miRNAkey and mireap; miRanalyzer; Mirena, and so on. Among this software, miRDeep and mireap are considered to be the best for prediction of novel miRNAs from mammalian data sets (Li *et al.*, 2012). Here, we will discuss miRDeep (Friedlander *et al.*, 2008, 2012). The codes and associated annotations have been taken from available guidances available online. In several cases, the explanations are verbatim with the source. The source URLs have been duly cited in this chapter.

miRDeep is a tool that helps in identifying miRNAs from the large pool of sequenced transcripts from a deep sequencing run. A probabilistic model is used to take into account the miRNA biogenesis for scoring fitness, and position the RNA sequence with the secondary structure of the miRNA precursor. miRDeep2 is an overhauled version of the original miRDeep algorithm, with added extensive new packages. The accuracy and sensitivity of miRDeep2 are estimated through its internal statistical controls.

Both the canonical and non-canonical miRNAs in deep sequencing data can be identified through miRDeep2. The miRNA expression profiling across samples can also be done using this tool. This includes: preprocessing of raw Illumina reads with mapper.pl script; quantification and expression profiling by quantifier.pl script; and miRNA identification by the miRDeep2.pl script.

45.2 PREPROCESSING OF READS

The reads are processed and mapped to the reference genome using mapper.pl script. This mapper module processes deep sequencing reads and/or maps them to the reference sequence.

Basic Applied Bioinformatics, First Edition. Chandra Sekhar Mukhopadhyay, Ratan Kumar Choudhary and Mir Asif Iquebal.

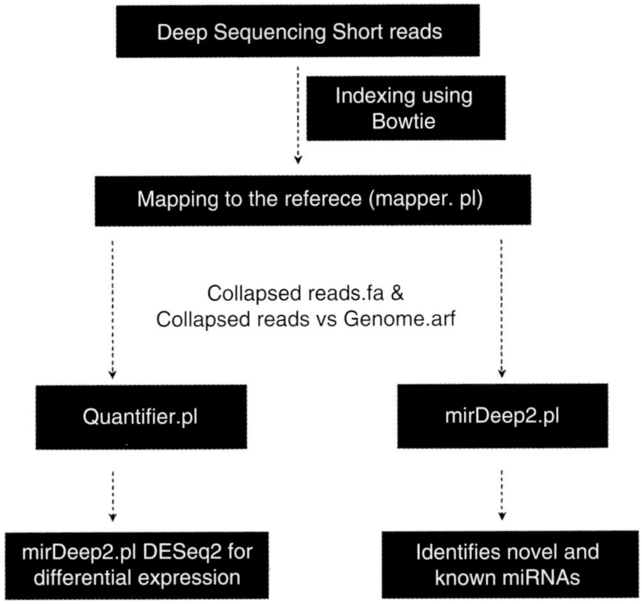

FIGURE 45.1

The module can process or map data that are in FASTA format, and can also handle sequence space data. It has a number of functions that can be implemented specifically with Illumina data. This entire chapter is explained using the datasets available in the miRDeep2 tutorial: (https://www.mdc-berlin.de/36105849/en/research/research_teams/systems_biology_of_gene_regulatory_elements/projects/miRDeep/documentation).

45.3 INPUT FORMATS OF THE DATA FILE

The default input file can be in FASTA, seq.txt or qseq.txt formats. For more options, please refer to https://www.mdc-berlin.de/36105849/en/research/research_teams/systems_biology_of_gene_regulatory_elements/projects/miRDeep/documentation

45.4 OUTPUT FORMATS THAT CAN BE GENERATED

The output depends on the options used. A *.fasta file containing the processed reads or an *.arf file with mapped reads (or both) can be generated as output. For example, we may say that the user generally wishes to analyze deep sequencing data mapping to a ≈ 6 kb region on *C. elegans* chromosome II for known and novel miRNA genes (this is as per the mirDeep2 tutorial at the address given previously).

45.5 PRELIMINARY FILES USED IN THE EXAMPLE

These are as per the miRdeep2 tutorial:

- cel_cluster.fa: a FASTA file with the reference genome.
- mature_ref_this_species.fa: *.fasta file containing reference miRBase mature-miRNA sequences for the species (*C. elegans* miRBase v.14 mature miRNAs) (http://petang.cgu.edu.tw/Bioinfomatics/Lecture/0_HTS/04/20120316.pdf).
- mature_ref_other_species.fa: *.fasta file harboring mature-miRNA sequences (from miRBase) for related species (*C. briggsae* and *D. melanogaster* miRBase v.14 mature miRNAs).
- precursors_ref_this_species.fa: Similarly, this is a FASTA file with the precursor miRNAs for the species (*C. elegans* miRBase v.14 precursor miRNAs, from miRBase).
- reads.fa: a FASTA file with the deep sequencing reads.

45.5.1 Step 1: Building index with bowtie

: >./bowtie-build cel_cluster.fa cel_cluster.

This command generates six files in the bowtie folder. Copy all the index files to the miRDeep2 folder.

45.5.2 Step 2: Process reads and map them to the genome

- The –c option designates that the input file is a FASTA file.
- The –j option: to remove the entries with non-canonical letters (characters other than a, c, g, t, u, n, A, C, G, T, U, N).
- The –k option: to clip the adapters.

```
APPLEs-Mac-Pro:~ appleserver$ cd downloads
APPLEs-Mac-Pro:downloads appleserver$ cd bowtie
APPLEs-Mac-Pro:bowtie appleserver$ ./bowtie-build cel_cluster.fa cel_cluster
Settings:
  Output files: "cel_cluster.*.ebwt"
  Line rate: 6 (line is 64 bytes)
  Lines per side: 1 (side is 64 bytes)
  Offset rate: 5 (one in 32)
  FTable chars: 10
  Strings: unpacked
  Max bucket size: default
  Max bucket size, sqrt multiplier: default
  Max bucket size, len divisor: 4
  Difference-cover sample period: 1024
  Endianness: little
  Actual local endianness: little
```

FIGURE 45.2

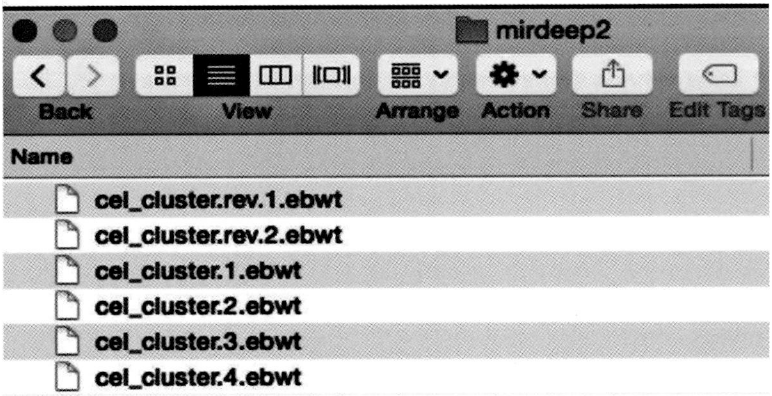

FIGURE 45.3

- The –l option: to discard the reads that are shorter than 18 nts.
- The –m option: to collapse the reads.
- The –p option: for mapping the processed reads against the previously indexed genome (cel_cluster).
- The –s option: to name the output file of processed reads.
- The –t option: for specifying the name of output file of genome mappings.
- –v gives verbose output to the screen.

Go to the mirdeep2 directory and type the following command:
mapper.pl reads.fa –c –j –k TCGTATGCCGTCTTCTGCTTGT –l 18 –m –p cel_cluster –s reads_collapsed.fa –t reads_collapsed_vs_genome.arf –v

```
APPLEs-Mac-Pro:mirdeep2 appleserver$ mapper.pl reads.fa -c -j -k TCGTATGCCGTCTTC
TGCTTGT  -l 18 -m -p cel_cluster -s reads_collapsed.fa -t reads_collapsed_vs_gen
ome.arf -v

grep: /proc/cpuinfo: No such file or directory
discarding sequences with non-canonical letters
clipping 3' adapters
discarding short reads
collapsing reads
mapping reads to genome index
trimming unmapped nts in the 3' ends
Mapping statistics

#desc    total    mapped   unmapped          %mapped %unmapped
total: 378333    330869   47464    0.875    0.125
seq: 378333      330869   47464    0.875_   0.125
```

FIGURE 45.4

The reads collapsed are those reads that are generated after clipping the adapter sequence. The collapsed reads mapped to the genome are given in the .arf file.

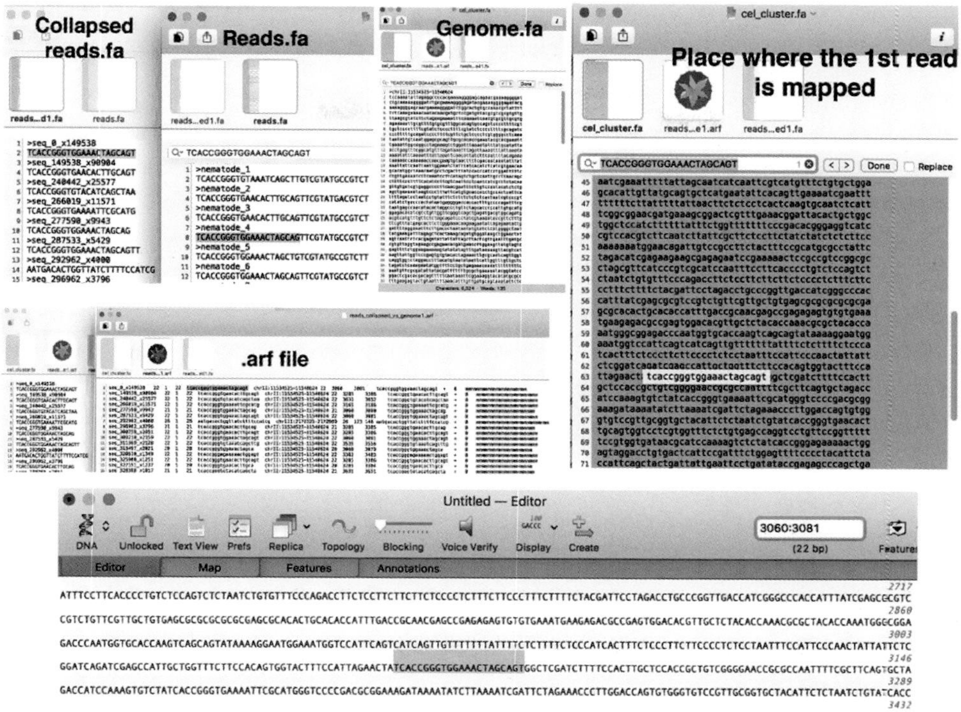

FIGURE 45.5

Figure 45.5 shows the reads in reads.fa that were collapsed to collapsedreads.fa. For example, the first read in collapsedreads.fa is obtained after clipping the adaptor sequence of sequence 4 (>nematiode_4) in the reads.fa file. The.arf file is the aligned reads file that shows the place where the reads match exactly in the genome. For example, the collapsed read one exactly matches with reference genome at positions 3060–3081.

45.5.3 Step 3: Fast quantitation of reads mapping to known miRBase precursors

Quantification of reads to known mirBase precursors is done using a quantifier.pl script. The deep sequencing reads are mapped to the predefined miRNA precursors by the quantifier module, to determine the expression of the corresponding miRNAs. Initially, the predefined mature miRNA sequences are mapped to the predefined precursors, followed by the mapping of the deep sequencing reads to the precursors.

- *Input*: this could be a FASTA file with precursor sequences, a FASTA file with mature miRNA sequences, a FASTA file with deep sequencing reads or, optionally, a FASTA file with star sequences and the three-letter code of the species of interest.
- *Output*: a tab-separated file called miRNAs_expressed_all_samples.csv with miRNA identifiers and its read counts, a signature file called miRBase.mrd, a file called expression.html that gives an overview of all miRNAs in the input data, a

directory called pdfs that contains for each miRNA, and a.pdf file showing its signature and structure (see the mirDeep2 tutorial at the address given previously).

The command is:

quantifier.pl –p precursors_ref_this_species.fa –m mature_ref_this_species.fa –r reads_collapsed.fa –t cel –y 16_19

The –p option denotes miRNA precursor sequences from miRBase database. The –m option designates miRNA sequences from miRBase database, the –t option designates the name of the species which is being analyzed, and the –y option designates the timestamp.

```
APPLEs-Mac-Pro:mirdeep2 appleserver$ quantifier.pl -p precursors_ref_this_specie
s.fa -m mature_ref_this_species.fa -r reads_collapsed.fa -t cel -y 16_19
getting samples and corresponding read numbers

Converting input files
building bowtie index
mapping mature sequences against index
mapping read sequences against index
Mapping statistics

#desc    total    mapped   unmapped          %mapped %unmapped
total: 374130     175870   198260   0.470    0.530
seq: 374130       175870   198260   0.470    0.530
analyzing data

7 mature mappings to precursors

Expressed miRNAs are written to expression_analyses/expression_analyses_16_19/mi
RNA_expressed.csv
not expressed miRNAs are written to expression_analyses/expression_analyses_16_1
9/miRNA_not_expressed.csv

Creating miRBase.mrd file

after READS READ IN thing

make_html2.pl -q expression_analyses/expression_analyses_16_19/miRBase.mrd -k ma
ture_ref_this_species.fa -t worm -y 16_19  -o -i expression_analyses/expression_
analyses_16_19/mature_ref_this_species.fa_mapped.arf  -l -m cel -M miRNAs_expres
sed_all_samples_16_19.csv
miRNAs_expressed_all_samples_16_19.csv file with miRNA expression values
parsing miRBase.mrd file finished
creating PDF files
creating pdf for cel-mir-36 finished
creating pdf for cel-mir-37 finished
creating pdf for cel-mir-40 finished
creating pdf for cel-mir-39 finished
creating pdf for cel-mir-41 finished
creating pdf for cel-mir-38 finished
creating pdf for cel-mir-229 finished_
```

FIGURE 45.6

The output is generated in the form of:

miRNAs_expressed _all_samples_16_19.csv, which gives the read counts of the reference miRNAs in the data in tabular format pdfs_16_19 – details of miRNA were identified.

expression_16_19. html – presents all the results in html format. This file is present in the expression analyses folder in the mirdeep2 directory

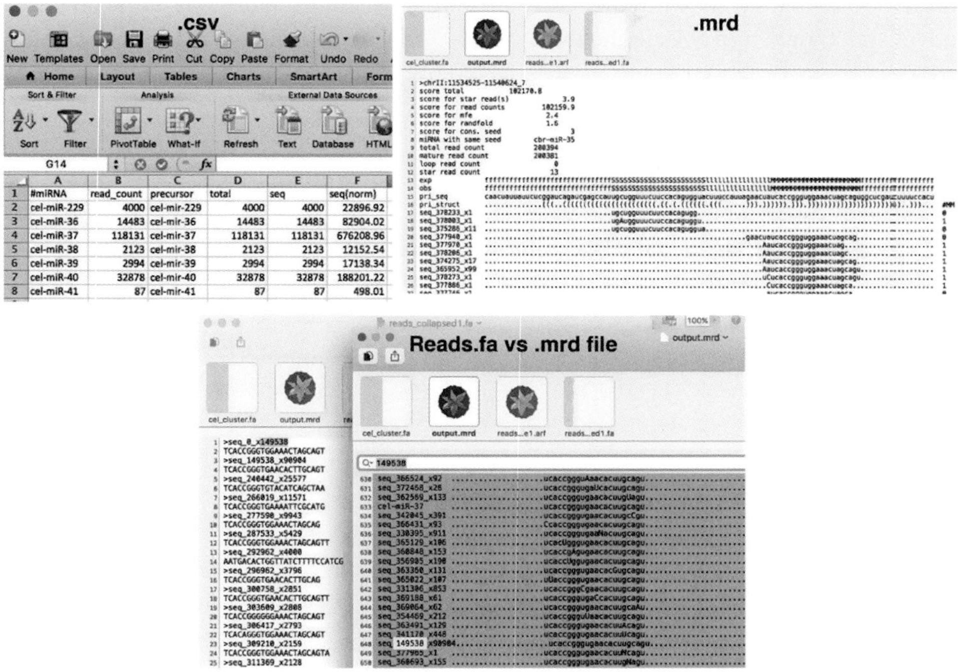

FIGURE 45.7

45.5.4 Step 4: Identification of known and novel miRNAs in the deep sequencing data

The novel and known miRNA detection can be done using the miRDeep2.pl script. The output from mapper module is used by the miRDeep2 module.

- Input: the input files for miRDeep2 are: a FASTA file with deep sequencing reads; a file of mapped reads to the genome in miRDeep2 arf format; a FASTA file of the corresponding genome; an optional FASTA file with known miRNAs of the analyzing species; and an optional FASTA file of known miRNAs of related species.
- *Output*: the output generated is a spreadsheet and an html file with an overview of all detected miRNAs.

Go to the miRDeep2 directory and type the following command:

>miRDeep2.pl reads_collapsed.fa cel_cluster.fa reads_collapsed_vs_genome.arf mature_ref_this_species.fa mature_ref_other_species.fa precursors_ref_this_ species.fa -t C.elegans 2>report.log

The file "mature_ref_this_species.fa" contains all mature miRNA of *C. elegans* species, while the "mature_ref_other_species.fa" file contains all mature miRNA of

C. briggsae and *D. melanogaster* species. By using “2>”, all progress output will be piped to the report.log file.

The results.html generated after running the above command contains all the results generated from miRDeep2.pl. In addition, the command will also generate a directory with .pdfs showing the read signatures, structures, and score breakdowns of novel and known miRNAs in the data.

45.6 QUESTIONS

1. Name two software programs that that can predict novel miRNA from RNA-Seq reads.
2. True or false? The .arf file is similar to the .sam file.
3. What is an .mrd file?

miRNA Target Prediction

CHAPTER 46

Sarika, Mir Asif Iquebal and D Kumar
CABiN, IASRI, New Delhi

46.1 INTRODUCTION

MicroRNAs (miRNAs) are short, non-coding RNA molecules that function to regulate gene expression post-transcriptionally. Due to the potential of one miRNA to target multiple gene transcripts, miRNAs are recognized as a primary mechanism to regulate gene expression. MicroRNA targets can be identified by pairing between the miRNA seed region and complementary sites within target mRNAs. Several miRNA target prediction tools, based on different computational approaches (e.g., modeling of physical interactions, machine learning techniques, etc.) are available. These tools include *TargetScan*, *miRanda*, and so on.

A common set of rules for miRNA target prediction includes complementarity, free energy calculations and evolutionary arguments and cooperativity of binding.

This chapter aims at predicting mRNA targets of the given microRNAs using various miRNA prediction tools.

46.2 miRNA TARGET PREDICTION BY TARGETSCAN (http://targetscan.org/)

This predicts the biological targets of miRNAs by searching for the presence of sites that match the seed region of each miRNA (Agarwal *et al*., 2015).

46.2.1 Objective

To search for predicted miRNA sequences in the human genome for the gene "HMGA2".

46.2.2 Procedure

a. Open the TargetScan web server: http://targetscan.org/
b. Select the species "Human" and enter the human gene symbol (i.e., HMGA2).

Basic Applied Bioinformatics, First Edition. Chandra Sekhar Mukhopadhyay, Ratan Kumar Choudhary and Mir Asif Iquebal.

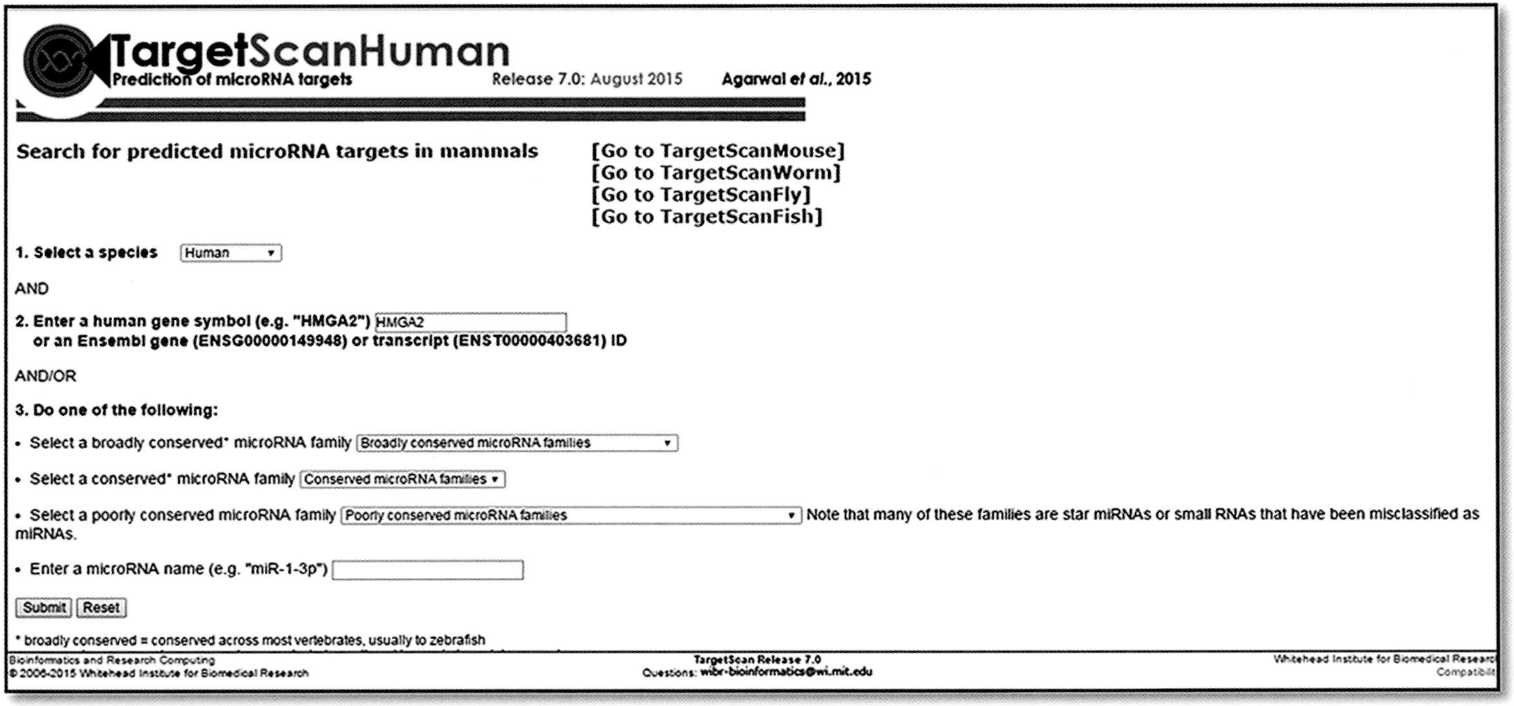

FIGURE 46.1 Home page of TargetScan tool.

c. Select the required conditions from the drop-down menu of given options.
d. Click on the "submit" button to obtain the result, or the "reset" button to perform a new prediction (Figure 46.1).

46.2.3 Results

a. For the gene "HMGA2", TargetScan gives multiple transcripts. Click on the transcript to see the details.
b. The resulting page gives the conserved sites for miRNA families corresponding to the gene HMGA2 (in all the Mammalia in this case).
c. The table has compiled detailed information (position, length, score, etc.) about the conserved sites for miRNA families corresponding to the gene HMGA2).

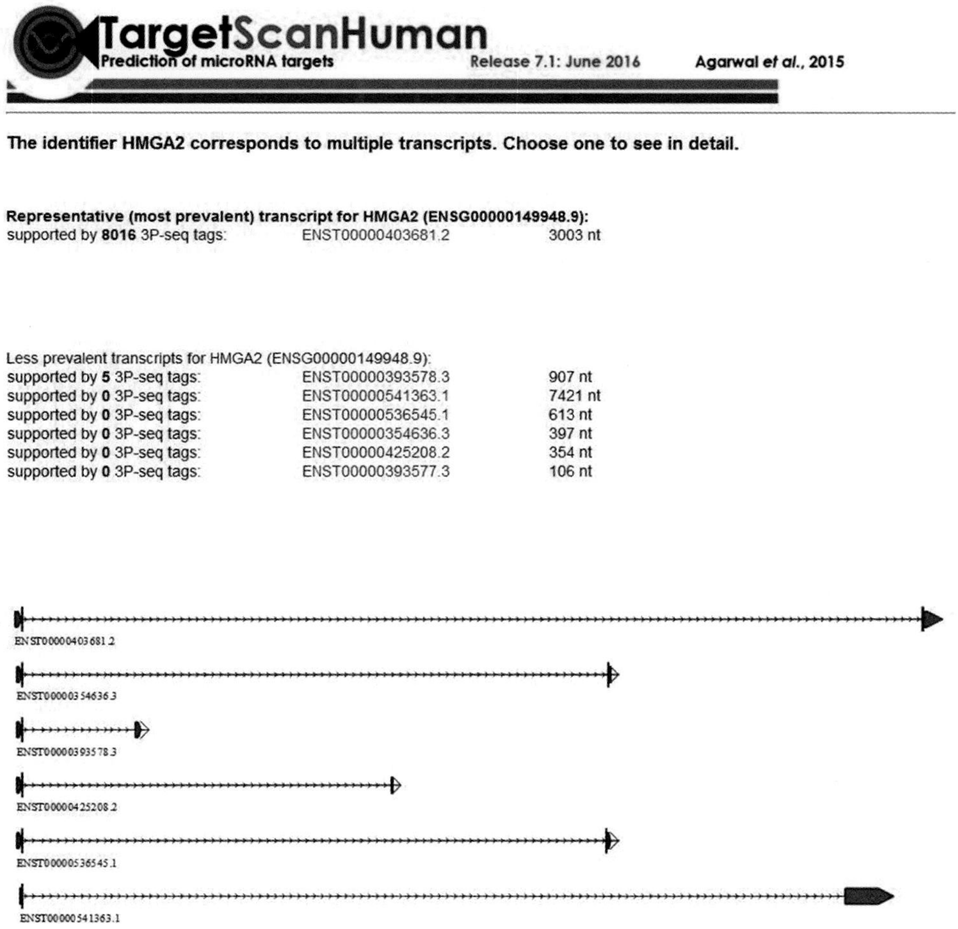

FIGURE 46.2 Output page showing multiple transcripts in the TargetScan tool. (*See insert for colour representation of the figure.*)

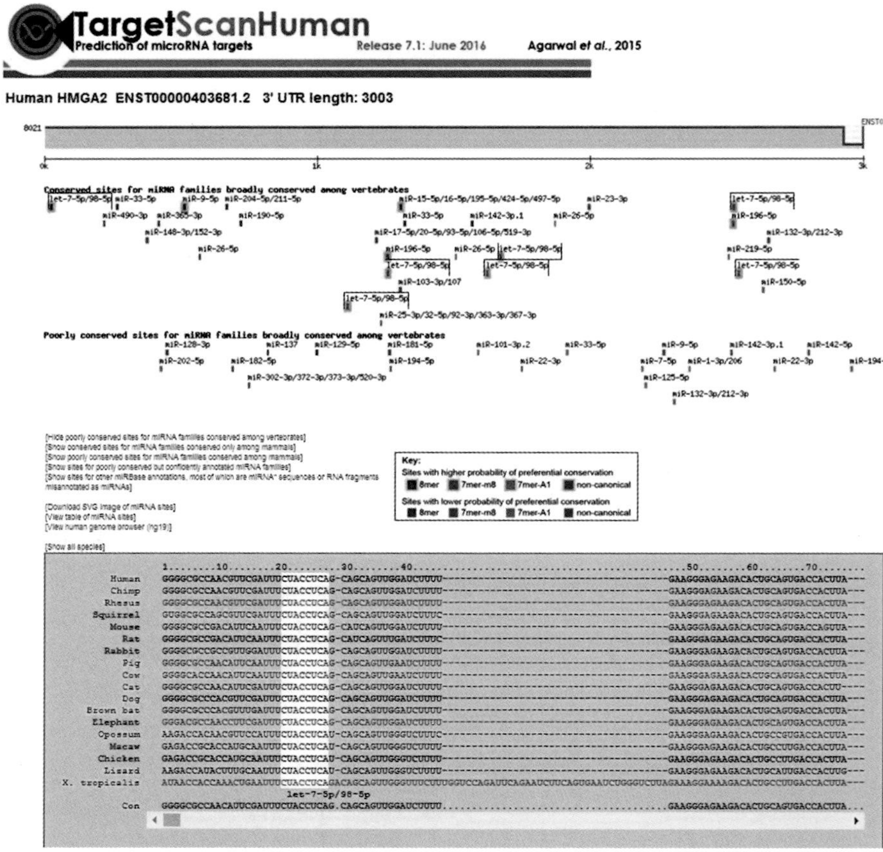

FIGURE 46.3 Output page of the TargetScan tool, showing conserved sites for miRNA families. (*See insert for colour representation of the figure.*)

46.3 miRNA TARGET PREDICTION BY TARGETSCAN IN HUMAN

46.3.1 Objective

To search for all the predicted miRNA targets for the miRNA "miR-1-3p" in the TargetScan tool.

46.3.2 Procedure

a. Select the species "Human" and enter the microRNA name (i.e., miR-1-3p).
b. Click on the "submit" button to obtain the result or the "reset" button to perform a new prediction.

46.3.3 Results

The resulting page gives a list of all the predicted targets for the human miRNA "miR-1-3p". The table has the information about the target gene symbol, transcripts with sites, gene name, score, and so on.

Conserved

	Predicted consequential pairing of target region (top) and miRNA (bottom)	Site type	Context++ score	Context++ score percentile	Weighted context++ score	Conserved branch length	P_{CT}
Position 21-28 of HMGA2 3' UTR hsa-let-7e-5p	5' ...GCCAACGUUCGAUUUCUACCUCA... ||||||| 3' UUGAUAUGUUGGAGGAUGGAGU	8mer	-0.63	99	-0.63	4.512	0.94
Position 21-28 of HMGA2 3' UTR hsa-let-7b-5p	5' ...GCCAACGUUCGAUUUCUACCUCA... ||||||| 3' UUGGUGUGUUGGAUGAUGGAGU	8mer	-0.63	99	-0.63	4.512	0.94
Position 21-28 of HMGA2 3' UTR hsa-let-7c-5p	5' ...GCCAACGUUCGAUUUCUACCUCA... ||||||| 3' UUGGUAUGUUGGAUGAUGGAGU	8mer	-0.63	99	-0.63	4.512	0.94
Position 21-28 of HMGA2 3' UTR hsa-let-7i-5p	5' ...GCCAACGUUCGAUUUCUACCUCA... ||||||| 3' UUGUCGUGUUUGAUGAUGGAGU	8mer	-0.63	99	-0.63	4.512	0.94
Position 21-28 of HMGA2 3' UTR hsa-let-7f-5p	5' ...GCCAACGUUCGAUUUCUACCUCA... ||||||| 3' UUGAUAUGUUAGAUGAUGGAGU	8mer	-0.63	99	-0.63	4.512	0.94
Position 21-28 of HMGA2 3' UTR hsa-let-7d-5p	5' ...GCCAACGUUCGAUUUCUACCUCA... ||||||| 3' UUGAUACGUUGGAUGAUGGAGA	8mer	-0.64	99	-0.64	4.512	0.94
Position 21-28 of HMGA2 3' UTR hsa-miR-4458	5' ...GCCAACGUUCGAUUUCUACCUCA... ||| ||||||| 3' AAGAAGGUGUGGAUGGAGA	8mer	-0.67	99	-0.67	4.512	0.94
Position 21-28 of HMGA2 3' UTR hsa-miR-4500	5' ...GCCAACGUUCGAUUUCUACCUCA... ||||||| 3' UUCUUUGAUGAUGGAGU	8mer	-0.63	99	-0.63	4.512	0.94
Position 21-28 of HMGA2 3' UTR hsa-let-7g-5p	5' ...GCCAACGUUCGAUUUCUACCUCA... ||||||| 3' UUGACAUGUUUGAUGAUGGAGU	8mer	-0.63	99	-0.63	4.512	0.94
Position 21-28 of HMGA2 3' UTR hsa-let-7a-5p	5' ...GCCAACGUUCGAUUUCUACCUCA... ||||||| 3' UUGAUAUGUUGGAUGAUGGAGU	8mer	-0.63	99	-0.63	4.512	0.94

FIGURE 46.4 Detailed table of all the conserved sites. (*See insert for colour representation of the figure.*)

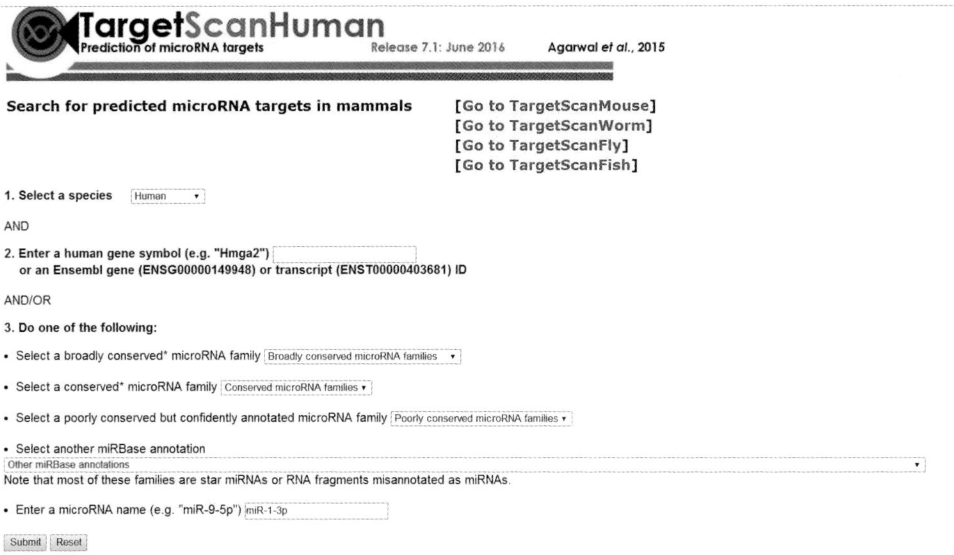

FIGURE 46.5 Input page of the TargetScan tool. (*See insert for colour representation of the figure.*)

46.4 miRNA TARGET PREDICTION BY psRNATARGET (http://plantgrn.noble.org/psRNATarget>/)

psRNATarget is a plant small RNA target analysis server, which provides reverse complementary matching between small RNA and target transcript, and target-site accessibility evaluation by calculating the unpaired energy essential to "open" secondary structure around the target site on mRNA of the small RNA (Dai and Zhao, 2011).

46.4.1 Objective

To search for the targets of *Arabidopsis thaliana* (denoted as *ath*) miRNA sequences as inputted by the user, against a genomic library of *Arabidopsis thaliana*.

46.4.2 Procedure

a. Open psRNATarget web server: http://plantgrn.noble.org/psRNATarget/.
b. Paste the following *Arabidopsis thaliana* miRNA sequences in the box provided:
   >ath-miR156a
   UGACAGAAGAGAGUGAGCAC
   >ath-miR157a
   UUGACAGAAGAUAGAGAGCAC
   >ath-miR158a
   UCCCAAAUGUAGACAAAGCA
   >ath-miR398a
   UGUGUUCUCAGGUCACCCCUU
   >ath-miR398b
   UGUGUUCUCAGGUCACCCCUG
   >ath-miR398c
   UGUGUUCUCAGGUCACCCCUG

TargetScanHuman
Prediction of microRNA targets   Release 7.1: June 2016   Agarwal *et al.*, 2015

Human | miR-1-3p/206

883 transcripts with conserved sites, containing a total of **976** conserved sites and **336** poorly conserved sites.

Genes with only poorly conserved sites are not shown.   [View top predicted targets, irrespective of site conservation]

Table sorted by cumulative weighted context++ score   [Sort table by aggregate P_{CT}]

The table shows at most one transcript per gene, selected for being the most prevalent, based on 3P-seq tags (or the one with the longest 3' UTR, in case of a tie).   [Download table]

Target gene	Representative transcript	Gene name	Number of 3P-seq tags supporting UTR + 5	Link to sites in UTRs	Conserved sites				Poorly conserved sites				6mer sites	Representative miRNA	Cumulative weighted context++ score	Total context++ score	Aggregate P_{CT}	Previous TargetScan publication(s)
					total	8mer	7mer-m8	7mer-A1	total	8mer	7mer-m8	7mer-A1						
CORO1C	ENST00000261401.3	coronin, actin binding protein, 1C	1970	Sites in UTR	2	2	0	0	0	0	0	0	2	hsa-miR-206	-1.08	-1.09	0.96	2005, 2007, 2009, 2011
SMIM14	ENST00000295958.5	small integral membrane protein 14	601	Sites in UTR	2	2	0	0	3	0	1	2	3	hsa-miR-206	-1.03	-1.38	0.98	2007, 2009, 2011
ARPC3	ENST00000228825.7	actin related protein 2/3 complex, subunit 3, 21kDa	323	Sites in UTR	1	1	0	0	2	1	1	0	0	hsa-miR-1-3p	-0.99	-1.30	< 0.1	
PTPLAD1	ENST00000261875.5	protein tyrosine phosphatase-like A domain containing 1	11697	Sites in UTR	2	1	1	0	1	0	1	0	1	hsa-miR-206	-0.97	-1.11	0.95	2003, 2007, 2009, 2011
ARCN1	ENST00000534182.2	archain 1	1139	Sites in UTR	1	1	0	0	1	0	1	0	0	hsa-miR-1-3p	-0.91	-0.91	0.95	2005, 2007, 2009, 2011
TAGLN2	ENST00000368096.1	transgelin 2	2192	Sites in UTR	2	0	2	0	1	0	0	1	1	hsa-miR-1-3p	-0.85	-0.85	0.89	2005, 2007, 2009, 2011
GJA1	ENST00000282561.3	gap junction protein, alpha 1, 43kDa	1732	Sites in UTR	2	2	0	0	0	0	0	0	1	hsa-miR-206	-0.84	-0.84	0.93	2003, 2005, 2007, 2009, 2011
ERMP1	ENST00000381506.3	endoplasmic reticulum metallopeptidase 1	53	Sites in UTR	2	0	2	0	1	1	0	0	0	hsa-miR-1-3p	-0.81	-0.81	ORF	
SERP1	ENST00000239944.2	stress-associated endoplasmic reticulum protein 1	3242	Sites in UTR	3	0	0	3	0	0	0	0	2	hsa-miR-1-3p	-0.77	-0.80	0.97	2005, 2007, 2009, 2011
TMSB4X	ENST00000451311.2	thymosin beta 4, X-linked	5	Sites in UTR	1	1	0	0	0	0	0	0	0	hsa-miR-206	-0.74	-0.74	0.77	2009, 2011
MMD2	ENST00000406755.1	monocyte to macrophage differentiation-associated 2	5	Sites in UTR	1	1	0	0	1	1	0	0	0	hsa-miR-1-3p	-0.69	-0.69	0.86	2005, 2007, 2009, 2011
BDNF	ENST00000439476.2	brain-derived neurotrophic factor	2696	Sites in UTR	3	1	2	0	0	0	0	0	0	hsa-miR-1-3p	-0.68	-0.95	0.98	2003, 2005, 2007, 2009, 2011
SLC10A7	ENST00000264986.3	solute carrier family 10, member 7	77	Sites in UTR	2	1	1	0	0	0	0	0	2	hsa-miR-1-3p	-0.68	-0.90	0.88	2009, 2011
G6PD	ENST00000393562.2	glucose-6-phosphate dehydrogenase	9	Sites in UTR	3	0	3	0	0	0	0	0	0	hsa-miR-1-3p	-0.68	-0.68	> 0.99	2011
SRI	ENST00000265729.2	sorcin	278	Sites in UTR	1	1	0	0	0	0	0	0	0	hsa-miR-1-3p	-0.65	-0.65	0.89	2011
GLCCI1	ENST00000223145.5	glucocorticoid induced transcript 1	654	Sites in UTR	2	2	0	0	0	0	0	0	2	hsa-miR-206	-0.65	-0.73	0.92	2005, 2007, 2009, 2011

FIGURE 46.6 Detailed information about the target gene symbol in the TargetScan tool. (*See insert for colour representation of the figure.*)

>ath-miR834
UGGUAGCAGUAGCGGUGGUAA
>ath-miR390a
AAGCUCAGGAGGGAUAGCGCC
>ath-miR390b
AAGCUCAGGAGGGAUAGCGCC

c. An existing library of transcripts or a genomic library for target search is to be selected from the list – namely, *Arabidopsis thaliana*, genomic DNA, 3.4 K segments from a strand with 0.4 K overlapped region, TAIR.
d. Add the required parameters given options, or use default values.
e. Click on the “submit” button to obtain the result or the “reset” button to perform a new prediction.

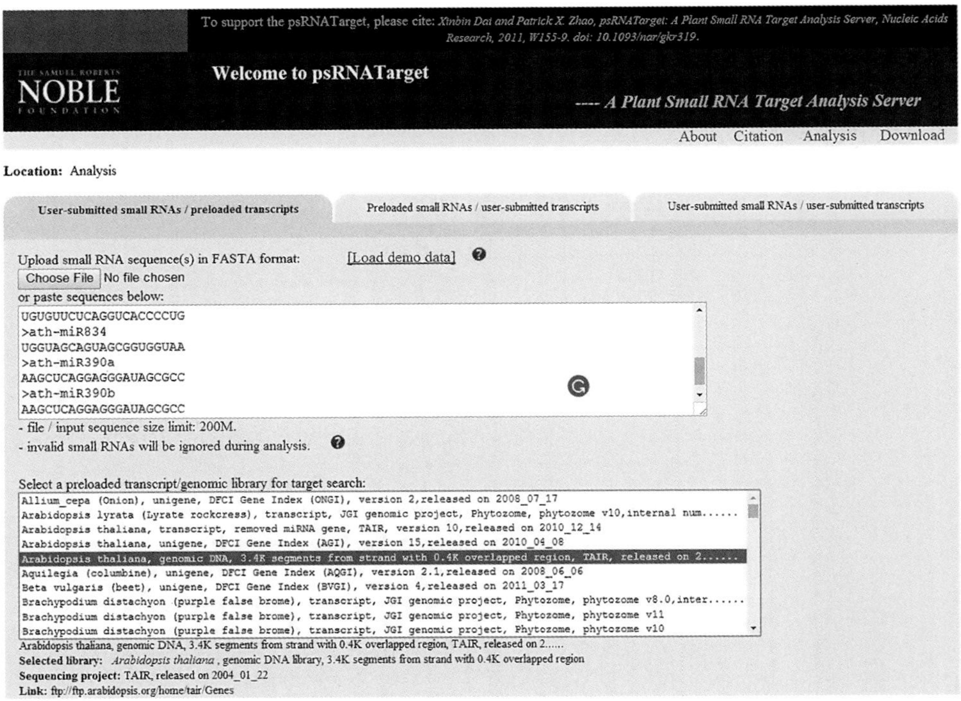

FIGURE 46.7 Input page of the psRNATarget tool. (*See insert for colour representation of the figure.*)

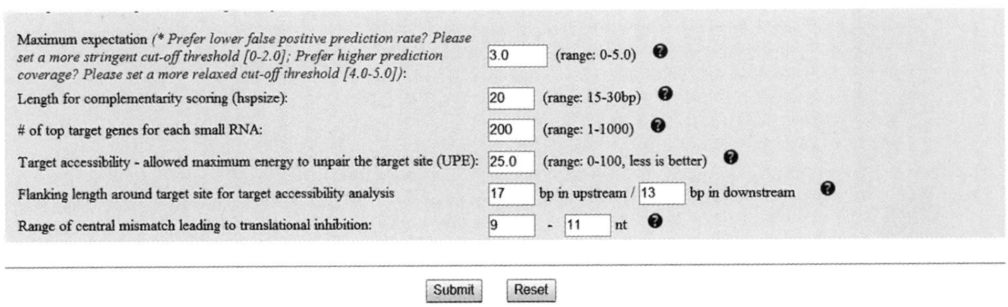

FIGURE 46.8 Input page of the psRNATarget tool for other parameters. (*See insert for colour representation of the figure.*)

FIGURE 46.9 Result page of the psRNATarget tool. (*See insert for colour representation of the figure.*)

46.4.3 Results

The result page gives a list of predicted miRNA/Target pairs for the input miRNA sequences against the selected genomic library. The result tabulates miRNA accession, target accession, expectation score, target accessibility, alignment, and also the target description. The server also has a facility for batch-downloading of the complete result.

46.5 miRNA TARGET PREDICTION BY miRANDA (http://www.microrna.org)

This is command-line software that runs on Linux and uses a weighted dynamic programming algorithm to obtain the candidate sequences. This algorithm uses a score to rank the predictions that consist of a weighted sum based on matches, mismatches, and G:U wobbles (Enright *et al.*, 2004).

46.5.1 Objective

To search for the targets of user submitted miRNA sequences against the RNA sequence.

46.5.2 Procedure

Here, sheep miRNA will be used to predict the target by search against the set of sheep RNA, using miRanda:

a. Paste the following sequence, create an input file named mature.fa. and keep it in the src directory of miRandatool:
   >oar-let-7b MIMAT0014963
   UGAGGUAGUAGGUUGUGUGGU
b. Download a coding set of sheep from NCBI from the link ftp://ftp.ncbi.nlm.nih.gov/genomes/all/GCF_000298735.2_Oar_v4.0/GCF_000298735.2_Oar_v4.0_rna.fna.gz, and keep in the src directory of miRanda software.
c. Sheep.fa is the coding sequence for sheep, as downloaded from NCBI.

```
root@wrk07 src]#./miranda mature.fa sheep.fa>sheep_target.out
root@wrk07 src]#
```

d. Type the command./mirandamature.fasheep.fa>sheep_target.out

46.5.3 Result

miRanda provides details on predicted miRNA/Target pairs for the input miRNA sequences against the coding sequence of sheep. Gene symbol, gene names, RefSeq id, alignment, score, and so on are stored in the output file.

```
Complete

Read Sequence:gi|426218486|ref|XM_004003429.1| PREDICTED: Ovis aries autophagy related 4B, cysteine peptidase (ATG4B), mRN
A(1365 nt)
=-=-=-=-=-=-=-=-=-=-=-=-=-=-=-=-=-=-=-=-=-=-=-=-=-=-=-=-=-=-=-=-=
Performing Scan: oar-let-7b vs gi|426218486|ref|XM_004003429.1|
=-=-=-=-=-=-=-=-=-=-=-=-=-=-=-=-=-=-=-=-=-=-=-=-=-=-=-=-=-=-=-=-=

   Forward:     Score: 146.000000  Q:2 to 19  R:872 to 892 Align Len (17) (70.59%) (82.35%)

   Query:    3' uggUGUGUUGGAUGAUGGAGu 5'
                   |||  :|| |||||:||
   Ref:      5' ccaACAGTGCCCACTACTTCa 3'

   Energy:  -20.620001 kCal/Mol

Scores for this hit:
>oar-let-7b     gi|426218486|ref|XM_004003429.1|        146.00  -20.62  2 19    872 892 17      70.59%  82.35%

Score for this Scan:
Seq1,Seq2,Tot Score,Tot Energy,Max Score,Max Energy,Strand,Len1,Len2,Positions
>>oar-let-7b    gi|426218486|ref|XM_004003429.1|        146.00  -20.62  146.00  -20.62  6       21      1365    872
Complete

Read Sequence:gi|426218488|ref|XM_004003430.1| PREDICTED: Ovis aries THAP domain containing 4 (THAP4), mRNA(1863 nt)
=-=-=-=-=-=-=-=-=-=-=-=-=-=-=-=-=-=-=-=-=-=-=-=-=-=-=-=-=-=-=-=-=
Performing Scan: oar-let-7b vs gi|426218488|ref|XM_004003430.1|
=-=-=-=-=-=-=-=-=-=-=-=-=-=-=-=-=-=-=-=-=-=-=-=-=-=-=-=-=-=-=-=-=

   Forward:     Score: 145.000000  Q:2 to 18  R:187 to 207 Align Len (16) (68.75%) (87.50%)
```

FIGURE 46.10 Results file of the miRanda tool.

46.6 QUESTIONS

1. Predict biological targets of the conserved miRNA family in mouse using the TargetScan web server with broadly conserved microRNA family "miR-21-5p/590-5p". How many target genes are predicted from this miRNA family?
2. From the above search made, how many are conserved sites, and how many poorly conserved sites are there?
3. Find the predicted target with conserved sites only in the mouse, for conserved miRNA family miR-21-5p/590-5p.
4. Find the predicted conserved miRNA target sites for gene Khc-73 in *D. melanogaster* using the TargetScanFly server. Discuss the output in detail.

Appendix A: Usage of Internet for Bioinformatics

RK Choudhary
School of Animal Biotechnology, GADVASU, Ludhiana

Bioinformatics is the branch of science that deals with the processing of biological data with the help of computer science. In sequencing gene or protein data, for example, voluminous amounts of data are processed, analyzed for deriving biological meaning and, ultimately, stored for further use with the help of bioinformatics. Therefore, it is a hybrid science, comprising biology and computer science, that is conceptualizing biology in terms of macromolecules (in the sense of physical chemistry) and then applying "informatics" techniques (derived from disciplines such as applied mathematics, computer science, and statistics) to understand and organize the information associated with these molecules on a large scale (Luscombe *et al.*, 2001).

The internet is an interconnected network of information across the globe that provides remote communications. The introduction of Transmission Control Protocol (TCP) and Internet Protocol (IP) (or, together, TCP/IP) by the Advanced Research Project Agency (ARPA) in 1969 evolved remote communication radically. The IP number (example; 96.47.32.230) that recognizes the server and the computer is unique. Since it is hard to recognize these long strings of numbers, IP addresses have been associated with a Fully Quantified Domain Name (FQDN). For example, the IP address given above is associated with www.gadvasu.in, a website belonging to the Guru Angad Dev Veterinary and Animal Science University. The top level of domain name includes .com, .edu, .gov, .org, and .in.

There are various facilities for the bioinformatics that are provided by the internet, such as:

- Electronic mail (email);
- Electronic journals;
- Educational resource materials on bioinformatics;
- Biological databases (NCBI, EBI, Genome browser, DDBJ, PDB, TIGR);
- Software tools;
- World Wide Web (www) searches.

Basic Applied Bioinformatics, First Edition. Chandra Sekhar Mukhopadhyay, Ratan Kumar Choudhary and Mir Asif Iquebal.

Email is the most convenient way of writing/replying and receiving mail electronically. In this, an email address is assigned to each sender or receiver, in the form *user@computer.domain*. The sender can attach files that can be sent with the email instantaneously. The speed of sending an email message with attachments varies with the speed of the local area network (LAN), the time of sending the email (because of network congestion), the size of attachments, and so on.

Despite several advantages offered by email in transmitting a message, users do experience difficulties in transmitting files, especially with attachments. Microsoft Exchange is one of the common platforms to send an email, but sending emails across the platform hinders decoding or detaching files by the receivers. Therefore, the urgency of a protocol has arisen, where files can be transferred to a remote server quickly. This protocol is called File Transfer Protocol (FTP). In FTP, a connection is made between the user's computer and a remote computer, files are transferred at a faster rate, and the connection remains effective until the session is over. TCP facilitates files back and forth between FTP-server and FTP-client.

In the case of public or anonymous FTP, server sign-in may not be required, but if using a private FTP server, sign-in with a valid username and password to initiate the data transfer is a must. Data can be transferred via FTP by three different ways: stream mode, block mode, and compressed mode. In stream mode, the file is transferred in a continuous stream from the port without any data formatting. This occurs when server and client have identical operating systems. In block mode, data are transferred into blocks of information, such as header, byte count and actual data. In compressed mode, large data files are compressed and modified by codes and then transferred. In response to the need to transfer sensitive data, the need for more security has arisen. In 1994, Netscape developed a Secure Sockets Layers (SSL) protocol and FTP transferred, now armed with SSL protection called SFTP.

Although FTP and SFTP have tremendous use in file transfer from one computer to another computer, there are a number of limitations. In FTP, the user has to enter a particular directory, and has access to only those files that are there in the appropriate directory of the server's computer. However, the user cannot access another user's directory located on the different server. This inherent drawback has led to the development of an interactive client application system called Distributed Document Delivery Systems (DDDS). DDBS, commonly known as the World Wide Web (www), enabled navigation to the Web without prior knowledge of the location of the server with the directory and information. The need to access files over the Web led to the development of Hyper Text Transfer Protocol (HTTP or http) as the means of obtaining information on the World Wide Web. The Web also provides room for keeping other information through the linkage (called hyperlinked files). The WWW is now the most popular method of using the internet, providing access to any web pages by entering http:// in front of the address. Today, browsers do not require the user to type "HTTP", because it is the default method of communication for accessing the Web.

Browsers are client-server applications, and are connected to a remote website to download requested information. Since the information is retrieved in a fast and continuous manner, a platform-independent format is required to display the information. The development of Hyper Text Markup Language (HTML), a text-based format, has allowed graphics, images and other information to be displayed in a separate file whose form is standard to most users.

Example of a small HTML document

```
<!DOCTYPE html>
<html>
<body>

<h1>GURU ANGAD DEV VETERINARY AND ANIMAL SCIENCE UNIVERSITY</h1>
<h2>School of Animal Biotechnology</h2>
<h3>Research Areas</h3>
<ol type="I">
  <li>Molecular Genetics</li>
  <li>Diagnostics and Vaccinology</li>
  <li>Stem and Cell Biology</li>
</body>
</html>
```

Information pertaining to the above HTML codes will appear on the website as:

GURU ANGAD DEV VETERINARY AND ANIMAL SCIENCE UNIVERSITY
School of Animal Biotechnology
Research Areas

I. Molecular Genetics
II. Diagnostics and Vaccinology
III. Stem and Cell Biology

Meaning of HTML notations

- All documents in HTML will start with a declaration as <!DOCTYPE html>.
- All HTML documents will begin with <html> and end with </html>.
- The body of information that is displayed is that in between <body> and </body>.
- <h1> describes the first heading, which ends with </h1>.
- Likewise, the order number will start with <o1> and ends with </o1>.

The Internet also a way to find information about bioinformatics, electronic journals and online tools on bioinformatics, as well as providing access to biological databases. Different types of online tools of bioinformatics, and various databases of genomics, transcriptomics and proteomics are elaborated elsewhere in this book.

Appendix B: Important Web Resources for Bioinformatics Databases and Tools

CS Mukhopadhyay and RK Choudhary
School of Animal Biotechnology, GADVASU, Ludhiana

INTRODUCTION

This chapter is a daily companion for bench workers who need to use web-based tools as well as the databases for various bioinformatics work. The major databases and tools required for basic bioinformatics jobs have been outlined, along with the uniform/universal resource locator (URL). It is not possible to provide an exhaustive list of the links to bioinformatics tools and databases in an appendix of this book. Some of the most well-known and frequently used sites have been covered here. Users are requested to bookmark the URLs in their systems and also keep abreast with the changes in the URLs of the pages, if applicable.

Site name	Universal Resource Locator (URL)	Description
NCBI, EMBL, DDBJ		
NCBI home page	http://www.ncbi.nlm.nih.gov/	A vast repository and a public database of nucleic acid sequences, literature and genome-specific resources. It also provides several biocomputational tools for sequence analysis and FTPs for sequence retrieval.
NCBI-dbVar	http://www.ncbi.nlm.nih.gov/dbvar/	dbVar is a database maintained by NCBI for structural variations at the genomic level.
GenBank	http://www.ncbi.nlm.nih.gov/genbank/	A public repository of nucleotide sequences provided by NCBI.
NCBI Human Genome Browser	http://www.ncbi.nlm.nih.gov/mapview/map_search.cgi?chr=hum_chr.inf&query	This map-viewer displays the human reference genome assembly. It depicts several components of the genome (genes, sequence tagged sites, expressed sequence tags, contigs, etc.) and the experimentally derived maps (BAC component map, cytogenetic and physical map, radiation hybrid map, etc.).

Basic Applied Bioinformatics, First Edition. Chandra Sekhar Mukhopadhyay, Ratan Kumar Choudhary and Mir Asif Iquebal.

Site name	Universal Resource Locator (URL)	Description
NIH Human Microbiome Project	http://www.hmpdacc.org/resources/data_browser.php/	The Human Microbiome Project (HMP) is the initiative of NIH, with an aim to explore the microbes in different organs and contents within the organs. The identified microbiome is characterized and associated with healthy and diseased states of human.
Entrez	http://www.ncbi.nlm.nih.gov/sites/entrez?db=pubmed	A search engine (global query search system against cross-databases) which is used to look for literature (journal articles: review or research category), book, documents in various sections of NCBI (e.g., OMIM, Genome, Structure, etc.). Users can also save the searched results in their NCBI account for referring later.
NCBI GenBank Taxonomy Database	http://www.ncbi.nlm.nih.gov/taxonomy/taxonomyhome.html/	Provides taxonomical information of an organism (used in molecular biology research work).
EMBL-EBI	http://www.ebi.ac.uk	The EBI, a part of EMBL, is an academic research institute located on the Wellcome Trust Genome Campus in Cambridge (UK). It serves as a public repository of molecular data. It also provides free online bioinformatic software and tools.
ENA-EMBL	http://www.ebi.ac.uk/embl/	This European Nucleotide Archive of EMBL-Bank is the repository of nucleotide sequences of various types, like NCBI GenBank. The latest release (ENA release 125: http://www.ebi.ac.uk/about/news/service-news/ena-release-125) maintains the annotated sequences of ENA.
DDBJ	http://www.ddbj.nig.ac.jp/	This nucleotide databank (DNA Databank of Japan) is similar to ENA-EMBL and GenBank-Nucleotide of NCBI. The International Nucleotide Sequence Database Collaboration (INSDC: http://insdc.org) links these three databanks to each other by computerized synchronization.
Protein databases		
PDB	http://www.rcsb.org/pdb/home/home.do	The RCSB-PDB is a repository of structural information and curated annotation of different types of experimentally determined structures, like protein, nucleic acids or other complex assemblies.
Pfam	http://pfam.sanger.ac.uk/	Information on protein families, characterized by alignment (by HMM-based algorithm) of amino acid sequences of the same family, annotation, multiple domain architecture analysis, etc. Links to protein structures are also provided.
PIR	http://pir.georgetown.edu/	A centralized resource for information on proteins in terms of sequence, function, resources for protein annotation (PIRSF, iProClass, iProLINK). In 2012, a single database called "UniProt" was created after merging the PIR, Swiss-Prot and TrEMBL databases.

Site name	Universal Resource Locator (URL)	Description
PROSITE	http://www.expasy.ch/prosite/	This database maintains information on domains, families and functional sites of proteins and the profiles for identifying proteins (based on a collection of rules called pro-rule). Tools for protein sequence analysis and detection of motifs are also provided by PROSITE.
SWISSPROT-TrEMBL	http://www.expasy.ch/sprot/	This is an official database that contains manually curated protein sequences with high-level annotation. Information on protein structure, post-translational modifications, etc. is available in this non-redundant database.
RCSB	http://home.rcsb.org/	The Research Collaboratory for Structural Bioinformatics (RCSB) undertakes research works to decipher the relationship between 3D-structural features of macromolecules and their functional aspects. RCSB is responsible for citation and annotation of PDB data.
NDB	http://ndbserver.rutgers.edu/	NDB maintains information about the three-dimensional structure of nucleic acids.
RNA Databases		
The RNAdb	http://research.imb.uq.edu.au/rnadb/	This is a popular non-coding RNA database (RNAdb) of mammals that harbors sequences and annotations for several noncoding RNAs, including microRNAs, snRNAs, and lncRNAs.
Comparative RNA database	http://www.rna.ccbb.utexas.edu/	This database maintains information about structural and evolutionary perspectives of RNAs, obtained through comparative analysis of RNA sequences.
European rRNA database		The related sequences (complete or partial) of small and large sub-units of ribosomal RNAs (rRNAs) are aligned and displayed in this database, along with secondary structure information.
miRNA Database	http://www.mirbase.org/	The miRBase is one of the most popular microRNA databases and archives the published miRNA sequences, position of each mature-miRNA in the respective pre-miRNA sequences and annotations. The nomenclature of the miRNAs is determined according to some set tenets.
Genome databases		
Genomes online database (GOLD)		Maintains information about the genome and metagenome sequencing projects operated around the world, plus the associated metadata.
A quick guide to sequenced genomes	http://www.genomenewsnetwork.org/resources/sequenced_genomes/genome_guide_p1.shtml	Describes the sequenced organisms, links to the published abstracts and provides the URL for (hyperlinks to) the sequencing centers/institutes.

Site name	Universal Resource Locator (URL)	Description
Completed genomes: Eukaryotes	http://www.bioinfbook.org/chapt16.htm	The web resources for completed eukaryotic genomes.
KEGG	http://www.genome.jp/kegg/	The Kyoto Encyclopedia of Genes and Genomes (KEGG) is a comprehensive collection of the database to assemble pertinent systems biology information viz. pathways (maps the cellular/organismal functions), complete genomes, chemical substances, drugs and diseases.
Metagenomics		
MEGAN4-MEtaGenome Analyzer	http://ab.inf.uni-tuebingen.de/software/megan/	A standalone tool for metagenomic analyses of short-read data.
MG-RAST	http://metagenomics.anl.gov/	An automated analysis platform for metagenomes. Use the Firefox web browser to use this server. The results quantitatively report the microbial populations from the analysis of the metagenomic data.
Terragenome	http://www.terragenome.org/	An international soil metagenome sequencing consortium.
R and PERL programming resources		
The Comprehensive R Archive Network	http://cran.r-project.org/bin/windows/base/	To download R for the specific operating system.
R and Data Mining	http://www.rdatamining.com/	This website is dedicated to R programming, and gives lots of examples of R code usage.
Bioconductor login page	https://stat.ethz.ch/mailman/options/bioconductor	An open source software project to enable development, sharing (codes and packages) of R packages for analysis of genomic data.
R Function Index	http://www.math.montana.edu/rweb/rhelp/00index.html	A list of R-hyperlinked function names. Each of the functions has been briefly discussed for usage, along with a description and example.
R Tutorial	http://heather.cs.ucdavis.edu/~matloff/r.old.html	Tutorial for R programming, package usage, etc.
Comprehensive Perl Archive Network	http://www.cpan.org/	A hub of Perl modules, Perl ports and source.
Perl-Meme	http://perlmeme.org/start_hereindex.html	The user will find standard Perl-codes, examples and Perl-meme on this page.
Perl-Learning site		This site contains all relevant information about Perl programming language, including books, basic aspect, module, etc.
List of Perl Functions	http://perldoc.perl.org/5.12.4/index-functions.html	Categorizes Perl functions either alphabetically or categorywise.

Site name	Universal Resource Locator (URL)	Description
NGS data analysis related		
FASTX Tool Kit	http://hannonlab.cshl.edu/fastx_toolkit/	A collection of command line tools for Short-Reads FASTA/FASTQ files preprocessing.
The Genome Analysis Toolkit (GATK)	http://bioops.info/2011/05/gatk-the-genome-analysis-toolkit/	A toolkit (a set of bioinformatics tools) that enables next-generation sequence data analysis, data quality checking, variant discovery, etc. The server is very fast in executing the codes to analyze the NGS data of genomes from a variety of organisms.
Genome-wide Complex Trait Analysis (GCTA)	http://www.complextraitgenomics.com/software/gcta/	A powerful tool to estimate various breeding-related parameters using genome-wide SNP data, such as inbreeding coefficient, chromosome-specific genetic variance, genetic analysis of complex traits to find out the proportion of phenotypic variance explained by genome- or chromosome-wide SNPs.
Burrows–Wheeler Algorithm Download	http://sourceforge.net/projects/bio-bwa/files/	The NGS-derived sequence-reads (short and long reads, separately) are aligned to the reference genome using BWA.
SAM Tools	http://samtools.sourceforge.net/	Alignment of SAM-formatted reads to reference sequence can be manipulated, including sorting, merging, indexing, etc.
Genome2Seq	http://agbase.msstate.edu/cgi-bin/tools/genome2seq.cgi	Using the genome coordinates of transcripts from the RNA-seq data, the transcript sequences are retrieved in a FASTA file.
Primer designing		
FastPCR	http://www.biocentr.helsinki.fi/bi/programs/fastpcr.html	Used for designing PCR primers or probe, oligonucleotide assembly and for repeat searching. This program can be downloaded and run in PCs.
Primer3 (version 0.4.0)	http://frodo.wi.mit.edu/	Freely available online software for designing primers and probe from a DNA sequence. A very popular software package, due to the availability of several parameters to design primers with high specificity and accuracy.
OligoAnalyzer 3.1	http://eu.idtdna.com/analyzer/applications/oligoanalyzer/	This online tool is provided by IDT for analyzing the properties of the oligos, as well as for predicting the likelihood of self- and heterodimer formation by oligos.
IDT Antisense Design	http://www.idtdna.com/scitools/applications/antisense/antisense.aspx	To synthesize antisense oligos for a specific target sequence of interest.
Oligonucleotide Properties Calculator	http://www.basic.northwestern.edu/biotools/oligocalc.html	A very useful oligonucleotide properties calculator. It displays the reverse complementary sequence, physical properties (length, molecular weight, GC%), T_m, thermodynamic constants, and hairpin and self-dimer production by a given primer/sequence.

Site name	Universal Resource Locator (URL)	Description
UnaFold	http://www.idtdna.com/scitools/applications/unafold/	The likelihood of secondary structure formation by the single-stranded target is checked by this software from IDT (freely available online).
Restriction digestion		
Restriction-Mapper	http://www.restrictionmapper.org/	Online, freely available tool for mapping restriction endonuclease sites on a DNA sequence.
Webcutter 2.0	http://rna.lundberg.gu.se/cutter2/	Another RE site detection program (online, free) for linear and circular DNA.
NEB Cutter	http://tools.neb.com/nebcutter2/index	An RE site mapper, hosted by New England Biolabs.
Sequence alignment		
Dotlet	http://myhits.isb-sib.ch/util/dotlet/doc/dotlet_about.html	Free online software used as a tool for diagonal plotting of sequences.
Dotplot(+)	http://www.hku.hk/bruhk/gcgdoc/dotplot.html	Used to identify the overlapping portions of two sequences and to identify the repeats and inverted repeats in a sequence.
Dotter	http://sonnhammer.sbc.su.se/dotter.html	A graphical dotplot program for detailed comparison of two sequences. It runs on MAC, Linux, Sun Solaris and Windows OS.
Clustal Omega	http://www.ebi.ac.uk/tools/msa/clustalo/	The latest form of the Clustal alignment program, it is online and command-line based. The distinguishing feature of Clustal Omega is its scalability, as several thousands of medium- to large-sized sequences can be aligned simultaneously. It will also make use of multiple processors where present. In addition, the quality of alignments is superior to the previous versions. The algorithm uses seeded guide trees and HMM profile-profile progressive alignments.
ClustalW	http://www.ebi.ac.uk/tools/clustalw2/index.html	A very popular site for pairwise and multiple sequence alignment. It runs in Windows, Linux/Unix and Mac operating systems.
ClustalX	http://bips.u-strasbg.fr/en/documentation/clustalx/	The latest version (v.2.0) is provided by "Plate-Forme Bio-Informatique de Strasbourge", along with detailed instructions (help) for operating ClustalX. This site also provides online tools (Actin Related Proteins Annotation server, EMBOSS, Gene Ontology Annotation, SAGE experiment parameters, GPAT, etc.), databases (SRS, BAliBase, InPACT) and documentation (tutorials to elucidate the parameters of Clustal, GCG, EMBOSS, Bioinformatics protocols, etc.).
LALIGN	http://www.ch.embnet.org/software/lalign_form.html	Online free tool for finding local alignment between two sequences (provided in stipulated input format, i.e., plain text without header line, Swiss-Prot ID, TrEMBL ID, EMBL ID, EST ID, etc.).

Site name	Universal Resource Locator (URL)	Description
FASTA	http://fasta.bioch.virginia.edu/fasta_www2/fasta_list2.shtml	This server is hosted by the University of Virginia, USA. It is a repository of online software for sequence (nucleic acid and amino acid) comparison, local and global alignment, Hydropathy plotting and protein secondary structure prediction.
MAFFT version 6	http://align.bmr.kyushu-u.ac.jp/mafft/software/	Another useful tool to perform MSA (online or offline) with precise scope to alter or modify the alignment parameters. The other facilities are Jalview depiction of whole alignment, Construction of NJ Tree and downloading the Newick file (*.NWK).
T-Coffee	http://www.es.embnet.org/services/molbio/t-coffee/	Tree-based Consistency Objective Function For alignment Evaluation (T-Coffee) is another popular multiple sequence alignment program, developed by Cedric Notredame, CRG Centro de Regulacio Genomica (Barcelona). It allows combining of results obtained from several alignment methods. The URL is http://www.ebi.ac.uk/Tools/msa/tcoffee/. The default output format is Clustal, but it also accepts sequences in PIR and FASTA format.
Online books		
Online Biology Book	http://www.estrellamountain.edu/faculty/farabee/biobk/biobooktoc.html	An online free book of biology covering the basic topics, including plant and anima cell, molecular genetics, muscular, reproduction, essential systems of animals, biological diversity, human evolution, etc.
MendelWeb	http://www.mendelweb.org/	A website containing resources for genetics studies. The site contains Mendel's papers, secondary sources of Mendel's paper, essays and commentary, etc.
The Biology Project	http://www.biology.arizona.edu/	Useful for learning the basic aspects of genetics and participating in discussions among teachers and the taught.
Molecular Biology Web Book	http://www.web-books.com/mobio/	A Web-book that covers molecular biology topics, including cell biology, structural and functional genetics, biotechnology and bioinformatics.
NCBI Bookshelf	http://www.ncbi.nlm.nih.gov/books	Online book repository (life science books) maintained by NCBI.
Bioinformatics and Functional Genomics	http://www.bioinfbook.org/index.html	A very useful site that "features a complete bioinformatics teaching curriculum: PowerPoints for an entire course taught at the Johns Hopkins School of Medicine, and 1100 website links organized by chapter in the new textbook, Bioinformatics and Functional Genomics (John Wiley, 2003)".

Site name	Universal Resource Locator (URL)	Description
Tutorials		
Notes on Population Genetics	http://darwin.eeb.uconn.edu/eeb348/lecture-notes/notes.html	Visit this item to get notes on population genetics.
Genetics Education Centre	http://www.kumc.edu/gec/	This website is hosted by University of Kansas Medical Center. It provides a link to Human Genome Project, Genetic Resources (books, videos, curricula), Lesson Plans, Networking, Genetic Conditions, Careers, Glossaries, etc.
Complete PCR Solution	http://www.pcrlinks.com/	A web guide to PCR, with several links to PCR-related topics, books and variants in PCR.
Biology Related Internet Sites	http://lib.berkeley.edu/bios/selected_sites.html	A link to selected listed biology-related sites.
Protocols		
The Electronic Protocol Book	http://www.changbioscience.com/protocols/	An online protocol link for molecular biology and bioinformatics works.
Protocol Online	http://www.protocol-online.org/	Protocols for molecular biology works.
Protocols in Cytogenetics	http://www.biologia.uniba.it/rmc/0-1a_pagina/2_2_protocols.html	The website contains protocols for cytogenetics and molecular genetics.

For more details, the user is requested to visit http://www.bioinformaticssoftwareandtools.co.in/

Appendix C: NCBI Database: A Brief Account

CS Mukhopadhyay and RK Choudhary
School of Animal Biotechnology, GADVASU, Ludhiana

The information on each of the databases listed below has been collected from NCBI. In several cases, the description of the databases will be verbatim to that available in the NCBI pages. The information regarding these databases has been taken from NCBI-guide (https://www.ncbi.nlm.nih.gov/guide/) and related sites.

SN	Databases	Features
1	Assembly	This database maintains and periodically updates organism-wise information on assembled genomes (WGS) or complete chromosome sequence of prokaryotic and eukaryotic organisms.
2	Bio project	This database holds data and information related to a single project or a consortium. It enables users to obtain voluminous data belonging to a project, in one place. The type of records maintained in Bioprojects are Genome sequencing and assembly; Metagenomes; Genetic or RH maps; Targeted locus sequencing; Epigenetics; Phenotype or Genotype and Variation detection, Transcriptome sequencing and expression.
3	Biosystems	A data repository of information (list, sequence, structure) regarding biological molecules (genes, proteins, small molecules) and pathways involved in biological systems. This includes data from BioCyc (including its Tier 1 EcoCyc and MetaCyc databases and its tier 2 databases), KEGG, Reactome, the National Cancer Institute's Pathway Interaction Database, WikiPathways and Gene Ontology (GO).
4	Bookshelf	A database of freely accessible electronic books and documents in life science and healthcare. It integrates NCBI resources such as PubMed, Gene, OMIM and Pubchem.
5	ClinVar	ClinVar is the public repository of sequence variation, and information about its relationship to human health. ClinVar maintains records on various medical conditions due to genetic aberration(s) collected from a number of distinct sources, including SNOMEDCT, MeSH & OMIM, etc.

Basic Applied Bioinformatics, First Edition. Chandra Sekhar Mukhopadhyay, Ratan Kumar Choudhary and Mir Asif Iquebal.

SN	Databases	Features
6	Clone DB	A public database that maintains information (sequence data, map positions, and distributor information) for clones associated with genomics, cDNA and cell-based libraries belonging to different eukaryotic organisms.
7	Biosample	A central repository of biological resources (including tissues, cell lines, experimental organisms) used in different assays.
8	Computational resources from NCBI's structure group	It maintains access and links to resources (databases and tools) developed by the division of Biocomputational Structure Group of NCBI that determines macromolecular structures and identifies conserved domains. This resource also maintains tools for classification of protein, for determining small molecular biological activity and pathways analysis, etc.
9	Consensus CDS (CCDS)	A consensual collaboration among NCBI, EBI, University of California at Santa Cruz (UCSC) and Wellcome Trust Sanger Institute (WTSI) to identify and annotate a core set of protein-coding regions.
10	Conserved Domains Database (CDD)	CDD, a protein annotation resource, holds models of well-annotated multiple sequence alignment about primal domains, as well as the complete peptides.
11	Database of Expressed Sequence Tags (dbEST)	This is the EST database that contains short single-read transcript sequences obtained from GenBank.
12	Database of Genome Survey Sequences (dbGSS)	This NCBI database contains comprehensively annotated short, single-pass reads obtained for genomic sequences (which could be cDNA or non-coding DNA) obtained from sources such as random survey sequences, clone-end sequences, artificial chromosomes (BAC/YAC) or cosmids and exon- and gene-trapped sequences.
13	Database of Genomic Structural variation (dbVar)	Maintains information regarding large-scale genomic variation, namely sizeable InDels, translocations and inversions with regard to the association of these variations with phenotypes.
14	Database of Genotypes and Phenotypes (dbGaP)	This database archives and distributes the results of studies on the interaction of genotype and phenotype. The information pertains to molecular diagnostics, genome-wide association studies (GWAS) concerning the association of genotype with non-clinical traits. The GaP database also offers cloud computing services.
15	Database of Major Histocompatibility Complex (dbMHC)	Information on gene and related clinical data associated with Major Histocompatibility Complex (MHC) of human are maintained here. The tool dbMHCms searches for the portrayal for reported short tandem repeats (STRs) belonging to MHC. It has a "Reagent Database" section (reagent data needed to trace DNA typing) and a "Clinical" section (maintains clinical data from anonymous individuals sharing their clinical data in the project).
16	Database of Short Genetic Variations (dbSNP)	A public database for obtaining information regarding genetic variation within and across different species. SNP data obtained from several experiments, starting from physical mapping and association studies, pharmacogenomics to evolutionary studies can be submitted to dbSNP.
17	Epigenomics	This database holds epigenomic data on a biological sample, and also serves as a tool (as genome browser) for selecting, downloading and viewing multiple sets of epigenomic data.

SN	Databases	Features
18	GenBank	A public repository of annotated DNA sequences. The International Nucleotide Sequence Database Collaboration maintains the collaborative liaison among the DNA data of NCBI, EMBL and DDBJ. The FTP is updated every two months.
19	Gene	This database integrates information on nomenclature, variations and reference sequences (RefSeqs), gene-maps, molecular-pathways regarding phenomes. This information is linked to genome-, phenotype-, and locus-specific resources, with regard to highly divergent species. "Gene" can be accessed by querying on any word, restricting the query term to a certain field, or applying filters or properties.
20	Gene Expression Omnibus (GEO) Database	A public repository of experimental data generated from microarray experiment and high-throughput genomic data like next generation sequencing (NGS).
21	Gene Expression Omnibus (GEO) Datasets	Stores compiled gene expression DataSets, and original series, samples and platform records in the Gene Expression Omnibus (GEO) repository. The differential expression pattern is collated and displayed along with clustered heatmaps for easy comprehension.
22	Gene Expression Omnibus (GEO) Profiles	Maintains the curated gene expression profiles belonging to the Gene Expression Omnibus (GEO) archive.
23	GeneReviews	This database, being a part of the GeneTests website, archives peer-reviewed descriptions (diagnosis, counseling, etc.) of inherited diseases.
24	GeneTests	The repository is a knowledge base of diagnosis of the management of inherited diseases and genetic testing.
25	Genes and Disease	This database contains the articles related to genetic diseases and the causative genes.
26	Genetic Testing Registry (GTR)	This acts as a repository of information on genetic tests, including premises, promises, methodology, validity, utility, challenges, etc. associated with the testing of inherited diseases which are submitted by the test providers voluntarily.
27	Genome	This database archives the sequences and related map data from the whole genomes of different organisms (bacteria, archaea, and eukaryota), including the genomes of completely sequenced organisms and not yet complete ones.
28	Genome Reference Consortium (GRC)	This international consortium includes the eminent research institutes working on unraveling the genomic information in terms of genome mapping, association studies, genome-informatics, etc. with an aim to improve the human and mouse genome reference assemblies.
29	HIV-1, Human Protein Interaction Database	This database harbors links to PubMed records on interactions between HIV-protein and human-protein vis-a-vis to relevant sequences.
30	HomoloGene	A tool to identify the possible orthologs by comparing the homologous nucleotide sequences from different species.
31	Influenza Virus	Holds the data from the National Institute of Allergy and Infectious Diseases (NIAID), Influenza Genome Sequencing Project and GenBank, and maintains the NCBI Influenza Virus Sequence Database. Another important use of this database is the analysis of flu sequences, which are then submitted to GenBank following annotation.

SN	Databases	Features
32	Journals in NCBI Databases	A subset of the NLM Catalog database that maintains information on journals cataloged in PubMed and other NCBI database records.
33	Medical Subject Headings (MeSH) db	A comprehensive catalog of medical vocabulary used for indexing journal papers and books in the life sciences. The database is used to search for MeSH terminologies, get their definition and pertinent information and strategy building for PubMed search.
34	NCBI C++ Toolkit Manual	A public domain library containing system-independent (mostly) useful libraries, development framework, demos, release notes, etc.
35	NCBI Glossary	Contains definitions/portrayal of the tools available at NCBI, explanation of bioinformatic terms and acronyms, etc.
36	NCBI Handbook	Includes exhaustive explanatory notes on NCBI databases and software, which can be accessed through NCBI Bookshelf.
37	NCBI Help Manual	A collection of Help documents (downloadable) on tools like BLAST, Entrez (search engine), GenBank (databank), PubMed and NLM, etc.
38	NCBI Website Search	A search tool provided by NCBI to search documents, newsletters, sample codes and other resources at NCBI.
39	National Library of Medicine (NLM) Catalog	An electronic library catalog that enables searching the bibliographic data for around 1.5 million journals, books, software, audiovisuals-documents, etc. at National Library of Medicine, the largest online library of medical science.
40	Nucleotide Database	This maintains a vast repository of nucleotide sequences (gene/transcript/genome data) obtained from sources like GenBank, RefSeq, TPA and PDB.
41	Online Mendelian Inheritance in Animals (OMIA)	Textual information and references related to inherited disorders and associated genes in about 200 animal species are cataloged in this database. However, human and mice are not covered. The genetic disorders are linked to genes, and relevant literature (Pubmed) is also linked.
42	Online Mendelian Inheritance in Man (OMIM)	This database was developed to supply comprehensive information and reference on Mendelian disorders in a human being. The related genes, the relationship between genotype and disease phenotype are also detailed here. Each entry is linked to multiple genetic databases (gene and protein sequences), literature, genetic tests, mutation databases, etc.
43	PopSet	A repository of DNA sequences obtained from the members of a population (composed of individuals from different species or multiple species) to study their evolutionary relationship. One can submit DNA sequences to PopSet via Sequin of NCBI.
44	Probe	A public database for maintaining detailed information on reagents used in nucleic acid experiments (RNAi, microarray, genotyping, gene expression, etc.) conducted for a vast array of biomedical research. This helps researchers from different parts of the globe to assess information about useful biochemicals, molecular probes, distributors, etc.
45	Protein Clusters	The protclustdb (protein cluster database) maintains the clusters of RefSeq proteins from a variety of sources, including prokaryotic genome and plasmid, viruses, organelles, protozoa, and plants. The database consists of uncurated and manually curated cluster data, and is updated every three months. Cross-references to related external links (NCBI-COG, KEGG, InterPro, etc.) are provided for proteins and protein clusters.

SN	Databases	Features
46	Protein Database	*In silico* translated amino acid sequences from annotated coding sequences obtained from NCBI RefSeq, GenBank, etc., along with records from external sources of protein sequences, including SwissProt, PDB, PIR, etc. are maintained by this database. The GenPept sequence provides cross-references to cds (if applicable), PubMed, etc.
47	PubChemBioAssay	PubChemBioAssay is one of the three components of NCBI PubChem (a search tool to determine chemical similarity). The PubChemBioAssay is a link to the PubChem compounds that elaborates their bioactivity, including describing the bioassays, screening conditions, etc.
48	PubChem Compound	This database depicts the structure of the validated substances of the PubChem substance page of NCBI. This page maintains pre-clustered compounds based on similarity and links to related databases and information (structure information, references).
49	PubChem Substance	Describes the contents of PubChem (structure, cross-references, etc.) and provides links to biological screening results.
50	PubMed	This is one of the most popular databases and repositories of NCBI-NLM. It maintains biomedical books, as well as a wide range (including bioengineering and chemical sciences) of literature from different sources, including biological journals and MEDLINE. Each record is given a unique PMID.
51	PubMed Central (PMC)	Freely available biomedical literature are maintained by PMC.
52	PubMed Health	This is an archive of clinical reviews with an aim to cater to the clinicians and end users, so that they have access to research works directed towards biomedical and clinical issues.
53	RefSeqGene	A subset of the RefSeq database, where the reference genomic sequences pertaining to human genes are maintained. The curations obtained from locus-specific data, as well as information available from the genetic testing community, are included.
53	Reference Sequence (RefSeq)	This curated, non-redundant database maintains naturally occurring nucleotide (DNA, RNA) and protein sequences from a large number of species regarding linked records, from genomes to transcripts and translation products.
54	Retrovirus Resources	A public resource of research works on retroviruses, this provides certain online tools (genotyping tool using BLAST algorithm; alignment tool for global alignment; annotated maps, etc.).
55	SARS-CoV	Data (regarding sequence, genome sequence alignments of various isolates) and information (publication) on the SARS coronavirus are maintained in this database.
56	Sequence Read Archive (SRA)	This database archives short sequences (<1000 bases) produced from high-throughput sequencing, from massive parallel sequencing platforms, including Roche, Illumina, ABI SOLiD System, etc.
57	Structure (Molecular Modeling Database)	Macromolecular structures (from PDB) and visualization tools are available here. The Molecular Modeling DataBase (MMDB) or Entrez Structure DataBase (ESDB) stores the experimentally determined 3D structures of biomolecules.

SN	Databases	Features
58	Taxonomy	Holds standard nomenclature and scientific classifications of taxa from prokaryotic and eukaryotic origin. The species names are manually compiled for each of the organisms linked to the entries of INSDC (International Nucleotide Sequence Database Collaboration: GenBank + EMBL + DDBJ).
59	Third Party Annotation Database	The TPA database aims at maintaining and providing experimental (peer-annotated from evidence of wet-lab experiment) or inferential (not from direct wet-lab experimentation) results. It derives the TPA-sequence from already-available GenBank sequence data, and also annotates the sequences.
60	Trace Archives	This public repository has three sections: Sequence read archive: to store NGS data from a variety of NGS platforms; Trace Archive: sequencing data from gel or capillary sequencer; Trace assembly archive: assembles the reads of sequencing by pairwise or multiple sequence alignment.
61	UniGene	A repository of transcriptome sequencing reads obtained from expressed genes or pseudogenes. Each entry links to all the encoded transcripts from the same locus, and provides information about gene expression and genomic location, complementary DNA, and protein similarity.
62	UniGene Library browser	A database that enables users to browse the expressed sequence tags with respect to the organisms, tissue type, and stages of biological development.
63	UniSTS	Experimentally derived sequence tagged sites (STS) are archived in this comprehensive database.
64	Viral Genomes	Curated virus genome sequences are maintained in this database.
65	Virus variation	An organized collection of viral genome sequences with an aim to extend facilities for easy search, retrieval, display, and analysis of virus genomes. It provides pipelines for analysis of viral genomes to assist discovery using the available sequence data.

Appendix D: EMBL Databases and Tools: An Overview

S Jain[1], S Panwar[2] and A Kumar[3]
[1] Department of Applied Sciences & Humanities, Jai Parkash Mukand Lal Innovative Engineering and Technology Institute, Haryana, India
[2] Department of Genetics and Plant Breeding, Chaudhary Charan Singh University, Uttar Pradesh, India
[3] Department of Nutrition Biology, Central University of Haryana, Haryana, India

INTRODUCTION

The European Bioinformatics Institute (EBI) is a constituent body of EMBL and is situated at the Wellcome Trust Genome Campus, Cambridge (UK). It provides all sorts of molecular data, as well as bioinformatics databases, software and tools, at no cost. It has all kinds of life sciences information, and helps in basic and advanced research. The information in the databases and tools described in this chapter is extracted from the EMBL-guide and related sites. Therefore, in several instances, the information given may be verbatim.

THE EMBL DATABASES

Information on each of the databases has been collected from EMBL. The databases available via dbfetch are listed in Table 1. An overview of each database is also provided, which includes a short description and link to the databases.

THE EMBL TOOLS

This is the access and analysis point for numerous data resources through Web Services technologies (Li *et al.*, 2015; Lopez *et al.*, 2014). The program basically works on integration and inter-operation technology and has been created from Representational state transfer (REST), Simple Object Access Protocol (SOAP) and Web Services Description Language (WSDL).

The details and description of EMBL services are given in Table 2.

Basic Applied Bioinformatics, First Edition. Chandra Sekhar Mukhopadhyay, Ratan Kumar Choudhary and Mir Asif Iquebal.

TABLE 1 Features and links of various EMBL databases.

S.N.	Databases	Features	Links
1.	EDAM	EMBRACE Data and Methods (EDAM) Ontology.	http://edamontology.sourceforge.net/
2.	ENA Coding	European Nucleotide Archive (ENA) Coding is a database of nucleotide sequences of the CDS (coding sequence) features, as annotated in the ENA Sequence database. ENA Coding records contain the nucleotide sequence of the CDS, along with annotated parent nucleotide, in addition to spontaneously produced annotation.	http://www.ebi.ac.uk/ena/
3.	ENA Geospatial	A database of nucleotide sequences of the ENA Geospatial Sequence.	http://www.ebi.ac.uk/ena/
4.	ENA Non-coding	A database of nucleotide sequences of the non-coding RNA features, as annotated in the ENA Sequence database. ENA Non-coding records contain the nucleotide sequence of the RNA feature, along with annotated parent nucleotide, in addition to spontaneously produced annotation.	http://www.ebi.ac.uk/ena/
5.	ENA Sequence	ENA Sequence (formerly known as EMBL-Bank) is Europe's primary nucleotide sequence resource. The main sources of the DNA and RNA sequences in the database are submissions from individual researchers, genome sequencing projects, and patent applications.	http://www.ebi.ac.uk/ena/
6.	ENA Sequence Constructed	The ENA Sequence Constructed database division represents complete genomes and other long sequences constructed from segment entries. Instead of containing the sequence, these entries detail how to assemble the sequence from other ENA Sequence entries.	http://www.ebi.ac.uk/ena/
7.	ENA Sequence Constructed Expanded	Expanded entries include the complete nucleotide sequence of the constructed entry.	http://www.ebi.ac.uk/ena/
8.	ENA/SVA	The ENA Sequence Version Archive (SVA) is a repository of all entries which have ever appeared in the EMBL Nucleotide Sequence Databank (EMBL-Bank) or ENA Sequence databases.	http://www.ebi.ac.uk/cgi-bin/sva/sva.pl
9.	Ensembl Gene	Ensembl genome databases for vertebrate species and model organisms. For other species, see below.	http://www.ensembl.org/
10.	Ensembl Genomes Gene	Genome databases for metazoa, plants, fungi, protists and bacteria.	http://www.ensemblgenomes.org/

TABLE 1 (Continued)

S.N.	Databases	Features	Links
11.	Ensembl Genomes Transcript	Genome databases for metazoa, plants, fungi, protists and bacteria.	http://www.ensemblgenomes.org/
12.	Ensembl Transcript	Ensembl genome databases for vertebrate species and model organisms. For other species, see Ensembl Genomes instead.	http://www.ensembl.org/
13.	European Patent Office (EPO) Proteins	Patented Protein present in the European Patent Office.	http://www.ebi.ac.uk/patentdata/proteins/
14.	HGNC	HUGO Gene Nomenclature Committee (HGNC) approved gene name and symbol (short-form abbreviation) for each human gene.	http://genenames.org/
15.	IMGT/HLA	The International ImMunoGeneTics (IMGT) database provides a specialist database for the sequences of the human major histocompatibility complex (HLA), including the official sequences for the WHO Nomenclature Committee For Factors of the HLA System.	http://www.ebi.ac.uk/imgt/hla/
16.	IMGT/LIGM-DB	A comprehensive database of immunoglobulins and T cell receptors (LIGM) from human and other vertebrates.	http://imgt.cines.fr/cgi-bin/IMGTlect.jv
17.	InterPro	The InterPro database (Integrated Resource of Protein Domains and Functional Sites) is an integrated documentation resource for protein families, domains, and functional sites. It was originally used to rationalize the complementary efforts of the PROSITE, PRINTS, Pfam and ProDom database projects, but now it also includes the SMART, TIGRFAMs, PIR SuperFamilies and most recently SUPERFAMILY databases.	http://www.ebi.ac.uk/interpro/
18.	IPD-KIR	A centralized repository for human Killer-cell Immunoglobulin-like Receptor (KIR) sequences.	http://www.ebi.ac.uk/ipd/kir/
19.	IPD-MHC	Sequences of the major histocompatibility complex (MHC) in a number of species.	http://www.ebi.ac.uk/ipd/mhc/
20.	IPRMC	InterPro Matches Complete (IPRMC) for UniProtKB proteins.	http://www.ebi.ac.uk/interpro/
21.	IPRMC UniParc	InterPro Matches Complete (IPRMC) for UniParc proteins.	http://www.ebi.ac.uk/interpro/
22.	JPO Proteins	Protein sequences are appearing in patents from the Japanese Patent Office (JPO).	http://www.ebi.ac.uk/patentdata/proteins/

(Continued)

TABLE 1 (Continued)

S.N.	Databases	Features	Links
23.	KIPO Proteins	Patented Protein present in the Korean Intellectual Property Office (KIPO).	http://www.ebi.ac.uk/patentdata/proteins/
24.	MEDLINE	Comprises citations and abstracts records of more than 5000 medically related journals published in the United States and 70 other countries. The files contain over 19 million citations, dating back to the mid-1940s, and are updated weekly.	http://www.nlm.nih.gov/pubs/factsheets/medline.html
25.	Patent DNA NRL1	Non-redundant patent nucleotides level 1 (NRL-1). Nucleotide sequences from patents clustered by 100% sequence identity over the whole length.	http://www.ebi.ac.uk/patentdata/nr/
26.	Patent DNA NRL2	Non-redundant patent nucleotides level 2 (NRL-2). Nucleotide sequences from patents clustered by patent family, and then by 100% sequence identity over the whole length.	http://www.ebi.ac.uk/patentdata/nr/
27.	Patent Protein NRL1	Non-redundant patent proteins level 1. Protein sequences from patents clustered by 100% sequence identity over the whole length.	http://www.ebi.ac.uk/patentdata/nr/
28.	Patent Protein NRL2	Non-redundant patent proteins level 2. Protein sequences from patents clustered by patent family and then by 100% sequence identity over the whole length.	http://www.ebi.ac.uk/patentdata/nr/
29.	Patent Equivalents	Patent number equivalents (families) and patent classifications for patents containing sequence data. The patent equivalents are obtained from the patent numbers cited in the major sequence databases (e.g., EMBL-Bank and Patent Proteins), which are then expanded into a set of patent equivalents forming a WIPO Simple Patent Family.	http://www.ebi.ac.uk/patentdata/
30.	PDB	Comprises structure and sequence information of proteins and nucleotides.	http://www.ebi.ac.uk/pdbe/
31.	Reference Sequence project (RefSeq)	All sorts of information on reference sequences of natural molecules.	http://www.ncbi.nlm.nih.gov/refseq/
32.	RefSeq (protein)	All sorts of information on reference sequences of natural molecules.	http://www.ncbi.nlm.nih.gov/refseq/
33.	SGT	Structural Genomics Targets (SGT) is a protein target registration database, providing information on the experimental progress and status of target amino acid sequences selected for structural determination.	http://targetdb.pdb.org/
34.	Taxonomy	Taxonomic classification of organisms for which there are sequences in the INSDC databases (i.e., DDBJ, EMBL-Bank, and GenBank) and many other biological databases.	http://www.ncbi.nlm.nih.gov/Taxonomy/

TABLE 1 (Continued)

S.N.	Databases	Features	Links
35.	Trace Archive	An archive of capillary electrophoresis trace data.	http://www.ebi.ac.uk/ena/
36.	UniParc	Protein sequences retrieval system.	http://www.uniprot.org/
37.	UniProtKB	Curated protein information retrieval system.	http://www.uniprot.org/
38.	The UniProt Reference Clusters UniRef100/ UniRef90/ UniRef50	Access point for combined resemble sequences. In UniRef100, UniRef90 and UniRef50, no sequence mutual pair identity exceeds > 100%, > 90% or > 50%.	http://www.uniprot.org/
39.	UniProtKB Sequence/ Annotation Version Archive (UniSave)	Access point for UniProtKB/Swiss-Prot and UniProtKB/TrEMBL admitted versions.	http://www.ebi.ac.uk/uniprot/unisave/
40.	United States Patent and Trademark Office (USPTO) Proteins	Patented Protein present in the USPTO.	http://www.ebi.ac.uk/patentdata/proteins/

TABLE 2 Description of various EMBL tools.

General Services
Including data retrieval, access various sequence, and structural databases

S.N.	Service	Description
1.	ArrayExpress	Microarray data searching with ArrayExpress.
2.	ChEBI Web Services	Entry retrieval from the ChEBI database.
3.	ChEMBL Web Services	Retrieval data system.
4.	EB-eye (SOAP)/(REST)	EBI search engine (EB-eye).
5.	ENA Browser	Access point for sequence retrieval .
6.	Gene Expression Atlas API	Access point for statistics data over a curated subset of ArrayExpress Archive.
7.	MartService	Searching and retrieving the data through BioMart.
8.	PDBe (REST)	Helps in gathering facts from PDB and EMDB.
9.	PSICQUIC	Information retrieval system for molecular interaction, comprising ChEMBL, Reactome, and IntAct.
10.	Rhea	Access point for manually annotated chemical reactions information.
11.	Universal Protein Resource UniProt.org	Protein sequence information including annotated.
12.	WSDbfetch (REST)/(SOAP)	Identifier entry retrieval system.

(*Continued*)

TABLE 2 (Continued)

Protein Functional Analysis (PFA)
Identifying protein-related information, i.e., sequences, motifs, conserved regions, etc.

	REST/SOAP Service	Description
13.	FingerPRINTScan	Recognizing the proximal matching fingerprints motif.
14.	InterProScan 5	This tool is used for bringing different protein signature recognition methods into one platform or page.
15.	HMMER hmmscan	Access point for Hidden Markov Models (HMMs) database.
16.	PfamScan	PfamScan is used to explore the similar sequences for a query FASTA sequence against a library of Pfam HMM.
17.	Phobius	Prediction of transmembrane topology and signal peptides from the amino acid sequences of protein.
18.	Pratt	Identifying conserved patterns in unaligned protein sequences.
19.	PROSITE Scan	Comparing a protein sequence against the signatures in PROSITE (both patterns and profiles).
20.	RADAR	Repeat identification and alignment system in protein sequences.

Sequence Similarity Search (SSS)
Provides the identification of homologous sequences.

	REST/SOAP Service	Description
21.	FASTA	Fast protein or nucleotide comparison access tool.
22.	FASTM	Peptide fragment access point from FASTA.
23.	NCBI BLAST	Nucleotide and protein sequence comparison system.
24.	PSI-BLAST	Position Specific Iterative BLAST (PSI-BLAST), guided mode
25.	PSI-Search	Iterative Smith and Waterman using a PSI-BLAST strategy

Multiple Sequence Alignment (MSA)
Alignment of a set of three or more, protein or nucleotide sequences.

	REST/SOAP Service	Description
26.	Clustal Omega	Sequence alignments tool.
27.	ClustalW2	Global multiple sequence alignment of DNA and protein sequences using ClustalW2.
28.	DbClustal	Global multiple sequence alignment of DNA or protein sequences using anchor regions from BLAST results
29.	Kalign	Sequence alignment system of large sequences.
30.	MAFFT	Sequence alignment using the MAFFT method. Fast, and capable of handling large sequences.
31.	Multiple Sequence Comparison by Log-Expectation (MUSCLE)	Sequence alignment tool.

TABLE 2 (Continued)

32.	MView	Reformat a multiple sequence alignment or create a multiple sequence alignment from a sequence similarity search result (e.g., BLAST or FASTA).
33.	PRANK	Sequence alignment using the PRANK method.
34.	T-Coffee	Sequence alignment using the T-Coffee method.

Phylogeny
Phylogenetic analysis

	REST/SOAP Service	Description
35.	ClustalW2 Phylogeny	Neighbor-joining or UPGMA phylogenetic trees access system.

Pairwise Sequence Alignment (PSA)
Alignment of two sequences

	REST/SOAP Service	Description
36.	EMBOSS matcher	Waterman–Eggert local alignment using EMBOSS matcher.
37.	EMBOSS needle	Needleman–Wunsch global alignment using EMBOSS needle.
38.	EMBOSS stretcher	Myers and Miller global alignment using EMBOSS stretcher.
39.	EMBOSS water	Smith–Waterman local alignment using EMBOSS water.
40.	GeneWise	Provides comparison of protein and genomic DNA sequence.
41.	lalign	Huang and Miller sim local alignment using lalign
42.	PromoterWise	Comparison of two DNA sequences, allowing for inversions and translocations.
43.	Wise2DBA	The Wise2 DNA Block Aligner (DBA) aligns two DNA sequences.

RNA
RNA Analysis

	REST/SOAP Service	Description
44.	Infernal cmscan	Searching system for CM-format Rfam database.
45.	MapMi	Accessing mapping and analysis of miRNA sequences.

Sequence Format Conversion
Convert between homologous sequences or confirm the formatting of a sequence.

	REST/SOAP Service	Description
46.	EMBOSS seqret	Accessing manipulated sequence entries.
47.	MView	Reformatting of multiple sequence alignment data.
48.	Readseq	Convert biosequences between a selection of common biological sequence formats.

(*Continued*)

TABLE 2 (Continued)

Sequence Statistics
Analyze a sequence to determine its properties and use statistics to assign significance.

	REST/SOAP Service	Description
49.	EMBOSS cpgplot	European Molecular Biology Open Software Suite (EMBOSS) cpgplot identifies and plots CpG islands in a nucleotide sequence.
50.	EMBOSS isochore	Plots isochores in DNA sequences.
51.	EMBOSS pepinfo	Plots amino acid properties.
52.	EMBOSS pepstats	Provides calculation of protein properties.
53.	EMBOSS pepwindow	Generates a hydropathy plot for protein.
54.	SAPS	Statistical Analysis of Protein Sequences.

Sequence Translation
Translate a coding nucleotide sequence into a protein sequence and vice versa.

	REST/SOAP Service	Description
55.	EMBOSS transeq	Translates the nucleiceotide sequences.
56.	EMBOSS sixpack	Displays DNA sequences with six-frame translation and ORFs.
57.	EMBOSS backtranseq	Back-translates the protein sequences.
58.	EMBOSS backtranambig	Back-translates protein sequences to ambiguous nucleotide sequences.

Structural Analysis
Analysis of macromolecular structures.

	REST/SOAP Service	Description
59.	DaliLite	Pairwise structure comparison.
60.	MaxSprout	Provides fast database algorithm for making protein backbone and side chain.

Literature and Ontologies
Look-up ontology terms and navigate ontology relationships.

	Service	Description
61.	BioModels	Access point for mathematical models of biological interest.
62.	PICR	Protein Identifier Cross-Reference Service.
63.	QuickGO	Gene Ontology (GO) and Gene Ontology Annotation (GOA) databases.
64.	Europe PMC Web Service	Provides searching access from Europe PubMed Central.
65.	WSMIRIAM	Web Services for the Minimal Information Requested In the Annotation of biochemical Models (MIRIAM).
66.	WSOntology Lookup	Search multiple ontologies from a single location.
67.	WSSBO	Web Services for the Systems Biology Ontology (SBO).
68.	WSWhatizit	permits text mining tasks.

Appendix E: Basics of Molecular Phylogeny

CS Mukhopadhyay and RK Choudhary
School of Animal Biotechnology, GADVASU, Ludhiana

Phylogenetic analysis indicates the splits and diversions of species within ancestral lines, leading to a procreation of a clade. The term "clade" means a cluster of two or more species related by a common ancestor. The principle of phylogeny is relatedness among various organisms, due to descending from a nearer or remote common ancestor (CA). Thus, phylogeny is the relationship among different organisms due to sharing of a recent common ancestor (Zimmermann, 1931). It is a method by which to obtain an idea of the evolution and origin of an organism. The term "phylogeny" originates from two Greek words: Phylon (Stem) and Genesis (Origin).

GEOLOGICAL CLOCK

This is based on the regularity of the decay process of radioactive elements. Suppose an ancient rock which has been lying undisturbed is tested, using a mass spectrophotometer, for the amount of radioactive uranium (^{235}U) and normal lead (^{207}Pb). The former is decayed into the latter, with a half-life of 710 million years (MY) (Guttman, 2007). The wider the ratio of uranium to lead, the older the rock is. Thus, the approximate time of fossilization of an individual can be estimated by geological study, and this forms the geological clock. It is thought-provoking to note that the first fossil evidence for many of the animal phyla is available from the rocks preserved since the Cambrian Period of the Paleozoic era (510–540 MY) (Benton 1993; Graham 1993).

Geological studies have revealed some geological events that are closely related to the evolution of plants and animals. The birds and mammals first appeared during the Jurassic period of the Mesozoic era (208 million years ago (Mya)), which was the time of the dinosaurs. The supercontinent Pangea (whole land areas of the earth lying together) first disintegrated into Gondwanaland (which included India, Australia, Africa, etc.) and Laurasia (North America and Greenland) during the Mesozoic era (i.e., 160–170 Mya). The first primates had appeared on the earth by the Paleocene epoch of the Tertiary period of the Cenozoic era (≈66.4 Mya). The earliest hominids date back to the

Basic Applied Bioinformatics, First Edition. Chandra Sekhar Mukhopadhyay, Ratan Kumar Choudhary and Mir Asif Iquebal.

Pliocene Epoch (5.3 Mya) (Guttman, 2007). Thus, the genealogical clock reflects on the evolutionary perspective of the earth and the origins of different species on it.

MORPHOLOGICAL PHYLOGENY TO MOLECULAR PHYLOGENY

Early phylogenetic studies (prior to the 1960s) were based on morphological (morphos (Gr.): form, logos (Gr.): study) similarity and dissimilarities only. Fossil records and anatomical measurements are the prime sources of data for determining ancestral lineages. However, the morphology-based approach has some inherent limitations, such as the fact that several morphological traits seems to be convergent and seem to overlap with each other. For example, different species of chickadee (*Poecile atricapillus*), a small North American songbird, have several apparently indistinguishable characters that can bewilder a skilled birder.

Morphological features are more qualitative than quantitative where the underlying inheritance pattern is not well established (http://www.life.umd.edu/classroom/bsci338m/Lectures/Systematics.html), and the limited availability of morphological data and fossil record makes it further challenging. No consistent results with genealogy or family pedigree can be obtained using morphological data but, nevertheless, the phenotypes of microbes hold little promise in depicting the evolutionary relationship among microbes, using morphology as a means.

Now the other side of the morphological systematic is the confounding resemblance between unrelated species, which could be due to convergent evolution (i.e., independent evolution in a similar environment, such as sharks and dolphins, or African euphorbias (*Euphorbia* spp.) and American cactus) (Ghosh and Mallick, 2008).

Adaptation to different ecological niche could also bring about strikingly different morphology among closely related species. The Hawaiian islands, an archipelago of eight major islands in the North Pacific Ocean, were formed about 0.5 to 0.8 Mya and became detached from the mainland. Hawaiian honeycreepers, which have descended from a common ancestor, exhibit different beak shapes due to their adaptation to varying ecological niches.

The limitations of morphology-based phylogeny have now been replaced by molecular phylogeny, which uses molecular data (DNA/RNA/amino acid sequences, enzymatic data, etc.) for constructing the phylogenetic tree. Frederick Sanger first did the sequencing of bovine insulin in 1953. Later, the RNA sequencing technique (Min-Jou *et al.*, 1972) and then DNA sequencing, using mainly Sanger's method (Sanger *et al.*, 1977), became available, enabling scientists to make use of these sequences in reconstructing molecular phylogeny. FHC Crick suggested (in 1958) using the molecular sequences for phylogenetic tree reconstruction. However, Emile Zuckerkandl and Linus Pauling used aligned amino acid sequence data to build the first ever phylogenetic tree in 1962 (Morgan *et al.*, 1998) and proposed the theory of the molecular clock (Morgan, 1998). The theory of molecular evolution then gained momentum. In 1967, Walter Fitch and Emanuel Margoliash designed the first algorithm (applying least squares) for phylogenetic tree reconstruction using protein sequences (Fitch and Margoliash, 1967; Fitch, 1970, 1971).

BASIS OF MOLECULAR PHYLOGENY

The phenotype (expression of a trait) of an individual is the result of its genotype (allelic combination(s) of a locus or multiple loci), modification of the genotypic effect by the environment in which it is raised, and the interaction between genotype and environment. The DNA sequence of the coding region of a gene is, to a great extent, the determinant of its phenotypic uniqueness. Genetic relationship among the close relatives confers similarity among them and discriminating uniqueness from unrelated individuals.

Traits can show homology as synapomorphies or as symplesiomorphies. Synapomorphies are the homologies that are derived from a common ancestor – in other words, ancestral homologies which are first observed in the ancestor of the clade. Thus, synapomorphies define a clade. On the other hand, symplesiomorphies are shared ancestral characters which have already arisen before the common ancestor of the clade. They are also passed on to the downstream taxa through the common ancestor. The phylogenetic tree is constructed on the basis of the evidence obtained from the synapomorphies only (http://biology.unm.edu/ccouncil/Biology_203/Summaries/Phylogeny.htm).

Shared characteristics among related individuals (having a common ancestor(s)) are the cornerstone of the theory of evolutionary phylogeny. The evolutionary process is depicted by the tree of life (TOL), where each species occupies a distinct position on the branch. The phylogeny represents the evolutionary process through the paths descending from the common ancestor(s) (CA), through the intermediate nodes to the ultimate terminal node or leaf, where the species/gene occupies its position. This path is known as the lineage. Molecular phylogeny, thus, uses the sequence data of DNA, RNA or protein. The accuracy of results depends on the types of input sequences (DNA, RNA, amino acid) and the divergence among the taxa incorporated in the study (Ghosh and Mallick, 2008):

- Amino acid sequences are applied efficiently for most remote homologies.
- DNA sequences are very sensitive, non-uniform rates of mutation.
- Coding DNA sequence (cds) are used to determine purifying selection in coding region.
- RNA sequences are useful for remote homologies.
- 16s rRNA: considered as the most suitable phylogenetic marker.

The primary mechanism of molecular evolution is nucleotide substitution during the process of DNA replication. Different types of mutations (gross or point mutation) contribute to different types of germ-line mutations that alter the phenotype. Among the point mutations, InDels (Insertions, Deletions) are frequently encountered. The types of point mutation vis-à-vis corresponding changes in the translated amino acid are shown in Figure E1. Besides, transposition, i.e. movement of the entire gene or non-coding regions, exon shuffling, i.e. duplication of exons, exchange of structural or functional domains between protein-coding genes (in multiple exons), transitions and transversions are also the underlying mechanisms.

Apart from mutations, natural selection of individual, genetic drift in a small population, bottleneck effects and so on play a significant role in the process of speciation.

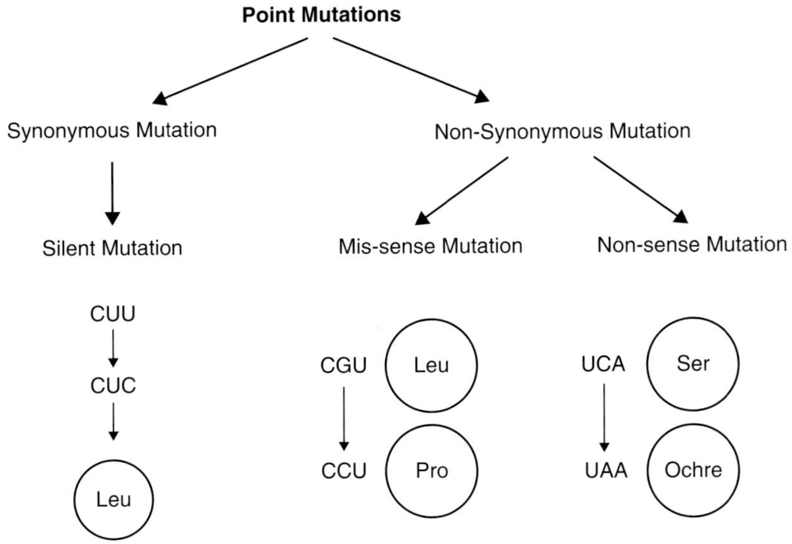

FIGURE E1 Different types of point mutations leading to codon change.

MUTATION RATE

This measures the tempo or pace of mutations occurring during one unit of time. The mutation rate varies with the type of gene, or the type of organism whose genome is being studied. It can be measured in terms of mutations per base pair per cell division, or per gene (or per genome) generation. The molecular clock studies a region with predictable mutation rate, to calculate the time of divergence of two species, in geologic history. The estimated mutation rates of different types of organisms are as follows:

- Unicellular eukaryotes and bacteria: ~0.003 mutations per genome per generation.
- DNA viruses: 10^{-6} to 10^{-8} mutations per base per generation.
- RNA viruses: 10^{-3} to 10^{-5} per base per generation.
- Human mitochondrial DNA: $\sim 3\times10^{-5}$ to $\sim 2.7\times10^{-5}$ per base per 20-year generation.
- Human genomic mutation: $\sim 2.5\times10^{-8}$ per base per generation.
- Human genome (WGS data): $\sim 1.1\times10^{-8}$ per site per generation.

COMPONENTS OF A PHYLOGENETIC TREE

A phylogenetic tree is a tree-like structure. However, this can be rooted or unrooted. Various terms used to specify the components of a tree are given below:

- **Terminals/leaves**: the species or the genes that have been sampled.
- **Internal nodes**: ancestral state reconstruction for the characters being studied.
- **Branches**: the relationships between the nodes. These can also represent the relative divergence among the terminal and nodes.

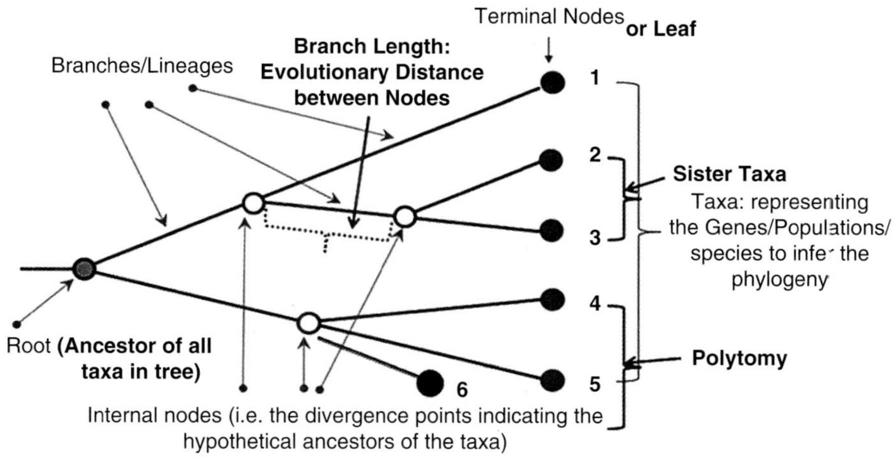

FIGURE E2 The components of a rooted phylogenetic tree.

- **Horizontal branch length** determines the time between speciation events according to the mutation rate or the mutation among the lineages, depending on the tree topology (Figure E2). Branch length is proportional to the evolutionary distance between the nodes (internal as well as external nodes), expressed as substitution or residue per site.
- **Distance scale**: A scale that assesses the distances between different nodes, expressed in terms of a number of differences. It is expressed in a range between 0 and 1, which can be inferred as differences for 0 to 100% of the residues.
- **Operational Taxonomic Unit (OTU)**: refers to the hierarchical groups comprising external/terminal and internal nodes. The element of OTUs is a group of either genes or species that are sufficiently distinguishable from others (Figure E2).
- **Analogs** refer to the traits which look similar as a result of convergent evolution, not due to inheritance from a common ancestor.
- **Taxa** is a general term applied to a taxonomic group (i.e., families, genera or species, etc.). The most closely related taxa are called "sister taxa" in a phylogenetic tree.

There are some terminologies which are frequently used in phylogeny:

A **clade** starts with a node (ancestor), and includes all taxa descending from the ancestor.

A monophyletic group is a good example of a clade. It is a group that includes a common ancestor along with all the descendants of that ancestor, but excluding all non-descendants.

A paraphyletic group is a group of taxa which contains a common ancestor and some (but not all) of the descendants of that ancestor.

The polyphyletic group includes multiple taxa, but not the common ancestors (Figure E3).

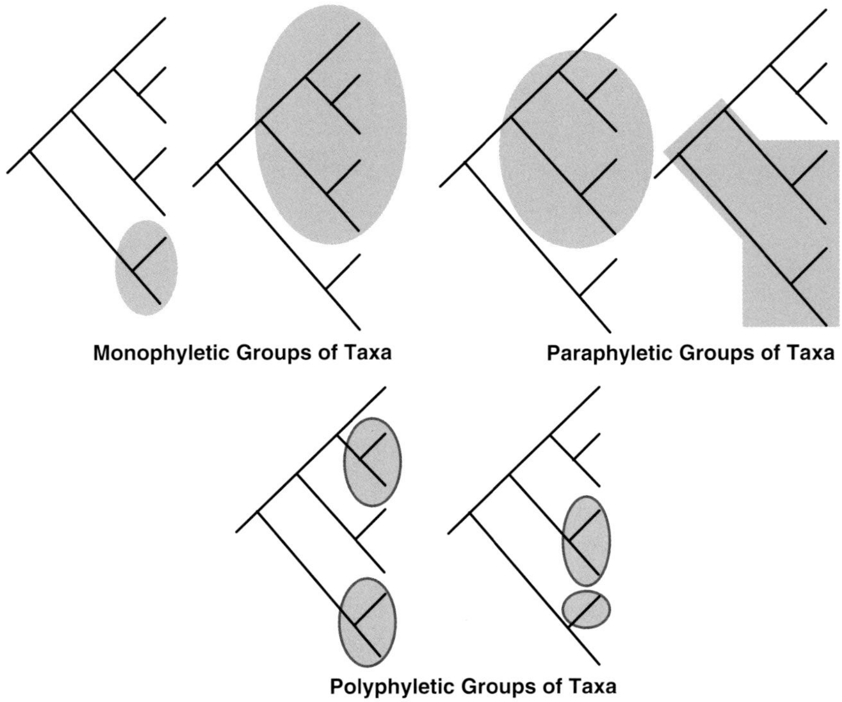

FIGURE E3 Diagrammatic representation of monophyletic, paraphyletic and polyphyletic groups of taxa.

TYPES OF PHYLOGENETIC TREES

Unrooted tree

This illustrates the relatedness of the OTUs without making any assumptions about ancestry. No ancestor is determined in this type of tree. Unrooted trees show the differences between the taxa (regarding distance or proportion of residue change). However, no time frame can be deduced from the orientation of taxa in an unrooted tree.

Rooted tree

This is a directed tree, characterized by the finally converged node signifying the most recent common ancestor of all the entities. In other words, the tree topology shows a common ancestor to all the involved taxa. The lineages/branches sprouting from the common ancestor determine the evolutionary path (and its direction). A rooted tree can be generated by introducing an outgroup as the root. The outgroup comprises one or more distantly related taxa, known to share a distant common ancestor. A rooted tree is also generated using a molecular clock where the evolutionary process is assumed to happen at a constant rate along the branches of a tree. The topology is rooted at a point where it splits the amount of character evolution in half.

Converting an unrooted tree into a rooted tree

The inclusion of an outgroup: "Outgroup" refers to the lineage (or taxon) in a phylogenetic analysis that is the least related to the rest of the taxa in the analysis. Thus, it branches off at the base of that phylogeny. An outgroup is remote with respect to the clade being studied, since the members of the clade exhibit closer relatedness to each other than to the outgroup (http://evolution.berkeley.edu/evolibrary/glossary/glossary_popup.php?word=outgroup). The outgroup taxon (or taxa) is known to be external to the group being analyzed. Thus, the root lies at the branch joining the outgroup to the original clade (i.e., the ingroup).

- **Choosing an outgroup**: the underlying assumption is that the inclusion of an outgroup does not alter (or influence) the relationship of the taxa of the original clade.
  - The inclusion of an incorrect outgroup may result in long branch attraction (LBA), a phenomenon which occurs if the distantly related clades cluster together, due to erroneous inferences drawn from shared homoplasies. This has been discussed in http://self.gutenberg.org/articles/Long_branch_attraction as "It is a result of the way clustering algorithms work: terminals or taxa with many autapomorphies (character states unique to a single branch) may by chance (convergence) exhibit the same states as those on another branch". As a result, rapidly evolving taxa may be interpreted as closely related. The alternative approach is to include a group of taxa as an outgroup, instead of a single OTU.
  - The chance of LBA can be minimized either by changing the phylogeny model or by splitting the long branches with more taxa. Fast-evolving taxa can also be removed from the set of the ingroup taxa.
  - It is better not to root an unrooted tree, if an erroneous inference is drawn due to the inclusion of an outgroup.
- **Using a molecular clock**: The assumption behind using a molecular clock is a similar rate of evolution for all the lineages since splitting from the common ancestor. The most distant taxa are selected based on the branch lengths. The tree root is selected at the mid-point between the two farthest taxa, so that the source is equidistant from all the external nodes.

Appendix F: Evolutionary Models of Molecular Phylogeny

CS Mukhopadhyay and RK Choudhary
School of Animal Biotechnology, GADVASU, Ludhiana

INTRODUCTION

Studies of molecular evolution and construction of phylogenetic trees are based on mathematical models that underpin the process of nucleotide substitution causing the process of speciation. A number of models have been hypothesized, based on certain assumptions of evolution. Some of the important models are described in this chapter.

Jukes–Cantor Model (1969) or JC69

This is the simplest DNA substitution model, which assumes equal base frequencies and a constant rate of evolution with a base substitution rate (Figure F1) of "α". The transition (Purine to Purine and Pyrimidine to Pyrimidine) frequency is assumed to be same as the transversion (Purine to Pyrimidine and vice versa) frequency; hence, the model has a single parameter "α". A substitution matrix (Figure F2) is generated for all possible base substitutions, assuming a fixed rate of changeover.

The substitution matrix is generated from a constant rate of base substitution per unit time, based on the JC69 model.

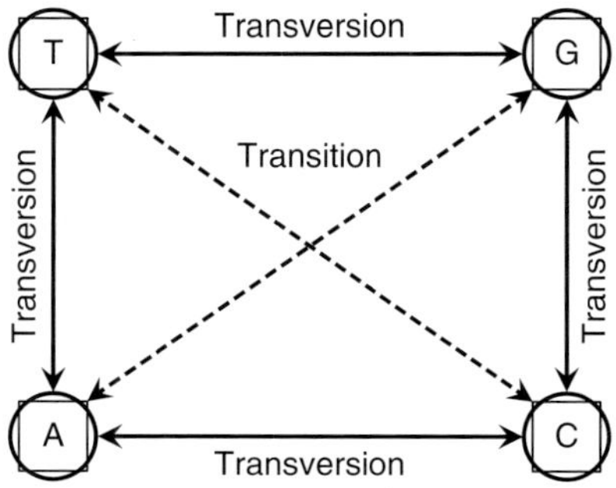

FIGURE F1 Substitution of nucleotides leading to transition and transversion.

Basic Applied Bioinformatics, First Edition. Chandra Sekhar Mukhopadhyay, Ratan Kumar Choudhary and Mir Asif Iquebal.

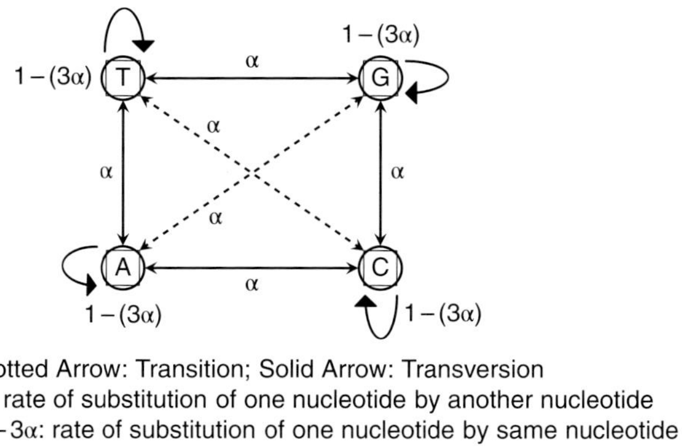

FIGURE F2 Jukes–Cantor one-parameter substitution model (Jukes and Cantor, 1969).

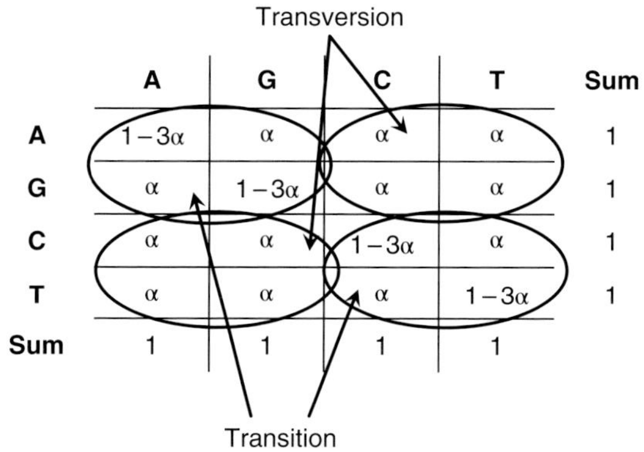

FIGURE F3 Rates of transition and transversion are the same (α).

	A	G	C	T	
A	$1-3\alpha\zeta$	$\alpha\zeta$	$\alpha\zeta$	$\alpha\zeta$	1
G	$\alpha\zeta$	$1-3\alpha\zeta$	$\alpha\zeta$	$\alpha\zeta$	1
C	$\alpha\zeta$	α	$1-3\alpha\zeta$	$\alpha\zeta$	1
T	$\alpha\zeta$	$\alpha\zeta$	$\alpha\zeta$	$1-3\alpha\zeta$	1
	1	1	1	1	

FIGURE F4 Amount of base-substitution in a period of time "ζ" and equal rates of transition and transversion (α).

The substitution matrix generated from a constant rate of base substitution for a specific time period (ζ), based on the JC69 model, is shown in Figure F4.

The amount of base substitution can be calculated, considering a stipulated lapse of time (ζ) from initiation of the evolutionary event in the study.

Mathematically, the amount of base substitution can be reduced to:

Branch length (for substitution of a base by the same base itself):=

$$\Theta(\zeta) = \frac{\left(1+3e^{(-4\alpha\zeta)}\right)}{4}$$

Branch length (for substitution of a base by any base other than itself):

$$\varphi(\zeta) = \frac{\left(1-3e^{(-4\alpha\zeta)}\right)}{4}$$

Estimate of evolutionary distance $(d) = (-3/4 \ln(1-4/3p))$ where p is the p-distance that estimates the proportion of sites that differ between two sequences under study.

The JC69 model is applicable under simple conditions, but is not at all suitable for a complex evolutionary model.

Kimura's Two-Parameter Model (1980) or K80 model

The K80 model is an extension of JC69. The model assumes different rates of transitions (α) and transversions (β).

Let:

α = probability of substitution due to a transition at time "t"
β = probability of substitution due to a transversion at time "t"

The substitution matrix at unit time ($t = 1$) will be as follows:
The rate of substitution can be approximated to the following:

- Branch length for transversion (other than itself) $= \psi(\zeta) = \frac{\left(1-e^{(-4\alpha\zeta)}\right)}{4}$
- Branch length for transition (other than itself) $= \varphi(\zeta) = \frac{\left(1+e^{(-4\alpha\zeta)} - e^{-2(\alpha+\beta)\zeta}\right)}{4}$
- Branch length for substitution of a base by the same base itself $= \Theta(\zeta) = 1-2\psi(\zeta) - \varphi(\zeta)$.

The distance obtained from the Kimura 2 parameter model is:

$$(d) = \left(-\frac{1}{2}\ln(1-2p-q) - \frac{1}{4}\ln(1-2q)\right)$$

	A	G	C	T	Sum
A	$1-\alpha-2\beta$	α	β	β	1
G	α	$1-\alpha-2\beta$	β	β	1
C	β	β	$1-\alpha-2\beta$	α	1
T	β	β	α	$1-\alpha-2\beta$	1
Sum	1	1	1	1	

FIGURE F5 K80 model: amount of base-substitution in unit time period ($t = 1$), assuming different rates of transition (α) and transversion (β).

	A	G	C	T	Sum
A	$1-3\rho_A$	ρ_A	ρ_A	ρ_A	1
G	ρ_G	$1-3\rho_G$	ρ_G	ρ_G	1
C	ρ_C	ρ_C	$1-3\rho_C$	ρ_C	1
T	ρ_T	ρ_T	ρ_T	$1-3\rho_T$	1
Sum	1	1	1	1	

FIGURE F6 F81 model: rate of base-substitution is different for four bases: adenine (ρ_A), guanine (ρ_G), cytosine (ρ_C) and thymine (ρ_T).

where:

p = proportion of sites undergoing transition;
q = proportion of sites undergoing transversion

Felenstein model (1981) or F81 model

The JC69 model assumes an equal rate of substitution among all nucleotide residues, which does not hold good for several practical and complex DNA evolution models. The F81 model modifies this assumption with unequal substitution frequencies for different bases. Thus, the substitution rate matrix has four different proportions for the four common bases:

$$\text{The expected number of substitutions per site } (\beta) = \frac{1}{\left(1-\rho_A^2-\rho_G^2-\rho_C^2-\rho_T^2\right)}.$$

The other models that have been developed to study DNA evolution are Hasegawa *et al.* (1985), Tamura (1992), Tamura and Nei (1993), the generalized time reversible model, etc. The scope of this chapter is too limited to discuss the models in length and breadth.

Glossary

Term	Meaning
ab initio	A Latin term that means starting "from the beginning" or initiation.
Accession number	The unique number assigned to an accepted submission (e.g., molecular sequence, genome project data, WGS, STs, etc.) by the database (NCBI, DDBJ, EMBL) to differentiate the submission from another similar type. The accession number is alphanumeric, and the format differs among molecular sequences (nucleotide and protein) as well as the type of database (NCBI, Swiss-Prot-UniProt, etc.)
Adapter	Priming site created by ligation of short oligonucleotide to the DNA which is to be sequenced or amplified
Algorithm	A set of rules set to complete an assignment or operation by a computer (in general). The term is derived from the name of Iraqi mathematician Mohammed ibn Musa al-Khwarizmi (9th century AD).
Allosteric Protein	A protein having multiple ligand binding sites, whose conformation changes upon ligand binding. The enzyme can be an allosteric protein.
Amplicon	Gene-specific nucleotides sequences amplified by PCR.
Annealing Temperature (T_m)	The temperature at which 50% of the DNA helices are dissociated during PCR amplification.
Annotation	Comments on or explanation of a text or data.
Barcode	Short sequences of typically six or more nucleotides that are used to identify/label individual samples when they are pooled in one sample.
Binary Tree	Tree-like data structures with two (binary) branches. The point from where each branch separates is called a node.
Binding site	A region of protein or DNA where the ligands bind.
Bioinformatics	A branch of science that interprets biological data with the help of statistics, computer science, mathematics and engineering.
Biostatistics	The application of statistics in biological science.

Basic Applied Bioinformatics, First Edition. Chandra Sekhar Mukhopadhyay, Ratan Kumar Choudhary and Mir Asif Iquebal.

Term	Meaning
Bit score (S′)	The similarity between two sequences by alignments, expressed by bit scores (denoted by "S"). The higher the scores are, the better the alignment is. It is calculated from the formula that considers conserved sequence, identical sequence and gaps therein.
BLASTn	Standard Nucleotide BLAST. Here, two nucleotide sequences are compared. The word BLAST (Basic Local Alignment Search Tool) is online software to compare query sequences from an online database.
BLASTp	The term BLASTp stands for protein BLAST. Here, two amino acid sequences are aligned and compared.
BLASTx	BLASTx aligns six conceptually translated DNA sequences from both the stands with a database of protein sequences.
Bridge amplification	Amplification of fragments attached on a chip by the adapter at both of its ends.
Burrows–Wheeler transform	An aligner that helps in reading large volumes of short-read data that have not been fully studied
Clustering	In gene expression analysis, a microarray cluster is the grouping together of genes of similar functions. In phylogenetic tree analysis, the data points having smaller or larger distances are connected and form different clusters. The distance matrix is calculated based on some algorithm, and there are more than 100 algorithms published. Hierarchical clustering is one of the common examples of connectivity-based clustering methods.
CpG Islands	Word CpG stands for Cytosine-phosphate diester-Guanine. CpG is an area of increased density C and P in the DNA (100–1000 bp long) at various places. CpG areas are usually non-methylated and present near 5′-end of gene at transcription initiation sites. In humans, there are around 45 000 CpG islands in the DNA. CpG sites are important, as they are involved in regulation of gene transcription.
C_t value	The cycle number in real time PCR when the fluorescent signal is above the threshold limit and can be detected by a machine. It is also called the Cp value.
de novo Assembly	Sequencing of genetic materials if the reference sequence is not available.
Deep sequencing	Repeated time sequencing of genetic material, measured in terms of coverage.
Delta BLAST	Domain Enhanced Lookup Time Accelerated (Delta) BLAST is a new algorithm to yield better homology of remote protein sequences. It searches a database of the pre-constructed position-specific scoring matrix (PSSM) before searching a protein sequence database. The web link for the paper describing Delta-BLAST for the first time is: http://www.ncbi.nlm.nih.gov/pmc/articles/PMC3438057/.
Delta C_t value	The difference between the two threshold cycles (C_t) of two genes (say, target and control genes, or target and reference genes).
Docking	Docking refers to a method that predicts the orientation of one molecular by binding to another molecular while making a complex. In bioinformatics, docking is a computational simulation of a ligand binding to its receptor.
Domain	A conserved part of a protein whose tertiary structure changes independently from that of the rest of the protein.
Dynamic Programming	A method of solving a complex problem by breaking it down into many sub-problems. DP in bioinformatics has been used in the sequence alignment, DNA-protein binding prediction, and protein structure prediction.

Term	Meaning
Edge Length	In the phylogenetic tree, an edge length is a number associated with an edge which represents either time or expected genetic distance from the other branches.
Energy Functions	
E-Value (Expectation value)	A way of representing the significance of the alignment. It is a probability of this alignment occurring with a particular bit score (S) or better in the database search. The lower the value, the better is the chance of getting this alignment.
Expressed sequence tag (EST)	A short sequence of cloned cDNA that is used to identify gene transcripts in gene discovery.
FASTA	FASTA is the first algorithm for searching database similarity in sequences. It is a text-based format for representing nucleotide or amino acid sequences. The sequence in FASTA format begins with ">" (greater than sign), followed by sequence description and sequence.
Fastq file	Result of primary analysis representing individual reads with quality indicators for each base of corresponding sequences
Functional annotation	Use of the analyzed output data of genomic and transcriptomic projects to describe gene/protein functions and interactions.
Gap-penalty	During sequence alignment, to compensate insertion or deletion of query sequences, gaps in the sequences are introduced. Introduction or extension of gap is penalized in the scoring of an alignment of nucleotide or protein sequences and is called the gap penalty.
GC-clamp	The presence of Guanine (Gs) and Cytosine (Cs) nucleotides at the 3′-end of the primer. More than three Cs should be avoided. GC-clamps help in the specific binding of primer with the DNA template.
Gene Identity Number (gi)	Sometimes written as "gi", this number is simply a series of digits assigned to each sequence of NCBI. It has been discontinued from September 2016.
Gene ontology	The bioinformatics process is to annotate, assimilate and disseminate information of gene and gene product across all species through a common platform.
Genetic Code	These are the triplets of three nucleotides that code for amino acids. Those triplets that do not code for any amino acid (UGA, UAG, and UAA) are called stop codons and, therefore, halt translation.
Genomic Survey Sequences (GSS)	Genome survey sequences are the short genomic DNA sequences from coding, non-coding and repetitive portions of genomic DNA that aid in rapid characterization of the unknown genome.
Genomics	Study of the whole DNA of an organism, e.g., genes, their structure, and organization, location in the chromosome, etc.
Gibb's free energy	The Gibbs free energy of a system at any time is defined as the enthalpy (H) of the system – the product of the entropy (S) of the system multiplied by the temperature (T), i.e., G = H – ST.
Global Alignment	The Needleman–Wunsch based algorithm dynamic programming methods of aligning two or more nucleotide sequences that are similar in nature.
Guide tree	This is constructed during multiple sequence alignment from the pair-wise distance scores. It is different from the phylogenetic tree that is constructed at the end of the MSA.

Term	Meaning
Hairpin loop (turn)	A hairpin loop is formed by single-stranded DNA or mRNA when a portion of strand folds up and pairs with another section of the same strand. In designing primers (short oligonucleotides) for PCR, the formation of the hairpin loop at the 3′ end is avoided because if affects PCR efficiency.
Heuristic program	A method of problem-solving that often involves experimentation on the basis of trial and error. Likewise, a heuristic program is an algorithm that produces an acceptable solution without formal proof of its correctness.
Hidden Markov Model (HMM)	HMM is used to present the probability distributions over the sequences of observations. It is a Markov model with a hidden (unobserved) state, where the state is not directly visible but the output is visible.
High-throughput genomic sequences (HTGS)	The division to accommodate rapidly growing unfinished genomic sequence databases of DDBJ, EMBL, and GenBank, where sequences are available for BLAST homology. When sequences are at the finished level (phase 3: finished with no gaps either with or without annotations), the data are moved from HTGS to the corresponding taxonomic division.
High-scoring segment pair (HSP)	HSP is the basic unit of BLAST algorithm output. It consists of two sequence fragments whose alignment is locally maximal, and for which the alignment score meets or exceeds a threshold or cut-off score.
Homology	Homology is the shared ancestry between a pair of the genes in different species.
InDel	An abbreviation of "insertion and deletion" of genes in mutation.
InDels	One or more Insertion or Deletion event detected in sequences of genetic materials.
Internal Node	The intermediate node between root node and leaf node in a phylogenetic tree.
International Nucleotide Sequence Database Collaboration (INSDC)	A long-standing foundational collaboration between DDBJ, EMBL, and NCBI in data raw reads, their alignment, assemblies and functional annotations, with related information on samples and experiments associated with the data.
Iteration	The process of repeating a process many times unless the desired results are achieved.
Leaf	In a phylogenetic tree, a leaf usually represents a single present-day taxon that is typically a DNA sequence whose genetic distance is measured with other taxa.
Library	This refers to a collection or pool of DNA or cDNA of an entire organism. A collection of the entire genome (exon and introns) is called a genomic DNA library, and a collection of all complementary DNA is called the cDNA library.
MegaBLAST	Alignment of larger DNA sequences that differ slightly as a result of sequencing. MegaBLAST is similar to BLASTn but able to efficiently handle longer DNA sequences.
Microarray	It is a set of DNA sequences representing the entire set of genes of an organism that are arranged (arrayed) in a grid pattern for use in gene expression analysis (cDNA microarray) or genetic testing (DNA microarray). A typical microarray experiment involves hybridization of mRNA molecules (called targets) to the DNA template (called probes) from which it is originated.

Term	Meaning
Mispriming	When primers of PCR anneal to non-specific sites, leading to the background or non-specific amplification, this is called mispriming.
Monte Carlo Simulation	A computerized mathematical technique to analyze risk assessment in quantitative analysis. It provides all possible outcomes of decisions and risk assessment, allowing scientists to make a better decision.
Motif	Motifs are the structural characteristics of a protein that are associated with a particular arrangement of amino acids. When such arrangements of amino acids are associated with a function like DNA binding or catalytic activity, then it is called a domain.
Multiple alignments	A computational method that lines up, as a set of three or more sequences in row, to identify overlapping positions with maximum accuracy and minimum mismatches and gaps.
Next-generation sequencing	High-throughput sequencing to sequence DNA and RNA much more quickly and cheaply than the previously used Sanger sequencing, by producing thousands or millions of sequences at once
Node	A node in a phylogeny represents the common ancestor of a set of taxa, from which different taxa are descended.
omics	The word "omics" is informally related to the field of biology such as genomics, proteomics or metabolomics, where the suffix -omics refers to the field of study of the genome, protein or metabolites, respectively.
Paired-end sequencing	The sequence of the DNA is obtained from the 5′ ends of both strands of the insert.
Palindrome	A sequence of the word (or nucleotide) that reads the same backwards or forwards. For example, in the word RACECAR, the arrangement of the word is the same forwards and backwards.
Phi angle	A torsion angle of right-handed rotation around the N-atom of the NH_2 group and the C-alpha atom of the Carboxyl group (N-Ca bond). The angle ranges from −180 to +180 degrees.
Phred scale	Measurement of base calling accuracy using the Phred quality score (Q score) for assessing the accuracy of a sequencing platform.
Position Hit Initiated BLAST (PHI-BLAST)	A variant of PSI-BLAST, based on the construction of Position-Specific Scoring Matrix (PSSM) around a motif of protein.
Position-Specific Iterative BLAST (PSI-BLAST)	An iterative search of the protein BLAST algorithm.
Position-Specific Scoring Matrix (PSSM)	A profile providing matching of an amino acid of a target sequence from a query sequence, estimated by log-odd scores.
Primary structure	The primary structure of a protein or polypeptide is a linear sequence of amino acids from the N-terminal to the C-terminal end.
Primer	18–25 bp of nucleotides sequences (in pairs usually) used to amplify specific genes in PCR.
Probe (microarray)	In a spotted microarray, probes refer to synthesize short oligonucleotides or DNA that is complementary to mRNA.

Term	Meaning
Prosthetic Group	"Prosthetic" means an external part that supports the functions of an organ. Similarly, a prosthetic group is a non-protein part, like vitamins or metal ions, that accelerates functions of an enzyme or protein.
Protein families	Like gene families, protein families are evolutionarily related proteins that share common features or functions.
Protein Isoelectric Point (pI)	The pH of a solution at which amino acid does not migrate in an electric field. For example, the pI of aspartic acid is 2.77, and of arginine is 10.76.
Proteomics	The entire set of proteins expressed by a genome of a cell/tissue/organism at a particular point in time.
Pseudo Count	In probability estimation of a model, an amount is added to the number of observed cases. Those priori counts, which might a subjective value, are called pseudo counts.
Psi angle	A torsion angle of right-handed rotation around the C-alpha atom of the carboxyl group and C-atom bond (Ca-C bond). The angle ranges from −180 to +180 degrees.
Query Coverage	The percentage of the query sequence that overlaps the subject sequence.
Ramachandran Plot	A diagrammatic visualization of protein structure by dihedral angles, psi (ψ) against phi (φ), against amino acid residues. It was originally developed by a team led by Ramachandran.
Raw alignment score (S)	A number used to assess the biological relevance of alignments of two sequences, where a higher score corresponds to a higher similarity of two sequences.
RCSB	Research Collaboratory for Structural Bioinformatics, founded in 1998 and responsible for maintaining protein data bank (PDB). PDB is the single worldwide repository maintaining the 3D structure of proteins and nucleic acids.
Real-time PCR	The real-time quantitative polymerase chain reaction (RT-qPCR), where the formation on amplicons can be visualized in real time on a monitor or screen. It is an advanced form of conventional PCR and utilizes a double-stranded DNA binding dye that combines with accumulated amplicon to be detected by the camera.
Reference genome	Reference assembly is a digital nucleic acid sequence database of set of genes assembled as a representative example of a species and can be retrieved using three different genome browsers.
RefSeq	"Reference Sequence" of either protein or nucleotide in a database of NCBI, derived from curation and computation of archived sequences.
Re-sequencing	Sequencing of genetic material with reference sequence available.
Restriction Enzyme	Also called "molecular scissors", used to chop DNA/plasmid sequences at specific sites in either a blunt or sticky end fashion to generate recombinant DNA.
Rn Value	An abbreviation of "normalized reporter value". The Rn value is the fluorescent signal of SYBR Green dye (DNA intercalating dye) normalized to (divided by) the signal of the passive reference dye (e.g., Rox). The delta Rn value is the Rn value of the reaction minus the Rn value of the baseline signal of the instrument.
Root	The root of a tree is the node of the phylogenetic tree that represents a common ancestor.

Term	Meaning
RSCB	A protein databank, an informative tool of predict molecular structure of proteins, genomic position and sequence alignments. The web link to the RSCB portal is: www.rscb.org/
SCOP	Standing for Structural Classification of Proteins, this is a manual classification of protein structural domains based on their amino acid sequences and structures. The SCOP database was discontinued in the year 2009, and a newer and better prototype is available, called SCOP2.
Secondary Structure	The second level of protein structure. The most common type of secondary structure in proteins is the alpha-helix. Beta-sheets are another type of secondary structure of protein.
Sequence format	The method of writing the nucleotide bases of a sequence is called the sequence format. There are various ways to write sequences, including: plain sequence format; EMBL format; FASTA format; GCG format; GenBank format; and IG format.
Sequence Similarity	Comparing sequences of either DNA, RNA or protein with each other for a degree of similarity is one of the most frequent tasks of computational biology. Two sequences showing a high degree of similarity often implies similar functions.
Sequence Tagged Sites (STS)	A 200–500 bp long DNA sequence that occurs singly (one copy) in a genome whose location and sequence are known. STS may contain repetitive sequences, but usually flanked by unique flanking regions (not present elsewhere in the genome). The microsatellite is a type of STS.
Short read	Single-End and Pair-End methods of sequencing of fragments of genetic material as per the specified read length.
SNP mining	The extraction of valuable information from single nucleotide polymorphism (SNP) data. SNP is a fast and cost-effective means of studying genetic variation.
Subtree	A part of the original tree, representing a fraction of the taxa being studied.
Taxa	The singular form of taxa is the taxon. This is a generic name for a taxonomic group, such as species. Taxon also represents genera, families, orders, phyla, and so on.
Taxonomy	Taxonomy is a branch of science that deals with the classification of new organisms and species systematically.
tBLASTn	Alignment of protein vs. translated nucleotide sequences for the identification of database sequences that encode proteins.
tBLASTx	Alignment of translated nucleotide vs. translated nucleotide sequences for identification of nucleotide sequences, based on their coding potential.
Tertiary Structure	The third level of protein structure, describing complex and irregular folding of peptide chains in three dimensions.
Third party annotation (TPA)	An annotated database derived from GenBank primary data or DDBJ/EMBL sequence databases. A TPA database could be experimental (if annotated from wet-lab experiment) or inferential (annotated by inference only).
Threading (protein sequence)	Protein threading refers to a method of protein modeling, where proteins may not be homologous but may have the same fold as a protein of known structure.
Topology	The physical layout of a gene or protein network is referred to as its topology. The three main topologies of a network are ring, bus, and star, which more likely exist as hybrid networks (combinations of ring and bus, or ring and star, or bus and star).

Term	Meaning
Torsion Angle	The angle of the geometric relation of two parts of a molecule joined by a chemical bond.
Transcriptome Shotgun Assembly (TSA)	An archived data of computationally assembled sequences derived from ESTs and next-generation sequencing.
Transcriptomics	Study of whole RNA profile (transcripts) of cells/tissue at a particular point in time (development stage, normal or diseased stage).
Tree	A phylogenetic tree, or simply "tree", is an evolutionary relationship among a set of organisms called a taxon.
Ultrametric Tree	It is a rooted tree with equal edge lengths from the root and represents an equal rate of mutation in all the lineages. It is also called a "dendrogram."
Whole Genome Shotgun Contigs	The sequence of the overlapping fragments of the whole genome
X-Ray crystallography	A tool to identify the atomic and molecular structure of a crystal by using X-rays.

References

Abajian C (1994). *Sputnik*. http://www.abajian.com/sputnik.

Agarwal V, Bell GW, Nam JW, Bartel DP (2015). Predicting effective microRNA target sites in mammalian mRNAs. *eLife digest* **4**, e05005.

Altschul SF, Gish W, Miller W, Myers EW, Lipman DJ (1990). Basic local alignment search tool. *Journal of Molecular Biology* **215**(3): 403–10.

Arquès DG, Lacan J, Michel CJ (2002). Identification of protein coding genes in genomes with statistical functions based on the circular code. *Biosystems* **66**(1–2): 73–92.

Ashburner M, Ball CA, Blake JA, Botstein D, Butler H, Cherry JM, Davis AP, Dolinski K, *et al.* (2000). Gene ontology: tool for the unification of biology. The Gene Ontology Consortium. *Nature Genetics* **25**: 25–9.

Bainbridge D (2003). *The X in Sex: How the X Chromosome Controls Our Lives*. Harvard University Press, Cambridge, MA.

Baxevanis AD, Ouellette BFF (2001). *Bioinformatics: A Practical Guide to the Analysis of Genes and Proteins*, 2nd edition. Wiley and Sons, Chichester, UK.

Benson G. (1999). Tandem repeats finder: a program to analyze DNA sequences. *Nucleic Acids Research* **27**(2): 573–580.

Benton MJ (1993). *The Fossil Record 2*. Chapman and Hall, New York.

Bindea G, Mlecnik B, Hackl H, Charoentong P, Tosolini M, Kirilovsky A, *et al.* (2009). ClueGO: a Cytoscape plug-in to decipher functionally grouped gene ontology and pathway annotation networks. *Bioinformatics* **25**: 1091–3.

Breslauer KJ, Frank R, Blöcker H, Marky LA (1986). Predicting DNA duplex stability from the base sequence. *Proceedings of the National Academy of Sciences of the United States of America* **83**(11): 3746–50.

Breslauer KJ, Frank R, Blöcker H, Marky LA (1986). Predicting DNA duplex stability from the base sequence. *Proceedings of the National Academy of Sciences of the United States of America* **83**: 3746–3750.

Basic Applied Bioinformatics, First Edition. Chandra Sekhar Mukhopadhyay, Ratan Kumar Choudhary and Mir Asif Iquebal.

Bujnicki JM (2006). *Practical Bioinformatics*, 1st edition. Springer (India) Private Limited, New Delhi.

Burge C, Karlin S (1997). Prediction of complete gene structures in human genomic DNA. *Journal of Molecular Biology* **268**(1): 78–94.

Burge CB (1998). Modeling dependencies in pre-mRNA splicing signals. In: Salsberg SL, Searls DB, Kasif S (eds). *Computational Methods in Molecular Biology*. Elsevier Science, Amsterdam.

Byrne KA, Wang YH, Lehnert SA, Harper GS. *et al.* (2005). Gene expression profiling of muscle tissue in Brahman steers during nutritional restriction. *Journal of Animal Science* **83**, 1–12.

Catchen J, Hohenlohe PA, Bassham S, Amores A, Cresko WA (2013). Stacks: an analysis tool set for population genomics. *Molecular Ecology* **22**(11): 3124–3140.

Choudhary RK, Li RW, Evock-Clover CM, Capuco AV (2013). Comparison of the transcriptomes of long-term label retaining-cells and control cells microdissected from mammary epithelium: an initial study to characterize potential stem/progenitor cells. *Frontiers in Oncology* **3**: 21.

Dai X, Zhao PX (2011). psRNATarget: a plant small RNA target analysis server. *Nucleic Acids Research* **39**(suppl 2), W155–W159.

Dayhoff MO, Schwartz R, Orcutt BC (1978). *A Model of Evolutionary Change in Proteins. Atlas of Protein Sequence and Structure*, 3rd edition. National Biomedical Research Foundation, Waltham, MA.

Dean J, Ghemawat S (2008). MapReduce: simplified data processing on large clusters. *Communications of the ACM* **51**(1): 107–113.

DePristo MA, Banks E, Poplin R, Garimella KV, Maguire JR, Hartl C, *et al.* (2011). A framework for variation discovery and genotyping using next-generation DNA sequencing data. *Nature Genetics* **43**(5): 491–498.

Desper R, Gascuel O (2005). The Minimum-Evolution Distance Based Approach to Phylogenetic Inference. In: Gascuel, O (ed). *Mathematics of Evolution and Phylogeny*. Oxford University Press, Oxford, UK.

Durbin RM, Eddy SR, Krogh A, Mitchison G (1998). *Biological Sequence Analysis: Probabilistic Models of Proteins and Nucleic Acids*, 1st edition. Cambridge University Press, Cambridge, UK.

Edgar RC (2004). MUSCLE: multiple sequence alignment with high accuracy and high throughput. *Nucleic Acids Research* **32**(5): 1792–97.

Enright AJ, John B, Gaul U, Tuschl T, Sander C, Marks DS (2004). MicroRNA targets in Drosophila. *Genome Biology* **5**(1), R1–R1.

Felsenstein J (1981). Evolutionary trees from DNA sequences: a maximum likelihood approach. *Journal of Molecular Evolution* **17**(6): 368–376.

Fetchko M, Kitts A (2011). *Users Guide to Bankit*: http://www.ncbi.nlm.nih.gov/books/NBK63586.

Fitch W (1969). Locating gaps in amino acid sequences to optimize the homology between two proteins. *Biochemical Genetics* **3**: 99–108.

Fitch WM (1970). Distinguishing homologous from analogous proteins. *Systematic Biology* **19**(2): 99–113.

Fitch WM (1971). Toward defining the course of evolution: minimum change for a specified tree topology. *Systematic Zoology* **20**(4): 406–416.

Fitch WM, Margoliash E (1967). Construction of phylogenetic trees. *Science* **155**: 279–284.

Frech K, Danescu-Mayer J, Werner T (1997). A novel method to develop highly specific models for regulatory units detects a new LTR in GenBank which contains a functional promoter. *Journal of Molecular Biology* **270**: 674–687.

Frech K, Quandt K, Werner T (1997). Finding protein-binding sites in DNA sequences: the next generation. *Trends in Biochemical Sciences* **22**: 103–104.

Friedlander MR, Chen W, Adamidi C, Maaskola J, Einspanier R, Knespel S, Rajewsky N (2008). Discovering microRNAs from deep sequencing data using miRDeep. *Nature Biotechnology* **26**: 407–15.

Friedlander MR, Mackowiak SD, Li N, Chen W, Rajewsky N (2012). miRDeep2 accurately identifies known and hundreds of novel microRNA genes in seven animal clades. *Nucleic Acids Research* **40**: 37–52.

Garber M, Grabherr MG, Guttman M, Trapnell C (2011). Computational methods for transcriptome annotation and quantification using RNA-seq. *Nature Methods* **8**: 469–77.

Gardiner-Garden M, Frommer M (1987). CpG islands in vertebrate genomes. *Journal of Molecular Biology* **196**(2): 261–82.

Gascuel O, Bryant D, Denis F (2001). Strengths and limitations of the minimum-evolution principle. *Systematic Biology* **50**: 621–627.

Ghosh J, Mallick B (2008). *Bioinformatics: Principles and Applications.* Oxford University Press, Oxford, UK.

Ghosh Z, Mallick B (2012). *Bioinformatics: Principles and Applications*, 3rd edition. Oxford University Press, Oxford, UK.

Gonnet GH, Cohen MA, Benner SA (1992). Exhaustive matching of the entire protein sequence database. *Science* **256**(5062): 1443–5.

Graham LE (1993). *Origin of Land Plants*. John Wiley, New York.

Grosdidier A, Zoete V, Michielin O (2011). SwissDock, a protein-small molecule docking web service based on EADock DSS. *Nucleic Acids Research* **39**(Web Server issue): W270–7.

Guttman BS (2007). *Evolution: A Beginner's Guide*. Oneworld Publications, Oxford, UK.

Hasegawa M, Kishino H, Yano T (1985). Dating of human-ape splitting by a molecular clock of mitochondrial DNA. *Journal of Molecular Evolution* **22**(2): 160–174.

Hu H, Wang J, Bu D, Wei H, Zhou L, Li F, Loor JJ (2009). *In vitro* culture and characterization of a mammary epithelial cell line from Chinese Holstein dairy cow. *PLoS One* **4**, e7636. doi: 10.1371/journal.pone.0007636

Huang da W, Sherman BT, Lempicki RA (2009). Systematic and integrative analysis of large gene lists using DAVID bioinformatics resources. *Nature Protocols* **4**: 44–57.

Iquebal MA, Jaiswal S, Mukhopadhyay CS, Sarkar C, Rai A, Kumar D (2015). *Applications of Bioinformatics in Plant and Agriculture*, 1st edition. Springer Publications, New York.

Jukes TH, Cantor CR (1969). *Evolution of Protein Molecules*. Academic Press, New York.

Karlin S, Altschul SF (1990). Methods for assessing the statistical significance of molecular sequence features by using general scoring schemes. *Proceedings of the National Academy of Sciences of the United States of America* **87**: 2264–2268.

Katoh K, Misawa K, Kuma K, Miyata T (2002). MAFFT: a novel method for rapid multiple sequence alignment based on fast Fourier transform. *Nucleic Acids Research* **30**(14): 3059–66.

Kerstens HH, Kollers S, Kommadath A, Del Rosari M, Dibbits B, Kinders SM, Groenen MA (2009). Mining for single nucleotide polymorphisms in pig genome sequence data. *BMC Genomics* **10**(1): 4.

Kimura M (1980). A simple method for estimating evolutionary rates of base substitutions through comparative studies of nucleotide sequences. *Journal of Molecular Evolution* **16**(2): 111–120.

Kozak M (1987). At least six nucleotides preceding the AUG initiator codon enhance translation in mammalian cells. *Molecular Biology* **196**: 947–950.

Kozak M (1989). The scanning model for translation: an update. *Cell Biology* **108**: 229–241.

Kumar S, Stecher G, Tamura K (2015). MEGA7: Molecular Evolutionary Genetics Analysis version 7.0 for bigger datasets. *Molecular Biology and Evolution* **33**(7):1870–4.

Kyte J, Doolittle RF (1982). A simple method for displaying the hydropathic character of a protein. *Journal of Molecular Biology* **157**(1): 105–32.

Lagesen K, Dave W, Ussery DW, Wassenaar TM (2010). Genome update: the 1000th genome – a cautionary tale. *Microbiology* **156**(3): 603–608.

Larkin MA, Blackshields G, Brown NP, Chenna R, McGettigan PA, McWilliam H, *et al.* (2007). ClustalW and ClustalX version 2. *Bioinformatics* **23**: 2947–2948.

Lassez J-L (1976). Circular codes and synchronization. *International Journal of Computer Systems Science* **5**: 201–208.

Leng N, Dawson JA, Thomson JA, Ruotti V, Rissman AI, Smits BM, *et al.* (2013). EBSeq: an empirical Bayes hierarchical model for inference in RNA-seq experiments. *Bioinformatics* **29**: 1035–43.

Li B, Dewey CN (2011). RSEM: accurate transcript quantification from RNA-Seq data with or without a reference genome. *BMC Bioinformatics* **12**: 323.

Li B, Ruotti V, Stewart RM, Thomson JA, Dewey CN (2010). RNA-Seq gene expression estimation with read mapping uncertainty. *Bioinformatics* **26**: 493–500.

Li H, Durbin R (2010). Fast and accurate long-read alignment with Burrows–Wheeler transform. *Bioinformatics* **26**(5): 589–595.

Li H, Jiang T (2004). *A class of edit kernels for SVMs to predict translation initiation sites in eukaryotic mRNAs*. Proceedings of the 8th International Conference on Research in Computational Molecular Biology, pp. 262–271.

Li H, Handsaker B, Wysoker A, Fennell T, Ruan J, Homer N, Durbin R (2009a). The sequence alignment/map format and SAMtools. *Bioinformatics* **25**(16): 2078–2079.

Li H, Handsaker B, Wysoker A, Fennell T, Ruan J, Homer N, Marth G, Abecasis G, Durbin R, Genome Project Data Processing S (2009b). The Sequence Alignment/Map format and SAMtools. *Bioinformatics* **25**: 2078–9.

Li W, Cowley A, Uludag M, Gur T, McWilliam H, Squizzato S, Park YM, Buso N, Lopez R (2015). The EMBL-EBI bioinformatics web and programmatic tools framework. *Nucleic Acids Research* **43**(W1): W580–4.

Li W-H (1997). *Molecular Evolution*. Sinauer Associates, Sunderland, MA. ISBN: 978-0878934638.

Li Y, Zhang Z, Liu F, Vongsangnak W, Jing Q, Shen B (2012). Performance comparison and evaluation of software tools for microRNA deep-sequencing data analysis. *Nucleic Acids Research* **40**: 4298–305.

Liu L, Qu C, Wittkop B, Yi B, Xiao Y, He Y, Li J (2013). A high-density SNP map for accurate mapping of seed fibre QTL in *Brassica napus* L. *PLoS One* **8**(12): e83052.

Livak KJ, Schmittgen TD (2001). Analysis of relative gene expression data using real-time quantitative PCR and the 2(-Delta Delta C(T)) Method. *Methods* **25**: 402–8.

Lopez R, Cowley A, Li W, McWilliam H (2014). Using EMBL-EBI Services via Web Interface and Programmatically via Web Services. *Current Protocols in Bioinformatics* **48**: 1–3.

Love MI, Huber W, Anders S (2014). Moderated estimation of fold change and dispersion for RNA-seq data with DESeq2. *Genome Biology* **15**: 550.

Lukashin A, Borodovsky M (1998). GeneMark.hmm: new solutions for gene finding, *Nucleic Acid Research* **26**(4): 1107–1115.

Luscombe NM, Greebaum D, Gerstein M (2001). What is bioinformatics? A proposed definition and overview of the field. *Methods of Information in Medicine* **40**(4): 346–358.

Magis C, Taly JF, Bussotti G, Chang JM, Di Tommaso P, Erb I, Espinosa-Carrasco J, Notredame C (2014). T-Coffee: Tree-based consistency objective function for alignment evaluation. *Methods in Molecular Biology* **1079**: 117–29.

Manikanandakuar K (2009). *Dictionary of Bioinformatics*, 1st Edition. MJP Publishers, Chennai, India.

Markoff A, Savov A, Vladimirov V, Bogdanova N, Kremensky I, Ganev V (1997). Optimization of single-strand conformation polymorphism analysis in the presence of polyethylene glycol. *Clinical Chemistry* **43**(1): 30–3.

Michel CJ, Pirillo G, Pirillo MA (2008). A relation between trinucleotide comma-free codes and trinucleotide circular codes. *Theoretical Computer Science* **401**: 17–26.

Miller SL (1953). Production of Amino Acids Under Possible Primitive Earth Conditions. *Science* **117**(3046): 528–529.

Miller SL, Urey HC (1959). Organic Compound Synthesis on the Primitive Earth. *Science* **130** (3370): 245–251.

Min-Jou W, Haegeman G, Ysebaert M, Fiers W (1972). Nucleotide sequence of the gene coding for the bacteriophage MS2 coat protein. *Nature* **237**(5350): 82–88.

Morgan GJ (1998). Emile Zuckerkandl, Linus Pauling, and the Molecular Evolutionary Clock, 1959–1965.*Journal of the History of Biology* **31**: 155–178.

Mortazavi A, Williams BA, McCue K, Schaeffer L, Wold B (2008). Mapping and quantifying mammalian transcriptomes by RNA-Seq. *Nature Methods* **5**: 621–8.

Moyes KM, Drackley JK, Morin DE, Rodriguez-Zas SL, Everts RE, Lewin HA, Loor JJ (2010). Mammary gene expression profiles during an intramammary challenge reveal potential mechanisms linking negative energy balance with impaired immune response. *Physiological Genomics* **41**, 161–70. doi: 10.1152/physiolgenomics.00197.2009

Mukhopadhyay CS, Osahan SS (2015a). Sequence alignment: concepts and methods. In: *Bioinformatic Approaches for Livestock Genome Analysis*. Satish Serial Publishing House, Delhi, India.

Mukhopadhyay CS, Osahan SS (2015b). Molecular phylogeny: basics, methods, and applications. In: *Bioinformatic Approaches for Livestock Genome Analysis*. Satish Serial Publishing House, Delhi, India.

Mukhopadhyay CS (2015c). Designing and *in silico* quality checking of PCR primers. In: *Bioinformatic Approaches for Livestock Genome Analysis*. Satish Serial Publishing House, Delhi, India.

Mukhopadhyay CS (2015d). Submitting Nucleotide sequence to Bankit. In: *Bioinformatic Approaches for Livestock Genome Analysis*. Satish Serial Publishing House, Delhi, India.

Needleman SB, Wunsch CD (1970). A general method applicable to the search for similarities in the amino acid sequences of two proteins. *Journal of Molecular Biology* **48**: 443–453.

Nielsen PH (1997). *Neural network prediction of translation initiation sites in eukaryotes: perspectives for EST and genome analysis*. In: Proceedings of the 5th International Conference on Intelligent Systems for Molecular Biology, pp. 226–233.

Notredame C, Higgins DG, Heringa J (2000). T-Coffee: A novel method for fast and accurate multiple sequence alignment. *Journal of Molecular Biology* **302**(1): 205–17.

Ogden R, Gharbi K, Mugue N, Martinsohn J, Senn H, Davey JW, Congiu L (2013). Sturgeon conservation genomics: SNP discovery and validation using RAD sequencing. *Molecular Ecology* **22**(11): 3112–3123.

Oparin AI (1924). *The Origin of Life*. Moscow Worker publisher, Moscow (in Russian: *Proiskhozhdenie zhizny*).

Pedersen AG, Nielsen H (1997). *Neural network prediction of translation initiation sites in eukaryotes: perspectives for EST and genome analysis*. Proceedings of the 2nd International Conference on Intelligent Systems for Molecular Biology, pp. 226–33.

Pfaffl MW (2001). A new mathematical model for relative quantification in real-time RT-PCR. *Nucleic Acids Research* **29**: e45.

Ravindran R, Saravanan BC, Rao JR, Mishra AK, Bansal GC, Ray D (2007). A PCR-RFLP method for the simultaneous detection of *Babesia bigemina* and *Theileria annulata* infections in cattle. *Current Science* **93**(12), pp. 1840–1843.

Reimand J, Kull M, Peterson H, Hansen J, Vilo J (2007). g:Profiler – a web-based toolset for functional profiling of gene lists from large-scale experiments. *Nucleic Acids Research* **35**: W193–200.

Reimand J, Arak T, Vilo J (2011). g: Profiler-a web server for functional interpretation of gene lists (2011 update). *Nucleic Acids Research* **39**: W307–15.

Robinson MD, McCarthy DJ, Smyth GK (2010). edgeR: a Bioconductor package for differential expression analysis of digital gene expression data. *Bioinformatics* **26**: 139–40.

Robinson MD, Oshlack A (2010). A scaling normalization method for differential expression analysis of RNA-seq data. *Genome Biology* **11**: R25.

Rozen S, Skaletsky HJ (2000). Primer3 on the WWW for general users and for biologist programmers. In: *Bioinformatics Methods and Protocols: Methods in Molecular Biology*. Humana Press, Totowa, NJ.

Rust AG, Mongin E, Birney E (2002). Genome annotation techniques: new approaches and challenges. *Drug Discovery Today* **7**(11): S70–6

Rzhetsky A, Nei M (1993). Theoretical foundation of the minimum-evolution method of phylogenetic inference. *Molecular Biology and Evolution* **10**(5): 1073–95.

Saitou N, Nei M (1987). The neighbor-joining method: a new method for reconstructing phylogenetic trees. *Molecular Biology and Evolution* **4**(4): 406–25.

Salamov T, Nishikawa MBS (1998). Assessing protein coding region integrity in cDNA sequencing projects. *Bioinformatics* **14**: 384–390.

Salzberg S (1997). A method for identifying splice sites and translational start sites in eukaryotic mRNA. *Computer Applications in Biosciences (CABIOS)* **13**: 365–376.

Sanger F, Nicklen S, Coulson AR (1977). DNA sequencing with chain-terminating inhibitors. *Proceedings of the National Academy of Sciences of the United States of America* **74**(12): 5463–5467.

SantaLucia J, Jr (1998). A unified view of polymer, dumbbell, and oligonucleotide DNA nearest-neighbor thermodynamics. *Proceedings of the National Academy of Sciences of the United States of America* **95**(4): 1460–1465.

Sarika MAI, Mukhopadhyay CS, Koringa PG, Rai A, Joshi CG, Kumar D (2015). Genome annotation in Prokaryotes and Eukaryotes. In: *Bioinformatic Approaches for Livestock Genome Analysis*. Satish Serial Publishing House, Delhi, India.

Schlee D (1978). In Memoriam Willi Hennig 1913–1976. Einebiographische Skizze. *Entomologica Germanica* **4**: 377–391.

Schmittgen TD, Livak KJ (2008). Analyzing real-time PCR data by the comparative CT method. *Nature Protocols* **3**: 1101–1108.

Schulz J (2008). *Inroduction to dot-plots.* Available online at http://www.code10.info/index.php?option=com_content&view=article&id=64:inroduction-to-dotplots&catid=52:cat_coding_algorithms_dotplots&Itemid=76.

Sievers F, Higgins DG (2014). Clustal omega. *Current Protocols in Bioinformatics* **48**: 1–16.

Sievers F, Wilm A, Dineen DG, Gibson TJ, Karplus K, Li W, *et al.* (2011). Fast, scalable generation of high-quality protein multiple sequence alignments using Clustal Omega. *Molecular Systems Biology* **7**: 539.

Simons A Tils D, von Wilcken-Bergmann B, Müller-Hill B (1984). Possible ideal lac operator: *Escherichia coli* lac operator-like sequences from eukaryotic genomes lack the central G X C pair. *Proceedings of the National Academy of Sciences of the USA* **81**, 1624–1628.

Smit AFA, Hubley R, Green P (1996). *RepeatMasker Open-3.0.* http://www.repeatmasker.org.

Smith TF, Waterman MS (1981). Identification of common molecular subsequences. *Journal of Molecular Biology* **147**: 195–197.

Sokal R, Michener C (1958). A statistical method for evaluating systematic relationships. *University of Kansas Science Bulletin* **38**: 1409–1438.

Stothard P (2000). The Sequence Manipulation Suite: JavaScript programs for analyzing and formatting protein and DNA sequences. *Biotechniques* **28**: 1102–1104.

Suchyta SP, Sipkovsky S, Halgren RG, *et al.* (2003). Bovine mammary gene expression profiling using a cDNA microarray enhanced for mammary-specific transcripts. *Physiological Genomics* **16**: 8–18.

Tamura K (1992). Estimation of the number of nucleotide substitutions when there are strong transition-transversion and G+C content biases. *Molecular Biology and Evolution* **9** (4): 678–687.

Tamura K, Nei M (1993). Estimation of the number of nucleotide substitutions in the control region of mitochondrial-DNA in humans and chimpanzees. *Molecular Biology and Evolution* **10**(3): 512–526.

Tamura K, Nei M, Kumar S (2004). Prospects for inferring very large phylogenies by using the neighbor-joining method. *Proceedings of the National Academy of Sciences (USA)* **101**: 11030–11035.

Tavaré S (1986). Some Probabilistic and Statistical Problems in the Analysis of DNA Sequences. *Lectures on Mathematics in the Life Sciences* (American Mathematical Society) **17**: 57–86.

Temnykh S, DeClerck G, Lukashova A, Lipovich L, Cartinhour S, McCouch S (2001). Computational and experimental analysis of microsatellites in rice (*Oryza sativa* L.): frequency, length variation, transposon associations, and genetic marker potential. *Genome Research* **11**(8): 1441–1452.

Thiel T, Michalek W, Varshney RK, Graner A. (2003). Exploiting EST databases for the development and characterization of gene-derived SSR-markers in barley (*Hordeum vulgare* L.). *Theoretical and Applied Genetics* **106**(3): 411–422.

Thompson JD, Gibson TJ, Plewniak F, Jeanmougin F, Higgins DG (1997). The CLUSTAL_X windows interface: flexible strategies for multiple sequence alignment aided by quality analysis tools. *Nucleic Acids Research* **25**: 4876–4882.

Thornton B, Basu C (2011). Real-Time PCR (qPCR) Primer Design Using Free Online Software. *Biochemistry & Molecular Biology Education* **39**: 145–154.

Trapnell C, Roberts A, Goff L, Pertea G, Kim D, Kelley DR, *et al.* (2012). Differential gene and transcript expression analysis of RNA-seq experiments with TopHat and Cufflinks. *Nature Protocols* **7**: 562–78.

Vincze T, Posfai J, Roberts RJ (2003). NEBcutter: A program to cleave DNA with restriction enzymes. *Nucleic Acids Research* **31**(13): 3688–91.

Wagner GP, Kin K, Lynch VJ (2012). Measurement of mRNA abundance using RNA-seq data: RPKM measure is inconsistent among samples. *Theory in Biosciences* **131**: 281–5.

Wang Z, Gerstein M, Snyder M (2009). RNA-Seq: a revolutionary tool for transcriptomics. *Nature Reviews Genetics* **10**: 57–63.

Wingender E, Chen X, Hehl R, Karas H, Liebich I, Matys V, *et al.* (2000). TRANSFAC: an integrated system for gene expression regulation. *Nucleic Acids Research* **28**: 316–319.

Wingender E, Chen X, Fricke E, Geffers R, Hehl R, Liebich I, Krull M, Matys V, Michael H, Ohnhauser R *et al.* (2001). The TRANSFAC system on gene expression regulation. *Nucleic Acids Research* **29**: 281–283.

Wu TD, Watanabe CK (2005). GMAP: a genomic mapping and alignment program for mRNA and EST sequences. *Bioinformatics* **21**: 1859–75.

Ye J, Coulouris G, Zaretskaya I, Cutcutache I, Rozen S, Madden T (2012). Primer-BLAST: A tool to design target-specific primers for polymerase chain reaction. *BMC Bioinformatics* **13**(1): 134.

Ying H (2013). MicroRNA and transcription factor mediated regulatory network for ovarian cancer: regulatory network of ovarian cancer. *Tumor Biology* **34**: 3219–3225.

Ying H, Lv J, Ying T, Li J, Yang Q, Ma Y (2013). MicroRNA and transcription factor mediated regulatory network for ovarian cancer: regulatory network of ovarian cancer. *Tumor Biology* **34**: 3219–3225.

Zimmermann W (1931). Arbeitsweise der botanischen Phylogenetik und anderer Gruppierungswissenschaften. In: Abderhalden E (ed). *Handbuch der biologischen Arbeilsmethoden*, **9**: 941–1053.

Zuker M (2003). Mfold web server for nucleic acid folding and hybridization prediction. *Nucleic Acids Research* **31**(13): 3406–3415.

Zvelebil, Marketa J (2013). *Study Guide for Understanding Bioinformatics*. Cram101 Publisher; ISBN-13 9781490216034.

Webliography

bioinformaticssoftwareandtools.co.in/bio_tools.php#restriction_table – acquires various online bioinformatics databases and tools

ftp://ftp.ncbi.nlm.nih.gov/genomes/all/GCF_000298735.2_Oar_v4.0/GCF_000298735.2_Oar_v4.0_rna.fna.gz

ftp://ftp.ncbi.nlm.nih.gov/pub/factsheets/HowTo_BLASTGuide.pdf – an enhanced report for search results generated by the BLAST web service

http://align.bmr.kyushu-u.ac.jp/mafft/software/ – another efficient MSA tool

http://biit.cs.ut.ee/gprofiler/ – web server link for characterizing and manipulating gene lists of high-throughput genomics

http://bioinf.cs.ucl.ac.uk/psipred/ – protein sequence analysis tool

http://bioinfogp.cnb.csic.es/tools/venny/ – tool for comparing lists with Venn Diagrams

http://biology.unm.edu/ccouncil/Biology_203/Summaries/Phylogeny.htm

http://blast.ncbi.nlm.nih.gov/Blast.cgi – Blast home page

http://catchenlab.life.illinois.edu/stacks/comp/denovo_map.php – the software pipeline "Stack" provides the scripts for necessary analysis of SNP mining

http://creskolab.uoregon.edu/stacks/ – software pipeline for building loci from short-read sequences

http://dnafsminer.bic.nus.edu.sg/Tis.html – link is used to predict translation initiation site(s) in vertebrate DNA/mRNA/cDNA sequences

http://emboss.sourceforge.net/docs/themes/SequenceFormats.html – an online sequence format conversion tool (along with an explanation of the basic features of sequence format conversion)

http://en.bio-soft.net/format/MACAW.html – link to Multiple Alignment Construction & Analysis Workbench (MACAW) software

http://espressosoftware.com/pages/sputnik.jsp

http://eu.idtdna.com/site – home page of integrated DNA technologies

http://eu.idtdna.com/UNAFold? – site contains information regarding various components (Resuspension Calculator; Dilution Calculator; qPCR Assay Design) of UNAFold Tool

Basic Applied Bioinformatics, First Edition. Chandra Sekhar Mukhopadhyay, Ratan Kumar Choudhary and Mir Asif Iquebal.

http://evolution.berkeley.edu/evolibrary/glossary/glossary_popup.php?word=outgroup – information about glossary of evolution of origin of life

http://exon.gatech.edu/gmhmmp.cgi – provides access to gene prediction program GeneMark.hmm prokaryotic (versions 3.25)

http://genes.mit.edu/GENSCAN.html – link provides access to the program GenScan for predicting the locations and exon-intron structure of genes in genomic sequences fron a variety of organisms

http://genome.crg.es/courses/Bioinformatics2003_genefinding/results/GENSCAN.html – explanation on GENSCAN output is available

http://genome.crg.es/courses/Bioinformatics2003_promoters/ – Online tutorial, "Regulation of Human obese protein geneRegulation of Human obese protein gene", which demonstrates the exercise on transcription binding site prediction

http://gor.bb.iastate.edu/ – tool for predicting secondary structure of proteins

http://lion.img.cas.cz/sms2/index.html/ – site contains a program suit for manipulating DNA and Protein sequences

http://molprobity.biochem.duke.edu/ – home page of molprobity

http://mordred.bioc.cam.ac.uk/~rapper/rampage.php – site used for Ramachandran plot analysis

http://myhits.isb-sib.ch/util/dotlet/doc/dotlet_about.html – Dotlet is online software used for diagonal plotting of sequences.

http://opal.biology.gatech.edu/GeneMark/ – online gene prediction programme

http://petang.cgu.edu.tw/Bioinfomatics/Lecture/0_HTS/04/20120316.pdf – guidelines for using miRDeep2

http://pgrc.ipkgatersleben.de/misa/ – a microsatellite identification tool

http://pgrc.ipk-gatersleben.de/misa/misa.html – a microsatellite identification tool

http://plantgrn.noble.org/psRNATarget/ – a Plant Small RNA Target Analysis tool

http://primer3.ut.ee/ – site used to screen candidate oligos against mispriming library

http://raptorx.uchicago.edu/StructurePrediction/predict/ – online protein sequence prediction (up to 20 proteins) tool

http://self.gutenberg.org/articles/Long_branch_attraction

http://sonnhammer.sbc.su.se/Dotter.html – Dotter is a graphical dotplot program for thorough comparison of two molecular sequences

http://swift.cmbi.ru.nl/servers/html/index.html – an online algorithm used to calculate data from symmetry, torsion angles, polar fraction through protein analysis and bond angles

http://swissmodel.expasy.org/workspace/index.php?func=show_workspace – an online analysis for SWISS-MODEL workspace

http://tandem.bu.edu/trf/trf.html – public database of tandem repeats for users to run their own sequences

http://targetscan.org/ – site belonging to TargetScan, used to predict biological targets of miRNAs by searching for the presence of 8mer, 7mer, and 6mer sites

http://tools.neb.com/NEBcutter2/index.webline – tool for the identification of restriction enzyme sites.

http://www.acdlabs.com/resources/freeware/chemsketch/ – ACD/Chemsketch

http://www.biobase-international.com/product/transcription-factor-binding-sites – online MATCH Analysis tool to Predict binding sites for Transcription Factors in a particular DNA sequence http://www.biochem.ucl.ac.uk/bsm/dbbrowser/CINEMA2.1/ – online multiple sequence alignment tool called Color INteractive Editor for Multiple Alignments (CINEMA)

http://www.bioinformatics.org/JaMBW/1/2/: a Java-based online sequence format conversion tool

http://www.biomedcentral.com/content/supplementary/1471-2164-11-156-s2/Additionalfile2/Genscan_output/GENSCAN%20output%20EG926217.htm – sequence output tool

http://www.cambridgesoft.com/software/overview.aspx: ChemDraw – site for desktop and enterprise software

http://www.cbi.pku.edu.cn/docs/faq/blastspecifics.html – a useful page that has frequently asked questions (FAQs) on BLAST

http://www.cbs.dtu.dk/services/NetStart/ – online server to produce neutral network predictions of translation start in vertebrate and *Arabidopsis thaliana* nucleotide sequences

http://www.deduveinstitute.be/~opperd/private/upgma.html – online tree construction tool

http://www.ebi.ac.uk/Tools/clustalw2/index.html (Clustal W) – site for online multiple sequence alignment

http://www.ebi.ac.uk/Tools/dbfetch/dbfetch/dbfetch.databases – provides information on a large number of dbfetch databases

http://www.ebi.ac.uk/Tools/msa/clustalo/ – link for online multiple sequence alignment tool Clustal Omega

http://www.ebi.ac.uk/Tools/sfc/emboss_seqret/ – sequence format conversion tool provided by EBI.

http://www.ebi.ac.uk/Tools/webservices/ – provides programmatic access to various data resources and analysis tools EMBL-EBI

http://www.es.embnet.org/Services/MolBio/t-coffee/ – online T-Coffee tool for MSA

http://www.gene-regulation.com/cgi-bin/pub/programs/match/bin/match.cgi – a site for Match tool that is used for *in silico* transcription factor binding studies

http://www.hhmi.umbc.edu/toolkit/ClustalWGuide.html – provides help and guidelines on Clustal W multiple sequence alignment program

http://www.hku.hk/bruhk/gcgdoc/dotplot.html – Dot-plot(+) is used to identify the overlapping portions of two sequences and to identify the repeats and inverted repeats of a particular sequence.

http://www.jalview.org/ – link to Java Alignment Viewer (JALVIEW) that uses a Java-based platform

http://www.kazusa.or.jp/codon/ – a codon usage database that holds records of thousands of organisms

http://www.life.umd.edu/classroom/bsci338m/Lectures/Systematics.html – provides information regarding importance of phylogenesis, biodiversity and evolution

http://www.life.umd.edu/classroom/bsci338m/Lectures/Systematics.html – provides information regarding importance of phylogenesis, biodiversity and evolution

http://www.megasoftware.net/ – an integrated tool for conducting automatic and manual sequence alignment

http://www.microrna.org – resource for miRNA target prediction and expression

http://www.ncbi.nlm.nih.gov/books/NBK153387/ – book source for blast sequence analysis tool

http://www.ncbi.nlm.nih.gov/protein/ – a protein database

http://www.ncbi.nlm.nih.gov/Structure/CN3D/cn3d.shtml – 3-dimensional structure viewer tool

http://www.ncbi.nlm.nih.gov/tools/primer-blast/index.cgi?LINK_LOC=BlastHome – primer designing too

http://www.ncbi.nlm.nih.gov/WebSub/?tool=genbank – site used for sequence submission

http://www.rcsb.org/pdb/ – Protein databank site

http://www.simsoup.info/Origin_Landmarks_Oparin_Haldane.html – webpage on Oparin-Haldane hypothesis

http://www.uniprot.org/uniprot/ – provides the scientific community with a comprehensive, high-quality and freely accessible resource of protein sequence and functional information

http://www.vcru.wisc.edu/simonlab/bioinformatics/programs/stacks/ref_map.pl.txt – explanation of the commands for Reference mapping (ref_map.pl), in Stacks pipeline

http://www.w3schools.com/html/html_intro.asp – a web developer's site that has a discussion on HTML documentation

http://www-bimas.cit.nih.gov/molbio/readseq/ – ReadSeq is an online sequence format conversion tool

http://zinc.docking.org/: ZINC – database used for virtual screening of commercially available compounds

https://david.ncifcrf.gov – link provides information regarding a comprehensive set of functional annotation tools for investigators to understand biological meaning behind large list of genes

https://eu.idtdna.com/analyzer/Applications/OligoAnalyzer/ – site used to examine an oligonucleotide for primer-dimer, self- dimer, hetero-dimer etc.

https://genome.ucsc.edu – site contains the reference sequence and working draft assemblies for a large collection of genomes

https://pubchem.ncbi.nlm.nih.gov/ – provides information on the biological activities of small molecules

https://software.broadinstitute.org/gatk/download/ – to download the latest version of GATK

https://sourceforge.net/projects/bowtie-bio/files/bowtie/1.1.1/ – a very fast short-read aligner for DNA-seq or RNA-seq studies

https://www.broadinstitute.org/gatk/index.php

https://www.cgl.ucsf.edu/chimera/ – program used for interactive visualization and analysis of molecular structures and related data

https://www.mdc-berlin.de/36105849/en/research/research_teams/systems_biology_of_gene_regulatory_elements/projects/miRDeep/documentation – miRDeep commands and explanation

rasmol.org/ – RasMol is a program for molecular graphics visualization

wssp.rutgers.edu/StudentScholars/WSSP08/DotPlotter/DotPlotPractice.html – site for dotplot analysis of amino acid and nucleic acid sequences. This site is not functional currently.

ww2.chemistry.gatech.edu/~williams/bCourse_Information/4581/labs/tbp/rasmol/rasmol_tbp_fset.html – site for RasMol tutorial

www.bioinformatics.org/sms2/mirror.html/ – site for online databases and bioinformatics tools for analysis.

www.computerhope.com/jargon – a dictionary and glossary for computer terms

www.expasy.org/ – maintains biological databases and online tools related to proteomics, genomics, phylogenetics, etc.

www.ncbi.nlm.nih.gov/nucleotide/ – very useful and popular primary repository of nucleotide sequences.

www.rcsb.org/pdb/home/home.do – database for biomacromolecular structures.

www.repeatmasker.org/ – program used to screen DNA sequences for interspersed repeats and low complexity

www.swissdock.ch/docking: SwissDock server – used to search target proteins or ligand

www.vivo.colostate.edu/molkit/dnadot/ – site for nucleic acid dot plot. Currently not available

Index

Basic Applied Bioinformatics, First Edition. Chandra Sekhar Mukhopadhyay, Ratan Kumar Choudhary and Mir Asif Iquebal.